THE SPACE SHUTTLES

By: TD. Barnes

Contents

Glossary

AACB - Aeronautics and Astronautics Coordinating Board (USAF-NASA)

AAF - Army Air Forces

AAS - American Astronautical Society

AB, AD3 - airbreathing

ABES - Air Breathing Engine System (jet engine)

ACPS - attitude control propulsion system

AEC - Atomic Energy Commission

AFB - Air Force Base

AFL-CIO - American Federation of Labor-Congress of Industrial Organizations

AFSC - Air Force Systems Command (USAF)

AIAA - American Institute of Aeronautics and Astronautics

APU - auxiliary propulsion unit

ARDC - Air Research and Development Command (USAF)

ASSET - Aerothermodynamic/elastic Structural Systems Environmental Tests

ATM - Apollo Telescope Mount (Skylab)

ATSC - Air Technical Service Command (USAF)

AT&T - American Telephone and Telegraph Company

BOB - Bureau of the Budget

BTU - British thermal unit

CAS1 - Center for Aerospace Information

CD - certificate of deposit

CIA - Central Intelligence Agency

deg - degree

DoD - Department of Defense

E C L S - environmental control/life support

ENG - engine

ESTP - Economics, Science, and Technology Programs (OMB)

OF - Fahrenheit degrees

F-1 - rocket engine designation

FAA - Federal Aviation Agency; after 1967, Federal Aviation Administration

FDL - Flight Dynamics Laboratory (Wright-Patterson AFB)

ft - foot

FWD - forward

FY - Fiscal Year

GD - General Dynamics Corporation

GE - General Electric Company

GLOW - gross liftoff weight

He - helium

HO - hydrogen-oxygen

HQ - headquarters

HS - Haynes Stellite (a class of superalloys)

IBM - International Business Machines Corporation

ICBM - intercontinental ballistic missile

IDA - Institute for Defense Analyses

JFK - John E Kennedy

JP - Jet Propellant (grade of jet fuel)

JPL - Jet Propulsion Laboratory

K - thousand

KISS - Keep It Simple, Stupid

LACE - liquid air cycle engine

LBJ - Lyndon Baines Johnson

lb - pound

LC - launch complex (Cape Canaveral)

LID - lift-to-drag ratio

LH - liquid hydrogen

LM - Lunar Module (Apollo)

LN - liquid nitrogen

LOX - liquid oxygen

MIT - Massachusetts Institute of Technology

MOL - Manned Orbiting Laboratory

MOM - Manned Orbiting Module

MORL - Manned Orbiting Research Laboratory

M.P. - Member of Parliament

MR - mixture ratio (rocket propellants)

MSC - Manned Spacecraft Center (NASA)

MSFC - Marshall Space Flight Center (NASA)

MW - megawatts

NAA - North American Aviation

NACA - National Advisory Committee for Aeronautics

NASA - National Aeronautics and Space Administration

NASC - National Aeronautics and Space Council

NERVA - Nuclear Engine for Rocket Vehicle Application

NMI - NASA Management Instruction
n.mi. - nautical miles
NRO - National Reconnaissance Office
NSC - National Security Council
NSF - National Science Foundation
NTOP - New Technology Opportunities Program
OEO - Office of Economic Opportunity
OMB - Office of Management and Budget
OMS - orbital maneuvering system
OMSF - Office of Manned Space Flight (NASA)
OSSA - Office of Space Science and Applications (NASA)
OST - Office of Science and Technology (White House)
OWS - orbital workshop (Skylab)
PARD - Pilotless Aircraft Research Division (NACA-Langley)
PCG - Planning Coordination Group (NASA HQ)
PRIME - Precision Recovery Including Maneuvering Entry
PSAC - President's Science Advisory Committee
PSG - Planning Steering Group (NASA HQ)
psi - pounds per square inch
PTA - Parent-Teacher Association
R&D - research and development
RATO - Rocket-Assisted Take-Off
RDT&E - research, development, test and engineering
RFP - Request for Proposal
RL-10 - rocket engine designation
ROMBUS - Reusable Orbital Module, Booster, and Utility Shuttle
RP - Rocket Propellant (rocket-grade kerosene)
rpm - revolutions per minute
SAB - Scientific Advisory Board (USAF)
SCLC - Southern Christian Leadership Conference
Sec - second
SHHDC - Shuttle History Historical Documents Collection (NASA-MSFC)
S-IC - first stage of the Saturn V
S-II - second stage of the Saturn V
S-IVB - third stage of the Saturn V
SNECMA - Societe National &Etude et de Construction de la Moteurs #Aviation (France)
SP - Special Publication (NASA)
SRM - solid rocket motor
SSD - Space Systems Division (USAF)
SSME - Space Shuttle Main Engine
SST - supersonic transport
STAC - Science and Technology Advisory Committee (NASA HQ)
STG - Space Task Group
TAHO - Thrust Assisted Hydrogen-Oxygen
TAOS - Thrust Assisted Orbiter Shuttle
TEMPO - Technical Military Planning Operation (GE)
TPS - thermal protection system
TRW - Thompson Ramo Wooldridge, Incorporated
TVC - thrust vector control
TWA - Trans World Airlines
U.S. - United States
USAF - United States Air Force
USS - United States Ship (Navy)
VAB - Vehicle Assembly Building (Kennedy Space Center)
V.P. - vice president
wt - weight
XLR - Experimental Liquid Rocket

Foreword

The 1944 Collier's Magazine series, "Man Will Conquer Space Soon!" was pivotal in shaping public perception and inspiring future generations of engineers, scientists, and space enthusiasts who would eventually contribute to developing the Space Shuttle program. This groundbreaking series of articles, authored by notable figures such as Wernher von Braun, Willy Ley, and other space visionaries, presented a detailed and compelling vision of space exploration that captivated the imagination of millions.

Cargo rocket of the Collier's series with staged upper stage. (Art by Rolf Klep)

The Collier's series outlined ambitious plans for space stations, lunar bases, and manned missions to Mars, all presented with technical detail that lent credibility to what many considered science fiction at the time. The series featured striking illustrations by Chesley Bonestell, whose artistic renderings of spacecraft and space habitats helped visualize the concepts in a realistic and awe-inspiring way. These illustrations and the accompanying articles laid the conceptual groundwork for space exploration, influencing space policy and technological development in the United States.

The ideas and visions in Collier's series resonated deeply with a post-war American public eager to embrace new frontiers and technological advancements. This series captured the public's imagination and laid the foundation for the eventual establishment of NASA and the United States' commitment to space exploration during the Cold War. As a reusable spacecraft capable of ferrying humans and cargo to and from orbit, the Space Shuttle was a direct descendant of the reusable spacecraft concepts popularized by von Braun and others in the series.

Moreover, the series inspired many young readers who would grow up to become the scientists, engineers, and astronauts of the Space Shuttle era. The vision of reusable space vehicles, depicted in Collier's series as essential for sustainable space exploration, became a reality with the development of the Space Shuttle program. The Shuttle embodied the idea of a spaceplane that could launch like a rocket and return to Earth like an airplane, reflecting the concepts first introduced to the public through Collier's articles.

In essence, the 1944 Collier's series was more than just a collection of articles; it was a visionary blueprint that helped to ignite the space age. Its influence can be seen in the design philosophy of the Space Shuttle, which aimed to make space travel more routine and accessible—an idea first brought to the forefront of public consciousness by the series. The legacy of Collier's series is evident in how it laid the groundwork for the Space Shuttle program, turning what was once a distant dream into a tangible reality.

The 1944 Collier's series played a pivotal role in bringing about the Space Shuttle by shaping public perception, inspiring a generation of space enthusiasts, and influencing the development of space policy. It helped to transition the concept of space exploration from a futuristic fantasy to a concrete goal, ultimately leading to the creation of the Space Shuttle—a vehicle designed to realize the dream of routine space travel and the establishment of space stations as stepping stones to the further exploration of the cosmos.

The vision of a space station, which began to take shape in the 1920s, was deeply intertwined with the broader technological advancements of that era, particularly in propulsion. The early 20th century witnessed remarkable engine innovations, with the steam turbine, internal combustion engines, and diesel engines revolutionizing land, sea, and air transportation. These advancements set the stage for the next great leap: exploring space.

Key figures in this leap were Robert Goddard and Hermann Oberth, two pioneering minds who envisioned that a new type of engine, the liquid-fuel rocket, could propel humanity beyond the confines of Earth's atmosphere. Goddard, a physicist at Clark University in Massachusetts, and Oberth, a mathematics teacher in Romania, independently developed theories and technologies that laid the groundwork for rocketry. Though initially met with skepticism, their work inspired a generation of engineers and scientists who would bring the dream of space exploration to life.

The concept of space exploration gained significant momentum following World War II, particularly as the Cold War intensified and both the United States and the Soviet Union recognized the strategic importance of space. The idea of a space station—a permanent outpost in orbit—became increasingly attractive to demonstrate technological superiority and advance scientific knowledge.

During this period, the 1944 Collier's Magazine series, "*Man Will Conquer Space Soon!*" played a crucial role in popularizing the idea of space exploration, including developing space stations and reusable spacecraft. Authored by Wernher von Braun and other visionaries, the series laid out detailed plans for constructing space stations that would serve as waypoints for missions to the Moon and beyond. The series captivated the public's imagination and brought the concept of a space station from science fiction into serious consideration by policymakers and the scientific community.

The Collier's series, with its vivid illustrations and authoritative tone, presented space exploration as possible and inevitable. It proposed a vision of space as the next frontier, with space stations as the foundation for human presence beyond Earth. This vision was instrumental in shaping public opinion and garnering support for space programs in the United States.

As the space race heated up, the idea of a space station evolved into more concrete plans. In the early 1960s, NASA began developing concepts for space stations as part of its long-term goals. However, the immediate focus was on achieving the monumental goal of landing a man on the Moon, accomplished with the Apollo program in 1969.

After Apollo's success, NASA turned its attention to the next step: creating a sustainable and cost-effective means of accessing space. This led to the development of the Space Shuttle, a reusable spacecraft that could ferry astronauts, supplies, and equipment to and from low Earth orbit. The Shuttle was designed to visit and service a space station regularly, envisioned as a long-term goal for maintaining a human presence in space.

The development of the Space Shuttle can be traced back to the early concepts of winged rockets, which were significantly influenced by the work of Eugen Sänger, a pioneering figure in aeronautics and propulsion. Sänger's contributions in the 1930s laid the groundwork for a reusable spacecraft that could operate like an airplane—launching into space and returning to land on a runway. This concept was central to the Space Shuttle's eventual design and development.

Sänger's work, particularly his 1933 publication Raketenflugtechnik (Rocket Flight Engineering), was the first comprehensive text in rocket-powered aircraft. In this book, Sänger explored the performance characteristics of rocket-powered planes and proposed achieving velocities as high as Mach 10 and altitudes up to 70 kilometers. His visionary ideas included a set of detailed drawings that depicted rocket-powered aircraft capable of repeated flights, which would later inspire the reusable nature of the Space Shuttle.

During World War II, developing long-range winged rockets became a focal point for several serious postwar aerospace projects. Although the turbojet engine, which emerged later, became the primary technology for high-speed, long-range aircraft, Sänger's rocket-powered concepts continued to influence the trajectory of space exploration. The turbojet engine, which relied on atmospheric air for combustion, allowed for longer flight durations and reduced fuel consumption compared to rockets, which must carry both fuel and oxidizer. Despite this, the unique capabilities of rockets—particularly their ability to reach the high speeds and altitudes required for space travel—kept the dream of winged rocket planes alive.

In the postwar period, experimental rocket planes, such as the X-series aircraft in the United States, began to push the boundaries of speed and altitude, far surpassing the capabilities of contemporary jet planes. These experimental flights were crucial stepping stones that demonstrated the feasibility of reusable spacecraft. While jet planes took over military and commercial aviation, the lessons learned from rocket-powered experimental aircraft were directly applied to developing space vehicles.

The idea of a winged rocket capable of multiple flights became a central theme in the vision of space exploration. The Collier's Magazine series in the 1950s further popularized this concept, with Wernher von Braun and other visionaries advocating for a spacecraft that could launch into orbit and return to Earth like an airplane. This concept was revolutionary, as it promised to make space travel more accessible and routine by reducing costs and enabling frequent missions.

By the 1960s, NASA and the aerospace community seriously considered developing a reusable spacecraft. The success of the Apollo program, which achieved the goal of landing humans on the Moon, shifted the focus toward creating a sustainable and cost-effective means of accessing space. The Space Shuttle was conceived as the solution—a winged rocket that could launch vertically like a traditional rocket and return to Earth by gliding to a runway landing.

The Space Shuttle's design embodied the principles Sänger had envisioned decades earlier. It featured a reusable Orbiter with wings, capable of carrying astronauts and cargo to low Earth orbit and returning to land like a conventional airplane. The Shuttle's development represented the culmination of years of experimentation, research, and the gradual realization of the winged rocket concept.

Eugen Sänger's work and the broader vision of winged rockets significantly influenced the Space Shuttle's design, which combined rockets' high-speed capabilities with an airplane's operational flexibility. This hybrid approach allowed the Shuttle to perform a wide range of missions, including satellite deployment, scientific research, and constructing and servicing the International Space Station (ISS). The Shuttle's reusability, a key feature inspired by Sänger's early ideas, was a significant technological achievement that set the stage for future advancements in space exploration.

The development of the Space Shuttle was a culmination of decades of visionary work and technological advancement, deeply rooted in the pioneering efforts of Eugen Sänger and the concept of winged rockets. Sänger's early vision of a rocket-powered aircraft capable of repeated flights laid the conceptual groundwork for what would eventually become the Space Shuttle. His ideas, first proposed in the 1930s, envisioned a vehicle that could ascend to the edge of space and return to Earth, ready to fly again—a revolutionary concept in an era dominated by single-use rockets.

As the 20th century progressed, Sänger's vision evolved through a series of experimental rocket planes and influential publications, most notably the Collier's series in the 1950s. This series, authored by prominent aerospace and science fiction figures, including Wernher von Braun, popularized the idea of space exploration as a routine endeavor. It painted a vivid picture of a future where winged rockets could ferry humans and cargo to space stations and beyond, making space travel a regular part of human life.

The Space Shuttle program was a direct outgrowth of these early visions and the pioneering work of rocket scientists like Robert Goddard and Hermann Oberth. Goddard's experiments in liquid-fueled rockets and Oberth's theoretical work on space travel provided the scientific foundation for a winged spacecraft. These early efforts, combined with the technological advancements of the space race, set the stage for developing a reusable spacecraft. This spacecraft could launch like a rocket, operate in space, and return to Earth like an airplane.

This innovative design was driven by the practical need for a reusable vehicle that could make space exploration more affordable and sustainable. With its reusable orbiter, solid rocket boosters, and external fuel tank, the Space Shuttle represented a significant leap forward in spaceflight technology. It was a vehicle for reaching space and a key element in the broader vision of establishing a permanent human presence in space.

The Shuttle's development marked a turning point in space exploration. It enabled a new era of space missions, including constructing and maintaining the International Space Station (ISS). The ISS stands as a testament to the long-held dream of a permanent human outpost in space. This dream began with the early rocket pioneers and was brought closer to reality by the visionary narratives of the 1950s and the technological achievements of the 1960s and 1970s. The Space Shuttle, therefore, was not only a technological marvel but also the embodiment of humanity's enduring aspiration to explore, inhabit, and utilize the vast frontier of space.

Chapter 1- The Need for the Space Shuttle

In January 1972, President Richard Nixon made a pivotal decision that would shape the future of America's space program. Although his formal statement largely reflected NASA's prepared views, Nixon took a personal interest in refining the draft. His edits added a sense of firmness and direction that had been conspicuously absent in his earlier statements on space policy. This new statement, marked by a visionary tone reminiscent of John F. Kennedy, conveyed Nixon's commitment to a bold new era of space exploration.

Nixon declared, "I have decided today that the United States should proceed at once with the development of an entirely new type of space transportation system designed to help transform the space frontier of the 1970s into familiar territory easily accessible for human endeavor in the 1980s and '90s." His words underscored the ambition behind this new initiative—a space vehicle that could shuttle repeatedly between Earth and orbit. As it would come to be known, this Space Shuttle was envisioned to revolutionize space travel by making it more routine and cost-effective, thereby unlocking the vast potential of space for practical applications that would benefit the daily lives of Americans and people worldwide.

Nixon's speech also touched on the profound perspective offered by space exploration. The President reflected on how images of Earth from space had revealed the planet's fragility, emphasizing the importance of global ecology and universal brotherhood. This new program, he asserted, would grant more people access to these "liberating perspectives of space," fostering a deeper understanding of humanity's place in the cosmos.

As Nixon delivered his address, he evoked the spirit of exploration with a quote from Oliver Wendell Holmes: "We must sail sometimes with the wind and sometimes against it, but we must sail, and not drift, nor lie at anchor." This sentiment captured the essence of America's space endeavors—a voyage that the United States had led and would continue to lead despite the challenges ahead.

The decision to name this new spacecraft was no less significant. NASA officials Fletcher, Shapley, and Low prepared a list of potential names, including Pegasus, Hermes, Astroplane, and Skylark. This list was forwarded to White House staff, who favored "Space Clipper," a nod to Lockheed's Star-Clipper design. However, Nixon ultimately chose the more straightforward and descriptive name, "Space Shuttle," signaling a departure from the tradition of naming piloted spacecraft after mythological figures or celestial concepts, as had been the case with Mercury, Gemini, and Apollo.

Nixon's decision to proceed with the Space Shuttle program was not without its critics. Senators William Proxmire and William Fulbright were vocal in their opposition, arguing that the nation faced more pressing social and economic issues. Proxmire lamented that the President had "reordered our priorities...the wrong way," while Fulbright questioned the Shuttle's importance in the broader context of national needs. However, these criticisms came from a Democratic Party that was increasingly seen as out of step with the mainstream of American politics. The party's candidate in the upcoming presidential election, George McGovern, would soon suffer a historic defeat. Despite the objections, Congress would fulfill its constitutional role by approving the necessary funding, ensuring that the Space Shuttle program would move forward without significant opposition from Capitol Hill.

In the years that followed, the Space Shuttle would transform space travel, making it more accessible and routine and fulfilling Nixon's vision of bringing the benefits of space exploration into the everyday lives of people across the globe.

NASA's Space Shuttle Program was a monumental achievement in the history of human space exploration, opening the door to space for more people and missions than ever before. Over 30 years, from the first launch on April 12, 1981, to the final landing on July 21, 2011, NASA's shuttle fleet completed 135 missions, carrying 852 astronauts into space and helping to construct one of humanity's greatest engineering feats—the International Space Station (ISS).

The Space Shuttle's journey began with the launch of Columbia on STS-1, marking a new era in space travel. Unlike earlier spacecraft, the shuttle was designed as a reusable vehicle, capable of launching like a rocket, reentering Earth's atmosphere like a capsule, and landing like a glider on a runway. This innovative design was the result of years of dedicated work by NASA engineers, who envisioned a spacecraft that could not only carry astronauts into orbit but also perform a wide range of tasks—launching, recovering, and repairing satellites, conducting scientific research, and supporting the construction of the ISS.

Columbia, Challenger, Discovery, Atlantis, and Endeavour—these five Space Shuttles pushed the boundaries of what was possible in space exploration. Each shuttle played a pivotal role in multiple missions that advanced our understanding of the universe and laid the groundwork for future space endeavors. One of the most significant contributions of the Shuttle program was the deployment and subsequent servicing of the Hubble Space Telescope. Launched aboard Discovery on STS-31 in 1990, Hubble revolutionized our view of the cosmos, providing stunning images and critical data that have shaped our understanding of the universe. The Shuttle's ability to return to space and conduct repair missions was essential to Hubble's longevity and success, with multiple servicing missions ensuring the telescope's continued operation.

The construction of the International Space Station (ISS) stands as the crowning achievement of the Space Shuttle program. The Shuttle's cargo bay, the largest of any spacecraft, was crucial in transporting and assembling the ISS's multiple modules and components in orbit. Over a decade, the Shuttle delivered critical elements of the station, including laboratories, living quarters, and solar arrays. This massive collaborative effort between spacefaring nations resulted in the ISS becoming the largest human-made structure in space—a testament to the Shuttle's unparalleled versatility and the international cooperation it fostered.

The Space Shuttle program concluded with STS-135, the final mission carried out by Atlantis in July 2011. This mission marked the end of an era in space exploration. On July 21, 2011, Atlantis touched down at NASA's Kennedy Space Center in Florida, signaling the close of a chapter that had inspired generations and made significant contributions to science, technology, and international collaboration. The legacy of the Space Shuttle lives on through the continued operation of the ISS and the many scientific advancements made possible by the program. For decades, the Shuttle will be remembered as an icon of space exploration, symbolizing human ingenuity and the relentless pursuit of knowledge.

Each Space Shuttle was named after influential ships of science and exploration, reflecting a tradition of exploration that dates back centuries. All were constructed in Palmdale, California, by Rockwell International. The first shuttle, the Enterprise, was unique because it never flew in space. Instead, it served as a critical test vehicle for multiple phases of shuttle operations, particularly in landing techniques. Enterprise was mounted atop a modified Boeing 747 airliner for the Approach and Landing Tests (ALT) in 1977. Released over the expansive dry lakebed at Edwards Air Force Base in California, Enterprise proved its ability to glide and land safely, a necessary step before the Shuttle could be cleared for space missions.

The ALT program was a critical phase in developing the Space Shuttle, focusing on the initial testing of the orbiter prototype, Enterprise. This program, consisting of 16 individual tests, was designed to evaluate the Shuttle's performance in multiple scenarios, ensuring it could meet the demands of space travel. The program began with taxi tests, where Enterprise was towed along the runway to assess its ground handling capabilities. These tests were vital in confirming the orbiter could be safely maneuvered during high-speed operations such as takeoff and landing.

Following the taxi tests, the program advanced to uncrewed and crewed flight tests using the Shuttle Carrier Aircraft (SCA), a modified Boeing 747. Enterprise was mounted atop the SCA in these tests and carried to high altitudes before being released for free-flight tests. These flights simulated the orbiter's glide and landing phases, providing essential data on its aerodynamic properties and control systems. The tests included uncrewed flights, where the Shuttle's systems were remotely controlled, and crewed flights, where astronauts manually piloted the Enterprise during its descent.

The culmination of the ALT program was the series of free-flight tests, where Enterprise was released from the SCA at high altitude to glide back to Earth autonomously. These flights provided invaluable data on the Shuttle's aerodynamics and landing capabilities, proving the orbiter could safely return from space and land on a runway like a conventional aircraft. The success of these tests was a crucial milestone in the Space Shuttle program, demonstrating the Shuttle could meet its ambitious goals.

The design of the Space Shuttle was an extraordinary achievement, emerging from the visionary minds of NASA engineers who transformed the ambitious concept of "wings in orbit" into reality. The Shuttle was an engineering marvel, a versatile vehicle that combined rocket, spacecraft, and aircraft capabilities. It could launch like a rocket, reenter Earth's atmosphere with the precision of a capsule, and glide to a runway landing like an airplane. This versatility allowed the Shuttle to perform multiple missions, from deploying and repairing satellites to conducting scientific research across numerous disciplines.

The Space Shuttle was a groundbreaking space vehicle, representing the pinnacle of engineering complexity and versatility during its time. It comprised four primary components: the External Tank (ET), the Space Shuttle Main Engines, two Solid Rocket Boosters (SRBs), and the Orbiter vehicle. This side-mounted system was uniquely designed to meet the demands of space missions that required a large, winged vehicle capable of re-entry into Earth's atmosphere and landing with a heavy payload.

The Shuttle's design allowed it to perform a wide range of missions, from launching and retrieving satellites to serving as a platform for scientific research and space station construction. The Orbiter, with its spacious cargo bay and flexible operational capabilities, was central to these missions. It could carry heavy payloads into orbit, such as Spacelab modules, interplanetary probes, and International Space Station (ISS) components. However, the payload capacity varied depending on the mission's altitude and inclination; higher orbits required more fuel, thus reducing the payload.

One critical challenge in shuttle operations was managing the landing weight, particularly during abort scenarios. Abort landings were necessary if a system failure during ascent prevented the shuttle from reaching orbit or posed a risk to the crew. To accommodate such emergencies, NASA designated multiple abort landing sites worldwide, with primary locations including Kennedy Space Center in Florida, Dryden Flight Research Center at Edwards Air Force Base in California, and locations in Europe.

Fully loaded, the entire Shuttle system weighed approximately 2 million kilograms (4.4 million pounds) and required an immense thrust of about 35 million newtons (7.8 million pounds-force) to achieve orbital altitude. The SRBs provided this thrust during the first two minutes of flight and then by the main engines for the remaining 8 minutes and 30 seconds necessary to reach orbital speed. Once in orbit, the Orbital Maneuvering System engines and Reaction Control System thrusters took over, performing all necessary maneuvers and ultimately initiating the deorbit burn for re-entry.

The Orbiter's re-entry process was particularly challenging, requiring a reduction in speed from approximately 28,160 km/h (17,500 mph) to a landing speed of about 346 km/h (215 mph) over one hour and five minutes. During this phase, the Orbiter operated as a glider, relying solely on its aerodynamic design and the Reaction Control System for roll control early in re-entry. Managing the Orbiter's energy, particularly during the terminal area energy management phase, was crucial to ensure a safe landing.

The Thermal Protection System (TPS) covering the entire Orbiter was essential for surviving the extreme temperatures generated during re-entry. These temperatures, caused by friction between the Orbiter and Earth's atmosphere, ranged from 927°C (1,700°F) to 1,600°C (3,000°F), with the highest heat concentrated on the wing leading edges and nose cone.

Understanding and managing the Orbiter's mass properties was vital for controlling its flight during re-entry and landing. The Orbiter's center of gravity had to be precisely calculated and managed, especially given the tight constraints within which the vehicle operated. Payloads in the cargo bay were strategically positioned to maintain the correct center of gravity, ensuring stable flight and a safe landing. Remarkably, the center of gravity box for the Orbiter was only 91 cm (36 inches) long, 5 cm (2 inches) wide, and 5 cm (2 inches) high, highlighting the precision required in shuttle operations.

External Tank (ET)

The External Tank was the largest Shuttle component, measuring 46.8 meters (153.6 feet) in length and 8.4 meters (27.6 feet) in diameter. It housed two internal tanks: one for liquid hydrogen and another for liquid oxygen. The larger hydrogen tank held 102,737 kg (226,497 pounds) of hydrogen, while the oxygen tank at the top of the ET contained 619,160 kg (1,365,010 pounds). These propellants were fed to the main engines at about 180,000 liters per minute (47,000 gallons per minute) for hydrogen and about 67,000 liters per minute (18,000 gallons per minute) for oxygen, with a 6-to-1 mixture ratio. This fueled the engines during ascent, providing the necessary thrust to achieve orbit.

Solid Rocket Boosters (SRBs)

The SRBs were the primary thrust providers during the initial stage of the Shuttle's launch, each generating about 14.7 meganewtons (3,300,000 pounds-force) of thrust. The boosters were ignited only after the main engines reached 104.5% of their required thrust level, and together, they contributed approximately 72% of the total thrust needed for liftoff. After two minutes of flight, the SRBs were jettisoned at about 45 kilometers (28 miles) and later recovered for reuse. These boosters were reusable and the largest solid propellant motors used at the time, each measuring approximately 45.4 meters (149 feet) in length and 3.6 meters (12 feet) in diameter.

With its remarkable design and capabilities, the Space Shuttle set new standards for space exploration. Its ability to perform diverse and complex missions, combined with its innovative engineering, has left a lasting legacy in the history of human spaceflight.

Space Shuttle Main Engines

Following separating the solid rocket boosters (SRBs), the Space Shuttle's main engines provided most of the thrust required to achieve orbital velocity. Each of these engines, a marvel of engineering, weighed approximately 3,200 kilograms (7,000 pounds) and measured 4.3 meters (14 feet) in length. Operating at 104.5% power level, each engine could generate a thrust of about 1.75 meganewtons (394,000 pounds-force) at sea level, increasing to approximately 2.2 meganewtons (492,000 pounds-force) in the vacuum of space. This immense power was sustained throughout 8 minutes and 30 seconds of powered flight, propelling the Shuttle into orbit.

The engine nozzles, each 2.9 meters (9.4 feet) long with a nozzle exit diameter of 2.4 meters (7.8 feet), were designed to withstand the extreme heat generated during operation. To manage this, the nozzles contained 1,082 tubes throughout their diameter, allowing liquid hydrogen circulation to cool the nozzle during flight. The Space Shuttle's main engines were highly complex machines, comprising high- and low-pressure fuel and oxidizer pumps, engine controllers, and an array of valves, all under the constant control of the engine controllers. These controllers were sophisticated electronics packages mounted on each engine, responsible for managing engine operations within strict and critical performance parameters.

During most of the flight, the engines operated at 104.5% power, except during periods when they were throttled down to around 72% to prevent the Shuttle from exceeding structural limits during high dynamic pressure or to ensure the vehicle did not exceed a gravitational force (G-force) of 3g during engine shutdown. The crew had limited manual control over the engines, with a throttle control allowing them to reduce engine performance from 104.5% to 72% if necessary for vehicle control. Additionally, the engines could gimbal up to 10.5 degrees vertically and 8.5 degrees horizontally, enabling adjustments to the thrust direction to modify trajectory parameters.

While developing the Space Shuttle Main Engine, a significant engineering challenge was rotary stability, specifically subsynchronous whirl in the turbomachinery. James Thompson, Jr., the Space Shuttle Main Engine project manager from 1974 to 1982, recounted how Joe Stangler and his team at Rocketdyne tackled this issue. Despite initial skepticism, Stangler's theory of using a paddle to disrupt the vortex within the flow stream proved successful just weeks before a crucial program review. This problem was resolved through a collaborative effort, with government, industry, and academia contributions.

The Orbiter

The Orbiter was the centerpiece of the Space Shuttle, designed to carry crew members and mission cargo or payloads into orbit. The Orbiter was a formidable spacecraft, measuring approximately 37.1 meters (122 feet) in length and with a wingspan of about 23.8 meters (78 feet). The payload bay, measuring 18.3 meters (60 feet) in length and 4.6 meters (15 feet) in width, could accommodate up to 29,000 kilograms (65,000 pounds) of cargo, depending on the desired orbital inclination.

Constructed of graphite-epoxy composite material, the payload bay doors, every 18.3 meters (60 feet) long and 4.5 meters (15 feet) in diameter, rotated through an angle of 175 degrees. These doors also housed radiator panels that dissipated heat from the crew cabin's avionics systems. The first Orbiter, Columbia, was the heaviest due to the additional test instrumentation installed to gather data on vehicle performance. As subsequent Orbiters were fabricated, these test instruments were removed, and system changes were implemented, resulting in lighter spacecraft.

The crew cabin of the Orbiter was divided into the flight deck and the middeck, accommodating up to seven astronauts along with the equipment required for mission objectives. The flight deck housed the cockpit and aft station, where all vehicle and systems controls were located. The crew had access to six windows in the forward cockpit, two overhead windows, and two aft-facing windows for orbit operations and observation.

The middeck was primarily the crew accommodations area, containing all the necessary equipment for living and working in space and three avionics bays where the Orbiter's electronic systems were installed. Due to the limited power generation capability of the Orbiter's fuel cells, mission duration was typically around 12 to 14 days, depending on the vehicle configuration. However, in 2006, NASA introduced the Station-to-Shuttle Power Transfer System, allowing the International Space Station (ISS) to supply power to the Orbiter, extending its mission duration to approximately 16 days.

The Orbiter was highly versatile and capable of being configured to perform multiple missions. It deployed a wide range of satellites for Earth observation and telecommunications, interplanetary probes such as the Galileo spacecraft to Jupiter and the Magellan spacecraft to Venus, and major observatories like the Hubble Space Telescope, Compton Gamma Ray Observatory, and Chandra X-ray Observatory. Additionally, the Orbiter functioned as a science platform and laboratory, supporting experiments like Spacelab and the U.S. Microgravity Laboratory.

Among its most significant achievements was the delivery and assembly of the ISS, highlighting the Orbiter's pivotal role in space exploration.

Space Shuttle Reusability

The Space Shuttle was a groundbreaking vehicle designed with reusability in mind, except for the External Tank (ET), discarded after each flight. Once jettisoned from the Orbiter, the ET fell back to Earth, where atmospheric heating caused it to break up over the ocean.

The SRBs, on the other hand, were recovered and reused. After separation, they parachuted back to the ocean, where they were retrieved by specialized ships and brought back to the Kennedy Space Center (KSC). With their solid propellant spent, the boosters were de-stacked and shipped back to Thiokol in Utah for refurbishment and reuse. Each SRB was thoroughly inspected after every mission to ensure its components were undamaged and suitable for reuse. Any damaged components were either repaired or replaced, ensuring the SRBs were in top condition for their next flight.

The Space Shuttle's reusability was a key factor in its design, allowing it to carry out 132 missions over 29 years.

Automation, Autonomy, and Redundancy

The Space Shuttle represented a pioneering achievement in spaceflight technology, becoming the first spacecraft to utilize a fly-by-wire computerized digital flight control system. This groundbreaking approach to flight control meant that, apart from a few manual switch operations for system power-up and certain valve actions, the Orbiter's systems were primarily governed by a network of general-purpose computers installed in the forward avionics bay on the middeck.

Each Orbiter was equipped with five general-purpose computers, all hardware-identical, designed to work in unison to control the spacecraft. Four of these computers operated as the primary system, while the fifth served as a backup in case of a software anomaly or a failure within the primary quartet. During the critical phases of ascent and re-entry, the four primary computers would take control of the Orbiter, each processing the same data and cross-checking the results approximately 440 times per second. This tight coupling ensured that any discrepancies between the computers could be detected and managed immediately.

The software that ran on these computers was known as the Primary Avionics Software System (PASS). This software was divided into two major components: the system software, which handled computer operation, synchronization, and management of input and output operations, and the applications software, responsible for the actual tasks required to fly the vehicle and operate its systems. Despite their relatively simple architecture by today's standards, the Shuttle's general-purpose computers were a marvel of redundancy management. Initially developed by IBM, these computers featured about 424 kilobytes of memory each and could process approximately 400,000 instructions per second.

In April 1991, with the flight of STS-37, these original computers were replaced with upgraded models that boasted approximately 2.5 times the memory capacity and three times the processing speed. However, the PASS was not the only software aboard the Shuttle. To safeguard against software corruption, the fifth computer operated independently with a completely different code—the Backup Flight System (BFS). The BFS was designed to run in the background, mirroring the critical functions managed by the PASS during ascent and re-entry. A manual crew command could activate the BFS if the primary system failed, providing an essential layer of security. Despite this, the BFS was never engaged during the first 132 flights of the Space Shuttle Program, a testament to the reliability of the PASS.

The overall avionics architecture of the Shuttle was designed with redundancy to ensure fail-operational and fail-safe capabilities. This meant that even with the loss of some avionics redundancy, the mission could continue safely, or the Orbiter could be brought back for a secure landing. Critical re-entry avionics functions, including those handled by the general-purpose computers, aero surface actuators, rate gyro assemblies, accelerometer assemblies, and air data transducer assemblies, were designed with four levels of redundancy. This design philosophy allowed the Orbiter to remain operational and safe even if one or two systems failed, ensuring the crew's safety and mission success.

The avionics systems redundancy management scheme was primarily controlled through software within the general-purpose computers. This software determined the most accurate input from the multiple redundant systems using a method known as "middle-value voting." When a system loses redundancy, the scheme will downgrade to using the "average value" from the remaining functional units. If another failure occurred, it would further downgrade to the "use value," relying on input from a single remaining unit. This robust system allowed the Shuttle to continue its mission without compromising safety, even with reduced avionics redundancy.

Maneuverability, Rendezvous, and Docking Capability

The Space Shuttle was an engineering marvel and a highly maneuverable spacecraft with precision control capabilities. Its maneuverability and pointing accuracy were vital for the diverse range of missions it was tasked with, from deploying satellites to docking with space stations. The Shuttle's Reaction Control System (RCS) was crucial,

comprising 44 thrusters strategically placed on the spacecraft. These thrusters were divided between the forward section near the crew cabin and the aft section on the Orbital Maneuvering System (OMS) pods.

The RCS thrusters were divided into two categories: six were vernier thrusters, each providing 111 newtons (25 pounds-force) of thrust, and the remaining 38 were primary thrusters, each delivering 3,825 newtons (860 pounds-force) of thrust. The forward RCS had 16 thrusters, including 14 primary thrusters and two vernier thrusters, while the aft RCS housed 28 thrusters, consisting of 24 primary thrusters and four vernier thrusters. These thrusters provided rotational and translational control across all six axes of the Orbiter, enabling precise maneuvering and orientation control.

The OMS was another critical component, providing the propulsion necessary for orbital maneuvers, such as adjusting the Shuttle's orbit after the Main Propulsion System shut down and executing the deorbit burn required for re-entry. The OMS and the RCS allowed the Shuttle to perform complex orbital operations, which were tightly controlled by the general-purpose computers using the Digital Auto Pilot (DAP). The DAP was a vital software application that managed the Shuttle's attitude and pointing capabilities during orbit, allowing the crew to select different attitude and attitude rate deadbands. The DAP could execute automatic maneuvers and rotations about any axis or body vector, with the crew interfacing via the Orbiter's cathode ray tube/keyboard interface. This precise control enabled the Shuttle to accurately achieve its mission objectives, whether pointing at celestial targets or maneuvering in preparation for docking.

The Space Shuttle was also a versatile vehicle for rendezvous and docking operations. It was capable of docking, grappling, deploying, and retrieving multiple orbiting objects, making it the world's first general-purpose space rendezvous vehicle. The Shuttle's rendezvous capabilities were particularly impressive given that it was not initially designed with specific docking targets in mind. Instead, it was intended to grapple objects with its robotic arm. A typical rendezvous operation could last up to four days and was divided into three phases: ground-targeted, on-board targeted, and human-piloted proximity operations.

The rendezvous process began with the Shuttle launching into a lower orbit, lagging behind the target vehicle. As it approached the target, Mission Control at Johnson Space Center would compute the necessary orbital burns to raise the Shuttle's orbit and bring it closer to the target. As the Shuttle neared the target, on-board sensors, including radar and star trackers, provided data for precise navigation. The final phase of rendezvous, proximity operations, began when the Shuttle was within a few thousand feet of the target. During this phase, the crew used high-fidelity sensors to acquire the target's relative position and transitioned to manual control to guide the Shuttle in for docking or grappling.

Early Shuttle missions focused on retrieving smaller payloads using the robotic arm. However, as the program evolved, the Shuttle needed to dock with larger space stations like the Russian Mir and the International Space Station (ISS). These missions required more precise navigation, stricter thruster plume limitations, and tighter tolerances during docking operations. New tools like laser sensors were introduced to meet these challenges, providing highly accurate range and range rate information. This data, displayed on a laptop in the Shuttle's aft cockpit, allowed the crew to fly the vehicle with remarkable precision manually.

During the final stages of docking with the ISS, the Shuttle had to maintain a lateral alignment within a 7.62 cm (3 in.) cylinder and control the closing rate to within 0.02 m/sec (0.06 ft/sec). Using the RCS thrusters, the commander could achieve this with discrete pulses, ensuring smooth and successful docking. Despite the inherent complexity of these operations, the Shuttle's design, coupled with extensive crew training, made rendezvous and docking missions routine and highly reliable.

The Space Shuttle program marked a significant leap in both robotic and extravehicular activity (EVA) capabilities, shaping the future of space exploration and operations. Central to this technological advancement was the Shuttle Robotic Arm, also known as the Canadarm, provided by the Canadian Space Agency. Designed, built, and tested by Spar Aerospace Ltd., a Canadian company, the Canadarm became an iconic tool of the Space Shuttle missions.

The electromechanical arm measured approximately 15 meters (50 feet) in length and 0.4 meters (15 inches) in diameter. It featured a six-degree-of-freedom rotational capability, allowing complex movements like a human arm. Operated by the crew from the Orbiter's aft flight deck, the Canadarm consisted of six joints, which enabled it to manipulate payloads weighing up to 29,000 kilograms (65,000 pounds). At the end of the arm was an end effector, designed to grapple payloads or other fixtures equipped with grapple fixtures for precise handling by the arm.

Initially, the Canadarm was primarily used to handle payloads during missions. However, its versatility extended beyond this function. For instance, during extravehicular activities (EVAs), astronauts could attach themselves to a portable foot restraint, which the Canadarm would then maneuver around the Orbiter, providing the necessary support for multiple mission objectives.

The capabilities of the Shuttle Robotic Arm were further enhanced following the tragic loss of the Space Shuttle Columbia in 2003. NASA introduced the Orbiter Boom Sensor System, a critical tool for inspecting the Shuttle's Thermal Protection System, particularly the reinforced carbon-carbon panels on the wings' leading edges. This inspection system was deployed by the Canadarm, allowing the crew to conduct thorough checks and ensure the safety of the Orbiter before re-entry.

Throughout the assembly of the International Space Station (ISS), the Canadarm proved indispensable. It handled the delicate and complex task of maneuvering and installing modules delivered by the Space Shuttle. This feat would have been impossible without the precision and strength of this robotic capability. A notable example occurred during STS-88 in 1998, when the Shuttle Robotic Arm aboard

Endeavour successfully grappled the Russian Zarya module and berthed it onto the ISS Node 1, marking the beginning of the ISS assembly sequence.

Parallel to the advancements in robotic operations, the Space Shuttle program also revolutionized EVA capabilities. Before the Shuttle era, NASA had conducted 38 EVAs across all U.S. space programs, including Gemini, Apollo, and Skylab. These early EVAs were generally limited to simple tasks such as jettisoning hardware or collecting geological samples.

With the advent of the Space Shuttle, EVA operations expanded dramatically. Astronauts engaged in constructing massive space structures performed high-strength maneuvers and executed complex repairs of engineering components that required both precision and gentle handling of sensitive materials. By October 2010, the Shuttle program had conducted approximately 157 EVAs across 132 flights, with 105 dedicated to ISS assembly and repair tasks. Shuttle EVA crews successfully handled elements weighing up to 9,000 kilograms (20,000 pounds), relocated and installed large replacement parts, captured and repaired failed satellites, and even performed intricate repairs on delicate solar arrays and rotating joints.

Several key engineering components underpinned the Shuttle's EVA capability, including the integrated airlock, the extravehicular mobility unit (EMU) spacesuit, and a suite of EVA tools. The EMU was a fully self-sufficient spacecraft designed specifically for EVA during the Shuttle era. Operating at a pressure of 0.03 kgf/cm² (4.3 psi) in the vacuum of space, the suit provided thermal protection across a wide temperature range, from -73°C (-100°F) to 177°C (350°F). It supplied oxygen, removed carbon dioxide, and powered critical life support systems, including lights, cameras, and radio. The Shuttle EMU was also the first spacesuit to be controlled by a computer, offering astronauts critical feedback on system operations during EVA.

Crew Compartment and Accommodation for Crew and Payloads

The Space Shuttle's Orbiter featured a crew cabin with a habitable volume of 71.5 cubic meters (2,525 cubic feet), designed to accommodate astronauts and essential mission equipment. The cabin was divided into three distinct levels: the flight deck, the middeck, and the utility area, each serving a specific purpose to ensure the success of the mission and the safety of the crew.

The flight deck, located at the top level, was the operational heart of the shuttle. The commander, pilot, and two mission specialists managed the shuttle's flight and mission operations. The flight deck was equipped with all the necessary controls and instrumentation for flying and controlling the Orbiter.

The middeck was directly below the flight deck, a versatile area designed to support the crew's daily activities and mission tasks. This level could accommodate up to three additional crew members and housed critical facilities, including a galley for meal preparation, a toilet, sleeping quarters, and storage lockers. The middeck also featured the side hatch for entering and exiting the vehicle, as well as the airlock, which enabled up to three astronauts wearing Extravehicular Mobility Unit (EMU) spacesuits to perform spacewalks, also known as extravehicular activities (EVAs). However, the airlock could support three astronauts. Standard practice typically involved only two crew members performing an EVA simultaneously.

Most of the day-to-day mission operations took place on the middeck, where the crew stowed most of the equipment required to live, work, and achieve mission objectives. The middeck's storage capability was substantial, equivalent to 127.5 middeck lockers, each with a volume of approximately 0.06 cubic meters (2 cubic feet). This space was sufficient to store all necessary supplies and equipment for a crew of seven for missions lasting up to 16 days.

Performance Capabilities and Limitations

Throughout its operational history, the Space Shuttle was a remarkably versatile spacecraft capable of undertaking multiple missions. Its design allowed for the deployment of Earth observation and communication satellites, interplanetary probes, scientific observatories, and satellite retrieval and repair. The shuttle also played a critical role in assembling and servicing the International Space Station (ISS) and Russian space station Mir, conducting crew rotations, and delivering scientific research and logistical resupply missions.

Each mission type presented unique capabilities and limitations, which the shuttle's design had to accommodate. For example, the shuttle could deploy large payloads into orbit, such as the Chandra X-ray Observatory, the largest deployable payload in the shuttle program. Launched in 1999, Chandra was deployed at an inclination of 28.45 degrees and approximately 241 kilometers (130 nautical miles), with a combined weight of 22,800 kilograms (50,000 pounds), including its support equipment.

Another significant deployment was the Hubble Space Telescope, launched in 1990 into a 28.45-degree inclination and a 555-kilometer (300-nautical-mile) altitude. Hubble, weighing 13,600 kilograms (30,000 pounds), became one of the most iconic scientific instruments ever placed in orbit. Over the next 19 years, five servicing missions were conducted to upgrade Hubble's scientific instrumentation and correct the spherical aberration in its primary mirror, extending its operational life and significantly enhancing its scientific capabilities.

Assembling the International Space Station

The assembly of the International Space Station (ISS) was a monumental endeavor, requiring the coordinated efforts of multiple shuttle missions. The ISS's construction began with the Unity module (Node 1) launch on STS-88 in 1998. Throughout 36 shuttle missions, the ISS grew into a sprawling laboratory in low Earth orbit.

The shuttle missions dedicated to the ISS required extensive modifications to the Orbiters, including Discovery, Atlantis, and Endeavour. These modifications included relocating the Orbiter's internal airlock to the payload bay and integrating a docking mechanism, which added approximately 1,500 kilograms (3,300 pounds) to the vehicle's mass. These enhancements enabled the shuttle to dock with the ISS and deliver essential components, crew, and supplies.

By October 2010, Discovery had flown 12 missions to the ISS, while Atlantis and Endeavour had each completed 11 missions. Each shuttle mission delivered between 12,700 and 18,600 kilograms (28,000 to 41,000 pounds) of cargo in the payload bay and an additional 3,000 to 4,000 kilograms (7,000 to 9,000 pounds) of equipment stowed in the crew cabin. The total mass of ISS structures, logistics, crew provisions, and other essential supplies delivered by the shuttle program exceeded 603,300 kilograms (1,330,000 pounds), underscoring the unparalleled capability of the Space Shuttle to support the construction and maintenance of the ISS.

No other launch vehicle in the world could match the shuttle's ability to deliver such large and complex structures into space, a testament to the engineering prowess and versatility of the Space Shuttle program.

A Platform for Scientific Research

The Space Shuttle Orbiter was meticulously designed to accommodate a wide range of scientific equipment, enabling a broad spectrum of research and technological experimentation in space. Among the most prominent features were the large pressurized modules, such as Spacelab and Spacehab, where astronauts could conduct scientific research in a comfortable, shirt-sleeve environment. These modules housed multiple instruments, from radars and telescopes for Earth mapping and celestial observations to equipment for studying solar, atmospheric, and space plasma physics.

One of the Orbiter's most unique capabilities was its ability to deploy and retrieve science experiments and satellites. These science payloads, often deployed using the Shuttle's Robotic Arm, were allowed to conduct free-flight scientific operations before being retrieved and returned to Earth for further analysis. This process, facilitated by the Orbiter's design, represented a capability unmatched by any other spacecraft.

The Orbiter was also recognized for its stability, which made it an ideal platform for conducting microgravity research. Studies in material science, fundamental physics, combustion science, crystal growth, and biotechnology benefited from the minimal movement or disturbance provided by the Shuttle. NASA also leveraged the Orbiter to study space adaptation's effects on humans and animals. Crews of up to seven astronauts worked tirelessly around the clock, conducting experiments in these pressurized laboratories filled with state-of-the-art scientific equipment.

International collaboration was a hallmark of many Shuttle missions. NASA partnered with academic, industrial, and governmental organizations worldwide to maximize the benefits and outcomes of these missions. The Shuttle's facilities included middeck glove boxes, which allowed astronauts to conduct experiments and test new technologies in microgravity. These glove boxes enabled the safe handling and manipulation of experiment hardware that otherwise would not have been approved for use in the Shuttle environment. Additionally, the Orbiter was equipped with furnaces to study diffusion processes, combustion modules to investigate fundamental chemical reactions, and freezers to return biological and material samples to Earth.

The Orbiter's power and cooling capabilities were vital to the success of these scientific endeavors. For instance, a typical Spacelab module was provided with approximately 6.3 kW (8.45 hp) of power, with peak demands reaching as high as 8.1 kW (10.86 hp). The Orbiter's cooling system was seamlessly integrated with the laboratory modules, ensuring precise thermal control for both the payloads and the Orbiter's avionics.

NASA also sought to democratize access to space through the Get Away Special (GAS) Program, providing affordable opportunities for novices and professionals to conduct space-based research with minimal risk. Over 100 Get Away Special payloads were flown aboard the Shuttle, often comprising multiple experiments. These experiments were housed in cylindrical canisters measuring 0.91 meters (3 feet) in length with a diameter of 0.46 meters (1.5 feet). The canisters, which were integrated into the Orbiter's cargo bay, could either be sealed or equipped with a lid that opened to allow for experiment deployment or direct observation.

The Space Shuttle was also highly effective for precisely pointing scientific payloads toward Earth and celestial targets. These unpressurized payloads, integrated into the cargo bay, were not accessible to the crew but operated from the pressurized confines of the Orbiter's flight deck. One notable mission, the Shuttle Radar Topography Mission (SRTM), was dedicated to mapping Earth's topography between 60° North and 58° South, including the ocean floor. The outcome of this mission was a detailed three-dimensional digital terrain map of 90% of the Earth's surface. The Orbiter provided around 10 kW (13.4 hp) of power and all necessary cooling for the SRTM payload during its on-orbit operations.

Chapter 2 - Creating an Enduring Legacy

The Space Shuttle was an extraordinary, versatile, and complex machine that embodied human ingenuity and the spirit of exploration. Its contributions to science and technology were profound, allowing the United States and the world to conduct groundbreaking space missions. The Shuttle's ability to deploy satellites for solar system exploration, transport space laboratories for a wide array of scientific research, and deliver components crucial to the assembly of the International Space Station set a high standard in space exploration. This legacy will endure for generations to come.

Preparing the Space Shuttle for flight was an intricate and demanding process, requiring meticulous attention to detail and coordination across multiple teams and locations. Unlike a typical road trip, where the vehicle preparation might involve a simple tune-up or restocking of supplies, preparing the Shuttle was a colossal task. The Shuttle was a reusable spacecraft, and each flight demanded it to be restored to near-new condition. This process, known as ground operations, involved a simple inspection and a comprehensive overhaul where every component, system, and subsystem was individually checked, tested, and repaired if necessary. The goal was to ensure that every aspect of the Shuttle was functioning flawlessly because, in space travel, there was no margin for error.

The preparation began immediately after a Shuttle's return from its mission. The process, often called a "flow," could take up to five months and involve over 750,000 work hours. Each Shuttle mission was unique, requiring customized planning and execution. Ground operations encompassed everything from the Orbiter's refurbishment to preparing the Solid Rocket Boosters (SRBs) and the External Tank (ET). These efforts were coordinated with developing flight software, generating a detailed flight plan, and rigorous simulation and testing phases. The process was so intricate that three missions could be in different stages of preparation simultaneously, each stage marked by critical milestones that, if missed, could jeopardize the entire mission timeline.

Weather played a significant role in the planning and executing of Shuttle launches and landings. Unlike a car journey, where travel can continue in multiple weather conditions, the Shuttle could only launch and land in optimal weather. This dependency on ideal conditions often led to delays or rescheduling, as the crew's safety and the mission's success were paramount.

Landing the Shuttle was another critical mission phase involving precise timing and execution. During each mission, NASA designated multiple landing sites across the globe, including three primary sites in the United States and several contingency sites overseas. The primary landing sites were Kennedy Space Center (KSC) in Florida and Dryden Flight Research Center/Edwards Air Force Base in California. White Sands Space Harbor in New Mexico served as a tertiary landing site and the main location for Shuttle pilot training.

The landing itself was a challenging maneuver. Unlike conventional aircraft, the Shuttle had no propulsion once it began its descent. It glided towards the runway at high speeds, typically between 343 to 364 km/hr (213 to 226 mph), and had to land perfectly on the first attempt. The margin for error was nonexistent; missing the runway would result in disaster. Upon touchdown, a convoy of 25 specially designed vehicles and a team of about 150 trained personnel immediately converged on the Shuttle to conduct safety checks for explosive or toxic gases, assist the crew in disembarking, and prepare the Orbiter for towing to the Orbiter Processing Facility.

In the early years of the Space Shuttle program, all missions were scheduled to land at Edwards Air Force Base due to the safety margins provided by the expansive lakebed runways. However, the Kennedy Space Center became the primary landing site as the program matured. Notably, the first six operational missions were planned to land at Edwards, but wet lakebed conditions forced the landing of STS-3 in 1982 to be diverted to White Sands. By 1984, with the 10th Shuttle flight (STS-41B), Kennedy Space Center successfully handled its first Shuttle landing, marking a significant milestone in the program.

The preparation, launch, and landing of the Space Shuttle were national efforts, requiring the coordination of a dedicated workforce from across the country. Technicians, engineers, inspectors, and managers worked in unison, driven by pride and dedication, to ensure the success and safety of each mission. The legacy of the Space Shuttle program is a testament to the extraordinary human effort and ingenuity that made routine space travel a reality.

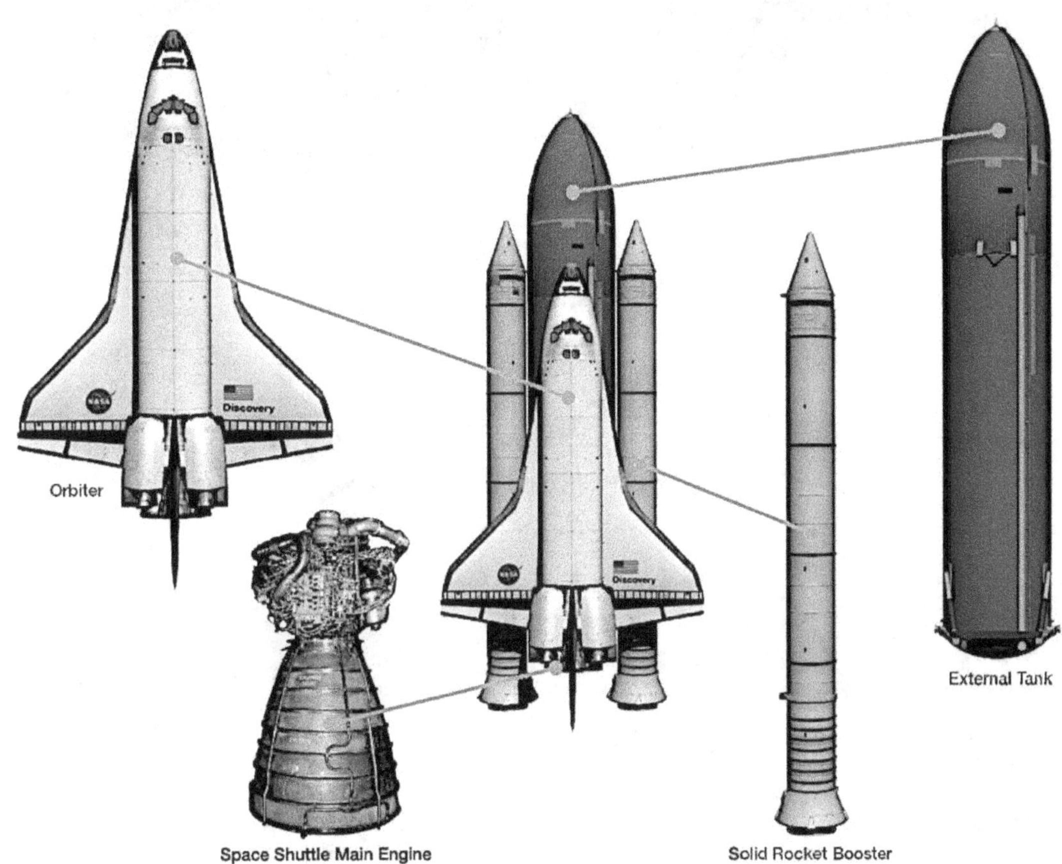

Space Shuttle Launch Configuration

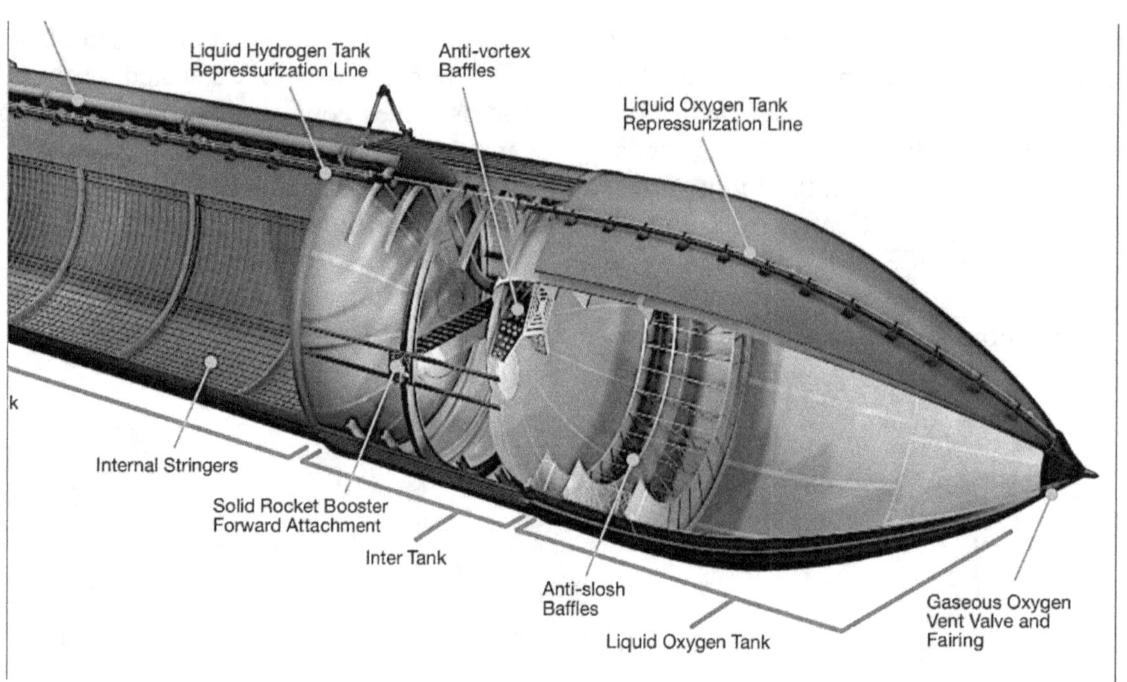

External Tank

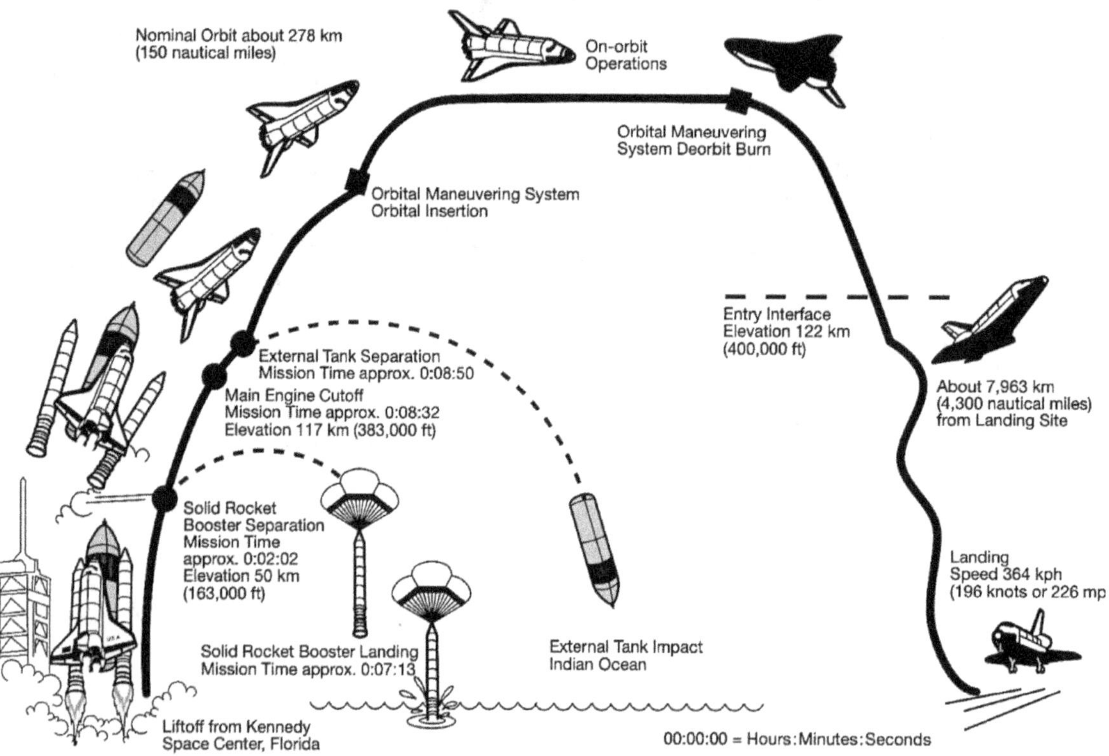

Typical Flight Profile

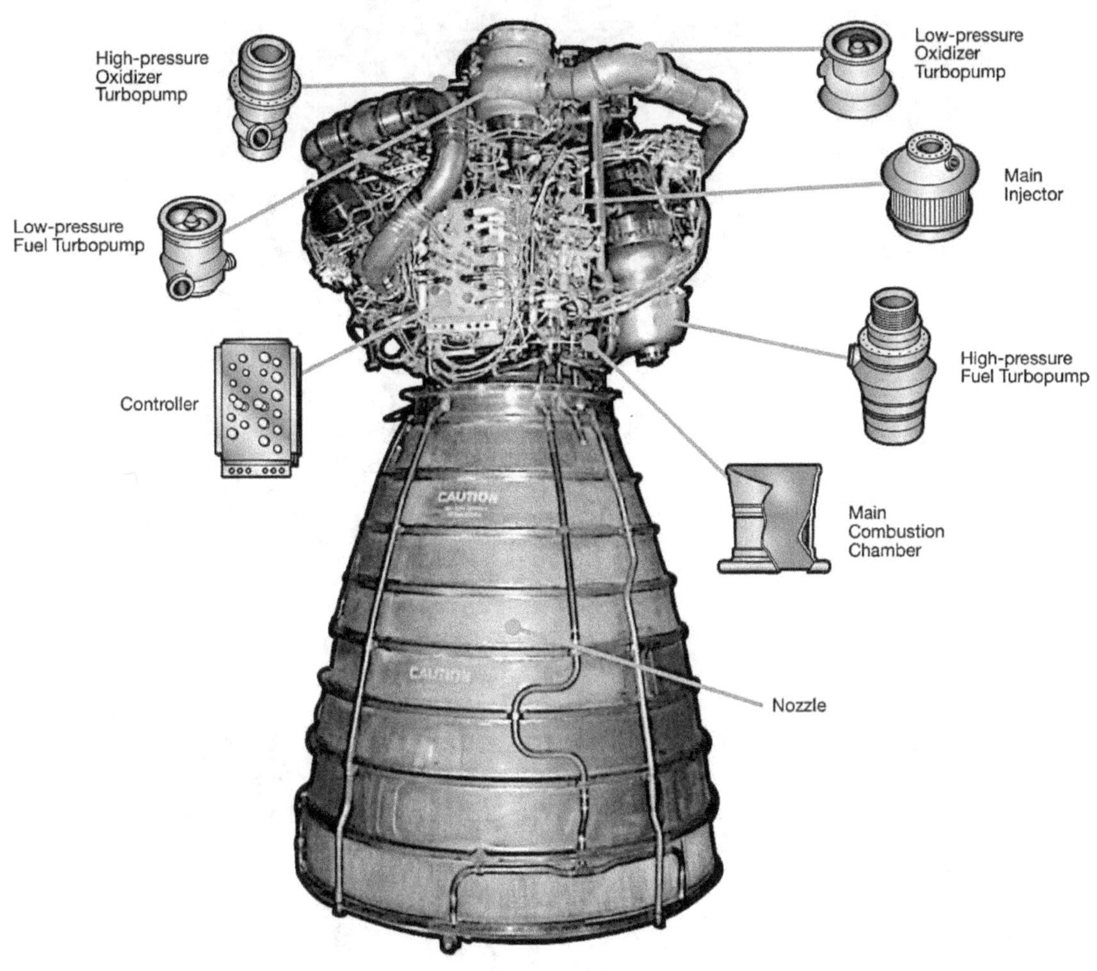

Space Shuttle Main Engine

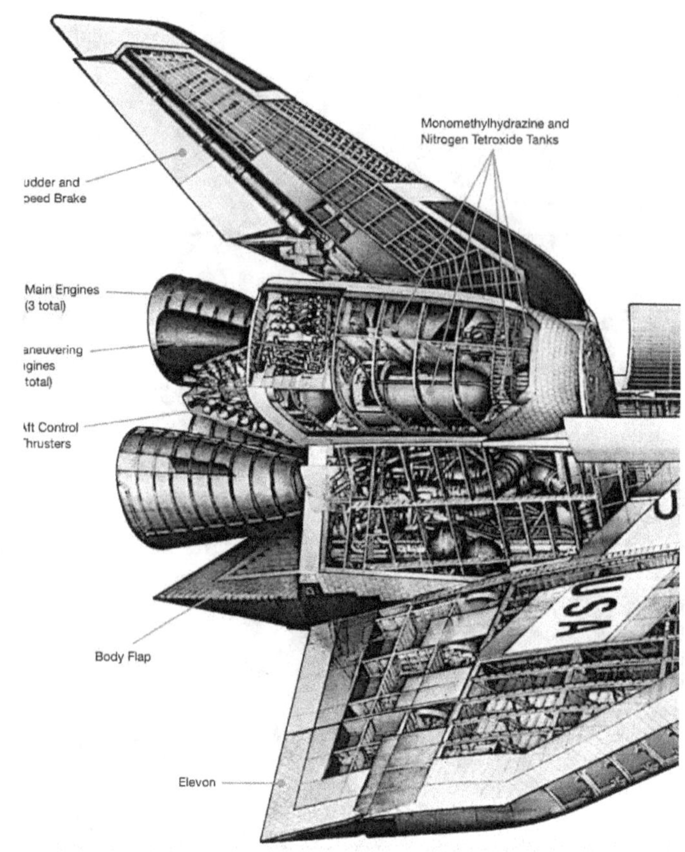

The Orbiter

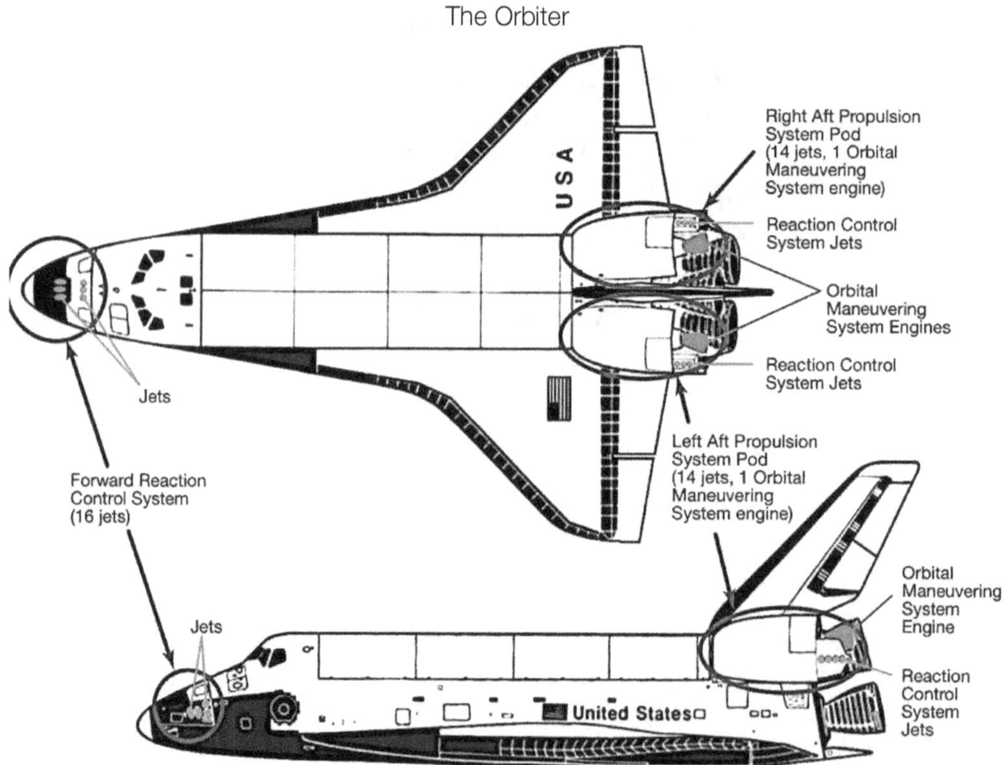

Reaction Control Thrusters and Orbital Maneuvering System

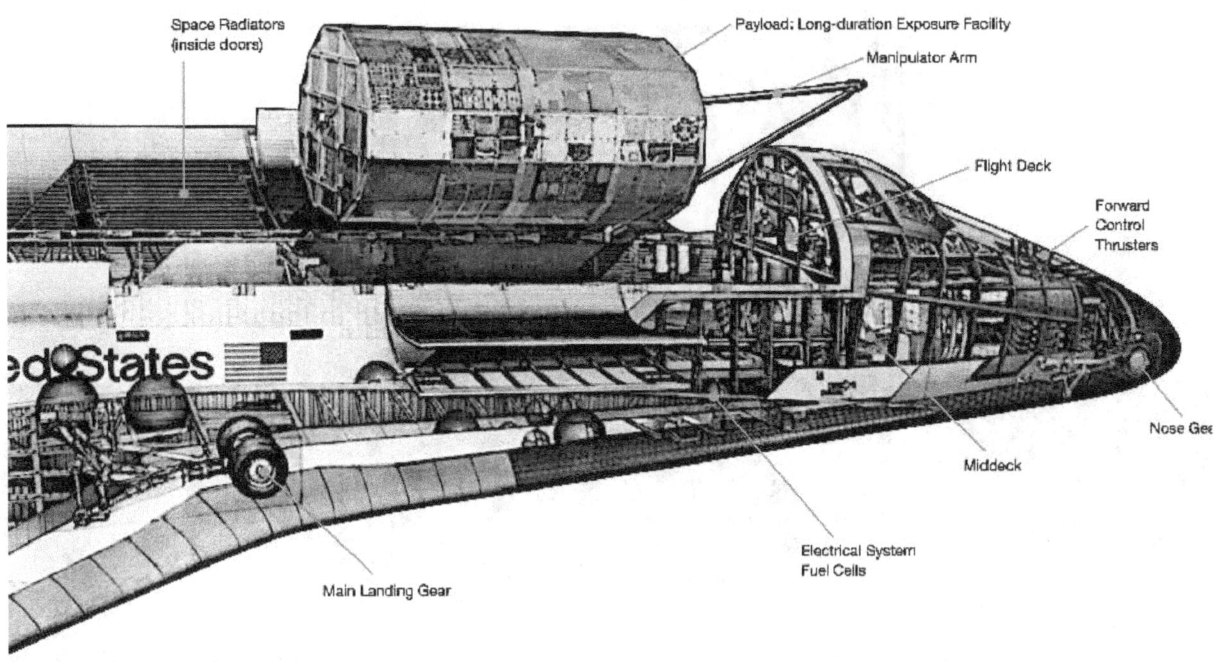

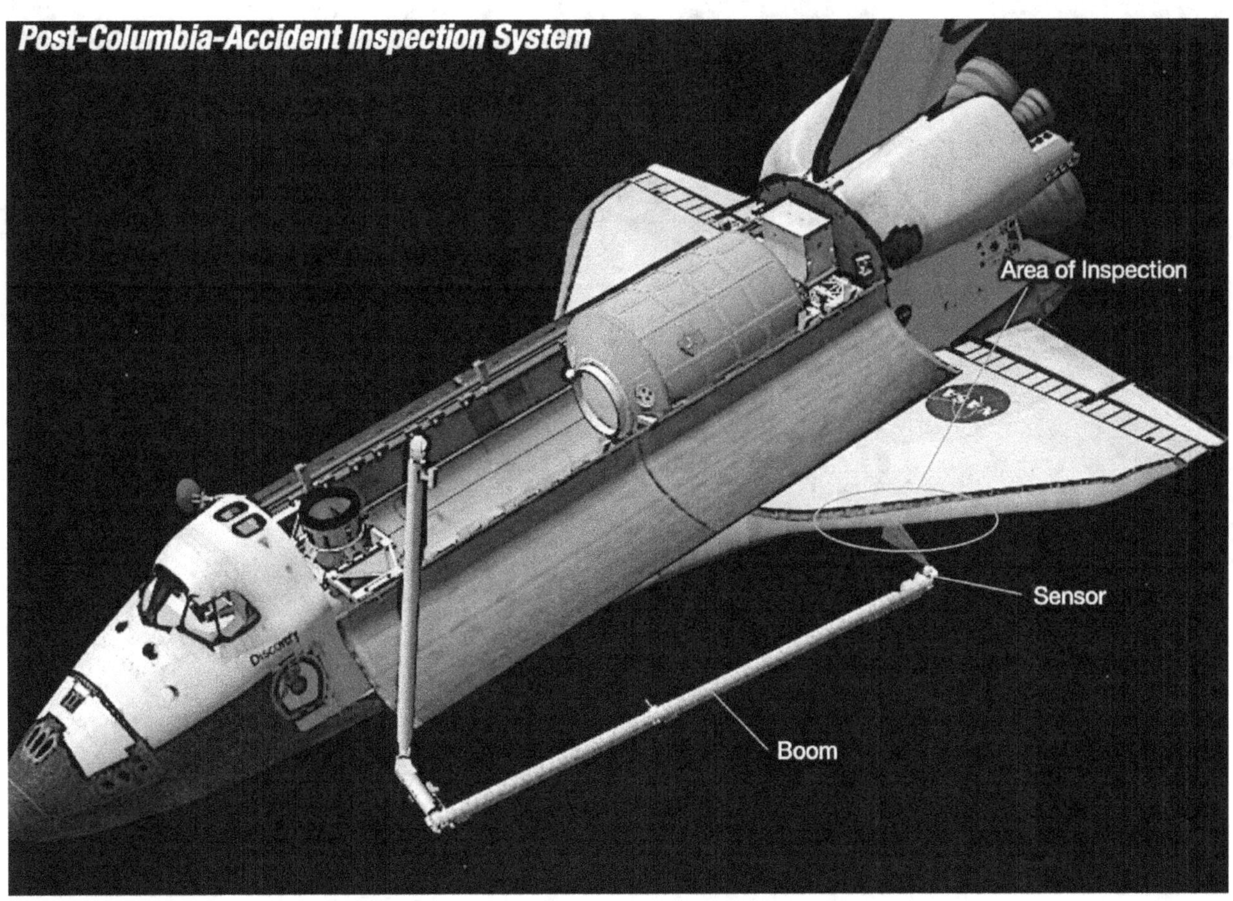

The Orbiter Boom Sensor System inspects the wing leading edge. This system was built for inspections after the Columbia accident

Space Shuttle Flight Deck

Space Shuttle Middeck

NA
ndeavour

Approach and Landing Tests

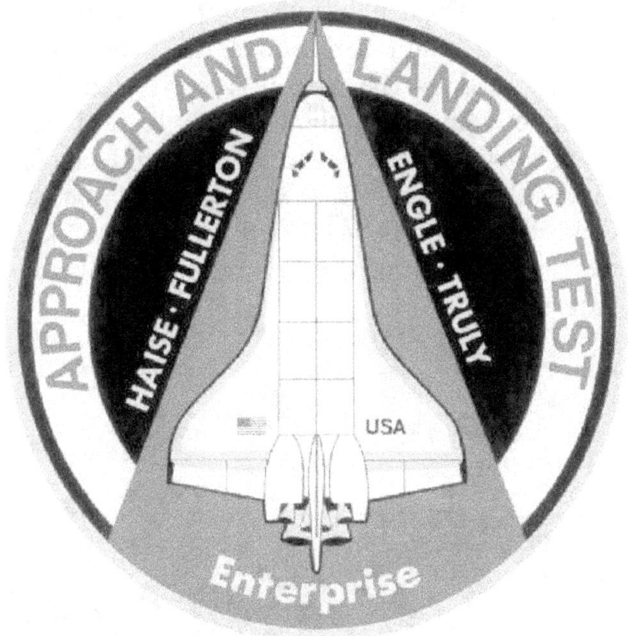

The Approach and Landing Tests (ALT) were a pivotal series of trials conducted in 1977 to evaluate the flight characteristics and overall systems of the Space Shuttle Enterprise, a prototype orbiter. These tests, crucial to the success of the Space Shuttle program, took place from February to October 1977 at Edwards Air Force Base in California. The ALT program was designed to ensure that the Shuttle could operate as intended in various phases of flight, particularly during its return to Earth from space.

The Enterprise was transported by road from the Rockwell plant in Palmdale, California, to the Dryden Flight Research Center at Edwards Air Force Base in January 1977, marking the beginning of the ALT program. The tests were divided into two primary phases: captive-carry flights and free-flight tests. In the first phase, Enterprise remained securely attached to the Shuttle Carrier Aircraft (SCA), a modified Boeing 747, for a series of eleven taxi-tests and flight trials. These captive-carry flights allowed engineers to study the Shuttle's aerodynamic properties while it was mated to the SCA, without subjecting the vehicle to the risks associated with independent flight.

The second phase of the ALT program was far more critical. It involved five free-flight tests in which the Enterprise was jettisoned from the SCA mid-air. During these tests, the Shuttle was piloted by a crew, who guided it through its descent and landing. These flights were essential for assessing the orbiter's ability to glide and land safely, simulating the final phase of an actual space mission.

NASA selected two pairs of astronauts to conduct these crucial tests. The first crew, consisting of Commander Fred W. Haise, Jr. and Pilot C. Gordon Fullerton, took the lead in piloting Enterprise. Haise, a veteran of the Apollo 13 mission, brought extensive experience to the program, while Fullerton would later fly as a pilot on STS-3 and command STS-51-F. The second crew included Commander Joe H. Engle and Pilot Richard H. Truly. Engle, who had previously earned astronaut wings as a pilot on the X-15 program, later flew on STS-2 and commanded STS-51-I. Truly, also a seasoned astronaut would go on to command STS-8.

In addition to the Shuttle crews, the program relied heavily on the expertise of the SCA flight crew, who were responsible for the safe transportation and release of Enterprise during the tests. This crew was led by Captain Fitzhugh L. Fulton, Jr., with Thomas C. McMurtry serving as co-pilot. They were supported by flight engineers Louis E. Guidry, Jr., Victor W. Horton, Vincent A. Alvarez, and William R. Young, who played vital roles in managing the complex systems of the SCA during the flights.

Throughout the ALT program, both ground-based and in-flight tests were conducted to validate the Shuttle's systems, including the launch pad procedures, aerodynamic properties, and landing capabilities. These tests were instrumental in confirming that the Shuttle's design was sound and that it could operate safely and effectively as intended. The success of the ALT program paved the way for the Space Shuttle's first orbital flights and ultimately the operational phase of the Space Shuttle program, marking a significant milestone in NASA's history.

The Approach and Landing Tests (ALT) program, which was integral to the development and success of the Space Shuttle, was meticulously divided into three distinct phases. Each phase was carefully designed to evaluate different aspects of the Shuttle's performance and the capability of the Shuttle Carrier Aircraft (SCA) to transport the orbiter safely.

The crew of NASA's 747 Shuttle Carrier Aircraft (SCA), seen mated with the Space Shuttle Columbia behind them, are from viewers left: Tom McMurtry, pilot; Vic Horton, flight engineer; Fitz Fulton, command pilot; and Ray Young, flight engineer. The SCA is used to ferry the shuttle between California and the Kennedy Space Center, Florida, and other destinations where ground transportation is impractical. The NASA 747 has special support struts atop the fuselage and internal strengthening to accommodate the additional weight of the orbiters. Small vertical fins have also been added to the tips of the horizontal stabilizers for additional stability due to air turbulence on the control surfaces caused by the orbiters.

ALT-1

The Space Shuttle Approach and Landing Test 1 (ALT-1) was pivotal in developing and testing NASA's Space Shuttle program. Conducted as part of a series of five approach and landing tests in 1977, this mission used the Space Shuttle Orbiter Enterprise—the first orbiter built for the program. These tests were designed to evaluate the shuttle's flight and landing capabilities rigorously, paving the way for its future operational missions.

The series began with ALT-1 on February 15, 1977, involving the first runway taxi test. In this test, the Shuttle Enterprise, mounted atop a modified Boeing 747 known as the Shuttle Carrier Aircraft (SCA), reached 89 mph (143 km/h). This low-speed test was crucial for verifying the structural integrity of the combined aircraft and shuttle before more advanced testing could proceed. The test was conducted on a concrete runway, with the shuttle's tailcone attached to ensure streamlined aerodynamics. The SCA crew of Fulton, McMurtry, Horton, and Guidry successfully executed this test, marking the first step in the shuttle's rigorous evaluation process.

left to right, Astronauts C. Gordon Fullerton, pilot of the first crew; Fred W. Haise Jr., commander of the first crew; Joe H. Engle, commander of the second crew; and Richard H. Truly, pilot of the second crew.

Following the taxi test, the ALT-1 mission progressed to the first of three planned "freeflights." In these tests, the Shuttle Enterprise was securely mounted atop the SCA, which carried the orbiter to approximately 24,000 feet (7,300 meters). At this height, the Enterprise was released to begin its brief glide, lasting just over two minutes. The mission's primary goal was to demonstrate the shuttle's ability to glide and land safely without propulsion, simulating the final approach phase of a typical shuttle mission.

During this critical test, astronauts Fred Haise and Gordon Fullerton expertly piloted the Enterprise. They executed a controlled descent, carefully aligning the shuttle with the runway at Edwards Air Force Base. The smooth touchdown of the Enterprise on the dry lakebed marked a significant achievement, providing NASA with essential data on the shuttle's aerodynamic performance and handling qualities. This test was a vital step in confirming that the shuttle could indeed glide to a precise and safe landing, as would be required in actual space missions.

The ALT-1 mission set the stage for subsequent tests, ALT-2 and ALT-3. Each test built upon the previous one by increasing the complexity and precision of the flight maneuvers. These tests were instrumental in validating the shuttle's design, ensuring its safety, and refining its performance characteristics.

The Space Shuttle prototype Enterprise rides smoothly atop NASA's first Shuttle Carrier Aircraft (SCA), NASA 905, during the first of the shuttle program's Approach and Landing Tests (ALT) at the Dryden Flight Research Center, Edwards, California, in 1977. During the nearly one year-long series of tests, Enterprise was taken aloft on the SCA to study the aerodynamics of the mated vehicles and, in a series of five freeflights, tested the glide and landing characteristics of the orbiter prototype. In this photo, the main engine area on the aft end of Enterprise is covered with a tail cone to reduce aerodynamic drag that affects the horizontal tail of the SCA, on which tip fins have been installed to increase stability when the aircraft carries an orbiter.

The success of ALT-1 and the subsequent missions in the series laid the essential groundwork for the Space Shuttle's transition from a concept to a fully operational spacecraft. The data gathered during these tests informed the final design adjustments, contributing significantly to the overall success of the Space Shuttle program. This program would eventually launch 135 missions over three decades, cementing its place as a cornerstone of human space exploration.

ALT-2

The Space Shuttle ALT-2 mission, the second in the series of Approach and Landing Tests (ALT), was conducted on September 13, 1977, at Edwards Air Force Base in California. Building on the success of ALT-1, this mission was a crucial step in further evaluating the aerodynamic and landing capabilities of the Space Shuttle Orbiter Enterprise. These tests were designed to confirm that the Shuttle could safely glide back to Earth and land like an aircraft, a distinctive feature set it apart from earlier spacecraft that relied on parachutes for descent and often splashed down in the ocean.

The orbiter is lowered onto the SCA in the Mate/Demate Device at Edwards AFB

As with the previous test, the ALT-2 mission began with the Enterprise securely mounted atop the modified Boeing 747 Shuttle Carrier Aircraft (SCA). The combined aircraft took off from Edwards Air Force Base and ascended to approximately 24,000 feet (7,300 meters). At this height, the Enterprise was released to begin its independent flight, marking the start of the mission's primary objective: to test the Shuttle's approach and landing characteristics under controlled conditions.

Astronauts Fred Haise and Gordon Fullerton, who had also piloted the ALT-1 mission, were once again in command of the Enterprise. While the overall mission profile for ALT-2 was similar to ALT-1, it introduced a key difference—this time, the glide phase included a more complex series of maneuvers. The Enterprise was required to perform controlled turns and approach patterns, simulating the maneuvers it would need to execute during an actual re-entry from space.

Upon release from the SCA, the Enterprise transitioned into a stable glide. Haise and Fullerton executed a series of S-turns during the descent, a maneuver critical for managing the shuttle's energy and aligning it with the landing site during re-entry. These turns also served to bleed off excess speed, ensuring that the orbiter approached the runway at a safe landing velocity.

In addition to the glide maneuvers, ALT-2 tested the Shuttle's handling and braking capabilities at higher speeds—approximately 140 mph (225 km/h). These tests were essential in verifying that the SCA, with the Shuttle Enterprise on board, could safely taxi under various conditions before taking to the skies. The data collected from these high-speed taxi tests were crucial for ensuring the safe operation of the Shuttle in future missions.

As the Enterprise descended, the crew skillfully guided it toward the designated landing area on the dry lakebed of Edwards Air Force Base. Engineers and controllers closely monitored the shuttle's performance throughout the flight, gathering valuable data on its handling characteristics and flight dynamics. The mission culminated in a successful landing, with the Enterprise touching down smoothly on the runway.

ALT-3

The Space Shuttle ALT-3 mission, conducted on September 23, 1977, marked the third and final free-flight in the series of Approach and Landing Tests (ALT) at Edwards Air Force Base in California. These tests were crucial for validating the Space Shuttle's unique capability to glide and land like an aircraft, a fundamental aspect of the program's reusability and safety.

Following the successful completion of ALT-1 and ALT-2, ALT-3 was designed to build upon the data and experience gained from those earlier tests. While ALT-1 primarily focused on basic glide and landing capabilities, and ALT-2 introduced more complex maneuvers, ALT-3 aimed to further challenge the Shuttle's aerodynamic performance by incorporating additional maneuvers and testing different approach profiles.

As in the previous missions, the Space Shuttle Orbiter Enterprise was mounted atop the Boeing 747 Shuttle Carrier Aircraft (SCA). The combined aircraft took off from Edwards Air Force Base and climbed to approximately 24,000 feet (7,300 meters). Upon reaching the designated release altitude, the Enterprise was separated from the SCA, initiating its freeflight.

Astronauts Fred Haise and Gordon Fullerton, who had also piloted the first two missions, once again controlled the Enterprise. The mission plan for ALT-3 included a more dynamic series of maneuvers than those executed in ALT-2. These maneuvers were designed to simulate the aerodynamic challenges the Shuttle would encounter during an actual re-entry from space, such as managing crosswinds, adjusting glide slopes, and executing precision turns.

Upon release, the Enterprise smoothly transitioned into its glide phase. The astronauts performed a series of S-turns to manage the orbiter's energy and refine its approach to the landing site. These S-turns were crucial for dissipating excess speed and ensuring the Shuttle aligned with the runway. In addition to the S-turns, the crew tested different approach angles and descent rates, simulating the Shuttle's conditions during various re-entry scenarios.

Conducted at higher speeds of 157 mph (253 km/h), ALT-3 continued to assess the SCA's handling and braking capabilities with the Shuttle Enterprise on board. These tests ensured the combined vehicle could safely taxi under various conditions before taking to the skies.

NASA engineers closely monitored the descent and gathered detailed data on the Shuttle's handling characteristics, control responses, and overall aerodynamic behavior. This data was invaluable for refining the Shuttle's flight control systems and ensuring that the orbiter could safely return to Earth under a wide range of conditions.

The mission concluded with a successful landing on the Edwards Air Force Base dry lakebed runway. The Enterprise touched down smoothly, demonstrating that the Space Shuttle's design could perform the complex and precise maneuvers required for a safe landing. The data collected during ALT-3 confirmed the orbiter's ability to handle a real mission's stresses and aerodynamic challenges, further validating the shuttle's design and operational concepts.

ALT-4

On February 18, 1977, NASA conducted the fourth test in the Approach and Landing Tests (ALT) series, ALT-4: Captive-Inert Flight #1. This mission marked the first captive-inert flight of the Space Shuttle Enterprise, a significant step in validating the Shuttle's design and aerodynamics. The Enterprise, securely mounted atop the Boeing 747 Shuttle Carrier Aircraft (SCA), ascended to 16,000 feet (4,877 meters) at 287 mph (462 km/h). The primary goal of this 2-hour, 5-minute flight was to gather data on the aerodynamics of the mated configuration, crucial for the Shuttle's future missions.

The test aimed to evaluate the aerodynamic and landing characteristics of the Space Shuttle Orbiter Enterprise. However, unlike its predecessors, ALT-4 featured a more challenging flight profile designed to simulate conditions closer to those expected during an actual re-entry from space. This test was critical in assessing the Shuttle's performance under more demanding circumstances, particularly its ability to handle the complex aerodynamic forces encountered during re-entry and landing.

As with earlier ALT missions, the Shuttle Enterprise was mounted atop the SCA for takeoff, which occurred over Edwards Air Force Base in California. The SCA carried the Enterprise to approximately 24,000 feet (7,300 meters). Once the release conditions were met, the Enterprise separated from the SCA, initiating its free-flight phase. This phase was crucial for testing the Shuttle's glide capabilities, a key feature of its design that allowed it to return to Earth as a glider after completing its missions in space.

The ALT-4 mission was piloted by experienced astronauts Fred Haise and Gordon Fullerton, who had also flown the previous ALT missions. Their expertise was invaluable as they guided the Enterprise through a series of demanding maneuvers designed to push the Shuttle's flight control systems to their limits. Upon release, the Enterprise entered a glide phase, performing complex maneuvers, including aggressive S-turns and a steeper descent rate. These maneuvers simulated the aerodynamic conditions the Shuttle would experience during the final stages of re-entry, where precise control and stability are critical for a safe landing.

One of the key objectives of ALT-4 was to test the Shuttle's ability to manage its energy and descent rate accurately. The orbiter's design required it to shed excess speed and altitude while maintaining a controlled approach to the runway. The crew executed precise adjustments to the Shuttle's flight path, ensuring it would land within the designated area on the dry lakebed runway. This test phase confirmed that the Shuttle could achieve a controlled and safe landing, even under challenging conditions.

Throughout the flight, NASA engineers and flight controllers closely monitored the Shuttle's performance, gathering vital data on its handling characteristics, stability, and control responses. This data was essential for refining the Shuttle's flight control systems and ensuring that the orbiter could perform as expected during an actual mission. The successful completion of ALT-4 provided NASA with the confidence that the Space Shuttle's innovative design was sound and could handle the complex aerodynamic forces it would face during re-entry and landing.

The mission culminated in a successful landing, with the Enterprise touching down smoothly on the runway. This achievement confirmed that the Shuttle could manage a more challenging flight profile while ensuring a safe and controlled descent. The success of ALT-4 was a significant milestone in the Space Shuttle program, demonstrating that the vehicle was ready for its upcoming orbital missions.

ALT-5

On February 18, 1977, NASA conducted the first captive-inert flight in the Approach and Landing Tests (ALT) series, ALT-4. This crucial test involved the Space Shuttle Enterprise, firmly attached to the Boeing 747 Shuttle Carrier Aircraft (SCA), flying at 16,000 feet (4,877 meters) at 287 mph (462 km/h). The 2-hour, 5-minute flight was designed to gather essential data on the aerodynamics of the mated configuration of the Shuttle and the SCA. The Enterprise's tailcone remained attached throughout the flight, which concluded with a safe landing on the runway, marking a successful step in validating the Shuttle's design.

Following the successful completion of ALT-4, NASA proceeded with the fifth and final mission in the ALT series, ALT-5. This mission culminated an extensive testing program to thoroughly validate the Space Shuttle's approach, glide, and landing capabilities. The data gathered from ALT-5 was critical in ensuring that the Shuttle was fully prepared for its operational role in spaceflight, providing NASA with the confidence to move forward with its first orbital flight.

The ALT-5 mission followed a procedure similar to the previous tests. The Space Shuttle Enterprise was again mounted atop the SCA at Edwards Air Force Base in California. The SCA, specially modified to carry the Shuttle, ascended to approximately 24,000 feet (7,300 meters) with the Enterprise securely attached to its back.

Astronauts Fred Haise and Gordon Fullerton, who had been at the helm for all the previous free-flight missions in the series, piloted the Enterprise for ALT-5. Their extensive experience and familiarity with the Shuttle's behavior were invaluable as they guided the Enterprise through the series' final and most challenging test.

Upon reaching the designated altitude, the Enterprise was released from the SCA and entered its free-flight phase. ALT-5 was meticulously designed to push the Shuttle's performance to its limits, testing its ability to handle a range of aerodynamic conditions that could be encountered during an actual re-entry from space.

During the ALT-5 mission, the Enterprise executed complex maneuvers, including tighter S-turns, varying descent rates, and a steeper approach angle, to simulate challenging conditions for the Shuttle's return to Earth. These maneuvers provided critical data on the vehicle's aerodynamic responses.

As the Enterprise descended, the crew adjusted its speed and altitude for a controlled approach. Accurate management was crucial for a safe landing amidst rapid deceleration and stresses.

NASA engineers and flight controllers monitored the Shuttle's performance, gathering data on its flight characteristics and control responses. This data finalized the Shuttle's flight control systems and ensured safe and reliable operations.

The mission concluded with a successful landing on Edwards Air Force Base's dry lakebed runway, confirming the Space Shuttle's handling of demanding flight conditions and safe descent.

ALT-6

On February 25, 1977, the Space Shuttle Enterprise took to the skies for its third captive-inert flight, ALT-6. This critical test, lasting 2 hours and 28 minutes, involved the Shuttle being securely attached to its carrier aircraft, the Boeing 747 Shuttle Carrier Aircraft (SCA). Flying at 26,600 feet (8,108 meters) and reaching 425 mph (684 km/h), the mission sought to gather crucial data on the Shuttle's aerodynamic properties at higher altitudes and speeds. This data would inch the program closer to the conditions expected during the Shuttle's operational flights.

The primary objective of the ALT-6 mission was to evaluate the orbiter's aerodynamic behavior, particularly during the landing phase—a critical aspect of future space missions. The crew for this mission consisted of two seasoned NASA astronauts: Fred Haise, a veteran of the Apollo 13 mission, and Gordon Fullerton, an accomplished test pilot. Haise served as the commander, while Fullerton took on the pilot role. Together, they managed the Shuttle's flight dynamics during its glide and landing, ensuring that the Enterprise adhered to the planned approach path.

As the mission progressed, the Enterprise's stable flight characteristics became evident. The orbiter responded accurately to control inputs as it glided toward the runway, demonstrating the effectiveness of its design. The mission's objectives included testing the Shuttle's handling qualities during approach and landing, validating the performance of its flight control systems, and assessing the capabilities of its thermal protection system during a simulated landing sequence.

Approaching the runway at Edwards Air Force Base, Haise and Fullerton skillfully managed the Shuttle's descent rate and speed. Their precise control guided the Enterprise to a smooth touchdown on the dry lakebed, a successful landing that underscored the soundness of the Space Shuttle's design. This achievement confirmed that the Shuttle could perform the critical landing phase of a mission, landing safely on a conventional runway after returning from space.

ALT-7

On February 28, 1977, the Space Shuttle Enterprise embarked on its fourth captive-inert flight, ALT-7. Continuing the series of critical tests, ALT-7 achieved 28,565 feet (8,707 meters) and maintained 425 mph (684 km/h). Over 2 hours and 11 minutes, this mission further validated the Shuttle's behavior in its mated configuration at higher altitudes, with all systems performing as expected.

As with previous missions, the Enterprise was carried to 24,000 feet by the Shuttle Carrier Aircraft (SCA) before being released over Rogers Dry Lake. This location provided an expansive and safe area for the Shuttle's unpowered glide and landing. The crew for ALT-7 included NASA astronauts Fred Haise, who served as the commander, and Gordon Fullerton, the pilot—both experienced from their earlier ALT missions.

Upon release from the SCA, the Enterprise smoothly transitioned into its glide phase. Haise and Fullerton closely monitored the orbiter's descent, skillfully managing its flight path, speed, and attitude as it approached the runway. The mission's objectives included fine-tuning the Shuttle's flight control software and gathering additional data on the orbiter's handling qualities under various flight conditions.

A key focus of ALT-7 was assessing the orbiter's behavior during the critical moments leading up to touchdown. The data collected during this flight was crucial in refining the procedures and systems that would later be employed during actual space missions. Haise and Fullerton executed a precise landing as the Enterprise neared the runway, bringing the orbiter to a smooth, controlled stop on the dry lakebed.

ALT-8

On March 2, 1977, the Space Shuttle Enterprise embarked on its fifth and final captive-inert flight, ALT-8. This crucial test marked the culmination of the captive-inert flight phase, a series of missions designed to validate the Shuttle's behavior while attached to the Shuttle Carrier Aircraft (SCA) at various altitudes and speeds. During ALT-8, the Shuttle and its carrier reached 30,000 feet (9,144 meters) and attained 474 mph (763 km/h), making it the highest and fastest flight in this phase.

The flight, which lasted an hour and 39 minutes, was a pivotal moment in the Space Shuttle program. It validated the orbiter's readiness to progress to more intricate phases, including captive-active and freeflight. Throughout the mission, the Shuttle's systems functioned flawlessly, providing the final data set to confirm its preparedness for the challenges ahead.

The success of ALT-8 marked the conclusion of the captive-inert flight series, paving the way for the subsequent critical stages of testing. These subsequent phases would entail more dynamic evaluations of the Shuttle's capabilities, ultimately culminating in its inaugural orbital flight.

ALT-9

On June 18, 1977, the Space Shuttle Enterprise entered a new testing phase with the commencement of ALT-9, the first captive-active flight. Building on the success of the previous captive-inert flights, this mission marked a significant step forward as the Shuttle's systems were powered on for the first time during flight, enabling it to control specific flight surfaces while remaining attached to the Shuttle Carrier Aircraft (SCA).

During ALT-9, the Shuttle, piloted by veteran astronauts Fred Haise and Gordon Fullerton, ascended to 14,970 feet (4,563 meters) and reached 208 mph (335 km/h). The flight lasted 55 minutes and 46 seconds, during which the Shuttle's ability to manipulate its aerodynamic surfaces, such as the elevons and rudder, was carefully tested and observed.

This mission demonstrated the Space Shuttle's capability to effectively manage its flight dynamics while simultaneously mating with the Space Carrier Aircraft (SCA). It provided valuable insights into the Space Shuttle's aerodynamic characteristics and the efficacy of its flight control systems when integrated with the carrier aircraft. The precise handling of the Enterprise by Haise and Fullerton during this flight further validated the Shuttle's design, demonstrating its ability to effectively respond to aerodynamic control inputs even in the captive-active configuration.

ALT-10

On June 28, 1977, the Space Shuttle Enterprise embarked on the ALT-10 mission, the second in a series of captive-active flights. This mission entailed astronauts Joe Engle and Richard Truly piloting the Enterprise through a pivotal testing phase. While still tethered to the Shuttle Carrier Aircraft (SCA), the Shuttle ascended to an altitude of 22,030 feet (6,715 meters) and attained 310 mph (499 km/h). The flight duration was 62 minutes, during which the Shuttle's handling characteristics under partial control of its aerodynamic surfaces were meticulously evaluated.

The crew members of the SCA, including pilots Fitz Fulton, Tom McMurtry, Vic Guidry, and NASA engineer Dick Young, played a pivotal role in ensuring the mission's success. Their seamless coordination with the shuttle crew facilitated the acquisition of data that would be instrumental in assessing the orbiter's performance under these specific flight conditions.

During ALT-10, the Space Shuttle Enterprise's flight control systems were partially activated, enabling Engle and Truly to manipulate the orbiter's aerodynamic surfaces, such as the elevons and rudder. Concurrently, the Shuttle remained connected to the SCA. This test provided invaluable data on the Shuttle's handling capabilities when it possessed partial control over its aerodynamics. It represented a significant milestone in validating the orbiter's design and ensuring its safe maneuverability during the final approach and landing phases.

ALT-11

On July 26, 1977, the Space Shuttle Enterprise progressed further in its testing with the ALT-11 mission, the third in the series of captive-active flights. This mission was piloted once again by seasoned astronauts Fred Haise and Gordon Fullerton, who had gained extensive familiarity with the Shuttle's capabilities through previous tests. During ALT-11, the Enterprise attained an altitude of 30,292 feet (9,233 meters) and 311 mph (501 km/h), significantly surpassing the Shuttle's performance in the captive-active configuration.

The flight, which lasted 59 minutes and 53 seconds, was immensely significant in validating the Shuttle's behavior at elevated altitudes and speeds. The mission's primary objective was to ascertain the Enterprise's ability to withstand the aerodynamic stresses it would encounter during actual space missions. By subjecting the Shuttle to these conditions, the team diligently collected crucial data on the orbiter's handling characteristics and the efficacy of its flight control systems under more demanding scenarios.

ALT 12

On August 12, 1977, a pivotal moment in the Space Shuttle program unfolded with the inaugural freeflight of the Shuttle Enterprise, designated ALT-12. This significant test marked the first instance of the Enterprise flying independently after being released from the Shuttle Carrier Aircraft (SCA). The mission, which took place over the expansive desert landscape of Edwards Air Force Base in California, was meticulously planned to evaluate the Shuttle's glide and landing capabilities.

Piloted by astronauts Fred Haise and Gordon Fullerton, the Shuttle Enterprise attained 310 mph (499 km/h) and ascended to an altitude of 24,100 feet (7,346 meters). These conditions closely simulated those anticipated during an actual re-entry from space. For nearly 54 minutes, the Shuttle remained aloft, although its independent glide lasted for just over five minutes. During this critical phase, the Enterprise demonstrated its ability to descend gracefully and land with precision, touching down smoothly on the dry lakebed of Rogers Dry Lake.

A notable feature of this test was the attachment of the tailcone, which served to reduce aerodynamic drag and create conditions analogous to those the Shuttle would encounter during re-entry into Earth's atmosphere. The successful completion of this flight provided invaluable data and bolstered confidence, paving the way for subsequent missions that would ultimately enable the Space Shuttle to transport astronauts into space and return them safely to Earth. This inaugural freeflight represented a monumental advancement, demonstrating the feasibility of the Shuttle's design and its potential as a reusable spacecraft, a cornerstone of NASA's vision for future space exploration.

ALT-13

On September 13, 1977, just a month after successfully completing the first free flight, the Space Shuttle Enterprise took to the skies once again for its second free flight, designated ALT-13. This mission was part of NASA's ongoing efforts to rigorously test and validate the Shuttle's ability to glide and land autonomously after release from the Shuttle Carrier Aircraft (SCA).

This time, the Enterprise was under the expert control of astronauts Joe Engle and Richard Truly, who would later command space shuttle missions in orbit. The flight followed a similar profile to its predecessor, with the Shuttle reaching 26,000 feet (7,925 meters) and maintaining 310 mph (499 km/h). The primary objective of ALT-13 was to further confirm the Shuttle's glide and landing capabilities under controlled conditions.

Enterprise on its approach during the second free-flight

Enterprise lands at the conclusion of freeflight #2

The Shuttle glided through the air for 5 minutes and 28 seconds, demonstrating its aerodynamic stability and precise control, before successfully landing on the dry lakebed at Edwards Air Force Base. The results of this flight reinforced the findings from the first freeflight, providing additional confidence in the Shuttle's design and performance. These early tests were crucial in building a foundation for the Space Shuttle program, ultimately leading to the Shuttle's role as a workhorse for NASA's manned space missions.

ALT-14

On September 23, 1977, the Space Shuttle Enterprise embarked on its third freeflight, designated ALT-14, further advancing NASA's ambitious Space Shuttle program. Piloted once again by astronauts Fred Haise and Gordon Fullerton, who had successfully conducted the first freeflight, this mission continued to build on the rigorous testing of the Shuttle's glide and landing capabilities.

During ALT-14, the Enterprise was released from the Shuttle Carrier Aircraft (SCA) at 24,700 feet (7,529 meters), slightly lower than the previous flight, and reached 290 mph (467 km/h). The primary objective of this flight was to continue validating the Shuttle's aerodynamic stability and landing precision under varying conditions.

The Shuttle glided smoothly for 5 minutes and 34 seconds, demonstrating consistent flight stability and control, before successfully landing on the dry lakebed at Edwards Air Force Base. The flight confirmed the Shuttle's ability to perform under different flight profiles and provided valuable data for future missions. These incremental tests brought NASA closer to realizing the Space Shuttle's potential as a reusable spacecraft capable of returning from space and landing safely, a critical component of the program's long-term goals.

ALT-15

On October 12, 1977, the Space Shuttle Enterprise completed its fourth free flight, ALT-15, marking a critical milestone in the Shuttle's testing phase. Piloted by astronauts Joe Engle and Richard Truly, the flight was significant as it was the first conducted without the tailcone, exposing the Shuttle's aft section and simulating the aerodynamic conditions it would experience during re-entry. Released from the Shuttle Carrier Aircraft (SCA) at 22,400 feet (6,828 meters), Enterprise reached 278 mph (447 km/h) and glided for 2 minutes and 34 seconds. This test provided NASA with crucial data on the Shuttle's stability and control in this configuration. The flight ended with a smooth landing on Edwards Air Force Base's dry lakebed, further validating the Shuttle's design.

On October 26, 1977, ALT-16 marked the final free flight of the Approach and Landing Tests (ALT). Piloted by Fred Haise and Gordon Fullerton, the Shuttle was released from the SCA at 19,000 feet (5,791 meters) and reached 283 mph (455 km/h). Like ALT-15, this flight was conducted without the tailcone, simulating re-entry conditions. Enterprise glided for 2 minutes and 1 second before successfully landing on the Edwards AFB runway, marking the first runway landing in the test series. This final test demonstrated the Shuttle's ability to glide through the atmosphere and land safely on a standard runway, simulating actual spaceflight conditions.

The successful completion of ALT-16 concluded the Approach and Landing Test program, confirming that the Shuttle could perform as intended: launch into orbit, re-enter the atmosphere, and land safely like an airplane. This validated the Shuttle's revolutionary concept of reusability and paved the way for the first orbital flight in 1981.

After completing the flight tests, Enterprise was repurposed for a series of ground tests to validate the Shuttle system. It was transported to the Marshall Space Flight Center in Alabama, where it underwent vibration tests to simulate the stresses of launch and ascent. These tests ensured the structural integrity of the fully assembled Shuttle stack, including the orbiter, external tank, and solid rocket boosters (SRBs).

Following the tests at Marshall, Enterprise was sent to the Kennedy Space Center (KSC) in Florida for fit checks of the Vehicle Assembly

Building (VAB) and the launch pad. These exercises confirmed the Shuttle system's readiness for launch, laying the groundwork for Columbia's first flight on STS-1 in April 1981.

Edwards Air Force Base (AFB) played a key role in the Shuttle program as the primary landing site during its early years. Its vast dry lakebeds provided a forgiving surface for landings, while its proximity to Plant 42 in Palmdale, California, where the Shuttle was serviced, allowed efficient processing between missions. Although Kennedy Space Center became the preferred landing site after 1991 due to cost savings, Edwards AFB remained a vital backup, hosting notable landings through the Shuttle program's end.

Enterprise mated to external tank and dummy SRBs stands on Kennedy Space Center Pad 39A during fit check tests twenty months prior to STS-1.

Chapter 2 - The Dawn of the Shuttle Era (1981–1982)

The Space Shuttle Mission Numbering

The United States Space Shuttle program, officially designated as the Space Transportation System (STS), employed an evolving mission designation system that mirrored its growth and challenges.

Initially, Shuttle missions were sequentially numbered based on their order of launch, with missions such as STS-7 indicating the seventh mission in the program. However, following the Apollo 13 mishap and due to NASA Administrator James M. Beggs's triskaidekaphobia—an aversion to the number 13—the system underwent a substantial transformation in 1984. NASA introduced a novel coding scheme to circumvent the numbering of a mission as STS-13.

This revised system assigned each mission a unique code, such as STS-41-B. The initial digit (or pair of digits for fiscal years 1990 and beyond) represented the fiscal year in which the program's schedule was planned. For instance, STS-41-B was scheduled for FY 1984, while missions scheduled for FY 1985 were numbered 51-A through 51-L. The second digit in the code denoted the launch site, where "1" signified Kennedy Space Center and "2" was reserved for Space Launch Complex 6 at Vandenberg Air Force Base, although the latter was never utilized for Shuttle launches. The letter in the code represented the sequence of the mission's scheduling.

These codes were assigned when launches were initially scheduled and remained unchanged even if missions were delayed or rescheduled. This coding system was used from STS-41-B through STS-51-L, although the highest code used was STS-61-C. Despite this external coding system, NASA used sequential numbers internally to process paperwork.

The tragic loss of Challenger during STS-51-L led NASA to abandon the coding scheme and return to a more straightforward sequential numbering system. This new numbering began with STS-26, the first "return to flight" mission after the Challenger disaster. To differentiate these flights from prior ones, an "R" suffix was added, starting with STS-26R. This suffix, indicating "reflight," was used through STS-33R before being dropped altogether.

Due to the different numbering systems, there were instances where missions under different systems shared the same number but with different letters, such as STS-51, which referred to a mission conducted in 1993, long after STS-51-A in 1984. This created some confusion, but it wasn't until STS-127 in 2009 that NASA returned to a standard, consistent order for numbering missions.

STS-1: A Historic Launch

The April 12 launch at Pad 39A of STS-1, just seconds past 7 a.m., carries astronauts John Young and Robert Crippen into an Earth orbital mission scheduled to last for 54 hours, ending with unpowered landing at Edwards Air Force Base in California.

STS-1, the first orbital mission of NASA's Space Shuttle program, marked a pivotal moment in space exploration, inaugurating a new era of reusable spacecraft. The mission, flown by the orbiter *Columbia*, launched on April 12, 1981, and returned safely to Earth on April 14, 1981, after 54.5 hours in space. During this time, *Columbia* completed 37 orbits around the Earth, demonstrating the capabilities of the shuttle and its systems. The flight was crewed by Commander John W. Young and Pilot Robert L. Crippen, making it the first American crewed spaceflight since the Apollo-Soyuz Test Project in 1975.

Columbia's launch coincidentally fell on the 20th anniversary of Yuri Gagarin's historic Vostok 1 mission, the Soviet Union's first human spaceflight. Initially planned for April 10, 1981, STS-1 was delayed due to a technical issue but successfully launched two days later, making the timing an unintended yet symbolic nod to the milestones in human space exploration.

The STS-1 crew members are: Commander, John W. Young and Pilot Robert L. Crippen.

John W. Young, the most experienced astronaut in NASA's history at the time, was selected as the mission commander in early 1978. Having flown twice on Project Gemini and twice on the Apollo program, including commanding *Apollo 16* and walking on the Moon in 1972, Young had unmatched expertise. He had also served as the Chief of the Astronaut Office since 1974, a role that allowed him to recommend himself for the historic mission. STS-1 was his fifth spaceflight, further solidifying his legendary status within NASA.

Robert L. Crippen, the mission's pilot, was a rookie astronaut, making his first spaceflight. Selected as part of NASA's Astronaut Group 7 after the cancellation of the Manned Orbiting Laboratory (MOL), Crippen had previously served in supporting roles, including as a capsule communicator (CAPCOM) during all three Skylab missions and the Apollo-Soyuz Test Project. His selection for STS-1 would make him the first member of his astronaut class to fly in space, and his partnership with Young provided a balanced crew of veteran leadership and fresh perspective.

The crew trained extensively for the mission, with both astronauts setting a record for the longest preflight training in NASA's history. Originally scheduled for launch in 1979, the mission's delay allowed Young and Crippen to rigorously prepare for the many challenges that could arise. They played a critical role in the design of the Shuttle's controls, which featured 2,214 switches and displays—three times more than the Apollo command module. The complex nature of the Shuttle's systems required extensive contingency planning, and the crew had 22 manuals, each three inches thick, detailing procedures for various potential failures. In one scenario involving an electronics failure due to a cooling system malfunction, the astronauts were prepared to execute a procedure involving 255 steps.

To ensure safety, *Columbia* was equipped with two Extravehicular Mobility Units (EMUs) in the event of an emergency spacewalk. In such a scenario, Crippen would be the first to exit the orbiter, while Young would stand by inside to assist if needed.

While Young and Crippen were the primary crew, a backup team consisting of Commander Joe H. Engle and Pilot Richard H. Truly was also prepared to fly the mission. This team would go on to fly STS-2 later in 1981. In addition, a support crew was integral to the mission's success. Daniel C. Brandenstein served as the ascent CAPCOM, overseeing communications during the critical launch phase, while Joseph P. Allen was the entry CAPCOM during re-entry. Henry W. Hartsfield also contributed to mission support, ensuring that the team on the ground could provide essential assistance to the crew in space.

During the planning stages of NASA's early Space Shuttle missions, the agency considered performing suborbital test flights before committing to a full orbital mission. Under the Carter Administration, NASA management, guided by Vice President Walter F. Mondale, chairman of the National Space Council, proposed a suborbital test that would land at an emergency site in Dakar, Senegal. Another suggestion was to utilize STS-1 to test the Return to Launch Site (RTLS) abort scenario, where the Shuttle would abort shortly after launch and return to Kennedy Space Center. The RTLS scenario required jettisoning the Solid Rocket Boosters (SRBs) and firing the Shuttle's main engines to reverse its trajectory back to the launch site.

This early abort test was considered high-risk, with potential dangers far exceeding the benefits of such a trial. John W. Young, the mission commander for STS-1, strongly opposed these proposals, citing the extreme risks involved. Young's extensive experience, which included walking on the Moon as Commander of *Apollo 16*, lent significant weight to his opinions. Ultimately, NASA followed Young's advice, and STS-1 proceeded as the first full orbital flight of the Space Shuttle. In Young's own words regarding the RTLS test: "Let's not practice Russian roulette, because you may have a loaded gun there." His caution proved wise, as STS-1 would go on to make history as a successful orbital mission.

On April 12, 1981, exactly 20 years after Yuri Gagarin's first human spaceflight, *Columbia* lifted off from Pad A at Launch Complex 39, Kennedy Space Center. The launch occurred at 12:00:04 UTC after a two-day delay caused by a technical issue with the Shuttle's four primary general-purpose computers. The computers had failed to provide proper timing to the backup flight system, which resulted in a software patch being installed to correct the problem.

STS-1 marked several firsts in spaceflight history: it was the first time solid-fuel rockets had been used for a NASA crewed launch and the first U.S. crewed spacecraft to fly without an uncrewed test flight. *Columbia* itself set a record, spending 610 days in the Orbiter Processing Facility (OPF) prior to launch, largely due to the time required to replace many of its heat shield tiles.

The mission's primary objective was to achieve a safe ascent into orbit, thoroughly test the Shuttle's systems, and ensure a safe return to Earth. The only payload aboard *Columbia* was a Development Flight Instrumentation (DFI) package that housed various sensors to record the Shuttle's performance throughout the mission. The mission met all 113 of its test objectives, verifying the Shuttle's spaceworthiness.

During the final countdown, Launch Director George Page delivered a message from President Ronald Reagan to the crew, expressing support and well wishes. At liftoff, the ignition of *Columbia*'s three RS-25 main engines was followed by the firing of the SRBs, which Crippen compared to being launched from a "steam catapult." As the Shuttle cleared the tower, it performed a right roll and pitched to a "heads-down" attitude to reduce wing stress while transitioning into orbit. Control was handed from the launch team in Florida to the Silver Team in Texas, led by Flight Director Neil Hutchinson and CAPCOM Dan Brandenstein.

As *Columbia* climbed, the Shuttle's engines were throttled down to 65% to mitigate aerodynamic stress during Max Q—occurring 56 seconds into the flight. The SRBs were jettisoned after 2 minutes and 12 seconds, at an altitude of 53,000 meters. The main engines shut down at 8 minutes and 34 seconds into the flight, with the external tank jettisoned soon after to break up over the Indian Ocean. Two Orbital Maneuvering System (OMS) burns positioned *Columbia* into a near-perfect orbit of 246 by 248 kilometers.

Once in orbit, Young and Crippen began testing the Shuttle's systems. Key tasks included calibrating the Crew Optical Alignment Sight (COAS), testing the Reaction Control System (RCS), and evaluating the performance of *Columbia*'s Inertial Measurement Unit (IMU) and star tracker. Opening the payload bay doors was critical, as the radiators on the doors allowed heat rejection from the Shuttle's systems. The crew noticed minor damage to thermal protection tiles on the OMS pods during this operation, which was televised to mission control. Despite the minor damage, *Columbia* performed as expected, and the scheduled systems tests were completed successfully.

Notably, the astronauts received a phone call from Vice President George H. W. Bush during their second day in orbit. President Reagan had planned to visit Mission Control but was recovering from an assassination attempt, making Bush's call a significant moment during the mission.

The mission's final phase began with the deorbit burn over the Indian Ocean. This maneuver reduced *Columbia*'s orbital altitude, allowing it to reenter the atmosphere. During reentry, *Columbia*'s aerodynamic performance differed slightly from preflight predictions, leading to higher roll oscillations and the need for larger body flap adjustments to maintain control. As *Columbia* streaked over the California coast, the Shuttle slowed and prepared for landing at Edwards Air Force Base. John Young manually piloted the Shuttle in its final approach, with T-38 chase planes flanking it. Touchdown occurred at 18:21 UTC on April 14, 1981, with *Columbia* rolling to a stop after a flawless reentry.

Despite approximately 70 anomalies observed during the mission—ranging from thermal protection tile loss to overpressure waves affecting the orbiter's structure—STS-1 was hailed as a success. Young and Crippen safely returned *Columbia* to Earth after 36 orbits, marking the beginning of a new era of space exploration with reusable spacecraft. Following some modifications, *Columbia* flew the next four Shuttle missions, cementing its place in history as NASA's pioneering orbiter.

STS-2: A Test of Reusability

On November 12, 1981, Space Shuttle Columbia lifted off from Kennedy Space Center for the second time, marking the historic STS-2 mission. This mission was significant as it was the first time in history that a spacecraft was reused for an orbital flight, emphasizing the Space Shuttle program's groundbreaking focus on reusability.

The primary objectives of STS-2 were to test the capabilities of the Space Shuttle further and to demonstrate the reusability of the orbiter, Columbia, which had completed its first mission, STS-1, just seven months earlier. The mission also aimed to conduct the first scientific research from the Shuttle, including Earth observation and space technology experiments.

Joseph H. Engle, an experienced test pilot who had previously flown the X-15 rocket plane, commanded the mission. Engle was joined by Pilot Richard H. Truly, who, like Engle, was making his first spaceflight. Both astronauts were integral to the Shuttle program's development, bringing extensive expertise to this critical mission.

STS-2 crewmembers Joe H. Engle (commander), left, and Richard H. Truly (pilot)

As with its predecessor, the launch of STS-2 was met with high anticipation. After several delays due to technical issues, Columbia finally lifted off from Launch Complex 39A at 10:10 a.m. EST. The launch was initially smooth, but one of the shuttle's three fuel cells failed approximately six minutes into the flight. The fuel cells were crucial for providing electrical power to the orbiter, and the failure prompted mission controllers to shorten the mission from its planned five days to just two.

Despite this setback, the mission proceeded with its scientific objectives. A major component of STS-2 was the deployment and operation of the Shuttle Imaging Radar-A (SIR-A), part of the first payload in the Shuttle's cargo bay. This radar was designed to capture high-resolution images of Earth's surface, providing valuable data for geological and environmental studies. Additionally, the mission carried the Office of Space Science-1 (OSS-1) package, which included a range of experiments designed to study solar physics, cosmic rays, and the effects of space on multiple materials.

Engle and Truly conducted these experiments while performing additional tests on the Shuttle's systems. These included evaluating the Shuttle's Remote Manipulator System (RMS), or robotic arm, in its first operational use. The RMS was designed to deploy, capture, and maneuver payloads in space, and its successful testing on STS-2 would pave the way for its critical role in future missions.

In addition to its primary objectives, the STS-1 mission included a series of scientific experiments and tests that would contribute to the growing body of knowledge in space science and engineering. Among these was the Shuttle Multispectral Infrared Radiometer, designed to capture data across different wavelengths of infrared light, which would be crucial for future environmental monitoring and remote sensing. The Feature Identification and Location Experiment aimed to enhance the accuracy of Earth observation from space by identifying specific geographical features and their locations. This experiment would lay the groundwork for more precise satellite mapping technologies.

Another significant experiment was the Measurement of Air Pollution from Satellites, which provided early insights into how space-based instruments could monitor air quality on Earth. The Ocean Color Experiment focused on analyzing the color of ocean waters from space, offering a new perspective on marine biology and oceanography. The mission also included the Night/Day Optical Survey of Lightning, which sought to improve the understanding of lightning patterns and their global distribution by observing them from space.

The Heflex Bioengineering Test was conducted to study the effects of the space environment on biological systems, providing valuable data for future bioengineering applications in space. Additionally, the Aerodynamic Coefficient Identification Package (ACIP) was used to refine the Shuttle's aerodynamic models by gathering data on the forces and moments acting on the vehicle during different phases of flight.

The STS-2 mission, originally planned for five days, was significantly shortened due to a critical technical failure. One of the three fuel cells responsible for generating electricity and drinking water malfunctioned early in the flight, forcing mission control to cut the mission to just two days. As a result, several planned activities, including testing of the Canadarm, were officially canceled. However, the crew, consisting of Commander Joe Engle and Pilot Richard Truly, took advantage of scheduled sleep periods and loss of signal (LOS) moments when communication with Mission Control was interrupted. During these times, they tested the Canadarm discreetly, demonstrating their initiative and dedication to the mission's objectives.

The deorbit and reentry phase of STS-2 differed from that of its predecessor, STS-1, in terms of guidance testing. While the first shuttle entry had followed a conservative approach to automatic guidance testing, the success of that mission allowed for more experimental maneuvers on STS-2. Engle, leveraging his extensive experience flying the X-15 rocket plane, manually executed 29 planned Programmed Test Inputs (PTIs) in Control Stick Steering (CSS) mode. These maneuvers were critical in assessing the Space Shuttle's stability margins, providing valuable data for future engineering modifications. Notably, Engle did not hand-fly the entire reentry phase, but contrary to some reports, he performed manual maneuvers throughout the entire speed range of reentry. Adjustments were also made to the elevon scheduling to relieve stress on the shuttle's body flap, addressing anomalies encountered during STS-1's entry.

The mission concluded with Columbia's landing on Runway 23 at Edwards Air Force Base at 21:23 UTC on November 14, 1981. The shuttle had completed 37 orbits around the Earth, covering 1,730,000 kilometers (1,070,000 miles) in just over two days—2 days, 6 hours, 13 minutes, and 12 seconds. Astronauts "Hoot" Gibson and Kathy Sullivan flew Chase 1, escorting Columbia during its final approach. Despite the shortened flight, more than 90% of the mission's objectives were successfully accomplished.

STS-2 also marked several key milestones in Space Shuttle history. Notably, modifications to the water sound suppression system at the launch pad, designed to absorb the overpressure wave from the solid rocket boosters, proved effective. No thermal protection tiles were lost during the launch, and only 12 tiles were damaged—a significant improvement over STS-1. Columbia was safely flown back to Kennedy Space Center on November 25, 1981, marking the mission's final milestone.

A significant issue uncovered during STS-2 was the first observation of O-ring blow-by in the solid rocket boosters. Though this was discovered post-flight, engineers conducted a series of tests, deliberately damaging another O-ring and subjecting it to three times the flight pressure. The O-ring survived, and the design was deemed flightworthy. Unfortunately, similar O-ring problems would reoccur in subsequent shuttle flights and eventually contribute to the Challenger disaster in 1986.

STS-2 was also the final mission where the Space Shuttle's external fuel tank was painted white. To reduce overall launch weight, NASA decided to forgo painting the tank starting with STS-3, saving approximately 272 kilograms (600 pounds). The unpainted external tank took on a distinct orange-brown color, becoming an iconic feature of the Space Shuttle in all subsequent missions.

Years later, in 2006, some members of the spaceflight community speculated whether the white paint might have prevented the foam shedding issue that led to the loss of Columbia during STS-107. However, NASA's consensus was that the paint would not have significantly prevented the disaster.

STS-3: A Test of Endurance and Ingenuity

STS-3, NASA's third Space Shuttle mission, marked another significant milestone in the early days of the Shuttle program. The mission, conducted by the Space Shuttle Columbia, launched on March 22, 1982, and landed eight days later on March 30, 1982. Crewed by Commander Jack R. Lousma and Pilot C. Gordon Fullerton, this mission was characterized by extensive orbital endurance testing of the Columbia orbiter and numerous scientific experiments.

This mission was notable for several firsts and unique challenges. It was the first Shuttle launch featuring an unpainted external tank, a change that reduced the vehicle's weight and improved its performance. Additionally, STS-3 was the only mission in the Shuttle program to land at White Sands Space Harbor near Alamogordo, New Mexico, a decision driven by the unexpected flooding at the originally planned landing site, Edwards Air Force Base in California.

STS-3 - Jack R. Lousma (commander), left, and C. Gordon Fullerton (pilot)

Jack R. Lousma, the mission commander, brought a wealth of experience to the flight, having previously served as the pilot of the second Skylab crew (Skylab 3) in 1973, where he spent 59 days aboard the space station. Lousma had also been selected as the pilot for STS-2, which was initially scheduled as a Skylab reboost mission. However, delays in the Shuttle's development meant Columbia could not be launched in time to rendezvous with Skylab in 1979. Following these delays and the retirement of STS-2 Commander Fred W. Haise Jr., Lousma was promoted to Commander of STS-3. His prior experience included serving on the support crews for Apollo 9, 10, and 13, where he famously acted as CAPCOM during the near-disastrous Apollo 13 mission. He was also the backup Docking Module Pilot for the Apollo-Soyuz Test Project in 1975.

Pilot C. Gordon Fullerton, a rookie on this flight, had transferred to NASA in 1969 following the cancellation of the U.S. Air Force's Manned Orbiting Laboratory program. Fullerton had previous experience with the Shuttle, having flown the Shuttle Enterprise as pilot alongside Haise during the Approach and Landing Tests (ALT) in 1977. He also served on the support crews for Apollo 14, 15, 16, and 17, gaining valuable experience that he brought to his first spaceflight on STS-3.

STS-3's mission objectives were ambitious, focusing on both testing and scientific research. The primary objective was the continued testing of the Shuttle's Remote Manipulator System (Canadarm), an essential tool for future missions. The mission also involved extensive thermal testing of Columbia by exposing different parts of the orbiter, including its tail, nose, and top, to the Sun for varying periods. This testing revealed a critical issue: prolonged exposure to the Sun caused the cargo bay doors to warp slightly, preventing them from closing fully. The crew ingeniously resolved this issue by rolling the orbiter to balance the temperatures around it.

The mission's payload bay carried several important scientific instruments. Columbia again carried the Development Flight Instrumentation (DFI) package and the OSS-1 (Office of Space Science and Applications) package, which was mounted on a Spacelab pallet. These instruments were designed to gather data on the near-Earth environment and study the extent of contamination caused by the orbiter itself. Notably, the OSS pallet included an X-ray detector for measuring the polarization of X-rays emitted by solar flares, providing valuable data for space weather research. A test canister for the Small Self-Contained Payload program, known as the Getaway Special (GAS), was mounted in the payload bay.

The Plasma Diagnostics Package (PDP) is grappled by the shuttle's Remote Manipulator System (Canadarm).

For the first time, several experiments were carried in the Shuttle's mid-deck lockers. These included an Electrophoresis Equipment Verification Test, which studied the separation of biological components in microgravity, and a Mono-disperse Latex Reactor experiment aimed at producing uniform micrometer-sized latex particles. The mission also carried the first Shuttle Student Involvement Project (SSIP), a study of insect motion in space, showcasing NASA's commitment to involving students in space research.

Despite the mission's successes, STS-3 was not without its challenges. A variety of minor problems occurred during the flight. The orbiter's toilet malfunctioned on its first use, leading to what Lousma humorously described as "eight days of colorful flushing." One of the Auxiliary Power Units (APU) overheated, though it functioned properly during descent. Both crew members experienced some degree of space sickness, and on March 26, 1982, the mission briefly lost three communication links. These issues, however, did not impede the mission's overall success.

Space Shuttle Columbia is shown seconds from touchdown with two T-38 chase planes following it in to the landing strip, following the completion of the STS-3 mission.

Originally planned as a seven-day flight, STS-3 was extended by one day due to weather-related delays at the landing site. The crew ultimately chose to land at White Sands Space Harbor instead of the Shuttle Landing Facility at Kennedy Space Center because they had trained extensively at White Sands. This decision necessitated a large-scale equipment movement from Edwards Air Force Base to White Sands, transporting "40 train carloads" of support equipment via the Santa Fe and Southern Pacific Railroads. This choice saved NASA approximately $2 million in transportation costs compared to using Air Force cargo planes.

High winds at White Sands reduced visibility and delayed the landing by a day, giving the crew what Lousma described as "an extra day in our world's favorite vacation spot." Despite the additional time in space, all mission objectives were accomplished. On March 30, 1982, Columbia re-entered the atmosphere and approached Northrop Strip (later renamed White Sands Space Harbor) for landing. Due to strong westerly winds, Columbia had to execute a high "right base" turn onto the final approach, a less desirable but necessary maneuver. At this stage in the Shuttle's development, the Orbiter had significantly less electronic energy management information available to the crew than on later missions, making the landing particularly challenging.

During the final approach, the autopilot engaged and caused fluctuations in speed by alternately closing and opening the speed brakes. Lousma, choosing to gather data on the autopilot's behavior, kept it engaged until the very late stages of the approach before manually touching down. The landing gear deployed at an altitude of 151 feet and locked just five seconds before touchdown. Despite these difficulties, Lousma successfully landed Columbia on runway 17 at 16:04:46 UTC, marking the completion of STS-3.

STS-3's landing at White Sands proved to be challenging and left Columbia coated in gypsum dust, requiring extensive cleaning and repair at Kennedy Space Center. The dust was so pervasive that traces of it were still found in the spacecraft during later missions. The mission completed 130 orbits, traveling a total of 5.3 million kilometers (3.3 million miles) over 8 days, 0 hours, 4 minutes, and 46 seconds. Despite losing 36 thermal protection tiles and damaging 19 others, Columbia returned safely to Earth.

STS-3 was the last mission for which NASA named a complete backup crew, a practice that would change in subsequent missions.

STS-4: The Final Test Flight

On June 27, 1982, the Space Shuttle Columbia launched from Kennedy Space Center, embarking on STS-4, the final test flight in the Shuttle program's orbital flight test phase. This mission lasted seven days and was critical in demonstrating that the Space Shuttle was ready to transition from its experimental phase to operational status. With STS-4, NASA sought to validate Columbia's capabilities as a reusable spacecraft and ensure the Shuttle could reliably support a broad range of missions in the future.

Commanded by Ken Mattingly, an experienced astronaut who had previously served as Command Module Pilot on the Apollo 16 mission, and piloted by Henry "Hank" Hartsfield, a seasoned test pilot making his first spaceflight, the crew of STS-4 brought a wealth of experience and technical expertise to the mission. Their primary objective was to conduct a series of rigorous tests that would evaluate Columbia's systems, payload capabilities, and overall performance under operational conditions.

One of the central goals of STS-4 was to test the Shuttle's thermal protection system further, particularly during reentry, when the orbiter would be subjected to intense heat. The mission was designed to push Columbia's heat-resistant tiles to their limits, ensuring the spacecraft could safely withstand the extreme temperatures encountered during the descent through Earth's atmosphere. These tests were vital for validating the durability of the thermal protection system, one of the most critical components of the Shuttle's design.

STS-4 - Henry W. Hartsfield Jr. (pilot), left, and Thomas K. (Ken) Mattingly II (commander)

In addition to thermal testing, STS-4 featured the deployment of scientific and commercial payloads, a first for the Shuttle program. The mission carried the first Department of Defense payload, though details of this payload remained classified. Including such a payload signaled the Shuttle's readiness to support national security missions, which would become a significant aspect of the program in the years ahead.

The Shuttle's cargo bay also housed the Continuous Flow Electrophoresis System (CFES), an experiment to refine the process of separating biological materials in the microgravity space environment. The success of this experiment demonstrated the Shuttle's potential for supporting scientific research and commercial ventures in low Earth orbit.

Furthermore, STS-4 included the first use of a Shuttle pallet to carry the Induced Environment Contaminant Monitor (IECM), an instrument designed to measure the gases and particles released by the Shuttle during flight. Understanding the contamination environment around the orbiter was crucial for ensuring the safety of future missions, particularly those involving sensitive scientific experiments or satellite deployments.

Columbia's mission also marked the debut of a new lightweight space suit, tested for flexibility and comfort during extravehicular activities (EVAs). Although no spacewalk was scheduled for STS-4, the suit's performance was evaluated to prepare for upcoming missions requiring astronauts to work outside the Shuttle.

The Induced Environment Contaminant Monitor (IECM) is grappled by the Canadarm.

Space Shuttle Columbia lands at Edwards Air Force Base runway 22.

President Reagan and his wife Nancy observe the shuttle's forward tiles and nosecone.

The reentry and landing phase of STS-4 was another significant milestone. For the first time in the Shuttle program, the mission was scheduled to land at Edwards Air Force Base in California with President Ronald Reagan and First Lady Nancy Reagan. This marked the first time a sitting U.S. President attended a Shuttle landing, underscoring the mission's importance and the Shuttle program's growing significance to the nation.

On July 4, 1982, Columbia reentered Earth's atmosphere, enduring the intense heat generated during reentry. Mattingly and Hartsfield skillfully guided the orbiter through its descent, landing smoothly on the dry lakebed runway at Edwards Air Force Base. The timing of the landing, coinciding with the United States' Independence Day celebrations, added a symbolic note to the event, highlighting the Shuttle program's role in advancing American leadership in space exploration.

The successful completion of STS-4 in July 1982 marked a pivotal moment in the Space Shuttle program, signaling the end of its test phase and ushering in a new era of operational missions. Columbia, the first orbiter in the fleet, had demonstrated its capability to perform a wide array of tasks, ranging from scientific experiments to national defense operations. This mission not only showcased the shuttle's versatility but also solidified the concept of a reusable spacecraft, a revolutionary idea in space

Throughout the 7-day, 1-hour, 9-minute, and 31-second flight, Columbia completed 112 orbits around the Earth, covering a total distance of 4.7 million kilometers (2.9 million miles). The mission objectives were nearly all achieved, except an Air Force payload that was not deployed as planned. However, a significant technical issue occurred when the main parachutes of the Solid Rocket Boosters (SRBs) failed, causing the empty casings to impact the ocean at high velocity and sink. This was only one of two instances—along with the later STS-51-L mission—where the SRBs were not recovered, a critical loss for NASA, which relied on recovering and reusing these boosters to support the shuttle's long-term cost-effectiveness.

Despite this setback, the mission was considered a success, achieving most of its scientific and technical objectives. Columbia touched down safely at Kennedy Space Center (KSC) on July 15, 1982, marking the end of an important chapter in the program's development. With the conclusion of STS-4, the Space Shuttle transitioned to operational status, ready to take on increasingly complex missions in support of NASA's ambitious goals. The Space Shuttle program was now poised to play a crucial role in advancing space exploration, forging partnerships, and supporting national security in the years to come.

STS-5

Columbia launched on schedule from Kennedy Space Center (KSC) at 07:19 a.m. EST on November 11, 1982, marking the fifth mission of the Space Shuttle program, designated STS-5. This mission was notable for carrying the largest crew aboard a spacecraft at the time—four astronauts: Commander Vance D. Brand, Pilot Robert F. Overmyer, and Mission Specialists Joseph P. Allen and William B. Lenoir. It was also the first mission to deploy commercial communications satellites from the Shuttle, a significant step in expanding the program's role beyond purely scientific and military purposes.

STS-5 mission crewmembers includes astronauts Vance D. Brand (second left), commander; Robert F. Overmyer (second right), pilot; and Joseph P. Allen (left) and William S. Lenoir, both mission specialists.

The Columbia payload bay contained two commercial satellites: SBS-3, owned by Satellite Business Systems, and Anik-C3, owned by Telesat Canada. Both were Hughes-built HS-376-series satellites designed for long-term geosynchronous operations. After deployment, McDonnell Douglas PAM-D kick motors successfully propelled these satellites into their designated orbits. This marked a new chapter in the commercialization of space, demonstrating the Shuttle's capability to carry and deploy satellites for private enterprises.

In addition to the satellites, STS-5 carried a West German-sponsored microgravity experiment known as a Getaway Special (GAS) canister, an early demonstration of international cooperation in space research. The crew also conducted three student-designed experiments, highlighting NASA's ongoing efforts to engage and inspire the next generation of scientists.

A pivotal moment of the mission was supposed to be a spacewalk, or extravehicular activity (EVA), the first in the Space Shuttle program. Mission Specialists Lenoir and Allen were scheduled to test newly developed space suits, designed as more efficient and cost-effective alternatives to those used in the Apollo program. However, the EVA was initially delayed when Lenoir experienced motion sickness—a common challenge for astronauts adjusting to the weightlessness of space. The following day, technical issues with the suits arose: Lenoir's oxygen regulator malfunctioned, and a broken recirculation fan in Allen's suit further complicated the situation. For the first time in spaceflight history, an EVA had to be canceled due to space suit issues, underscoring the complexities and risks of working in the harsh space environment.

After completing 81 orbits and traveling 3,397,082 kilometers (2,110,849 miles), Columbia returned to Earth, landing safely on Runway 22 at Edwards Air Force Base on November 16, 1982, at 06:33 a.m. PST. The mission lasted 5 days, 2 hours, 14 minutes, and 26 seconds. Notably, STS-5 was the first Shuttle mission in which the crew did not wear pressure suits during launch, reentry, or landing—a departure from previous procedures and a similarity to the Soviet Voskhod and Soyuz missions before the tragic Soyuz 11 disaster in 1971. Columbia was transported back to Kennedy Space Center on November 22, 1982, to prepare for future missions.

The success of STS-5 reinforced the belief that the Shuttle had moved beyond its test phase and into operational status, following the declaration made after STS-4. However, the tragic loss of Columbia and its crew during STS-107 in 2003 would later call this status into question. The Columbia Accident Investigation Board (CAIB) argued that the Shuttle had never truly been an "operational" vehicle in the sense that civilian and military aircraft are. While the Shuttle was not inherently unsafe, it was, in the CAIB's assessment, still an experimental vehicle, as evidenced by the ongoing modifications and the relatively small number of flights—less than 200—at the time of the disaster. NASA continued to operate the Shuttle under this understanding for the remainder of the program.

STS-6

STS-6, the sixth flight of NASA's Space Shuttle program and the maiden flight of the orbiter Challenger, was a landmark mission from April 4 to April 9, 1983. This mission was pivotal not only because it introduced Challenger to the fleet but also because it included the first spacewalk of the Shuttle program, marking a significant step forward in the United States' space exploration capabilities.

Challenger, OV-099, was named after the British Naval research vessel HMS Challenger, which sailed the Atlantic and Pacific oceans during the 1870s. Challenger was the second operational shuttle and made its first flight, STS-6, on April 4, 1983. Challenger hosted missions that saw astronauts take the first-ever spacewalks with jetpacks, including the first mission to pull a satellite out of orbit, fix it, and return it to service. Challenger and its seven astronauts were lost on January 28, 1986, when a seal on one of its boosters failed and hot gas burned through the external tank, igniting the propellants and causing the shuttle to break up in the resulting

explosion. That flight, STS-51L, was Challenger's 10th mission.

Paul J. Weitz (left), crew commander, and Karol J. Bobko, pilot. Standing are Donald H. Peterson (left), and Story Musgrave, both mission specalists.

The crew of STS-6 consisted of four astronauts: Commander Paul J. Weitz, Pilot Karol J. Bobko, and Mission Specialists Donald H. Peterson and Story Musgrave. Their mission objectives included deploying the first Tracking and Data Relay Satellite (TDRS-1). This satellite would revolutionize communications between the Space Shuttle, the International Space Station, other spacecraft, and ground control stations. The TDRS network would eventually become the backbone of NASA's space communication system, enabling near-continuous contact with orbiting spacecraft.

Challenger lifted off from Kennedy Space Center's Launch Complex 39A at 1:30 PM EST on April 4, 1983. The launch was nearly flawless, with Challenger performing as expected. However, the mission faced a significant challenge when the deployment of TDRS-1 encountered an issue. After the satellite was released from the Shuttle's payload bay, its Inertial Upper Stage (IUS) rocket malfunctioned, placing TDRS-1 in a lower-than-planned orbit. Despite this setback, ground controllers worked tirelessly over the next few months to raise the satellite to its correct geosynchronous orbit using its onboard thrusters, ultimately ensuring the success of its mission.

The highlight of STS-6 was the historic

extravehicular activity (EVA), or spacewalk, conducted by Peterson and Musgrave on April 7, 1983. This EVA, the first of the Space Shuttle program, was crucial for testing the new spacesuits and tools that would be used in future Shuttle missions, particularly those involving satellite repairs and the construction of space stations. The spacewalk lasted for four hours and ten minutes, during which the astronauts successfully tested the mobility of their suits and practiced multiple tasks essential for assembling and maintaining large structures in space.

STS-6 also included several scientific experiments, such as the Continuous Flow Electrophoresis System (CFES), which aimed to separate biological materials in microgravity, and the Monodisperse Latex Reactor (MLR), which investigated the formation of uniform-sized latex particles in space. These experiments contributed valuable data to NASA's understanding of the behavior of materials in the space environment.

Challenger's mission concluded with a successful landing at Edwards Air Force Base in California on April 9, 1983. The orbiter touched down at 10:53 AM PST, marking the completion of a mission that had achieved several important milestones despite its challenges. The success of STS-6 demonstrated the versatility and reliability of the Space Shuttle and set the stage for future missions that would further expand the frontiers of human space exploration.54

Chapter 3 - Exploring New Frontiers (1983–1985)

STS-7

Sally K. Ride (mission specialist), Robert L. Crippen (commander), Frederick H. Hauck (pilot); rear row: John M. Fabian (left) and Norman E. Thagard (mission specialists). STS-7 launched the first five-member crew and the first American female astronaut into space on June 18, 1983.

STS-7, the seventh flight of NASA's Space Shuttle program and the second mission for the orbiter Challenger, was significant and historic for several reasons. Launched on June 18, 1983, this mission was marked by the first flight of an American woman in space, Dr. Sally K. Ride, and the successful deployment of multiple satellites, demonstrating the Shuttle's capability to handle complex payload operations in space.

The crew of STS-7 consisted of five astronauts: Commander Robert L. Crippen, Pilot Frederick H. Hauck, and Mission Specialists John M. Fabian, Sally K. Ride, and Norman E. Thagard. Dr. Ride's participation in the mission made her a trailblazer, inspiring generations of women and girls to pursue careers in science, technology, engineering, and mathematics (STEM). Her role as a mission specialist included operating the Shuttle's robotic arm, known as the Canadarm, to deploy and retrieve satellites, showcasing her technical expertise and the vital role of the robotic arm in Shuttle missions.

Challenger lifted off from Kennedy Space Center's Launch Complex 39A at 7:33 AM EDT on June 18, 1983. The mission's primary objectives included the deployment of two communications satellites: the Canadian Anik C2 and the Indonesian Palapa B1. These satellites were crucial for expanding global communications networks and were successfully deployed using the Payload Assist Module (PAM-D) systems.

In addition to these satellite deployments, STS-7 included releasing and retrieving the Shuttle Pallet Satellite (SPAS-1), a platform designed to carry scientific experiments into orbit. SPAS-1 was released into space and later retrieved by the Shuttle, marking the first time a satellite had been deployed and brought back aboard the Shuttle. This capability demonstrated the Shuttle's unique ability to carry out missions that required both the deployment and recovery of payloads. This feature would be vital for future satellite servicing missions and the International Space Station (ISS) construction.

Dr. Sally Ride's operation of the Canadarm was a key component of these activities. Her skillful manipulation of the arm during the retrieval of SPAS-1 was a critical test of the Shuttle's ability to conduct complex in-orbit operations. Her performance during the mission demonstrated the versatility of the Shuttle's robotic systems and underscored the growing role of women in space exploration.

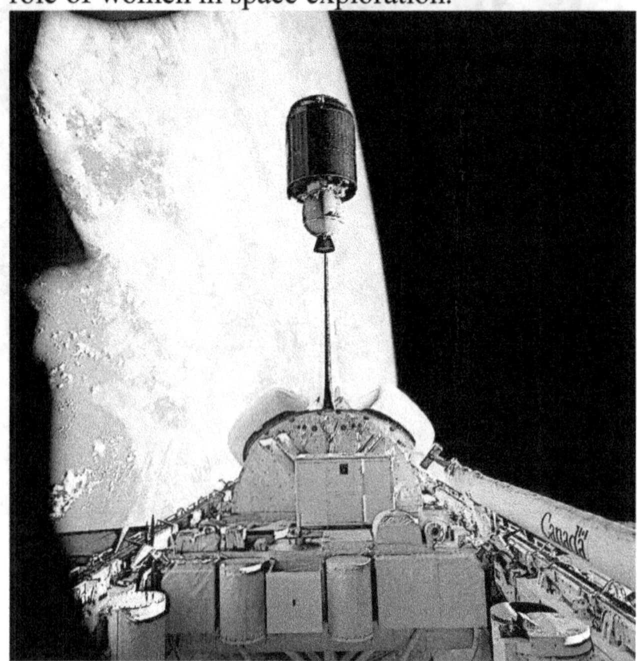

Deploy of PALABA-B1 Satellite durig Shuttle Mission STS-7 The Indonesian Palapa B communications satellite is just about to clear the vertical stabilizer of the Earth-orbiting Space Shuttle Challenger

The mission also included multiple scientific experiments, such as investigations into the effects of microgravity on materials processing and life sciences. These experiments provided valuable data that contributed to NASA's understanding of how different materials and biological systems behave in the space environment.

STS-7 was originally planned to extend the mission through an extravehicular activity (EVA) or spacewalk, but this was postponed to later missions due to time constraints and the complexity of other mission objectives. The crew thoroughly inspected the Challenger's systems, ensuring that all aspects of the mission were successful and safe.

After completing the mission objectives, Challenger returned to Earth, landing at Edwards Air Force Base in California on June 24, 1983, at 6:57 AM PDT. The landing marked the successful conclusion of a mission that had achieved numerous milestones, including breaking gender barriers and demonstrating the Shuttle's robust capabilities in deploying and retrieving payloads. STS-7 solidified the Space Shuttle's reputation as a versatile and powerful tool for space exploration and laid the groundwork for future missions that would continue to push the boundaries of human spaceflight.

Challenger's remote manipulator system (RMS) arm grasps Shuttle pallet satellite (SPAS-01) during proximity operations.

STS-8

STS-8 marked a significant milestone in the history of space exploration as it was the first nighttime launch and landing of a Space Shuttle. The mission, flown by the Space Shuttle Challenger, launched from Kennedy Space Center on August 30, 1983, at 2:32 AM EDT. This mission was notable for being the first to carry an African American astronaut, Guion S. Bluford, Jr., into space, further emphasizing the Space Shuttle program's role in breaking barriers and advancing human spaceflight.

In early planning for STS-8, the mission was scheduled for July 1983 as a three-day flight with four crew members. Its primary objectives included deploying two satellites: INSAT-1B, an Indian communications satellite, and TDRS-B, a NASA Tracking and Data Relay Satellite. However, due to issues with the Inertial Upper Stage (IUS) that occurred during the deployment of TDRS-A on STS-6, NASA made significant changes. In May 1983, the decision was made to remove TDRS-B from the STS-8 manifest, replacing it with the Payload Flight Test Article (PFTA). This shift delayed TDRS-B's launch until STS-51-L, the ill-fated mission that ended with losing Space Shuttle Challenger and its crew in January 1986.

Left to right are Daniel C. Brandenstein, pilot; Richard H. Truly, commander; and Guion S. Bluford Jr., Mission specialist. Standing from left to right are Dale A. Gardner, mission specialist; and William E. Thornton, mission specialist.

The main payload for STS-8 was INSAT-1B, the second satellite in a series operated by the Indian Space Research Organization (ISRO). It followed the unsuccessful INSAT-1A, launched by a Delta rocket in April 1982 but failed due to reaction control system malfunctions. INSAT-1B was carried in the rear of Challenger's payload bay and boosted into a Geostationary Transfer Orbit (GTO) by a Payload Assist Module (PAM-D) after deployment. Along with its upper stage, the satellite weighed 3,377 kilograms and was expected to enhance India's communication capabilities significantly.

The Payload Flight Test Article (PFTA), originally scheduled for STS-16 in June 1984,

was brought forward to STS-8 to fill the gap left by TDRS-B. The PFTA was an aluminum structure with two wheels connected by a central axle, designed to be manipulated by the Shuttle's Canadarm. Weighing 3,855 kilograms, it allowed the astronauts to practice moving bulky objects in space, providing essential experience for future missions requiring this robotic arm.

In addition to its primary payloads, STS-8 carried several other important experiments. The Development Flight Instrumentation (DFI) pallet, located in the forward section of the payload bay, mounted two experiments. The first studied the interaction between atomic oxygen and the Shuttle's structural materials, while the second tested a heat pipe intended for future spacecraft cooling systems. Four Getaway Special (GAS) payloads were also included, with experiments ranging from studying cosmic rays' effects on electronic equipment to an attempt to create snow crystals using water vapor.

Another unique aspect of the mission was its partnership with the United States Postal Service (USPS). The Shuttle carried 260,000 postal covers, each stamped with a special $9.35 express postage stamp. These covers were later sold to collectors, with the profits shared between NASA and the USPS.

Inside the orbiter, additional experiments were conducted. One significant experiment was the Continuous Flow Electrophoresis System (CFES), which separated biological materials using electric fields. This experiment was aimed at advancing research into treatments for diseases like diabetes. Additionally, a small animal cage containing six rats was flown to test its suitability for future student-involved experiments. Although no animal testing occurred on this mission, the flight provided vital data for future studies. Another experiment involved Guion Bluford, who used biofeedback techniques to determine their effectiveness in microgravity. The mission also featured a photography experiment designed to capture a spectrum of atmospheric luminosities observed around the orbiter.

Another key element of the mission was testing the TDRS-1 satellite, deployed during STS-6. This test ensured the satellite's operational readiness for the upcoming Spacelab program. Additionally, the Shuttle orbiter carried equipment for encrypted transmissions, a system intended for future classified missions.

STS-8 launched on August 30, 1983, into a circular orbit at 296 kilometers. One of the primary mission events occurred on August 31, when INSAT-1B was successfully deployed. Shortly after deployment, Challenger maneuvered to a safe distance as the satellite's booster motor ignited. However, the mission encountered several challenges, including temporary telemetry loss through TDRS-1, a fire alarm suspected to be false, and a minor cabin pressure leak, quickly resolved.

During the mission, the crew also tested the Canadarm by moving the PFTA around the payload bay, providing critical experience with the robotic arm. Communications tests through TDRS-1 were successful, although ground station issues at White Sands caused occasional problems.

On Challenger's middeck, Mission Specialist (MS) Guion Bluford, restrained by harness and wearing blood pressure cuff on his left arm, exercises on the treadmill. Forward lockers with data recording units and checklist notebooks are to Bluford's left.

On September 5, 1983, after six days in space, Challenger made history by performing the first night landing of the Shuttle program at Edwards Air Force Base in California. The orbiter had no onboard lights, so high-intensity xenon arc lamps illuminated the runway.

Despite the mission's success, post-flight analysis revealed several issues, most notably severe corrosion on the left Solid Rocket Booster (SRB). This corrosion was later found to be due to a faulty batch of resin used to line the booster nozzles. The defect was a near-disaster, with the potential to cause the Shuttle to lose control during ascent. Although this issue did not garner significant media attention at the time, it became more concerning after the Challenger disaster in 1986. Additionally, post-flight inspection found debris impacts on the Thermal Protection System (TPS) tiles, though the number and severity of these impacts were relatively low compared to other missions.

STS-9

Left to right are Owen Garriott, mission specialist; Brewster Shaw, pilot; John Young, commander; and Robert Parker, mission Specialist. Standing from left to right are the payload specialists, Byron Lichtenberg and Ulf Merbold.

STS-9, also known as Spacelab 1, was a landmark mission in NASA's Space Shuttle program. Launched on November 28, 1983, aboard the Space Shuttle *Columbia*, it marked the ninth shuttle mission and the sixth for *Columbia*. This mission was notable for carrying the first Spacelab laboratory module into space, a collaborative project between NASA and the European Space Agency (ESA), designed to facilitate advanced scientific research in microgravity.

The mission's crew was the largest to fly aboard the Space Shuttle up to that time, comprising six members: Commander John W. Young, Pilot Brewster H. Shaw, Mission

Specialists Owen Garriott and Robert A. Parker, and Payload Specialists Byron K. Lichtenberg and Ulf Merbold. Merbold, a physicist from West Germany, made history as the first non-U.S. citizen to fly on the Space Shuttle, representing ESA. His inclusion highlighted the growing international cooperation in space exploration during the early 1980s.

Young, a veteran of six spaceflights, commanded the mission in what would be his final spaceflight, cementing his status as one of NASA's most experienced astronauts. Young had previously flown on Project Gemini and Apollo, including walking on the Moon during Apollo 16 and commanding *Columbia* on its maiden flight, STS-1. Alongside him, Garriott, who had spent 56 days aboard *Skylab* in 1973, contributed his extensive experience to the mission.

View of the Spacelab module in the payload bay of the Columbia during STS-9 View taken through aft window on the flight deck of the Spacelab module in the payload bay of the Columbia during STS-9. The docking tunnel, leading from the environment of the orbiter to the Spacelab, is in the foreground. European Space Agency insignias are visible on the sides of the Spacelab.

STS-9 was primarily focused on Spacelab 1, a versatile research module that transformed the Shuttle's cargo bay into an orbiting laboratory. During the mission, 72 experiments were conducted in a wide array of scientific disciplines, including atmospheric and plasma physics, solar observations, astronomy, materials science, and biological research. These experiments showcased the potential of Spacelab to support complex research in a microgravity environment, with scientists on the ground at the Marshall Space Flight Center (MSFC) in Alabama coordinating with the crew in space.

Notably, Garriott made history by becoming the first amateur radio operator to transmit from space, using ham radio to communicate with operators on Earth. This marked the beginning of amateur radio operations as a valuable educational tool in space missions.

The road to launch was not without obstacles. The mission's original launch date of October 29, 1983, was scrubbed due to concerns over a defect in the right Solid Rocket Booster (SRB). For the first time in shuttle history, the entire shuttle stack was rolled back to the Vehicle Assembly Building (VAB), where the orbiter was destacked to address the issue. The orbiter was reassembled and returned to the launch pad in early November, paving the way for the successful launch later that month.

Throughout the mission, the shuttle's crew operated in two 12-hour shifts, with the Red Team (Young, Parker, and Merbold) alternating with the Blue Team (Shaw, Garriott, and Lichtenberg). This division of labor allowed for continuous operations throughout the mission's extended 10-day duration, which at the time was the longest shuttle mission.

However, STS-9 was not without technical difficulties. As the crew prepared for re-entry, two of the shuttle's General Purpose Computers (GPCs) malfunctioned due to a soldering defect that was triggered when the Reaction Control System (RCS) thrusters were fired. Fortunately, Commander Young's decision to delay re-entry and manually reboot the computers averted a potential disaster. Post-flight analysis revealed that the malfunction could have led to the loss of the vehicle and crew had the issue not been resolved in time.

Additionally, as *Columbia* approached landing, two of its three auxiliary power units (APUs) caught fire due to a hydrazine leak.

Despite this, the shuttle successfully landed at Edwards Air Force Base on December 8, 1983, after completing 167 orbits and traveling over 4.3 million miles (6.9 million kilometers). The fire was discovered only after landing, having caused significant damage to the APU compartment.

STS-9's success solidified Spacelab's role in future shuttle missions, proving the feasibility of conducting complex scientific research in space with non-NASA payload specialists. The mission also marked a turning point for the *Columbia* orbiter. Following the flight, *Columbia* was removed from service for extensive renovations, including the installation of upgraded engines and systems to bring it in line with newer orbiters like *Challenger* and *Discovery*. The shuttle would not fly again until STS-61-C in January 1986.

STS-41-B

Vance D. Brand, commander; and Robert L. Gibson, pilot. Standing left to right are mission specialists Robert L. Stewart, Ronald E. McNair, and Bruce McCandless.

STS-41-B, NASA's tenth Space Shuttle mission and the fourth flight of the Space Shuttle Challenger, holds a significant place in the history of space exploration. Launching on February 3, 1984, and concluding on February 11, 1984, the mission successfully deployed two communications satellites and made history with the first untethered spacewalk.

The mission was also notable for the Space Shuttle program's flight numbering system changes. Following the STS-9 mission, NASA adopted a new system to reflect better the complexity of the shuttle's missions and payloads. Originally designated as STS-11, this mission was reclassified as STS-41-B under the new scheme after the cancellation of STS-10 due to payload delays.

The crew of STS-41-B consisted of five astronauts: Commander Vance D. Brand, Pilot Robert L. Gibson, and Mission Specialists Bruce McCandless II, Ronald E. McNair, and Robert L. Stewart. Each member brought a wealth of experience and expertise to the mission, with Bruce McCandless II and Robert L. Stewart poised to make history with the first untethered spacewalks.

STS-41B launch

On the morning of February 3, 1984, at precisely 08:00 a.m. EST, the Space Shuttle *Challenger* lifted off from Kennedy Space Center amidst a crowd of approximately 100,000 spectators. The shuttle's mission, STS-41-B, aimed to deploy two communications satellites: Westar 6, built for America's Western Union, and Palapa B2, for Indonesia. Both satellites were Hughes-built HS-376-series models, released from the shuttle's payload bay approximately eight hours after launch. However, the deployment process encountered complications when the Payload Assist Modules (PAM) failed, leaving both satellites in orbits significantly lower than planned. This malfunction required future intervention. In November 1984, during STS-51-A, the orbiter *Discovery* successfully retrieved both satellites, demonstrating the shuttle's versatility in satellite recovery and repair operations.

One of the defining moments of STS-41-B occurred on February 7, the fourth day of the mission, when astronauts Bruce McCandless and Robert Stewart made history with the first untethered spacewalk. McCandless, using the Manned Maneuvering Unit (MMU), ventured out of *Challenger*'s payload bay, freely floating in the vacuum of space without the safety of a tether. At 8:25 a.m. EST, he activated the MMU's thrusters and maneuvered to a distance of 98 meters (322 feet) from the orbiter, an unprecedented feat in spaceflight history. Meanwhile, Stewart tested the "work station" foot restraint mounted on the shuttle's Canadarm (Remote Manipulator System), further advancing NASA's capabilities for space operations.

Later in the mission, both astronauts conducted another spacewalk to rehearse procedures for the upcoming retrieval and repair of the Solar Maximum Mission (SMM) satellite, a complex task planned for STS-41-C. These extravehicular activities not only demonstrated the MMU's potential but also laid the groundwork for future satellite servicing missions.

STS-41-B also marked the reflight of the West German-sponsored SPAS-1 pallet/satellite, which had first flown on STS-7. Unfortunately, due to an electrical malfunction in the Canadarm, SPAS-1 remained in *Challenger*'s payload bay throughout the mission. Despite this setback, the mission continued with a variety of scientific experiments. Five Get Away Special (GAS) canisters were deployed, carrying diverse research projects, including the Continuous Flow Electrophoresis System and the Monodisperse Latex Reactor experiments.

In a notable achievement for science education, one of the GAS canisters housed the first experiment designed and constructed by high school students to fly in space. In collaboration with Utah State University, a team of four students from Brighton High School in Utah developed a project on seed germination and growth in microgravity, exemplifying the educational impact of NASA's outreach programs.

Additionally, the mission included six live rats housed in the middeck area, as part of a study on the effects of microgravity on biological organisms. The payload also carried a Cinema-360 camera, allowing for immersive filming of the mission, further advancing public engagement with space exploration.

Throughout the mission, *Challenger* faced several technical challenges, particularly with its environmental control systems. The nozzles of its supply and wastewater venting systems encountered below-freezing temperatures, leading to a failure in the supply water dump valve. As a result, excess water had to be dumped via the shuttle's flash evaporator system for the remainder of the mission.

A more concerning issue arose during re-entry. Ice had formed around the wastewater dump valves near the orbiter's nose, and upon re-entering Earth's atmosphere, the ice broke free, striking the left Orbital Maneuvering System (OMS) pod. This impact damaged three Thermal Protection System (TPS) tiles, leading to a minor burn-through on the orbiter's surface. Post-flight inspections revealed that the wastewater dump line had ruptured due to freezing, and insulation around the nozzles was missing. Additionally, discolored TPS tiles indicated that ice buildup had occurred before re-entry. Fortunately, the damage was minor enough that *Challenger* and its crew returned safely.

After spending 7 days, 23 hours, 15 minutes, and 55 seconds in space, *Challenger* successfully completed its mission with a historic landing on February 11, 1984, at Kennedy Space Center's

Shuttle Landing Facility. This was the first time a spacecraft had returned to land at its launch site, demonstrating the shuttle's ability to perform runway landings. By the end of the mission, *Challenger* had completed 128 orbits around Earth, traveling a total distance of 5,329,150 kilometers (3,311,380 miles).

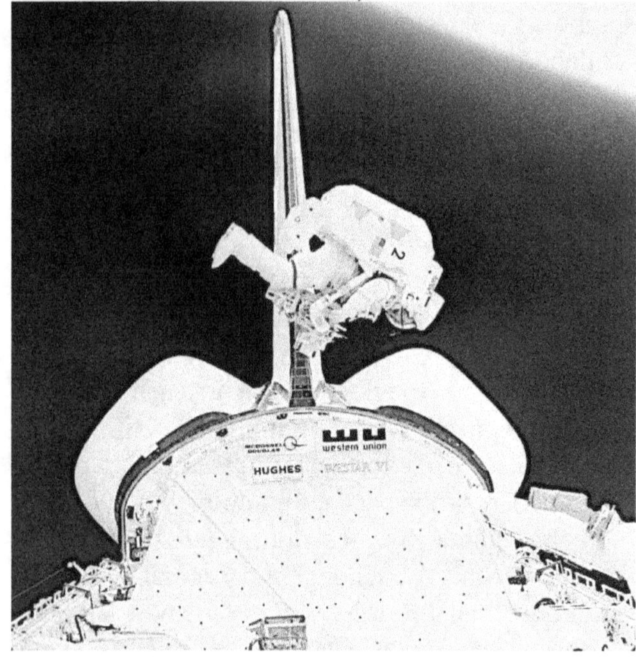

Astronaut Bruce McCandless, II, 41-B mission specialist, tests a the manned maneuvering unit (MMU) during an a test involving the trunion pin attachmet device (TPAD) he carries and the shuttle pallet satellite (SPAS-01A), partially visible at bottom of the frame. The Challenger was flying with its aft end aimed toward the Earth.

STS-41-C

STS-41-C, originally designated STS-13, marked NASA's eleventh Space Shuttle mission and the fifth flight of Space Shuttle Challenger. Launched on April 6, 1984, at 8:58 a.m. EST from Kennedy Space Center, the mission made history as the first Shuttle flight to use a direct ascent trajectory. This trajectory allowed Challenger to reach its intended orbit of 533 km (331 mi) using only its Orbital Maneuvering System (OMS) engines for circularization. However, the ascent was not without challenges—both the primary and backup computers at Mission Control Center failed, leaving controllers without telemetry data from the Shuttle for nearly an hour.

Robert L. Crippen, commander; Terry J. Hart, mission specialist; James D. Van-Hoften, mission specialist; George D. Nelson, mission specialist; and Francis R. (Dick) Scobee, pilot.

The experienced crew of STS-41-C was led by Commander Robert Crippen, making his third spaceflight. Francis R. "Dick" Scobee served as the pilot, marking his only spaceflight. Terry Hart, James van Hoften, and George Nelson completed the crew as mission specialists, with van Hoften and Nelson conducting spacewalks to repair Solar Max. Both van Hoften and Nelson were on their first spaceflights, while Hart was also making his only spaceflight.

The mission's two primary objectives were to deploy the Long Duration Exposure Facility (LDEF) and to capture, repair, and redeploy the malfunctioning Solar Maximum Mission satellite, commonly called "Solar Max." LDEF, a 9,700 kg (21,400 lb) cylindrical structure measuring 4.3 meters in diameter and 9.1 meters in length, carried 57 scientific experiments contributed by 200 researchers from eight countries. These experiments, housed in 86 removable trays, were designed to study the effects of long-term exposure to the space environment. The LDEF was successfully deployed on the mission's second day using the Shuttle's Remote Manipulator System (RMS), also known as the Canadarm. Originally scheduled for retrieval in 1985, delays, including the Challenger disaster in 1986, pushed the recovery to January 1990, when Space Shuttle

Columbia retrieved it during STS-32.

The second objective, the repair of Solar Max, proved more complicated. Launched in 1980, the satellite had malfunctioned and lost the ability to perform its solar observations. On the third day of the mission, Challenger's crew raised the Shuttle's orbit to 560 km (350 mi) and maneuvered within 61 meters (200 feet) of the satellite. Astronaut George Nelson, equipped with a Manned Maneuvering Unit (MMU), attempted to capture Solar Max using the Trunnion Pin Acquisition Device (TPAD). After several failed attempts, during which the satellite began to tumble uncontrollably, the effort was temporarily abandoned. Commander Robert Crippen had to carefully maneuver Challenger to keep up with both Nelson and the erratic satellite, nearly exhausting the Shuttle's Reaction Control System (RCS) fuel.

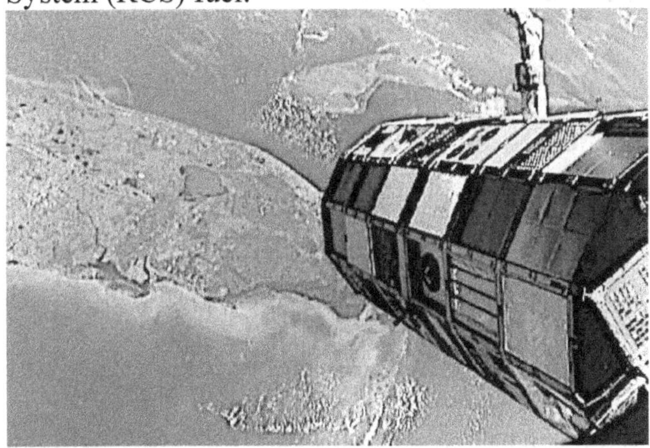

The deployed Long Duration Exposure Facility (LDEF), which became an important source of information on the small-particle space debris environment.

That night, engineers at the Solar Max Payload Operations Control Center (POCC) at Goddard Space Flight Center stabilized the satellite using its onboard magnetorquers. The next day, Challenger re-approached Solar Max, and Terry Hart successfully grappled the satellite with the RMS. Solar Max was then secured in a special cradle within the Shuttle's payload bay. Astronauts Nelson and James van Hoften conducted two spacewalks to replace the satellite's attitude control mechanism and repair its coronagraph instrument, restoring its functionality. Solar Max was redeployed into orbit on the fifth day of the mission, and after a

30-day checkout, resumed full operation.

George Nelson attempts to capture the Solar Maximum Mission satellite for repairs.

In addition to these high-profile tasks, STS-41-C carried out a unique student experiment demonstrating honeybees' ability to construct honeycomb cells in microgravity. The mission's highlights, including the deployment of LDEF and the repair of Solar Max, were captured using an IMAX camera, with the footage later featured in the 1985 film The Dream is Alive.

After six days, 23 hours, 40 minutes, and 7 seconds in space, Challenger completed 108 orbits, traveling 4,620,000 kilometers (2,870,000 miles). The mission concluded on April 13, 1984, with a pre-dawn landing at Edwards Air Force Base in California. The Shuttle touched down at 5:38 a.m. PST on Runway 17, marking another successful chapter in NASA's expanding Space Shuttle program. Challenger was ferried back to Kennedy Space Center on April 18, 1984, where it was prepared for future missions.

STS-41-D

STS-41-D, the twelfth mission of NASA's Space Shuttle program and the maiden flight of the Space Shuttle Discovery, marked a new chapter in the history of space exploration. Launched on August 30, 1984, from Kennedy Space Center's Launch Complex 39A, this mission introduced Discovery, the third orbiter in NASA's shuttle fleet, to operational service. The mission also set the stage for subsequent shuttle flights, showcasing the shuttle's capabilities in deploying satellites and conducting scientific experiments.

Discovery, OV-103, was named after one of the two ships used by the British explorer Captain James Cook when he discovered Hawaii and explored Alaska and northwestern Canada in the 1770s. Discovery was the third operational shuttle and made its first flight, STS-41D, in August 1984. Discovery has flown more than any other shuttle, with 39 missions. Discovery's noteworthy career also includes both Return to Flight missions after the Challenger and Columbia accidents. Discovery deployed NASA's Hubble Space Telescope, which has altered how we see and think of our universe. Discovery was the first space shuttle to retire from NASA's fleet, following its STS-133 mission to the ISS in February/March 2011.

Richard M. (Mike) Mullane, mission specialist; Steven A. Hawley, mission specialist; Henry W. Hartsfield, commander; and Michael L. (Mike) Coats, pilot. Standing in the rear are Charles D. Walker, payload specialist; and Judith A. (Judy) Resnik, mission specialist.

The crew of STS-41-D consisted of Commander Henry W. "Hank" Hartsfield, Pilot Michael L. Coats, and Mission Specialists Judith A. Resnik, Steven A. Hawley, Richard M. Mullane, and Charles D. Gemar. Additionally, the flight included Payload Specialist Charles D. Walker, the first non-government industrial astronaut, representing McDonnell Douglas. Walker's inclusion marked a milestone in the commercial use of space, as his role involved operating a continuous-flow electrophoresis system (CFES) experiment aimed at manufacturing high-purity pharmaceuticals in microgravity.

One of the primary objectives of STS-41-D was the deployment of three commercial communications satellites: SBS-D, SYNCOM IV-2 (also known as Leasat-2), and Telstar 3C. Each of these satellites was critical for enhancing global communications infrastructure. The deployment process involved using the shuttle's robotic arm to carefully place the satellites into their respective orbits, activating their onboard propulsion systems to reach their final geostationary positions.

The launch of STS-41-D was initially scheduled for June 1984 but was delayed multiple times due to technical issues, including a liquid hydrogen leak. These delays underscored the complexities and challenges of operating the Space Shuttle, mainly when introducing a new orbiter like Discovery.

One of the most memorable moments of STS-41-D was the unintentional yet iconic deployment of a large thermal blanket from the shuttle's cargo bay. The blanket, intended to shield part of the orbiter from the extreme temperatures of space, came loose and floated away, creating an unexpected yet visually striking scene. This incident highlighted the challenges of operating in space, where even small oversights can lead to unanticipated outcomes.

In its early years, the Space Shuttle program demonstrated the versatility of reusable spacecraft for deploying satellites and conducting scientific research. One of the hallmark missions, STS-41-D, saw Space Shuttle *Discovery* launch on August 30, 1984, carrying a varied payload. The primary cargo included three commercial communications satellites: SBS-4 for Satellite Business Systems, Telstar 302 for Telesat of Canada, and Syncom IV-2, or Leasat-2, a satellite leased to the U.S. Navy. All three satellites, manufactured by Hughes, represented a significant step forward in satellite technology. Notably, Leasat-2 was the first large communications satellite specifically designed for deployment from the Space Shuttle, underscoring the Shuttle's emerging role in supporting national defense and commercial communications. Each satellite was successfully deployed and became operational, highlighting the Shuttle's reliability in delivering high-value payloads to orbit.

Another significant payload on STS-41-D was the OAST-1 solar array, a groundbreaking experiment in solar technology. Measuring 4 meters (13 feet) wide and 31 meters (102 feet) high, the array demonstrated lightweight, expandable structures for space use. It folded into a compact package only 18 centimeters (7.1 inches) deep, and during the mission, it was deployed and extended to its full height multiple times. At the time, it was the largest structure ever extended from a crewed spacecraft and laid the groundwork for future applications, including solar arrays on the International Space Station (ISS). The array carried a variety of experimental solar cells, testing the feasibility of generating power for long-duration missions and future space stations.

STS-41-D also featured a McDonnell Douglas-sponsored experiment, the Continuous Flow Electrophoresis System (CFES), which built on previous missions. Using living cells, payload specialist Charles Walker operated the CFES for over 100 hours, refining techniques that could later be applied to pharmaceutical manufacturing in space. In addition, the mission hosted a student experiment on crystal growth in microgravity, further contributing to the understanding of space environments on scientific processes.

The mission was visually documented with an IMAX motion picture camera, capturing breathtaking footage of the crew's activities and the deployment of satellites. These scenes later appeared in the 1985 documentary *The Dream is Alive*, bringing the experience of spaceflight to audiences around the world.

However, the mission also faced challenges. On September 3, 1984, the crew encountered a peculiar issue when a 61-centimeter (24-inch) icicle, dubbed the "pee-sicle," formed on the shuttle's waste dump nozzle due to an obstruction in the external wastewater system. Astronaut Henry Hartsfield skillfully removed the ice using the Remote Manipulator System (Canadarm), ensuring that operations could continue without further incident.

STS-41-D lasted six days, with *Discovery* orbiting Earth 97 times and covering a total distance of 4,010,000 kilometers (2,490,000 miles). The mission concluded successfully with a landing at Edwards Air Force Base on September 5, 1984, at 06:37:54 a.m. PDT. Yet, the mission was marked by an ominous discovery: it was the first Shuttle flight in which blow-by damage to the solid rocket booster (SRB) O-rings was observed. A small amount of soot had passed beyond the primary O-ring, signaling a serious flaw in SRB design. This issue, later identified by Morton Thiokol engineer Brian Russell, was described as the first "big red flag" concerning the safety of SRB joints and O-rings, a warning that tragically preceded the Challenger disaster in 1986.

In retrospect, STS-41-D was a critical mission in the Space Shuttle program, balancing the success of satellite deployments and technological demonstrations with early signs of the risks leading to major safety overhauls.

STS-41-G

STS-41-G, the thirteenth mission of NASA's Space Shuttle program and the sixth flight of the Space Shuttle Challenger, was a historic and multifaceted mission from October 5 to October 13, 1984. This flight carried seven crew members—the largest crew to fly on a single spacecraft up to that time. Commander Robert L. Crippen led the mission, a veteran astronaut making his fourth Shuttle flight and his second in just six months. Crippen became the first American astronaut to complete two space missions within the same calendar year. Joining him was Pilot Jon A. McBride, along with Mission Specialists David C. Leestma, Sally K. Ride, and Kathryn D. Sullivan. Two Payload Specialists also flew on the mission: Paul D.

Scully-Power of the U.S. Naval Research Laboratory and Marc Garneau, who made history as the first Canadian astronaut in space.

Jon A. McBride, pilot; mission specialists Sally K. Ride, Kathryn D. Sullivan, and David C. Leestma. Standing in the rear, left to right, are payload specialists Paul D. Scully-Power and Marc Garneau with crew commander Robert L. Crippen in the

STS-41-G was notable for breaking new ground in space exploration. It was the first mission to fly two female astronauts together— Sally Ride, making her second spaceflight, and Kathryn Sullivan, who made history as the first American woman to walk in space. On October 11, 1984, Sullivan and Leestma conducted a three-hour Extravehicular Activity (EVA), during which they tested the Orbital Refueling System (ORS), proving the feasibility of refueling satellites in orbit. This demonstration marked an important step in the development of long-term orbital infrastructure.

Another major mission highlight was deploying the Earth Radiation Budget Satellite (ERBS) from the shuttle's payload bay. Using the Canadarm robotic arm, the crew released the 5,086-pound satellite nine hours after liftoff, and its onboard thrusters boosted it into a 350-mile-high orbit. ERBS was designed to measure the energy received from the Sun and radiated back into space, contributing to the study of Earth's climate systems. It was the first of three planned satellites to monitor the distribution of solar energy across the globe and its seasonal variations.

The mission also carried an advanced suite of Earth observation instruments, including the Shuttle Imaging Radar-B (SIR-B). Part of the OSTA-3 experiment package, SIR-B improved upon a similar radar system flown on STS-2 in 1981. With its large, 36-foot antenna array, SIR-B was designed to capture detailed radar images of Earth's surface, offering new insights into geological formations, deforestation, and changes in land use. Although some data transmission issues arose due to a malfunction with Challenger's Ku-band antenna, much of the valuable radar imagery was successfully recorded onboard.

In addition to SIR-B, the OSTA-3 payload included several other scientific instruments. The Large Format Camera (LFC) captured high-resolution photographs of Earth, while the MAPS (Measurement of Air Pollution from Satellites) system tracked atmospheric pollutants. The FILE (Feature Identification and Location Experiment) used television and still cameras to collect visual data for terrain analysis.

During the mission, Payload Specialist Paul Scully-Power conducted oceanographic observations, contributing to his expertise as a civilian oceanographer. Meanwhile, Marc Garneau led the CANEX experiments, a series of scientific investigations sponsored by the Canadian government. These experiments focused on materials science, atmospheric studies, and medical research, further cementing the international collaboration aboard the Shuttle.

The mission wasn't without controversy. On October 10, 1984, reports surfaced alleging that the Soviet Union's Terra-3 laser tracking center had targeted Challenger with a low-power laser, causing temporary malfunctions and disorientation among the crew. While these claims led to a U.S. diplomatic protest, the crew members aboard Challenger later denied the incident, and no definitive evidence emerged to support the allegations. However, in 2022, a former Soviet official revealed that the Soviets had indeed locked a laser tracker onto the Shuttle for a period, though the effect on the mission remains unclear.

Throughout its 8-day mission, *Challenger* traveled over 5.2 million kilometers and completed 133 orbits around the Earth. On

October 13, 1984, the Shuttle touched down at Kennedy Space Center's Shuttle Landing Facility at 12:26 p.m. EDT, marking only the second time a Shuttle had landed at its launch site.

STS-41-G's significance extended beyond its immediate achievements. Almost forty years later, one of *Challenger*'s OMS (Orbital Maneuvering System) engines from this mission was repurposed for NASA's Artemis 1 mission. The engine played a crucial role in propelling the uncrewed Orion spacecraft around the Moon, demonstrating the enduring legacy of Shuttle technology in shaping the future of human space exploration.

STS-51-A: The First Satellite Salvage Mission

STS-51-A, the 14th flight of NASA's Space Shuttle program, launched from Kennedy Space Center (KSC) in Florida at precisely 7:15 a.m. EST on November 8, 1984. This mission came just weeks after the successful STS-41-G flight, continuing a period of rapid shuttle deployment. Originally scheduled for the previous day, the launch was postponed due to concerns over high shear winds in the upper atmosphere, with the countdown halted at T-minus 20 minutes.

The five-person crew was led by Commander Frederick H. Hauck, embarking on his second spaceflight. Alongside him was pilot David M. Walker, and three mission specialists: Anna Lee Fisher, Dale A. Gardner, and Joseph P. Allen. Notably, Hauck was the first commander from NASA's 1978 astronaut class, marking a generational shift in leadership from the Apollo era. Both Gardner and Allen were veterans, each making their second shuttle flight, while Fisher, a medical doctor, became the first mother to fly in space.

STS-51-A's primary objectives included the deployment of two communications satellites and the unprecedented recovery of two malfunctioning satellites. The Hughes-built satellites, Anik D2 and Syncom IV-1 (also known as Leasat 1), were successfully deployed on the second and third days of the mission.

Following these deployments, the crew turned their attention to the recovery operations.

STS-51A mission included Frederick H. Hauck, commander,who is seated to the right. Standing, left to right, are Dale A. Gardner, mission specialist; David M. Walker, pilot; and mission specialists Anna L. Fisher, and Joseph P. Allen.

The two satellites targeted for recovery, Palapa B2 and Westar 6, had failed to reach their intended orbits following their launch aboard STS-41-B earlier in 1984. Ground controllers had lowered their orbits from around 600 miles (970 kilometers) to approximately 210 miles (340 kilometers) to facilitate retrieval. On the fifth day of the mission, Discovery rendezvoused with Palapa B2, initiating a complex extravehicular activity (EVA). Using the Manned Maneuvering Unit (MMU), Mission specialists Allen and Gardner approached the spinning satellite. Allen inserted the Apogee Capture Device (ACD), or "Stinger," into the satellite's apogee motor nozzle, enabling Gardner to attempt a grapple using the Remote Manipulator System (RMS), or Canadarm. After an initial failure to secure the satellite, Allen manually steadied Palapa B2 with Gardner's assistance, and with Fisher operating the Canadarm, they successfully maneuvered the satellite into the shuttle's cargo bay. The entire EVA took two hours, demonstrating the crew's ingenuity in overcoming technical challenges in space.

The recovery of Westar 6, the second satellite, occurred the following day and proved less complicated. Gardner used the same manual technique that Allen had employed, swiftly capturing the satellite and securing it in the shuttle's payload bay. In a moment of levity, Gardner held up a "For Sale" sign after the recovery, humorously referencing the troubled satellites. Ironically, Westar 6 was later sold to the Hong Kong-based AsiaSat.

In addition to the satellite operations, STS-51-A conducted important scientific experiments. The Diffused Mixing of Organic Solutions (DMOS) experiment, sponsored by 3M, was the first in a series of experiments exploring organic and polymer science in microgravity. The proprietary results were turned over to the company after the mission's success. A secondary experiment involved radiation monitoring, contributing to the growing body of data on space radiation exposure.

The Westar 6 satellite while Dale Gardner retrieves it during STS-51-A. Astronaut Dale A. Gardner, wearing the manned maneuvering unit (MMU) approaching the spinng Westar VI satellite over Bahama Banks. Gardner uses a large tool called the apogee kick motor capture device (ACD) to enter the nozzle of the spent Westar engine and stabilize the satellite to capture it for return to Earth.

STS-51-A is distinguished for performing the final untethered spacewalks of the Shuttle program. The mission marked the last use of the MMU, as subsequent spacewalks would utilize the Simplified Aid For EVA Rescue (SAFER)

system, tested a decade later on STS-64. In all future extravehicular activities (EVAs) by NASA and the Russian space program, astronauts would be tethered to their spacecraft.

After completing 127 orbits around Earth, Discovery returned safely to Kennedy Space Center, touching down on Runway 15 at 6:59 a.m. EST on November 16, 1984. The mission lasted 7 days, 23 hours, 44 minutes, and 56 seconds, and its successful conclusion was captured in footage featured in the 1985 IMAX film *The Dream is Alive*. The landing was only the third at KSC and marked the fifth and final shuttle mission of 1984, a year of tremendous achievement for NASA's burgeoning Space Shuttle program.

STS-51-C: A Milestone in National Security and Space Technology

Space Shuttle Discovery's mission STS-51-C was a significant chapter in the Space Shuttle program, marked by its classified nature and the expertise of its crew. The mission, which launched on January 24, 1985, was commanded by Colonel J. Dennis Hutchings, a seasoned Air Force pilot with extensive experience in aerospace operations. Serving alongside him was Pilot James D. Wetherbee, a Naval aviator who would go on to have a distinguished career in NASA, eventually becoming the only astronaut to command five Space Shuttle missions.

The mission specialists on board were William F. Readdy, Gary E. Payton, and Robert L. Stewart. Readdy, a Naval aviator and test pilot, brought a wealth of technical knowledge to the crew, having been selected as an astronaut in 1987. Gary E. Payton, a Colonel in the United States Air Force, was the first and only payload specialist on this mission, specifically chosen for his expertise in military space operations. His inclusion in the crew underscored the mission's importance to national security, as STS-51-C was the first dedicated Department of Defense mission in the Space Shuttle program. Finally, Robert L. Stewart, a veteran astronaut who had previously flown on STS-41-B, contributed his extensive experience in space operations. Stewart was known for his role in pioneering extravehicular activities (EVAs) using the Manned Maneuvering Unit (MMU), making him an invaluable asset to the crew.

The crew assigned to the STS-51C mission included (kneeling in front left to right) Loren J. Schriver, pilot; and Thomas K. Mattingly, II, commander. Standing, left to right, are Gary E. Payton, payload specialist; and mission specialists James F. Buchli, and Ellison L. Onizuka. Launched aboard the Space Shuttle Discovery on January 24, 1985 at 2:50:00 pm (EST), the STS-51C was the first mission dedicated to the Department of Defense (DOD).

STS-51-C launched from Kennedy Space Center (KSC) on January 24, 1985, at 19:50 UTC, marking the 100th human spaceflight to achieve orbit. Originally scheduled for January 23, the launch was delayed due to freezing weather conditions. Although the shuttle *Challenger* had initially been assigned to the mission, Discovery replaced it due to issues with *Challenger's* thermal protection tiles. This mission was the first of nine shuttle missions in 1985.

Unlike the longer civilian flights that often lasted a week or more, STS-51-C was a brief, three-day mission dedicated entirely to the U.S. Department of Defense (DoD). In a break from typical NASA procedures, pre-launch commentary was withheld until just nine minutes before liftoff, and much of the mission's details remain classified. The U.S. Air Force only disclosed that *Discovery* successfully deployed its payload using an Inertial Upper Stage (IUS)

on its seventh orbit. Although the exact nature of the payload has never been confirmed, it is widely believed to have been a Magnum SIGINT satellite placed into geosynchronous orbit. According to USAF Manned Spaceflight Engineer Gary Payton, this payload was "still up there, and still operating" as of 2009. Other DoD shuttle missions, such as STS-33 and STS-38, likely carried similar payloads.

Discovery initially entered an orbit of 204 × 519 km (127 × 322 mi) at an inclination of 28.45 degrees to the equator. The shuttle executed three Orbital Maneuvering System (OMS) burns, the last of which took place on the fourth orbit, to circularize its altitude at 519 km (322 mi). After completing its primary mission objectives, *Discovery* returned to Earth, landing on Runway 15 at KSC's Shuttle Landing Facility on January 27, 1985, at 21:23 UTC. The mission lasted a total of 3 days, 1 hour, 33 minutes, and 23 seconds. Footage from the STS-51-C launch was featured in the 1985 IMAX film *The Dream is Alive*, offering the public a rare glimpse of the mission's lift-off.

The importance of STS-51-C went beyond its classified payload, as it is indirectly connected to the tragic *Challenger* disaster that would occur just a year later during the STS-51-L mission. Among the crew of STS-51-L was astronaut Ellison Onizuka, who had also flown on STS-51-C. As part of the Rogers Commission's investigation into the *Challenger* disaster, it was revealed that during the launch of STS-51-C, there had been significant blow-by effects from the Space Shuttle's Solid Rocket Boosters (SRBs). The mission had experienced the worst blow-by to that point, indicating that the Viton O-rings used to seal the SRBs were not adequately containing the hot gases inside the combustion chambers. Post-flight analysis showed that both the right and left SRB O-rings displayed charring, with the primary O-ring in the right booster suffering unprecedented penetration and the secondary O-ring exhibiting heavy charring.

This data was crucial in establishing that low launch temperatures played a key role in the *Challenger* disaster. The temperature during the launch of STS-51-C was among the coldest recorded for a shuttle launch, at 12°C (54°F), providing early warnings of the risks posed by the compromised performance of the O-rings in cold conditions.

STS-51-D: A Mission of Diverse Objectives and Unexpected Challenges

Mission Specialist (MS) Karol J. Bobko, Crew Commander: Donald E. Williams, Pilot: Rhea Seddon and MS Jeffrey A. Hoffman, (back row) MS S. David Griggs, and Charles D. Walker and U.S. Senator Jake Garn (R-Utah) both Payload Specialists.

In the early hours of April 12, 1985, the Space Shuttle Discovery lifted off from Kennedy Space Center on mission STS-51-D, marking its fourth flight and the sixteenth mission of the Space Shuttle program. The launch was flawless, with Discovery soaring into the clear Florida sky, carrying a diverse payload and a crew of seven astronauts. The mission's ambitious objectives were deploying two communications satellites to conduct scientific experiments and showcasing the Shuttle's versatility as a platform for multiple space activities.

The crew of STS-51-D was a mix of experienced astronauts and rookies, led by Commander Karol J. "Bo" Bobko, a veteran of the Shuttle program. Pilot Donald E. Williams and Mission Specialists M. Rhea Seddon, Jeffrey A. Hoffman, and David Griggs were alongside him. Also aboard were two Payload Specialists: Charles D. "Charlie" Walker, a representative of

McDonnell Douglas, and U.S. Senator Jake Garn, a civilian with a unique background in politics and military service. Senator Garn's inclusion in the mission was significant, as he was the first sitting member of the United States Congress to fly into space, part of NASA's effort to broaden public understanding and support for the Shuttle program.

During STS-51-D, launched on April 12, 1985, Space Shuttle *Discovery* carried out a dual satellite deployment mission and performed several scientific and educational experiments. The mission's primary objective was to deploy two Hughes-built communications satellites: Telesat-I (Anik C1) and Syncom IV-3 (Leasat-3). The deployment of Telesat-I was successful, as it was released with the assistance of a Payload Assist Module (PAM-D) motor. However, the deployment of Syncom IV-3 encountered complications. Upon release, the satellite failed to initiate critical operations, such as antenna deployment, spin-up, and the ignition of its perigee kick motor, which was essential to place the satellite into its intended orbit.

Hoffman and Griggs attach the flyswatter device to the end of the Canadarm.

In an attempt to troubleshoot the satellite's malfunction, the mission was extended by two days. After reviewing the situation, NASA engineers determined that the satellite's spacecraft sequencer start lever needed to be manually adjusted. Astronauts Jeffrey Hoffman and David Griggs performed an unplanned Extravehicular Activity (EVA), during which they attached makeshift "Flyswatter" devices to the Shuttle's Remote Manipulator System (RMS), commonly known as the Canadarm. Mission Specialist Dr. Margaret Rhea Seddon then used the Canadarm to attempt to engage the satellite's start lever. Despite these efforts, the satellite's post-deployment sequence still failed to initiate. The malfunctioning satellite was later retrieved and successfully repaired and redeployed during the STS-51-I mission later that year.

In addition to the satellite deployment, *Discovery* carried a variety of payloads designed for scientific research and education. One of the key payloads was the Continuous Flow Electrophoresis System III (CFES-III), a device designed to investigate the separation of biological materials in microgravity, flying for its sixth mission. Other payloads included two Shuttle Student Involvement Program (SSIP) experiments, the American Flight Echocardiograph (AFE), two Getaway Special (GAS) experiments, and a set of Phase Partitioning Experiments (PPE). Additionally, an astronomical photography verification test was conducted, alongside various medical experiments. A notable educational experiment, "Toys in Space," studied the behavior of simple toys in microgravity, with the results shared with school students upon the shuttle's return to Earth.

Discovery landed on April 19, 1985, on Runway 33 at Kennedy Space Center's Shuttle Landing Facility. The landing, however, was not without incident. As the orbiter approached the runway, it encountered a 10-mile-per-hour crosswind from the right, which pushed the vehicle slightly off course. When the rear landing gear touched down, the orbiter had drifted 19 feet to the left of the centerline. As the orbiter continued to roll out, the crosswind increased its drift to 65 feet. Commander Karol J. Bobko applied differential braking, a technique used to steer the orbiter by applying more pressure to the brakes on one side. This maneuver successfully brought *Discovery* back toward the centerline,

but as the orbiter slowed, the right-side inboard tire locked up, followed by the outboard tire, which caused the inboard tire to blow out just before the orbiter came to a stop.

This incident prompted NASA to implement a new steering mechanism for the Shuttle fleet. Nose wheel steering was introduced to provide better control during landings, and by the end of 1985, this modification was completed. In the interim, NASA opted to land Shuttles at Edwards Air Force Base, where wide dry lakebeds provided a safer and more forgiving landing environment. The new nose wheel steering system was installed just before the Challenger disaster in January 1986, after which the Shuttle program was grounded for nearly three years. It would only be in STS-38 in 1990 that another Shuttle would land at Kennedy Space Center, marking a return to normal operations.

STS-51B: Advancing Space Science in Spacelab

On April 29, 1985, at precisely 12:02:18 p.m. EDT, Space Shuttle *Challenger* lifted off from Kennedy Space Center's Launch Pad 39A, marking the beginning of mission STS-51-B. This was *Challenger*'s seventh flight, carrying a distinguished crew led by Commander Robert F. Overmyer and Pilot Frederick D. Gregory. The mission specialists included Don L. Lind, Norman E. Thagard, and William E. Thornton. In contrast, payload specialists Lodewijk van den Berg from EG&G Energy Management, Inc., and Taylor G. Wang from NASA's Jet Propulsion Laboratory (JPL) also joined the team. With an average crew age of 48.6 years, STS-51-B held the record for the oldest American spaceflight crew at the time.

The mission's primary focus was the operation of the European Space Agency's (ESA) Spacelab module, marking its second flight aboard the Space Shuttle. Unlike its initial flight on STS-9, this mission saw the Spacelab in a fully operational configuration, allowing for comprehensive multidisciplinary research in the microgravity environment of space. The mission was divided into two 12-hour shifts to allow around-the-clock operation, with the Gold team comprising Overmyer, Lind, Thornton, and Wang, and the Silver team consisting of Gregory, Thagard, and van den Berg. This shift arrangement was essential in maintaining continuous oversight of the experiments aboard the shuttle.

The orbiter maintained a gravity gradient attitude, ensuring a stable environment for the delicate experiments being conducted, particularly in materials processing and fluid mechanics. Spacelab carried 15 primary experiments, 14 of which were successfully completed. The scientific payload also included two squirrel monkeys and 24 rats, housed in special cages for biomedical studies, marking the second time American astronauts had flown live non-human mammals aboard the shuttle. This research, supported by a dedicated Payload Operations Control Center at Johnson Space Center, added valuable insight into how biological systems respond to spaceflight.

Robert F. Overmyer, commander; and Frederick D. Gregory, pilot. Standing, left to right, are Don L. Lind, mission specialist; Taylor G. Wang, payload specialist; Norman E. Thagard, mission specialist; William E. Thornton, mission specialist; and Lodewijk van den Berg, payload specialist.

In addition to the Spacelab experiments, two Getaway Special (GAS) payloads were deployed for the first time during the mission. The Northern Utah Satellite (NUSAT) was successfully deployed, but the second satellite,

the Global Low Orbiting Message Relay (GLOMR), encountered deployment issues and was returned to Earth.

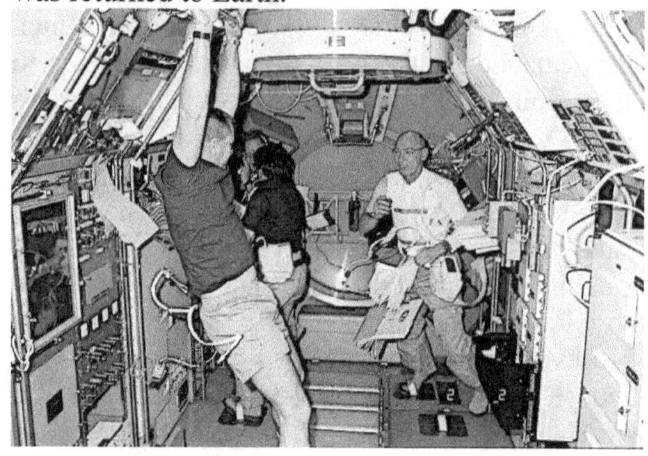

Crew members of the STS-51-B (Spacelab-3) mission inside the Spacelab laboratory module aboard the Space Shuttle Challenger. Left to right: STS-51-B Commander Robert F. Overmyer; Mission Specialist Don L. Lind (obscured); Payload Specialist Lodewijk van den Berg (obscured); and Mission Specialist William Thornton.

A notable experiment on the mission was India's Cosmic Ray Experiment, Anuradha. This instrument, developed by the Indian Space Research Organization (ISRO), was designed to measure the ionization states of low-energy cosmic rays in near-Earth space. Consisting of a barrel-shaped recorder with plastic sheets, Anuradha detected cosmic rays at a rate of seven per minute for 64 continuous hours, producing 10,000 sheets of data. This experiment provided important insights into cosmic radiation, contributing to the growing body of knowledge regarding the space environment.

After a successful mission, *Challenger* re-entered Earth's atmosphere and landed at Edwards Air Force Base at 12:11:04 p.m. EDT on May 6, 1985. The mission lasted a total of 7 days, 8 minutes, and 46 seconds, marking another triumph for the Space Shuttle program and advancing human knowledge in space science and technology.

STS-51G: A Global Endeavor in Space

On June 17, 1985, the Space Shuttle

Discovery lifted off from Kennedy Space Center on mission STS-51G, marking another significant chapter in the Space Shuttle program. This mission, the eighteenth of the program and the fifth for Discovery showcased the Shuttle's growing role as a versatile platform for international collaboration and diverse payload operations. The mission was notable for its multinational crew and its primary objective of deploying three communication satellites, underscoring the Shuttle's importance in supporting global communications infrastructure.

The crew of STS-51G was an international team led by Commander Daniel C. Brandenstein and Pilot John O. Creighton, both Shuttle program veterans. Four Mission Specialists joined them: Shannon W. Lucid, John M. Fabian, Steven R. Nagel, and Patrick Baudry, a French astronaut representing the French space agency CNES (Centre National d'Études Spatiales). Additionally, Sultan bin Salman Al Saud, a Payload Specialist from Saudi Arabia, became the first Arab and the first Muslim to fly in space, making this mission a symbol of international cooperation and achievement.

Astronauts Daniel C. Brandenstein (left) and John O. Creighton, commander and pilot, respectively. Astronauts Shannon W. Lucid, Steven R. Nagel, and John M. Fabian, mission specialist (l.-r.) joing Payload specialists Sultan Salman Abdelazize Al-Saud (second right) and Patrick Baudry on the back row.

The primary task of STS-51G was the deployment of three communications satellites, each representing a different region of the world.

The first satellite deployed was the Mexican Morelos-1, intended to enhance telecommunications across Mexico. Shortly afterward, the Arabsat-1B satellite was released, part of a regional communications system serving the Arab League nations. The final satellite, Telstar 3D, was an American commercial communications satellite designed to provide television and data transmission services.

Each satellite deployment required precise coordination and timing. The crew worked seamlessly to ensure the satellites were released into their correct orbits, using Discovery's sophisticated payload deployment systems. The successful deployment of these satellites highlighted the Shuttle's ability to serve as a critical link in global telecommunications, facilitating international cooperation and development.

In addition to the satellite deployments, STS-51G carried several secondary payloads and experiments. Among them was the SPARTAN-1 (Shuttle Pointed Autonomous Research Tool for Astronomy), an experiment package designed to conduct astronomical observations. SPARTAN-1 was released from the Shuttle's payload bay, operated independently for a period, and then retrieved by the crew using the Shuttle's robotic arm. This operation demonstrated the Shuttle's capability to deploy, operate, and recover free-flying scientific instruments in space.

The mission also included experiments in materials science and technology demonstrations. Notably, the Automated Directional Solidification Furnace (ADSF) experiment was conducted to study the effects of microgravity on the solidification of materials, with implications for manufacturing processes both in space and on Earth.

Another interesting aspect of the mission was Sultan bin Salman Al Saud's participation. He conducted experiments designed to observe how the human body adapts to spaceflight, including the effects of weightlessness on the cardiovascular system. His presence on the mission also had symbolic importance, representing the growing international interest and involvement in space exploration.

The crew's return to Earth on June 24, 1985,

was as smooth as the mission. Discovery landed safely at Edwards Air Force Base in California, completing a highly successful mission on all fronts. The crew's international nature and the mission's diverse objectives demonstrated the Shuttle's unique role in bringing nations together to advance science and technology.

STS-51G exemplified the Space Shuttle program's ability to serve as a platform for international cooperation. The deployment of communications satellites for Mexico, the Arab League, and the United States underscored the Shuttle's contribution to global communications infrastructure. The mission's scientific experiments and technology demonstrations also provided valuable insights that would influence future space missions.

In the broader context of space exploration, STS-51G highlighted the importance of collaboration among nations in the pursuit of knowledge and technological advancement. The mission's success reinforced the Space Shuttle's reputation as a versatile and reliable vehicle capable of supporting a wide range of objectives, from satellite deployment to scientific research, and set a precedent for future international partnerships in space.

Arabsat communications satellite deploying from Discovery's payload bay.

STS-51F: A Mission of Scientific Discovery and Adaptation

Gordon Fullerton, commander; and Roy D. Bridges, pilot. Standing, left to right, are mission specialists Anthony W. England, Karl G. Henize, and F. Story Musgrave; and payload specialists Loren W. Acton, and John-David F. Bartoe.

STS-51-F, also known as Spacelab 2, marked the 19th flight of NASA's Space Shuttle program and the eighth mission of *Space Shuttle Challenger*. Launched from Kennedy Space Center on July 29, 1985, and returning on August 6, 1985, this mission carried a range of scientific experiments. It was notable for several unique occurrences, including the first-ever in-flight abort.

The mission's primary objective was to conduct various scientific experiments using the *Spacelab 2* laboratory module. However, the launch of STS-51-F was not without challenges. The first launch attempt on July 12, 1985, was aborted at T−3 seconds due to a malfunction in one of the Space Shuttle Main Engines (SSMEs). After a delay of 17 days, *Challenger* successfully lifted off on July 29, 1985, though not without further complications.

At 3 minutes and 31 seconds into the ascent, one of the sensors in *Challenger's* center engine failed, leading to an engine shutdown. The mission shifted into an "Abort to Orbit" (ATO) procedure—the first and only time such a procedure was executed in the Shuttle program—resulting in a lower-than-planned orbit of approximately 265 km, rather than the intended 385 km. The quick thinking of Booster Systems Engineer Jenny M. Howard, who recommended inhibiting further automatic shutdowns based on faulty sensor readings, prevented a second engine shutdown and potentially catastrophic consequences for the mission and crew.

The seven-member crew was split into two 12-hour shifts, known as the "Red Team" and "Blue Team," to ensure continuous operations throughout the mission. Commander C. Gordon Fullerton led the mission, with Pilot Roy D. Bridges Jr. and Mission Specialists Karl G. Henize, F. Story Musgrave, and Anthony W. England forming the core crew. Payload Specialists Loren W. Acton and John-David F. Bartoe joined them. Henize, Acton, and Bridges comprised the Red Team, while Musgrave, England, and Bartoe made up the Blue Team.

Challenger was prepared for emergency scenarios and carried two Extravehicular Mobility Units (EMUs) for spacewalks, though they were not utilized during the mission. England and Musgrave were trained for this potential contingency.

Close-up view of a portion of the F6 forward flight station panel. Photo focuses on the sub-panel controlling the shuttle ascent abort mode, RCS firing command indicators and the Range Safety Arm/Safe light. The picture highlights the abort selector in the ATO position, as it was used in a successful Abort to Orbit on this flight, STS-51-F.

The centerpiece of the mission was the *Spacelab 2* module, which housed numerous experiments across a range of scientific fields, including plasma physics, astronomy, solar physics, and atmospheric studies. The mission aimed to verify the performance of Spacelab systems and measure the environment created by the Shuttle in orbit. The European Space Agency's (ESA) Instrument Pointing System (IPS) was tested for the first time in space, designed to track celestial objects with an accuracy of one arcsecond. Although initial problems arose with tracking the Sun, software adjustments enabled the system to function correctly.

One of the primary scientific instruments aboard was the *Spacelab Infrared Telescope* (IRT), a helium-cooled infrared telescope with a 15.2 cm aperture, designed to observe light between wavelengths of 1.7 to 118 micrometers. Despite concerns that heat emissions from the Shuttle might affect its long-wavelength observations, the IRT successfully collected valuable data on 60% of the galactic plane. However, a piece of insulation briefly obstructed the telescope's view, which was later rectified.

In addition, the Plasma Diagnostics Package (PDP), first flown on STS-3, returned for further study of the Earth's ionosphere. On the mission's third day, it was deployed from the payload bay using the Shuttle's robotic arm (Canadarm) and released for six hours. During this time, *Challenger* maneuvered around the PDP as part of a targeted proximity operations exercise. The PDP was then retrieved and stowed back in the payload bay.

Among the more publicized elements of the mission was a marketing experiment involving the two major soda brands, Coca-Cola and Pepsi. The astronauts tested the feasibility of consuming carbonated beverages in space using specially designed cans. The experiment drew significant media attention, though the results were mixed. While Coca-Cola's dispenser was relatively effective, Pepsi's system, described as resembling a shaving cream can, struggled to function properly in microgravity. Despite the novelty, the astronauts later preferred Tang, which could be mixed with chilled water, unlike the warm, excessively fizzy sodas.

Another significant experiment involved firing the Shuttle's thruster rockets over Tasmania and Boston to create plasma depletion regions—areas in the ionosphere where the plasma density was reduced. Geophysicists worldwide collaborated with NASA on this experiment, utilizing the data collected by Spacelab 2 to study the effects of these "holes" in the ionosphere.

Despite the complications resulting from the in-flight abort, STS-51-F succeeded. 17 orbits extended the mission to allow additional time for experiments affected by the lower-than-expected orbit. After eight days in space, *Challenger* touched down at Edwards Air Force Base, California, on August 6, 1985, at 12:45 p.m. PDT.

Upon its return to Kennedy Space Center on August 11, 1985, *Challenger* and its crew were celebrated for their accomplishments despite the technical hurdles faced during the mission.

STS-51I: A Mission of Repair and Deployment

On August 27, 1985, the Space Shuttle Discovery launched from Kennedy Space Center on mission STS-51I, the twentieth flight of the Space Shuttle program and the sixth for Discovery. This mission was distinguished by its dual objectives: deploying three communications satellites and performing an ambitious in-orbit repair of a previously deployed satellite. The mission highlighted the Shuttle's versatility as both a launch platform and a space-based service vehicle capable of addressing unforeseen issues with valuable space assets.

The crew of STS-51I consisted of five astronauts, led by Commander Joe H. Engle, a veteran test pilot and astronaut. Pilot Richard O. Covey and Mission Specialists James D. van Hoften, John M. Lounge, and William F. Fisher joined him. This experienced team was tasked with executing a complex mission that required precision, technical expertise, and adaptability.

The primary task of STS-51I was the deployment of three communications satellites crucial for expanding global communications networks and supporting both commercial and military operations: ASC-1 (American Satellite Company 1), Aussat-1 (the first satellite for Australia's national satellite system), and Syncom IV-4, also known as LEASAT-4, a satellite intended for use by the U.S. Navy.

STS-51I mission included (front row left to right) Joe H. Engle, commander, and Richard O. Covey, pilot. In the center is John M. (Mike) Lounge, mission specialist. On the back row, from left to right, are mission specialists James D. van Hoften, and William

The satellite deployments proceeded smoothly. Each satellite was released from Discovery's payload bay and boosted into its designated orbit using an attached perigee kick motor. The crew's ability to carry out these deployments with precision underscored the Shuttle's effectiveness as a satellite delivery system capable of launching multiple payloads on a single mission.

However, the most challenging aspect of STS-51I was repairing the Syncom IV-3 satellite, also known as LEASAT-3, which had failed to activate after being deployed during the STS-51D mission earlier that year. LEASAT-3 was stuck in a dormant state, unable to provide the critical communications capabilities it was designed for. NASA determined that an in-orbit repair was necessary, and STS-51I was tasked with this ambitious objective.

Mission Specialists James van Hoften and William Fisher performed two spacewalks, or Extravehicular Activities (EVAs) to accomplish this repair. On the first EVA, van Hoften and Fisher manually attached a custom-built switch cover to the satellite to bypass the malfunctioning electronics and activate the satellite's systems.

The spacewalkers displayed remarkable skill and dexterity as they worked in the challenging space environment, using specialized tools and techniques to address the problem.

The second EVA was equally critical, as van Hoften and Fisher completed the final steps needed to activate LEASAT-3 fully. Their efforts were successful, and the satellite was brought back to life, allowing it to fulfill its intended role in military communications. This repair marked a significant achievement for the Space Shuttle program, demonstrating the Shuttle could not only deploy satellites but also service and repair them in orbit, extending their operational lifetimes and saving valuable space assets.

In addition to the satellite deployments and repair, the STS-51I mission included several secondary payloads and experiments. Among them was the Automated Directional Solidification Furnace (ADSF), used to study the solidification of materials in microgravity, with potential applications for manufacturing processes both in space and on Earth.

the AUSSAT-1 satellite after deployment on STS-51-I. Australia's AUSSAT communications satellite is deployed from the payload bay of the Shuttle Discovery.

After completing all mission objectives, Discovery and her crew returned to Earth on September 3, 1985, landing safely at Edwards Air Force Base in California. The mission was hailed as a success, particularly for repairing LEASAT-3, which had initially been considered a lost asset. The ability to conduct such a complex repair in space underscored the growing maturity of the Space Shuttle program and its capacity to handle a wide range of tasks.

STS-51I demonstrated the Space Shuttle's role as a versatile and capable spacecraft for launching and deploying satellites and performing critical repairs that would otherwise be impossible. The mission's success reinforced the value of the Shuttle as a tool for maintaining and servicing space infrastructure, ensuring the continued operation of satellites essential for global communications and military operations. The achievements of STS-51I contributed to the growing body of knowledge and experience that would guide future space missions, solidifying the Shuttle's place as a cornerstone of NASA's space exploration efforts in the 1980s.

STS-51-J

STS-51-J, the 21st flight of NASA's Space Shuttle program and the second mission for Space Shuttle Atlantis, was a pivotal yet highly classified mission. It marked the shuttle's first mission dedicated entirely to the U.S. Department of Defense. Launched on October 3, 1985, from Kennedy Space Center's Launch Complex 39A, STS-51-J was shrouded in secrecy, with limited details released to the public due to the sensitive nature of its payload.

Atlantis, OV-104, was named after the primary research vessel for the Woods Hole Oceanographic Institute in Massachusetts from 1930 to 1966. Atlantis was the fourth operational shuttle and made its first flight, STS-51J, on October 3, 1985. The shuttle sent probes to Venus and Jupiter and carried NASA's Destiny laboratory to the International Space Station. Atlantis also served as the final shuttle servicing mission, STS-125, for NASA's Hubble Space Telescope. STS-135 will mark the 33rd mission and final flight for Atlantis. This will also be the final flight of the Space Shuttle Program.

The mission crew consisted of five astronauts

selected for their expertise and experience in handling classified operations. The commander was Karol J. "Bo" Bobko, a seasoned astronaut who had previously flown on STS-6 and STS-51-D. Alongside him was pilot Ronald J. Grabe, who was making his first spaceflight. Mission specialists included David C. Hilmers, Robert L. Stewart, and William A. Pailes, a payload specialist representing the U.S. Air Force.

Robert L. Stewart, mission specialist; Karol J. Bobko, commander; and Ronald J. Grabe, pilot. On the back row, left to right, are mission specialists David C. Hilmers, and Major William A. Pailes (USAF).

Atlantis lifting off the pad.

After a flawless countdown, Atlantis lifted off at 11:15 AM EDT, ascending into a clear Florida sky. The launch was routine, but the mission's true objectives remained closely guarded. Atlantis carried two military communications satellites, part of the Defense Satellite Communications System (DSCS). These satellites were crucial for secure, long-distance communication between U.S. military forces worldwide, ensuring that critical information could be relayed across continents without interception.

The deployment of these satellites was conducted smoothly, utilizing the shuttle's payload bay. Although specific details of their operation were classified, it is known these satellites were designed to operate in geostationary orbit, providing continuous coverage over specific areas of the globe.

The mission also tested classified Department of Defense systems and conducted experiments not disclosed to the public. These activities were part of the broader strategy of the U.S. military to leverage the unique capabilities of the Space Shuttle for national security purposes. The Shuttle's ability to deploy, repair, and retrieve satellites was a significant advantage in maintaining and expanding the United States' strategic capabilities in space.

Atlantis remained in orbit for nearly four days, circling the Earth 64 times before returning to Kennedy Space Center on October 7, 1985. The landing was precise, with Atlantis touching down at 1:00 PM EDT on Runway 15 of the Shuttle Landing Facility. The mission was deemed a success, having achieved all its objectives, though the specific outcomes of the experiments and satellite operations remained classified.

STS-61A

STS-61A, also known as Spacelab D1, was a historic mission during the final months of 1985. It was the first and only Space Shuttle mission funded and managed by a foreign country, West Germany, under the European Space Agency (ESA). This mission marked the first time that eight astronauts flew aboard a single spacecraft, setting a record for the largest crew in space to that date. The mission was a testament to international cooperation in space exploration and showcased the Shuttle's versatility in supporting complex scientific endeavors.

Reinhard Furrer, German payload specialist; Bonnie J. Dunbar, mission specialist; James F. Buchli, mission specialist; and Henry W. Hartsfield, Jr., commander. On the back row, left to right, are Steven R. Nagel, pilot; Guion S. Bluford, mission specialist.

Launched aboard the Space Shuttle Challenger from Kennedy Space Center's Launch Complex 39A on October 30, 1985, at noon EST, STS-61A was commanded by NASA astronaut Henry W. "Hank" Hartsfield, with Steven R. Nagel as the pilot. The mission's crew included NASA Mission Specialists James F. Buchli and Guion S. Bluford, as well as four payload specialists: Ernst Messerschmid and Reinhard Furrer from West Germany, Wubbo Ockels from the Netherlands, and NASA's Bonnie J. Dunbar. The diverse crew reflected the international nature of the mission and its scientific goals.

The primary focus of STS-61A was operating the Spacelab D1 module, a European-built laboratory housed in the Shuttle's payload bay. Spacelab D1 was equipped with a wide array of scientific instruments, allowing the crew to conduct materials science, life sciences, fluid physics, and atmospheric physics experiments. The mission's experiments were designed to take advantage of the microgravity space environment, providing insights that could not be obtained through Earth-based research.

The crew worked in two shifts throughout the mission, ensuring the Spacelab D1 operated continuously during the eight-day mission. The rigorous schedule allowed for the completion of over 75 experiments, many of which had been developed by German scientists. These experiments included studies on the behavior of fluids in microgravity, the growth of protein crystals, and the effects of spaceflight on the human body. The mission also provided valuable data on how the human body adapts to long-duration spaceflight, information that would prove critical for future missions, including those to the International Space Station.

STS-61A was also notable for its smooth operation and high coordination between NASA and the German Space Operations Center (GSOC), which managed the mission's payload operations from Oberpfaffenhofen, West Germany. This cooperation exemplified the potential for international partnerships in space exploration, setting a precedent for future collaborative missions.

Challenger returned to Earth on November 6, 1985, landing at Edwards Air Force Base in California at 9:44 a.m. PST after a mission duration of 7 days, 44 minutes, and 51 seconds. The Shuttle orbited the Earth 111 times, covering approximately 2.9 million miles. The successful completion of STS-61A not only achieved its scientific objectives but also demonstrated the effectiveness of international collaboration in space exploration, paving the way for future partnerships, including those that would later support the International Space Station.

STS-61-B

STS-61-B, the twenty-third flight of NASA's Space Shuttle program and the second-to-last mission of 1985, was a pivotal mission that demonstrated the Space Shuttle's capability to deploy multiple satellites in a single flight and tested construction techniques that would be essential for building structures in space, such as the International Space Station.

Launched aboard the Space Shuttle Atlantis from Kennedy Space Center's Launch Complex

39A on November 26, 1985, at 7:29 p.m. EST, STS-61B was commanded by Brewster H. Shaw, with Bryan D. O'Connor serving as the pilot. The mission's crew included Mission Specialists Jerry L. Ross, Mary L. Cleave, Sherwood C. Spring, Payload Specialists Rodolfo Neri Vela of Mexico, and Charles D. Gemar of the United States. Rodolfo Neri Vela became the first Mexican national to fly in space, highlighting the mission's international significance.

Back row: Walker, Ross, Cleave, Spring and Neri Vela
Front row: O'Connor and Shaw

The primary objectives of STS-61B were deploying three communications satellites and conducting experiments related to space construction. The three satellites deployed during the mission were MORELOS-B for Mexico, AUSSAT-2 for Australia, and SATCOM K2 for the United States. These satellites were critical for enhancing communication capabilities in their respective countries, and their successful deployment showcased the Shuttle's versatility in supporting commercial and governmental satellite operations.

In addition to satellite deployment, STS-61B featured the EASE/ACCESS experiment (Experimental Assembly of Structures in Extravehicular Activity/Assembly Concept for Construction of Erectable Space Structures), which tested techniques for assembling large structures in space. This experiment was particularly significant as it provided vital data and hands-on experience to be used later in constructing the International Space Station.

Two spacewalks, or extravehicular activities (EVAs), were conducted by Mission Specialists Jerry Ross and Sherwood Spring during the mission. The first EVA, lasting 5 hours and 32 minutes, took place on November 29, and the second, lasting 6 hours and 41 minutes, occurred on December 1. During these EVAs, the astronauts assembled a series of structural components, simulating the construction of a space station or other large structures in orbit. The success of these activities validated the feasibility of on-orbit construction and assembly, an essential capability for future space missions, including the assembly of the ISS.

STS-61B also carried the OAST-1 (Office of Aeronautics and Space Technology 1) solar array experiment, which tested the deployment of large solar arrays in space. The experiment provided valuable insights into the dynamics of deploying and operating solar arrays, which would become a crucial power source for future space stations.

After completing its objectives, Atlantis returned to Earth on December 3, 1985, landing at Edwards Air Force Base in California at 4:33 p.m. PST. The mission lasted six days, 21 hours, and 4 minutes, during which the Shuttle orbited the Earth 109 times, covering approximately 2.9 million miles.

STS-61B was a landmark mission that underscored the Space Shuttle's versatility and its vital role in advancing space exploration capabilities. The successful deployment of multiple satellites and the pioneering work in space construction during this mission laid the groundwork for future endeavors, including the assembly of the International Space Station. The mission also highlighted the growing importance of international collaboration in space exploration, with the participation of Mexico's first astronaut serving as a symbol of the increasingly global nature of space endeavors.

STS-61-C, the twenty-fourth mission of NASA's Space Shuttle program and the final flight before the tragic Challenger disaster focused on satellite deployment and scientific research. Launched aboard the Space Shuttle Columbia, STS-61-C was notable not only for its objectives but also for the diversity of its crew and the high-profile nature of one of its payloads.

Columbia lifted off from Kennedy Space

Center's Launch Complex 39A on January 12, 1986, at 6:55 a.m. EST. The mission's crew, commanded by Robert L. "Hoot" Gibson, included Pilot Charles F. Bolden, who would later become NASA's Administrator, and Mission Specialists Franklin R. Chang-Díaz, Steven A. Hawley, and George D. "Pinky" Nelson. The mission also carried two Payload Specialists: Robert J. Cenker, an engineer for RCA who was responsible for overseeing the deployment of the company's satellite, and Bill Nelson, a U.S. Congressman from Florida who flew as a part of NASA's efforts to involve policymakers in space missions.

STS-61-C

The primary objective of STS-61-C was deploying the SATCOM KU-1 communications satellite for RCA Americom. The satellite was successfully deployed and later used to enhance television broadcasting and communication services. This deployment continued the Shuttle's role in supporting the burgeoning commercial satellite industry, showcasing the vehicle's versatility in carrying out diverse missions.

Charles F. Bolden, Jr., pilot; and Robert L. (Hoot) Gibson, commander. On the back row, left to right, are payload specialists Robert J. Cenker, and Congressman Bill Nelson. To the right of Nelson are mission specialists Steven A. Hawley, George D. Nelson, and Franklin R. Chang-Diaz

In addition to the satellite deployment, STS-61-C carried several secondary payloads, including the Materials Science Laboratory-2 (MSL-2) experiment. MSL-2 was designed to study the behavior of multiple materials in microgravity, providing valuable data that would inform future materials science research and the development of new technologies.

One of the most interesting aspects of the mission was Congressman Bill Nelson's presence. Nelson became the second sitting member of the U.S. Congress to fly in space, following Senator Jake Garn's flight on STS-51-D. Nelson's participation was intended to give him firsthand experience with the Shuttle program and space exploration, which he could use to inform his work in Congress. His presence on the mission also reflected NASA's broader effort to engage the public and government in its space activities.

STS-61-C faced several challenges, including multiple delays before launch and a shortened mission duration due to concerns about the Shuttle's thermal protection system. Despite these challenges, the crew successfully completed the mission's objectives. The Shuttle's reentry and landing were complicated by adverse weather conditions, leading to several delays and wave-offs before Columbia finally touched down at Edwards Air Force Base in California on January 18, 1986, at 5:59 a.m. PST. The mission lasted 6 days, 2 hours, and 3 minutes, during which Columbia orbited the Earth 96 times, traveling approximately 2.5 million miles.

While successful in its primary objectives, STS-61-C is often remembered as the last mission before the Challenger disaster, which occurred just ten days later. The disaster brought the Shuttle program to a standstill and led to a comprehensive reevaluation of NASA's operations and safety protocols. In hindsight, STS-61-C serves as a poignant reminder of the inherent risks of human spaceflight and the need for constant vigilance in ensuring the safety of astronauts and the success of missions.

The mission's successful satellite deployment and scientific research contributions are a testament to the versatility and capability of the Space Shuttle program, even as it faced the challenges and uncertainties of the era. Bill Nelson's participation also highlighted the importance of fostering connections between space exploration and governmental policymaking, a relationship that continues to shape the future of space exploration.

STS-61-C lifts off from Launch Complex 39A at Kennedy Space Center

Chapter 4 - Tragedy and Triumph: The Challenger Era (1986 - 1993)

STS-51-L

Ellison S. Onizuka, Sharon Christa McAuliffe, Greg Jarvis, and Judy Resnik. In the front row from left to right: Michael J. Smith, Dick Scobee, and Ron McNair.

STS-51-L was the 25th mission of NASA's Space Shuttle program and the 10th flight of the Space Shuttle Challenger. Launched on January 28, 1986, this mission tragically became one of the most infamous in space exploration history due to the catastrophic loss of the shuttle and its seven-member crew just 73 seconds after liftoff.

The crew of STS-51-L included seven astronauts: Francis R. Scobee, the mission commander; Michael J. Smith, the pilot; and mission specialists Ronald McNair, Ellison Onizuka, and Judith Resnik. Also on board were payload specialist Gregory Jarvis and Christa McAuliffe, a New Hampshire teacher selected as the first civilian to fly in space as part of NASA's Teacher in Space Project.

The mission objectives for STS-51-L were ambitious. They included deploying the Tracking and Data Relay Satellite-B (TDRS-B) to enhance communications between orbiting spacecraft and ground stations. Another significant part of the mission was deploying the Spartan Halley spacecraft, designed to observe Halley's Comet as it approached the Sun. Additionally, Christa McAuliffe was scheduled to conduct two live lessons from space, part of an educational outreach program to inspire students across the United States.

On January 28, 1986, Challenger lifted off from Kennedy Space Center's Launch Complex 39B at 11:38 AM EST. The launch had been delayed multiple times due to weather conditions and technical issues. Despite concerns about the unusually cold temperatures in Florida that morning, the decision was made to proceed with the launch.

Challenger after the explosion 73 seconds after launch

Tragically, just 73 seconds into the flight, Challenger disintegrated over the Atlantic Ocean, resulting in the deaths of all seven crew members. The cause of the disaster was later determined to be the failure of an O-ring seal in its right solid rocket booster. The O-ring, designed to prevent hot gases from escaping the booster, had become brittle in the cold temperatures and failed to seal properly, allowing hot gases to escape and eventually causing the external fuel tank to explode.

The loss of Challenger and its crew was a devastating blow to NASA and the entire space community. The disaster led to the suspension of the Space Shuttle program for nearly three years while an extensive investigation and redesign of the shuttle's systems were undertaken. The Rogers Commission, appointed by President Ronald Reagan to investigate the accident, uncovered severe flaws in NASA's decision-making processes, leading to significant changes in the agency's safety protocols and organizational culture.

On March 7, 1986, divers from the USS *Preserver* made a significant discovery on the ocean floor—what they believed to be the Space Shuttle Challenger crew cabin. The following day, a dive confirmed their grim finding: it was indeed the crew cabin, and the remains of the seven astronauts were inside. This marked a poignant and sorrowful moment in the investigation of the *Challenger* disaster, as efforts continued to piece together the events that led to the tragedy.

Despite extensive investigations, no official inquiry has definitively determined the exact cause of death for the astronauts. However, it is widely believed that the explosion during the launch did not immediately kill the entire crew. Evidence for this comes from the recovery of four Personal Egress Air Packs (PEAPs), three of which had been manually activated. These PEAPs, designed to provide emergency oxygen, would only have been used in a critical situation such as loss of cabin pressure. The activation of the PEAPs suggests that at least some of the crew were conscious and attempting to respond to the emergency.

However, PEAPs do not provide pressurized air, and even with their activation, the astronauts would have lost consciousness within seconds due to the rapid depressurization of the cabin at such a high altitude. The crew would not have survived long in the unpressurized environment.

Media reports at the time claimed that NASA had a tape recording of the astronauts' voices during the 2-minute 45-second free fall after the explosion. Allegedly, this recording captured the crew's panic and final moments before the crew cabin impacted the Atlantic Ocean off the coast of Florida. However, these claims are widely regarded as fabrications. The crew did not wear individual voice recorders, and it is unlikely that they would have been able to communicate under the circumstances, as they were likely unconscious due to the depressurization.

This chapter in the *Challenger* investigation reflects the sorrow, mystery, and solemn respect surrounding the loss of the crew. Though the exact moments before their deaths remain unclear, the tragedy of the *Challenger* disaster continues to resonate as one of the most profound moments in space exploration history, highlighting both the inherent dangers of spaceflight and the courage of those who embark on such missions.

In the wake of the tragedy, the legacy of STS-51-L became one of both sorrow and determination. Christa McAuliffe's planned educational lessons were never broadcast, but her mission to inspire a new generation of explorers and dreamers lived on. The Challenger disaster served as a stark reminder of the inherent risks of space exploration, but it also spurred improvements that would make future missions safer.

The crew of STS-51-L is remembered as pioneers who paid the ultimate price in pursuing knowledge and expanding human presence in space. Their names are memorialized on monuments, in scholarships, and in the hearts of all who continue to look to the stars.

STS-26

STS-26, the 26th NASA Space Shuttle mission, marked a pivotal moment in the history of U.S. space exploration as the first mission following the *Challenger* disaster. Known as the "Return to Flight" mission, it was the seventh flight of the *Discovery* orbiter, launching from Kennedy Space Center on September 29, 1988, and landing four days later on October 3, 1988. This mission carried the hopes of NASA, its partners, and the nation, symbolizing the resilience of human spaceflight after the tragic loss of *Challenger* on January 28, 1986.

The mission was designated STS-26R, as the original STS-26 number had previously been assigned to STS-51-F, also known as Spacelab-2. Including the "R" in its designation and other missions up to STS-33 was necessary to avoid confusion with earlier tracking and data records assignments.

The STS-26 crew was notable for being an all-veteran team, the first such U.S. space mission since Apollo 11. Commander Frederick H. Hauck, Pilot Richard O. Covey, and Mission Specialists John M. Lounge, George D. Nelson, and David C. Hilmers had all flown in space previously. This experienced crew successfully executed the mission based on the original crew assignments for the planned *Challenger* mission, STS-61-F, scheduled to launch the Ulysses probe in 1986.

Mission Specialist Mike Lounge, Mission Specialist David C. Hilmers, Mission Specialist George D. Nelson. Front row: Pilot Richard O. Covey and Commander Frederick H. Hauck.

Covey had a particularly significant connection to the *Challenger* tragedy. As CAPCOM for the ill-fated STS-51-L launch, he had uttered the words, "Challenger, go at throttle up," moments before the orbiter disintegrated. His presence on STS-26 added a deeply personal layer to the mission's success.

At 11:37 a.m. EDT on September 29, 1988, *Discovery* lifted off from Launch Complex 39B, 975 days after the *Challenger* accident. The launch faced minor delays due to unseasonable light winds and technical issues with two crew members' flight suit cooling systems. Additionally, a brief concern over cabin air pressure, caused by the activation of oxygen systems in the crew's flight suits, was resolved just in time for a successful launch.

The primary payload of STS-26 was the deployment of the TDRS-C satellite, later known as TDRS-3, which played a crucial role in NASA's Tracking and Data Relay Satellite System (TDRSS). TDRS-C was deployed from *Discovery*'s cargo bay six hours into the flight, where its Inertial Upper Stage (IUS) successfully placed it into a transfer orbit, and a subsequent stage propelled it into a geosynchronous orbit over the Pacific Ocean. TDRS-3 became part of the satellite network, enhancing NASA's ability to track and communicate with orbiting spacecraft.

In addition to deploying TDRS-C, STS-26 conducted 11 scientific and technological experiments on the mid-deck. These included materials science experiments such as the Protein Crystal Growth (PCG) experiment, which sought to advance the understanding of protein structures, including an enzyme critical to AIDS research. However, two 11 proteins did not produce crystals suitable for analysis. The mission also included life sciences experiments like the Aggregation of Red Blood Cells study, which investigated the effects of microgravity on medical research and diagnostic tests.

An innovative aspect of STS-26 was using the Voice Control Unit (VCU), a speech recognition system controlling the cameras monitoring the Canadarm robotic arm. Although this was an experimental technology and initially faced problems due to changes in speech caused by weightlessness, the system was successfully retrained during the mission, achieving a recognition rate of over 96%.

During the mission, the crew donned newly designed partial-pressure flight suits for launch and landing, enhancing safety. These suits and a redesigned crew escape system underscored the significant advancements made to improve astronaut safety following the *Challenger* disaster.

Despite its successes, STS-26 was not without challenges. Shortly after ascent, the Flash Evaporator System, responsible for cooling the orbiter, iced up, raising the cabin temperature to an uncomfortable 31°C (88°F). The issue was resolved by Flight Day 4, and the temperature returned to normal. Additionally, a Ku-band antenna malfunctioned on Flight Day 2 and had to be stowed for the remainder of the mission.

In a poignant moment on October 2, 1988, the crew paid tribute to the *Challenger* astronauts, acknowledging the loss that had profoundly impacted the space program. The following day, *Discovery* touched down at Edwards Air Force Base in California at 12:37 p.m. EDT, concluding the mission with 4 days, 1 hour, and 11 seconds.

STS-26 marked a new chapter in NASA's Shuttle program. It successfully demonstrated the reliability of the redesigned solid rocket boosters, which showed no signs of leakage or overheating, addressing the flaws that had caused the *Challenger* accident. However, post-flight analysis revealed severe damage to *Discovery*'s thermal protection system. A piece of cork insulation from the solid rocket booster had struck the orbiter during ascent, causing erosion to the tiles on the underwing—an issue that would tragically resurface during the *Columbia* disaster in 2003.

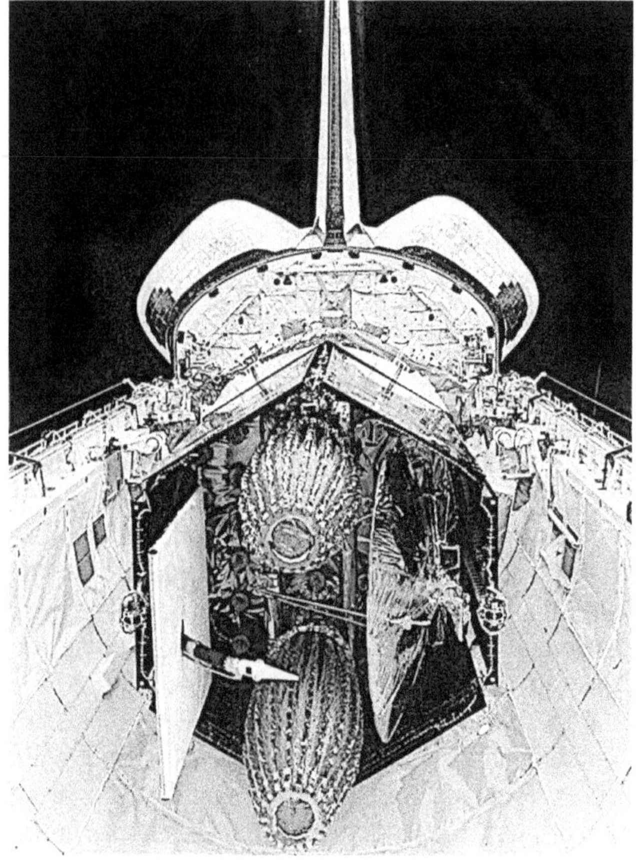

The TDRS-C in Discovery's payload bay

STS-27

STS-27, the 27th NASA Space Shuttle mission, marked the third flight of *Space Shuttle Atlantis* (OV-104) and was launched on December 2, 1988. This mission held great significance; it was only the second shuttle flight following the *Challenger* disaster of January 1986. Atlantis embarked on a classified mission for the U.S. Department of Defense (DoD), carrying a payload later identified as the Lacrosse 1 surveillance satellite. This satellite, equipped with side-looking radar, was deployed for all-weather reconnaissance, serving the National Reconnaissance Office (NRO) and the Central Intelligence Agency (CIA).

Originally scheduled for December 1, 1988, the launch was delayed by one day due to unfavorable weather conditions, including cloud cover and strong winds at Kennedy Space Center. At 09:30 a.m. EST on December 2, 1988, Atlantis lifted off from Launch Complex 39B. The four-day mission culminated in a safe landing at Edwards Air Force Base in California on December 6, 1988, at 6:36 p.m. EST. The total mission duration was four days, 9 hours, 5 minutes, and 37 seconds. Following the mission, Atlantis was transported back to Kennedy Space Center on December 13, 1988, for repairs and processing.

Guy S. Gardner, pilot; Robert L. Gibson, commander and Jerry L. Ross, mission specialist. On the back row, left to right, are mission specialists William M. Shepherd and Richard M. Mullane.

Commander Robert L. Gibson led the crew of STS-27 on his third spaceflight. Pilot Guy S. Gardner and Mission Specialists Richard M. Mullane, Jerry L. Ross, and William M. Shepherd joined him. This was their first spaceflight for Gardner and Shepherd, while Mullane and Ross were on their second.

During ascent, the shuttle sustained significant damage to its Thermal Protection System (TPS) tiles, a crucial layer that shields the vehicle from the extreme heat of reentry. About 85 seconds into the flight, ablative insulating material from the right solid rocket booster struck the orbiter, resulting in what was later revealed to be the most severe tile damage of any mission that safely returned to Earth. Over 700 tiles were damaged, with one completely missing. Remarkably, the missing tile was located over an antenna mounting plate made of aluminum, which offered additional protection to the shuttle's structure. This fortunate circumstance likely prevented a catastrophic outcome, akin to the one that befell *Space Shuttle Columbia* during STS-107 in 2003.

Due to the mission's classified nature, the crew could not send detailed images of the damage to Mission Control. They were forced to transmit low-resolution, encrypted images, which led engineers on the ground to underestimate the severity of the damage. Despite the crew's concerns—Commander Gibson believed they would not survive reentry—they were reassured by ground control that the damage was not critical. This communication disconnect left the crew frustrated and fearing for their lives. Gibson, known for his leadership, advised his crew to stay calm and, in a darkly humorous moment, mentioned that if the shuttle began to disintegrate, he would use his final moments to share his thoughts with Mission Control.

The shuttle's reentry was tense. Mission Specialist Mullane, filming through the upper deck windows, feared molten aluminum would stream across the vehicle's surface, reminiscent of rain on a windshield. When Atlantis touched down, the extent of the damage shocked NASA engineers. The safe return of the heavily damaged shuttle became a testament to the robustness of its design, but also a stark reminder of the inherent risks of space travel.

Speculation persisted about an extravehicular activity (EVA) during this mission. Years later, crew members suggested that repairs to the satellite after its deployment might have required a spacewalk, potentially by Mission Specialists Jerry L. Ross and William M. Shepherd. However, given the mission's classified nature, details about this possible EVA remain undisclosed.

In addition to its technical and operational challenges, the mission coincided with a significant global event. On December 7, 1988, the day after Atlantis' return, a devastating earthquake struck Armenia, killing tens of thousands. In a somber astronaut meeting, Commander Gibson made a darkly humorous reference to the classified nature of the mission, quipping that Armenia had been Atlantis' first target, although only set to "stun."

The incident with Atlantis' tile damage led to an intensive investigation. A review team conducted detailed inspections of the Thermal Protection System, examining the damaged tiles and compiling a fault tree to explore failure scenarios. Their findings led to recommendations to prevent similar occurrences in future missions, highlighting the importance of continuous improvement in shuttle safety protocols.

In retrospect, had STS-27 resulted in disaster, the shuttle program might have faced severe political repercussions. With only one successful mission since the loss of *Challenger*, the destruction of Atlantis could have ended the shuttle program entirely, altering the course of U.S. space exploration.

STS-29

STS-29, the 28th NASA Space Shuttle mission, was a significant step in the post-Challenger era of space exploration, marked by the deployment of a critical satellite and various experiments designed to advance human knowledge of spaceflight. Space Shuttle Discovery, making its eighth flight, launched from Kennedy Space Center, Florida, on March 13, 1989, after multiple delays due to technical issues. The mission played an important role in rebuilding confidence in the shuttle program following the Challenger disaster in 1986.

Standing (left ot right) are James P. Bagian, mission specialist 1; Robert C. Springer, mission specialist 3; and James F. (Jim) Buchli, mission specialist 2. Seated (left to right) are John E. Blaha, pilot, and Michael L. Coats, commander.

Initially planned for February 18, 1989, the launch was delayed due to the replacement of faulty liquid oxygen turbopumps in Discovery's three main engines. Further postponements occurred when a master event controller failed during prelaunch checkout and a fuel preburner oxidizer valve required replacement. Finally, on March 13, 1989, despite ground fog and high upper winds, Discovery lifted off at 9:57 a.m. EST, after a waiver was granted for wing loads.

The primary objective of STS-29 was the deployment of the Tracking and Data Relay Satellite (TDRS-D), the third and final satellite in NASA's Tracking and Data Relay Satellite System (TDRSS) constellation. This system, positioned in geosynchronous orbit 35,900 km (22,300 mi) above Earth, revolutionized communication with spacecraft, significantly reducing the gaps in contact with ground control. After its deployment, TDRS-D became TDRS-4, joining two other satellites to provide near-constant communication coverage for spacecraft.

Commander Michael Coats led the crew of five astronauts on his second spaceflight. Pilot John E. Blaha joined him, and three Mission Specialists: Robert C. Springer, James Buchli,

and James P. Bagian, each making their first spaceflight, except for Buchli, who was on his third mission. Together, they performed a series of critical tasks, including successfully deploying TDRS-D just six hours into the mission. The satellite was released from the shuttle's payload bay at 3:12 a.m. EST, and the Inertial Upper Stage (IUS) booster carried it to its final orbit, 12 hours and 30 minutes into the mission.

Onboard, Discovery encountered a minor technical issue when one of its cryogenic hydrogen tanks, essential for powering the shuttle's fuel cells, showed erratic pressure readings. This tank was deactivated, and the crew was instructed to conserve power. However, it was successfully reactivated on Flight Day 3 and operated without further issue for the remainder of the mission.

STS-29 was not solely about satellite deployment; it carried a suite of secondary payloads, including eight experiments. Among them were two Shuttle Student Involvement Program (SSIP) experiments: one involving four live rats to study the effects of spaceflight on bone healing and another involving 32 chicken eggs to examine the impact of spaceflight on fertilized embryos. Both experiments aimed to explore how living organisms respond to the unique conditions of microgravity, contributing to our understanding of biology beyond Earth.

One of the mission's notable experiments, the Space Station Heat Pipe Advanced Radiator Element (SHARE), was designed to test a potential cooling system for the proposed Space Station Freedom. Unfortunately, the experiment was only partially successful due to a design flaw in the manifold section, limiting its operation to just 30 minutes under powered electrical loads.

Other experiments, however, were more successful. The Protein Crystal Growth (PCG) experiment yielded crystals from all proteins studied, potentially aiding in pharmaceutical research. The Chromosomes and Plant Cell Division in Space (CHROMEX) experiment provided valuable insights into how microgravity affects plant root development, an important consideration for future long-term space missions where growing food in space could become essential.

Additionally, Discovery served as a calibration target for the U.S. Air Force Maui Optical Site (AMOS) in Hawaii, contributing to ground-based optical studies. An IMAX camera was also onboard to capture high-definition footage of Earth, including scenes of natural disasters such as floods, hurricanes, and volcanic eruptions, for the 1990 documentary *Blue Planet*.

After five days in orbit, Discovery's mission concluded on March 18, 1989, one orbit earlier than planned to avoid high winds at the landing site. The shuttle touched down smoothly on Runway 22 at Edwards Air Force Base, California, at 9:35 a.m. EST. STS-29 marked another successful step in NASA's journey to restore the Space Shuttle program, reaffirming the shuttle's reliability and contributing to advancements in both satellite communication and scientific research in space. The total mission duration was 4 days, 23 hours, 38 minutes, and 52 seconds, and Discovery completed 80 orbits around the Earth.

STS-30

Ronald J. Grabe, pilot; David M. Walker, commander; and mission specialists Norman E. Thagard, Mary L. Cleave, and Mark C. Lee.

STS-30 marked the 29th NASA Space Shuttle mission and the fourth for *Atlantis*. It was a significant flight for multiple reasons: it was the fourth shuttle mission since the Challenger disaster and the first since then to carry a female astronaut. On May 4, 1989, *Atlantis* launched from Kennedy Space Center, Florida, on a mission that would culminate in the deployment of the Magellan probe, bound for Venus. After four days in space, the shuttle landed safely at Edwards Air Force Base, California, on May 8, 1989.

Officially designated STS-30R, the mission's numbering reflected NASA's internal tracking system, which had changed after the Challenger accident. The "R" suffix was added to avoid conflicts with previous missions that had been renumbered. This system ensured clarity in mission documentation and tracking for STS-26 through STS-33.

The STS-30 crew consisted of five astronauts: Commander David M. Walker, Pilot Ronald J. Grabe, and Mission Specialists Norman E. Thagard, Mary L. Cleave, and Mark C. Lee. Each crew member brought specialized skills to the mission, ensuring the successful deployment of Magellan and the completion of other mission objectives.

After Atlantis reached orbit, the crew quickly began deploying the Magellan spacecraft. On May 4, just hours after launch, Magellan was successfully released from the Shuttle's payload bay. The spacecraft then fired its solid rocket motor, placing it on a trajectory toward Venus. The successful deployment was a momentous achievement, as it marked the first time since the end of the Apollo program that NASA had launched a planetary mission.

Preparation for STS-30R began soon after *Atlantis* completed STS-27. The orbiter underwent a three-month refurbishment in the Orbiter Processing Facility (OPF-2) at Kennedy Space Center, during which damaged Thermal Protection System (TPS) tiles were replaced. The shuttle was carefully inspected and prepared for its next mission. By March 11, 1989, *Atlantis* was mated with its External Tank (ET-29) and Solid Rocket Boosters (SRBs) in the Vehicle Assembly Building, before rolling out to Launch Pad 39B on March 22.

On May 4, 1989, at 14:46:59 EDT, *Atlantis* lifted off from Launch Complex 39B. The mission's primary objective was to deploy the Magellan spacecraft, which carried the Inertial

Upper Stage (IUS). Magellan was destined for Venus and represented the first American planetary mission in over a decade. Six hours into the flight, the crew successfully released the probe from the shuttle's payload bay, initiating a series of IUS burns that sent Magellan on its trajectory toward Venus. Magellan would reach its target in August 1990 and spend 243 days mapping Venus' surface using radar, revolutionizing our understanding of the planet's topography.

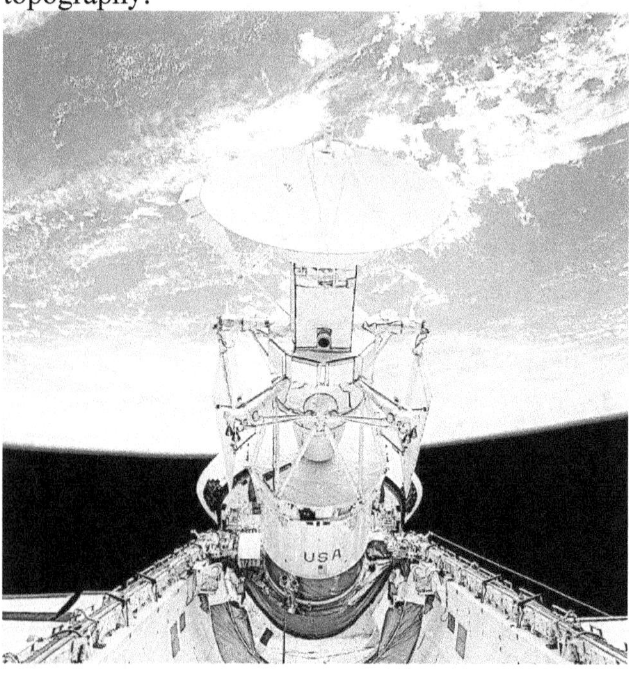

The Magellan spacecraft and its Inertial Upper Stage in the payload bay of Space Shuttle Atlantis during the STS-30 mission.

Although the mission's launch was originally scheduled for April 28, a series of technical issues delayed liftoff. On the first attempt, a problem with the liquid hydrogen recirculation pump in one of the Space Shuttle Main Engines (SSMEs) and a vapor leak in the hydrogen recirculation line forced NASA to scrub the launch just 31 seconds before ignition. When the launch finally occurred on May 4, poor weather at Kennedy Space Center caused another brief delay, as mission controllers waited for cloud cover and crosswinds to improve to ensure safe conditions for a potential Return To Launch Site (RTLS) abort.

The mission was largely successful, though not without a few minor setbacks. On May 7, one of the shuttle's four general-purpose computers failed and had to be replaced—a first for a Space Shuttle mission while in orbit. The redundancy of the onboard computer systems meant that the crew's safety was never in jeopardy, but the malfunction led to the cancellation of some planned experiments.

The crew encountered additional, albeit minor, technical issues. A Hasselblad camera intended for Earth observations malfunctioned, forcing it to be stowed. The Text and Graphics System (TAGS), a device designed to send visual data to the orbiter from Mission Control, jammed on the second day and was shut down. There were also problems with a medical device that measured central venous pressure, part of an experiment to study microgravity's effects on the cardiovascular system. The water dispenser in the shuttle's galley malfunctioned, complicating the preparation of meals for the crew.

On May 8, 1989, *Atlantis* re-entered Earth's atmosphere and landed at Edwards Air Force Base. Strong crosswinds at the initial runway forced mission controllers to switch from Runway 17 to Runway 22 just before landing. The mission had lasted 4 days, 56 minutes, and 28 seconds.

The Magellan spacecraft was the most significant payload on STS-30. Its mission, beginning six hours and 14 minutes after launch, was to map the surface of Venus, providing humanity's first detailed radar images of the planet. By the time Magellan arrived at Venus in August 1990, it had traveled 804 million kilometers (500 million miles), ready to begin its 243-day mapping mission. The data it gathered revolutionized scientific understanding of Venus, revealing a planet dominated by volcanic plains and shield volcanoes.

In addition to Magellan, the crew conducted three mid-deck experiments, all of which had flown on previous missions. One of the key experiments, the Fluids Experiment Apparatus (FEA), was a joint project between NASA and Rockwell International. Operated by Mission Specialist Mary Cleave using a portable laptop, the FEA allowed the crew to investigate the behavior of fluids in microgravity, contributing to research with direct applications for spacecraft design. For the first time on a shuttle mission, the crew used an 8mm video camcorder to record their work, including participating in the Mesoscale Lightning Experiment, which used cameras in the payload bay to capture lightning storms from space.

STS-30 was a pivotal mission for NASA, as it marked both a return to planetary exploration and the continued recovery of the Space Shuttle program after the Challenger disaster. The deployment of Magellan and the experiments conducted aboard *Atlantis* demonstrated NASA's commitment to advancing both human spaceflight and scientific knowledge.

STS-28

STS-28 marked the 30th mission of NASA's Space Shuttle program and was the fourth mission dedicated to the United States Department of Defense (DoD). Launched on August 8, 1989, this mission was the eighth flight of the Space Shuttle *Columbia*, and it traveled 3.4 million kilometers (2.1 million miles) in 81 orbits around the Earth. The mission concluded with a successful landing on Runway 17 at Edwards Air Force Base, California, on August 13, 1989.

Richard N. (Dick) Richards, pilot; Brewster H. Shaw, commander; and David C. Leestma, mission specialist 2. Standing, from left to right , are Mark N. Brown, mission specialist 3; and James C. (Jim) Adamson, mission specialist 1.

This flight was *Columbia*'s first since January 1986, when it flew on the STS-61-C mission, immediately preceding the Challenger disaster of STS-51-L. Given the classified nature of STS-28, much of its details remain undisclosed, but it is widely believed that the mission deployed the first Satellite Data System-2 (SDS-2) communications satellite. The spacecraft's altitude varied between 295 km (183 mi) and 307 km (191 mi).

The crew of STS-28 consisted of Commander Brewster H. Shaw, in his third and final spaceflight; Pilot Richard N. Richards, on his first mission; and three Mission Specialists: James C. Adamson, David Leestma, and Mark N. Brown. Adamson and Brown were both embarking on their first flights, while Leestma was making his second.

Columbia launched two satellites, designated USA-40 and USA-41, from Kennedy Space Center's Pad 39B at 8:37 a.m. EDT. Initial speculation suggested that one of these was an Advanced KH-11 photo-reconnaissance satellite. However, subsequent reports and amateur satellite observations indicate that USA-40 was likely a second-generation SDS relay satellite. These satellites shared the same bus design as the LEASAT series, which had been deployed on other Shuttle missions, and it is presumed that they were released similarly.

In addition to its primary objectives, STS-28 also carried a unique scientific payload. A human skull, weighing 5 kg (11 lbs), was a crucial part of the In-Flight Radiation Dose Distribution (IDRD) experiment, a joint NASA/DoD project. The experiment aimed to assess radiation penetration in the human cranium during spaceflight. The skull, encased in a plastic matrix representing human tissue, was sliced into ten layers, with thermoluminescent dosimeters embedded to measure radiation levels at various depths. This experiment would also fly on future missions STS-36 and STS-31, providing valuable data on space radiation exposure.

During the mission, the crew encountered technical challenges, including shutting down a thruster in the Reaction Control System (RCS) due to indications of a leak. Additionally, an RCS heater malfunctioned. Upon re-entry, the

Shuttle's thermal protection system (TPS) exhibited unusual heating, attributed to early turbulence in the plasma flow around the vehicle. A post-flight investigation identified protruding gap filler material as the likely cause, the same material that would later be removed during STS-114 in 2005, after the Columbia disaster.

Another important component of STS-28 was the Shuttle Infrared Lee-side Temperature Sensing (SILTS) camera, which made its second flight aboard *Columbia*. Mounted on the orbiter's vertical stabilizer, the SILTS system captured thermodynamic data during re-entry, mapping heat distribution across surfaces visible from the top of the Shuttle's tail. Tragically, the port wing, which the camera monitored, would be breached by superheated plasma during *Columbia*'s final flight in 2003, leading to the orbiter's destruction.

Before STS-28, *Columbia* modified its thermal protection system, aligning it with upgrades made to *Discovery* and *Atlantis*. These upgrades included replacing many of the white Low-temperature Reusable Surface Insulation (LRSI) tiles with felt insulation blankets, a change aimed at reducing weight and improving turnaround time between missions. One minor but notable change on STS-28 was the relocation of *Columbia*'s name from the payload bay doors to the fuselage, allowing for easier identification while in orbit.

STS-28's landing at Edwards Air Force Base on August 13, 1989, was notable for its exceptionally low touchdown speed. Due to a software glitch with the Shuttle's weight-on-wheels sensors, the crew was instructed to land as softly as possible. As a result, *Columbia* touched down at 154 knots (177 mph), the slowest landing of the entire Shuttle program, and just above the orbiter's stall speed. The mission lasted five days, 1 hour, and 8 seconds, contributing critical data to both defense and space science initiatives.

STS-34: The Journey of Galileo

On October 18, 1989, the Space Shuttle Atlantis roared to life at Kennedy Space Center's Launch Complex 39B, embarking on a mission that would take humanity one step closer to understanding the mysteries of our solar system. Mission STS-34, commanded by Donald E. Williams, was a pivotal moment in NASA's exploration efforts, marking the launch of the Galileo spacecraft, a sophisticated probe designed to study the planet Jupiter and its moons.

Shannon W. Lucid, mission specialist; Donald E. Williams, commander; Franklin R. Chang-Diaz, mission specialist; Michael J. McCulley, pilot; and Ellen S. Baker, mission specialist.

STS-34's primary objective was the successful deployment of Galileo, an ambitious project to deliver unprecedented data about the largest planet in our solar system. Named after the famed astronomer Galileo Galilei, the spacecraft was the first to orbit Jupiter and extensively survey its moons, atmospheric conditions, and magnetic environment. The mission was a testament to NASA's commitment to deep space exploration, building on the success of previous interplanetary missions like Voyager.

The STS-34 crew comprised five astronauts, each bringing specialized skills to ensure the mission's success. Alongside Commander Donald E. Williams was pilot Michael J. McCulley, and mission specialists Shannon W. Lucid, Franklin R. Chang-Díaz, and Ellen S. Baker. This diverse team was responsible for the safe deployment of Galileo and the operation of multiple experiments aboard Atlantis.

Ellen S. Baker, the crew's medical doctor,

was crucial in monitoring the team's health and well-being during the mission. Shannon W. Lucid, a biochemist, contributed her expertise to the scientific aspects of the flight. Franklin R. Chang-Díaz, a physicist and veteran astronaut, was instrumental in operating the onboard systems and contributed to the mission's overall success.

On October 18, 1989, Space Shuttle *Atlantis* lifted off from Launch Complex 39B at the Kennedy Space Center (KSC) at precisely 12:53:40 EDT. Its mission, STS-34, carried the Galileo spacecraft, bound for Jupiter, nestled securely in the orbiter's payload bay. The launch countdown experienced a brief delay of 3 minutes and 40 seconds at T-minus 5 minutes to allow for an update to the onboard computer regarding the Transoceanic Abort Landing (TAL) site. Initially planned for Ben Guerir Air Base in Morocco, the TAL site was changed to Zaragoza Air Base in Spain due to inclement weather conditions at the original location.

The launch of *Atlantis* was initially scheduled for October 12, 1989, during a 41-day launch window when the planets were aligned favorably for a direct flight to Jupiter. However, due to technical issues, including the need to replace a faulty controller on Space Shuttle Main Engine No. 2, the launch was rescheduled for October 17. Further delays on that day resulted from rain showers within 32 kilometers (20 miles) of the Shuttle Landing Facility at KSC, violating the launch commit criteria in the event of a Return to Launch Site (RTLS) abort. Ultimately, *Atlantis* lifted off on October 18, embarking on a journey to deliver Galileo, one of NASA's most ambitious planetary exploration missions.

Galileo's deployment occurred without issue approximately six hours after liftoff, at 19:15 EDT. Powered by an attached Inertial Upper Stage (IUS) booster, the spacecraft was boosted toward Venus on the first leg of its complex, six-year journey to Jupiter. At 20:20 EDT, the IUS successfully injected Galileo onto a Venus transfer orbit, and the spacecraft separated from the booster 47 minutes later.

Galileo's mission required a series of gravity assists—slingshots around Venus and Earth twice—to gain enough velocity to reach Jupiter,

located in the outer Solar System. The spacecraft's journey provided additional opportunities to study two asteroids, 951 Gaspra and 243 Ida, marking the first close observations of these celestial bodies. Galileo was equipped with two primary components: an orbiter designed to study Jupiter and its largest moons over an eight-year period and a probe intended to descend into the planet's atmosphere, capturing groundbreaking data before being destroyed by the immense heat and pressure of the gas giant.

Galileo being prepared prior to launch.

STS-34 was only the second shuttle flight to deploy a planetary spacecraft, following the deployment of the Magellan probe to Venus on STS-30. Galileo would go on to make history as the first spacecraft to orbit an outer Solar System planet and perform direct atmospheric sampling.

Aside from deploying Galileo, *Atlantis* carried additional experiments. The Shuttle Solar Backscatter Ultraviolet (SSBUV) experiment, making its first flight, was designed to calibrate ozone sounders on satellites and verify data on atmospheric ozone levels and solar irradiance. This experiment operated successfully throughout the mission. In the payload bay, other scientific endeavors included the Polymer Morphology (PM) experiment, sponsored by 3M, which studied the melting and resolidification of polymers in microgravity. The Mesoscale Lightning Experiment (MLE) continued its observations of lightning in the upper atmosphere, an effort begun on previous shuttle missions.

The crew also worked on a student experiment focused on ice crystal growth, overcoming an initial failure caused by ice formation on the cooling plate. By adjusting the experiment, the crew successfully grew several crystals. In addition, Lucid and Baker completed the Growth Hormone Concentration and Distribution in Plants experiment, which involved freezing samples of corn seedlings grown aboard the shuttle.

In the cabin, the crew operated an IMAX camera to capture footage for educational and documentary purposes. This footage later featured in Werner Herzog's 2005 film *The Wild Blue Yonder*. Another experiment focused on human health in space, as astronaut Franklin Chang-Díaz and Dr. Shannon Lucid used a high-resolution camera to photograph and videotape the retinal veins and arteries in Lucid's eyes. This data aimed to explore potential links between cranial pressure and motion sickness, a condition often experienced by astronauts. Dr. Baker also conducted tests of anti-motion sickness medication in microgravity conditions.

Throughout the mission, several minor anomalies occurred. An issue with the Auxiliary Power Unit (APU) 2's gas generator fuel pump system A led to an alarm on October 22. Problems with the Flash Evaporator System, which cooled the orbiter, and a cryogenic oxygen manifold valve left closed for the remainder of the mission also surfaced. One of the Hasselblad cameras jammed twice, prompting the crew to switch to a backup camera.

The mission also encountered external challenges. Demonstrators protested the launch due to the presence of a nuclear-powered device onboard Galileo. Chang-Díaz later remarked that they came close to aborting the mission three times because of technical malfunctions. However, he noted that the decision to continue was made in part to avoid the risk of an international incident that could have arisen from landing with the nuclear device at an airport in Senegal.

On October 21, 1989, Costa Rican President Dr. Óscar Arias spoke with Chang-Díaz, a native of Costa Rica, via a special telephone linkup, congratulating him on his achievements and praising his representation of Costa Rica and Latin America.

As the mission drew to a close, high winds predicted at the planned landing site led to an early reentry, two orbits ahead of schedule. On October 23, 1989, *Atlantis* touched down smoothly on Runway 23 at Edwards Air Force Base in California at 16:33:00 UTC, marking the successful completion of a mission that lasted 4 days, 23 hours, 39 minutes, and 20 seconds.

STS-34 was a milestone for planetary exploration and a demonstration of the shuttle program's versatility in advancing scientific knowledge across multiple fields, from space-based polymer studies to planetary exploration.

The Space Shuttle Columbia, returning to KSC after the successful STS-32 mission, is poised atop the Shuttle Carrier Aircraft (SCA) as the duo fly by the Vehicle Assembly Building (VAB) at KSC January 26. Columbia, carrying the Long Duration Exposure Facility (LDEF) in its payload bay, was compleitng a two-day ferry flight from Edwards Air Force Base, California. Landing at the Shuttle Landing Facility occurred a few moments later at 3:30 p.m.

After a successful mission, *Columbia* executed the third night landing in shuttle history, touching down at Edwards Air Force Base on January 20, 1990, at 1:35 a.m. PST. The orbiter's landing weight was 228,335 pounds, and it rolled out for 62 seconds over 10,732 feet on Runway 22. *Columbia* was then ferried back to Kennedy Space Center atop the Shuttle Carrier Aircraft, concluding a mission that underscored the versatility and endurance of the Space Shuttle as NASA entered the 1990s.

STS-33: A Critical National Security Endeavor

STS-33, flown by Space Shuttle *Discovery*, marked a significant milestone in NASA's Space Shuttle program as the final mission of the 1980s. Launching from Kennedy Space Center's Launch Complex 39B on November 22, 1989, at 7:23 p.m. EST, *Discovery* carried a highly classified payload for the United States Department of Defense (DoD). As the 32nd shuttle mission overall and the ninth flight of *Discovery*, this mission was the fifth shuttle mission directly supporting the DoD. Given its secretive nature, much of the mission's specifics remain classified.

The mission, originally designated STS-33R, followed a renaming protocol after the ill-fated Challenger STS-51-L mission. NASA adopted this designation system to avoid confusion using the same number with previous missions. Notably, the "R" suffix was used to distinguish STS-33 from the ill-fated Challenger mission in official documentation.

The crew of STS-33 included Commander Frederick D. Gregory, Pilot John E. Blaha, Mission Specialists Sonny Carter, Story Musgrave, and Kathryn C. Thornton. This mission was a mix of veteran and first-time flyers, with Gregory and Blaha both on their second spaceflights, Musgrave on his third, while Thornton was making her space debut. Tragically, Sonny Carter would later perish in a commercial plane crash in 1991, while S. David Griggs, originally slated to pilot the mission, died in a World War II aircraft accident in June 1989 during training for this mission. His legacy is honored with a single gold star on the mission insignia.

Kathryn C. Thornton, mission specialist 3; Manley L. (Sonny) Carter, mission specialist 2; Frederick D. Gregory, commander; John E. Blaha, pilot; and F. Story Musgrave, mission specialist 1

Originally set for launch on November 20, 1989, the mission faced delays due to issues with the shuttle's solid rocket booster electronics. Once launched, STS-33 became the third night launch in shuttle program history and the first since flights resumed after the 1986 Challenger disaster.

Discovery carried and deployed a single satellite, USA-48, believed to be a classified Magnum electronic intelligence (ELINT) satellite, similar to those deployed in previous shuttle missions, such as STS-51-C in 1985. The satellite was suspected to be tasked with intercepting military and diplomatic communications from adversaries like the Soviet Union, China, and other communist states. The satellite deployed by STS-33 was thought to replace an earlier one running low on maneuvering fuel. However, reports later indicated that the previous satellite was still operational.

During the mission, the shuttle orbited at an inclination of 28.45 degrees, with three Orbital Maneuvering System (OMS) burns that positioned the shuttle into a circular orbit of 519 km (322 miles). The satellite was successfully deployed on the seventh orbit, and its Inertial Upper Stage (IUS) booster was activated, which placed it into a geostationary transfer orbit (GTO). This was the eighth IUS launched aboard the shuttle and the seventh successfully deployed.

STS-33 also contributed to atmospheric studies. The U.S. Air Force's Maui Optical and Supercomputing Observatory (AMOS) used advanced spectrographic and infrared imaging to observe the shuttle's interactions with residual gases in orbit.

Discovery's return was initially scheduled for November 26, 1989, but strong winds at the landing site delayed it by a day. The shuttle landed safely at Edwards Air Force Base in California on November 27, 1989, at 7:30 p.m. EST. The mission lasted 5 days, 0 hours, 6 minutes, and 46 seconds, contributing to the growing legacy of the Space Shuttle program and the exploration of space for both civilian and defense purposes.

STS-32

Daniel C. Brandenstein, commander; and James D. Weatherbee, pilot. Pictured rear left to right are mission specialists Marsha S. Ivins, G. David Low, and Bonnie J. Dunbar.

STS-32, the 33rd mission of NASA's Space Shuttle program, marked several significant milestones in space exploration history. Launched on January 9, 1990, aboard Space Shuttle *Columbia*, this mission was notable as the first use of Launch Complex 39A since 1986 and the debut of Mobile Launcher Platform-3 (MLP-3) in the Shuttle program. These launch facilities, integral to the program's operations, had

undergone extensive modifications after the hiatus. The refurbishment of Launch Complex 39A included enhanced safety features, such as upgrades to the emergency egress system, improved payload bay capabilities, and additional protection for the platform's water and fuel systems. Integrating debris traps and weather protection measures further bolstered launch reliability, while a new umbilical system provided power and instrumentation for the solid rocket boosters.

The STS-32 crew consisted of Commander Daniel Brandenstein, Pilot Jim Wetherbee, and Mission Specialists Bonnie J. Dunbar, Marsha Ivins, and G. David Low.

Columbia lifted off from Kennedy Space Center at 7:35 a.m. EST, after two prior postponements—first from December 18, 1989, and again from January 8, 1990, due to inclement weather. Weighing 255,994 pounds at launch, *Columbia* embarked on a mission that would become the longest shuttle flight to date, lasting nearly 11 days and surpassing the previous record set by STS-9 in 1983. The mission's primary objectives were to deploy the Syncom IV-F5 military communications satellite, known as Leasat-5, and to retrieve NASA's Long Duration Exposure Facility (LDEF). This satellite had remained in orbit for over four years.

On the mission's second day, the crew successfully deployed Syncom IV-F5. The satellite was propelled into geosynchronous orbit by a Minuteman solid apogee kick motor, fulfilling its role in providing communications for the U.S. Department of Defense. However, the retrieval of LDEF, scheduled for the fourth day, was the more critical objective. LDEF had been launched in 1984 to collect data on the long-term effects of space exposure on various materials. Its retrieval had been delayed due to the Challenger disaster, and by 1990, the satellite's orbit had decayed to a precarious position, requiring prompt recovery to prevent atmospheric re-entry.

Using the shuttle's Remote Manipulator System (Canadarm), Mission Specialist Bonnie J. Dunbar carefully retrieved the 12-sided, 12,000-pound structure. The retrieval was meticulously planned, as heightened solar activity had increased drag on the satellite, accelerating its orbital decay. Had LDEF not been recovered in time, it would have descended beyond the shuttle's reach and been destroyed during re-entry. The crew conducted a 4.5-hour photographic survey of the satellite before securing it in the payload bay for its return to Earth. LDEF's recovery provided NASA with invaluable data from its 57 onboard experiments, which studied the effects of space on materials, biological samples, and electronic systems.

Aside from the satellite operations, the crew of STS-32 conducted various scientific experiments, many of which had flown on previous shuttle missions. These included studies on protein crystal growth, fluid dynamics, and neurospora circadian rhythms. The American Flight Echocardiograph experiment monitored the cardiovascular effects of spaceflight, while Earth observation experiments, such as the Mesoscale Lightning Experiment, sought to capture natural phenomena from orbit. Notably, footage of the Earth and the LDEF retrieval was filmed with an IMAX camera and later featured in the documentary *Destiny in Space.*

STS-32 also set a new endurance record for the shuttle program, as NASA tested the orbiter's capacity for extended missions. This mission paved the way for future long-duration flights, as new systems allowed the shuttle to operate for up to 16 days in orbit. These advancements would later be implemented during Columbia's STS-50 mission in 1992.

STS-36

On February 28, 1990, Space Shuttle Atlantis launched on STS-36, marking the 34th mission of the shuttle program and the sixth flight for Atlantis. The mission, which carried a classified payload for the U.S. Department of Defense (DoD), is believed to have deployed the Misty reconnaissance satellite into orbit. As one of the few national security missions, much of the mission's details remain officially classified, with the satellite deployment and its objectives shrouded in secrecy.

STS-36 was unique in many respects,

particularly in its launch trajectory. Originally scheduled for February 22, 1990, the launch faced multiple delays due to poor weather conditions and a minor respiratory illness affecting the mission's commander, John Oliver Creighton.

After several postponements, the launch occurred on February 28 at 07:50:22 UTC (2:50:22 am EST), from Kennedy Space Center in Florida. The payload's classified nature necessitated a unique "dog-leg" trajectory to achieve an orbital inclination of 62 degrees— higher than the typical shuttle mission's maximum of 57 degrees.

This trajectory allowed Atlantis to reach a required orbit while minimizing performance reductions. The mission originally intended for Vandenberg Air Force Base was instead launched from Florida after shuttle operations at Vandenberg were canceled. This unusual trajectory caused the shuttle to pass over regions typically avoided, including Cape Hatteras, Cape Cod, and parts of Canada.

The STS-36 crew portrait features 5 astronauts who served in the 6th Department of Defense (DOD) mission. Posed near the Space Shuttle Orbiter Discovery are (left to right) Pierre J. Thuot, mission specialist 3; John H. Caster, pilot; John H. Creighton, commander; Richard M. (Mike) Mullane, mission specialist 1; and David. C. Hilmers, mission specialist 2.

The STS-36 crew consisted of five astronauts. Commander John Oliver Creighton, on his second spaceflight, led the mission. He was joined by Pilot John Casper, who was flying for the first time, along with Mission Specialists Pierre J. Thuot, David C. Hilmers, and Mike Mullane. For Mullane, it marked his third and final spaceflight. Hilmers, the flight engineer, was embarking on his third mission while Thuot was on his first.

Due to the sensitive nature of the mission, the payload, designated AFP-731, remains largely unknown. However, it is widely speculated that Atlantis deployed a KH-11 Advanced photo-reconnaissance satellite, USA-53 or "Misty." KH-11 satellites, known for resembling the Hubble Space Telescope, are equipped with advanced digital imaging systems. Amateur satellite observers briefly tracked Misty in late 1990, suggesting its capabilities as a critical asset for U.S. national security.

In addition it carried a secondary scientific experiment involving a human skull. As part of the Detailed Secondary Objective 469, or the In-flight Radiation Dose Distribution Experiment (IDRD), the skull, seated in a plastic matrix, was equipped with thermo-luminescent dosimeters to measure radiation levels penetrating the human cranium during spaceflight. This experiment had flown previously on STS-28 and would later be conducted again on STS-31.

Atlantis returned to Earth on March 4, 1990, landing at Edwards Air Force Base, California, at 18:08:44 UTC (10:08:44 am PST). The landing marked the completion of the classified mission after a 2.41-kilometer rollout on runway 23. Post-mission inspections revealed 62 impacts to the orbiter's thermal protection system tiles, though only one tile required replacement. While minor issues were observed with hydraulic fluid leaks, the orbiter and its systems performed nominally during the mission's reentry and landing.

Despite the classified nature of its objectives, STS-36 stands as a significant mission within the shuttle program, highlighting the versatility of the shuttle in both civilian and defense-related operations.

STS-31

STS-31 marked the 35th mission of NASA's Space Shuttle program and played a pivotal role in space exploration with the Hubble Space Telescope (HST) deployment. The mission utilized Space Shuttle *Discovery*, making its tenth flight, and was launched from Launch Complex 39B at Kennedy Space Center, Florida, on April 24, 1990. This mission represented a significant milestone for NASA, as it was designed to place the Hubble Space Telescope into low Earth orbit, enabling groundbreaking astronomical research that continues to this day.

Charles F. Bolden, pilot; Steven A. Hawley, mission specialist; Loren J. Shriver, commander; Bruce McCandless, mission specialist; and Kathryn D. Sullivan, mission specialist.Charles F. Bolden Jr., pilot; Steven A. Hawley, Bruce McCandless II, and Kathryn D. Sullivan, all mission specialists.

Hubble Space Telescope in the cargo bay of Discovery

The mission was originally slated to launch in August 1986 as STS-61-J, using *Atlantis*. However, the tragic *Challenger* disaster postponed the mission, and by 1988, *Discovery* was designated to carry out this important task, with Loren Shriver as the commander replacing the originally assigned John W. Young. The significance of this launch was amplified by the fact that it marked the first time since January 1986 that two Space Shuttles were on launch pads simultaneously, with *Discovery* on LC-39B and *Columbia* on LC-39A preparing for STS-35.

The launch on April 24 was not without complications. A previous launch attempt on April 10 was scrubbed at T-minus 4 minutes due to a faulty valve in Auxiliary Power Unit (APU) number one. The issue was resolved, and the Hubble Space Telescope's batteries were recharged, ready for the next attempt. On launch day, the countdown was halted at T-minus 31 seconds due to a computer issue with a fuel valve line. Engineers quickly resolved the issue, allowing the launch to proceed successfully at 12:33:51 UTC (8:33:51 AM EDT).

The Hubble Space Telescope, designed to operate above Earth's atmosphere, was intended to observe the universe in ultraviolet, visible, and near-infrared wavelengths. This joint effort between NASA and the European Space Agency (ESA) began in the late 1970s, to revolutionize our understanding of the cosmos. On April 25, 1990, the *Discovery* crew successfully deployed Hubble into orbit. Positioned at approximately 613 × 615 kilometers (381 × 382 miles), *Discovery* briefly reached an apogee of 621 kilometers (386 miles), setting the record for the highest altitude ever attained by a Shuttle orbiter.

Aside from the Hubble deployment, the mission also focused on scientific experiments and documentation of the mission. Motion pictures were recorded using two IMAX cameras, later showcased in the 1994 film *Destiny in Space*. Among the scientific payloads were the Protein Crystal Growth (PCG) experiment, designed to explore protein crystal formation in microgravity, and the Radiation Monitoring Equipment III (RME III), which measured gamma radiation levels in the crew cabin. The Shuttle also carried a unique payload—an experiment involving a human skull to examine radiation penetration during spaceflight. Known as the In-flight Radiation Dose Distribution (IDRD) experiment, this joint NASA/DoD study had been previously flown on STS-28 and STS-36.

Most of the giant Hubble Space Telescope (HST) can be seen as it is suspended in space by Discovery's Remote Manipulator System (RMS) following the deployment of part of its solar panels and antennae

While deploying the Hubble Space Telescope, the crew encountered a potential issue when one of the observatory's solar arrays failed to unfurl properly. Mission Specialists Bruce McCandless and Kathryn Sullivan prepared for a contingency spacewalk to fix the array manually. However, before they could begin, ground controllers were able to resolve the problem remotely, successfully deploying the array without the need for extravehicular activity (EVA).

The mission concluded on April 29, 1990, when *Discovery* landed on Runway 22 at Edwards Air Force Base in California at 13:49:57 UTC (6:49:57 AM PDT). The landing rollout took 61 seconds, covering 2,705 meters (8,875 feet), marking the first use of carbon brakes on a Shuttle. *Discovery* orbited Earth 80 times during the mission, and its re-entry required the longest deorbit burn in Shuttle history up to that time, lasting 4 minutes and 58 seconds.

Following the mission, *Discovery* was ferried back to Kennedy Space Center on May 7, 1990, where it was prepared for future flights. The success of STS-31 and the deployment of the Hubble Space Telescope marked a new chapter in NASA's space exploration efforts, ushering in a golden era of space-based astronomy that continues to shape our understanding of the universe.

STS-41: A Mission of Scientific and Strategic Importance

Bruce E. Melnick, mission specialist 2; Robert D. Cabana, pilot; Thomas D. Akers, mission specialist 3; Richard N. Richards, commander; and William M. Shepherd, mission specialist 1.

The STS-41, launched on October 6, 1990, from Kennedy Space Center, marked another significant step in NASA's ambitious space exploration program. Aboard the Space Shuttle Discovery was a crew of five astronauts led by Commander Richard N. Richards, a seasoned pilot with extensive experience in naval aviation and test piloting. Alongside him were Pilot Robert D. Cabana and mission Specialists William M. Shepherd, Bruce E. Melnick, and Thomas D. Akers, each bringing their unique expertise to the mission.

The primary objective of STS-41 was deploying the Ulysses spacecraft, a joint venture between NASA and the European Space Agency (ESA). Ulysses was designed to study the Sun's polar regions, an area previously unexplored by spacecraft. This ambitious mission required Ulysses to be placed into a trajectory that would take it out of the solar system's ecliptic plane, utilizing a gravity assist from Jupiter to achieve this unusual orbit. The data gathered by Ulysses would provide unprecedented insights into the Sun's atmosphere and the solar wind, contributing significantly to our understanding of the heliosphere and its impact on the solar system.

Ulysses spacecraft and its upper stage system are deployed during STS-41.

On October 6, 1990, at precisely 7:47:16 a.m. EDT, Space Shuttle *Discovery* lifted off from Kennedy Space Center for the STS-41 mission. The launch occurred 12 minutes after the opening of a two-and-a-half-hour launch window that had begun at 7:35 a.m. EDT. *Discovery* weighed a remarkable 117,749 kilograms (259,592 pounds) at liftoff, carrying the heaviest payload in Shuttle history at the time.

The mission's primary objective was deploying the European Space Agency's (ESA) *Ulysses* spacecraft, designed to explore the Sun's polar regions. Ulysses was equipped with two upper stages: the Inertial Upper Stage (IUS) and the Payload Assist Module-S (PAM-S) to achieve its unique out-of-ecliptic trajectory. This was the first time these two stages were combined to send a payload on a path beyond the plane of the planets. The payload bay also carried various other scientific experiments, including the Shuttle Solar Backscatter Ultraviolet (SSBUV) experiment, Intelsat Solar Array

Coupon (ISAC), and several biological and combustion studies.

Approximately six hours after liftoff, *Ulysses* was deployed from the Shuttle's payload bay, marking the start of its mission to study the Sun's poles. A joint project between NASA and ESA, *Ulysses* first embarked on a 16-month journey to Jupiter. Upon reaching the gas giant, the spacecraft utilized Jupiter's gravity to slingshot itself out of the orbital plane and toward the Sun's southern polar region, which it reached in 1994. After crossing the Sun's southern pole, *Ulysses* passed over the solar north pole in 1995, fulfilling its unprecedented mission to study the Sun's polar environment.

Once *Ulysses* was safely on its way, the STS-41 crew focused on a series of scientific investigations. One of the key experiments was CHROMEX-2, which examined plant growth in space. Flowering plant samples grown in the Discovery's microgravity environment were used to investigate the effects of space on chromosome behavior. This research aimed to expand understanding of genetic responses to space travel, particularly for long-duration missions aboard space stations such as the then-planned Space Station Freedom.

Another critical study on board was the Solid Surface Combustion Experiment (SSCE), designed to analyze how fire behaves in a microgravity environment. Sponsored by the Lewis Research Center and Mississippi State University, the experiment involved burning a strip of paper in a sealed chamber to observe the development and movement of flames without the influence of convection currents. This research was essential for improving fire safety aboard the Shuttle and future space habitats.

The Shuttle Solar Backscatter Ultraviolet (SSBUV) instrument also played a pivotal role during STS-41, helping to calibrate ozone detection satellites in orbit. Atmospheric ozone depletion was a growing environmental concern at the time, and the data from *Discovery*'s instruments, combined with NASA's Nimbus 7 and NOAA's TIROS satellites, provided critical information to ensure accurate satellite ozone readings.

One of the mission's more commercially focused investigations involved solar arrays, tested in space to determine their durability in low Earth orbit. This experiment was a preparatory measure for the eventual repair of an Intelsat VI communications satellite that had been stranded in orbit. The findings from this test indicated that the solar arrays showed no significant damage from atomic oxygen exposure, which helped pave the way for NASA's successful repair mission, STS-49, in 1992.

The STS-41 mission also contributed to biomedical research with the Physiological Systems Experiment (PSE), which studied the effects of microgravity on bone and muscle loss. Pharmacological treatments, including proteins developed by Genentech, were administered to rats aboard *Discovery* to assess whether these treatments could counteract the negative effects of space travel on the musculoskeletal system. This research was vital for developing medical countermeasures for future long-duration space missions.

The Investigations into Polymer Membrane Processing (IPMP) experiment, sponsored by the Battelle Advanced Materials Center, sought to understand how microgravity affects the formation of membranes used in medical and industrial processes such as water desalination and kidney dialysis. Conducted during the mission, this research contributed to the broader understanding of material science in space.

In addition to the scientific experiments, the astronauts onboard STS-41 engaged in outreach efforts. They filmed educational demonstrations, which were later compiled into a video distributed to middle school students through NASA's Teacher Resource Center network. The crew also conducted a detailed evaluation of graphical user interfaces, using a Macintosh Portable laptop to test the suitability of trackball devices in a zero-gravity environment.

After a highly successful mission, *Discovery* returned to Earth, landing at Edwards Air Force Base in California on October 10, 1990, at 6:57:19 a.m. PDT. The Shuttle touched down on runway 22, rolling out over 2,523 meters (8,278 feet) in 49 seconds. Following its landing, *Discovery* was ferried back to Kennedy Space

Center on October 16, 1990, where it underwent preparations for its next mission.

The STS-41 mission not only accomplished its primary objective with the successful deployment of *Ulysses*, but also significantly advanced scientific research across a range of disciplines, from biology and material science to combustion and atmospheric studies. These experiments and the continued evaluation of spaceflight technologies contributed to the broader knowledge required for long-duration space exploration and the eventual construction and operation of the International Space Station.

STS-38: A Classified Mission with Strategic Importance

Frank L. Culbertson, pilot; and Richard O. Covey, commander. Standing (left to right) are mission specialists (MS) Charles D. (Sam) Gemar, (MS-3), Robert C. Springer, (MS-1), and Carl J. Meade, (MS-2).

STS-38, launched from Kennedy Space Center on November 15, 1990, was a significant yet shrouded chapter in the Space Shuttle program. This mission, flown by the Space Shuttle Atlantis, was the seventh flight of Atlantis and the 37th mission overall in the Shuttle program. Unlike most other missions, STS-38 was dedicated to the United States Department of Defense, and much of its details remain classified even today.

The crew of STS-38 comprised five astronauts: Commander Richard O. Covey, Pilot Frank L. Culbertson Jr., Mission Specialist 1 Charles D. Gemar, Mission Specialist 2 Robert C. Springer, and Mission Specialist 3 Carl J. Meade. Each astronaut was highly experienced and selected for their expertise and ability to manage the mission's classified objectives.

As a DoD mission, much of STS-38's payload was classified. Reports suggest that the Shuttle initially entered a 204 km x 519 km orbit with an inclination of 28.45° to the equator. Following three orbital maneuvers, it deployed its primary payload, believed to be a secret satellite codenamed *USA-67*. This satellite, thought to be an ELINT (Electronic Intelligence) gathering asset, likely headed for geosynchronous orbit. The satellite's deployment coincided with the ongoing Gulf War, and it was reportedly tasked with monitoring events during this critical period.

Further analysis indicates that *USA-67* might have been a Satellite Data System (SDS-2) military communications satellite, similar to others deployed in previous Shuttle missions. Additionally, amateur satellite observers speculated that a second, stealth payload—rumored to be a satellite known as *Prowler*—was also deployed during the mission. This satellite, allegedly designed to inspect other nations' geostationary satellites covertly, added another layer of intrigue to STS-38's secretive objectives.

As the mission neared its conclusion, adverse weather at Edwards Air Force Base, the intended landing site, forced NASA to extend the mission by one day. With continued unfavorable conditions, the landing was ultimately shifted to Kennedy Space Center's Shuttle Landing Facility. This marked a significant event, as it was the first landing at KSC for *Atlantis* and the first Shuttle return to the facility since *Discovery's* problematic touchdown five years earlier.

On November 20, 1990, at 21:42:46 UTC (4:42:42 p.m. EST), *Atlantis* touched down on KSC's Runway 33, rolling for 2,753 meters (9,032 feet) over 57 seconds. The orbiter weighed 86,677 kg (191,090 lbs) at landing, concluding its classified mission. Despite the secrecy surrounding the payload, STS-38 represented another successful demonstration of the Shuttle's ability to support national defense objectives, while also making history by returning to the Kennedy Space Center for the first time in years.

STS-35: A Voyage to the Stars

STS-35, the tenth flight of the Space Shuttle *Columbia*, marked the 38th mission of the Space Shuttle program and was dedicated to astronomical observations using the ASTRO-1 observatory, a Spacelab configuration with four telescopes. Launched on December 2, 1990, from Kennedy Space Center in Florida, this mission was originally slated for March 1986 as STS-61-E but was delayed due to the *Challenger* disaster. The mission's primary objective was to conduct round-the-clock ultraviolet and X-ray observations of celestial objects.

The crew of STS-35 included seven astronauts: Commander Vance D. Brand, Pilot Guy S. Gardner, Mission Specialists Jeffrey A. Hoffman, John M. Lounge, and Robert A. Parker, and Payload Specialists Samuel T. Durrance and Ronald A. Parise. Brand, at 59, became the oldest astronaut to fly into space until F. Story Musgrave flew at age 61 on STS-80 in 1996. The mission also marked the final spaceflight for both Brand and Lounge.

The ASTRO-1 observatory was initially scheduled to fly shortly after the *Challenger* accident. After the disaster, the mission was reconfigured as STS-35, incorporating the Broad-Band X-Ray Telescope (BBXRT-01) into the payload. *Columbia* rolled out to Pad 39A in April 1990 for a scheduled May launch, but persistent hydrogen leaks in the orbiter's aft compartment forced multiple delays. Attempts to repair the leaks required *Columbia* to be rolled back to the Vehicle Assembly Building (VAB) for repairs. Engineers replaced the orbiter-side umbilical assembly with one borrowed from the still-under-construction *Endeavour* and fixed the external tank hardware. Despite these efforts, further hydrogen leaks were detected during subsequent tanking tests in June and September 1990, forcing additional postponements.

Astronaut Vance D. Brand mission commander, Pilot Guy S. Gardner and Mission Specialist (MS) John M. Lounge. MS Robert A.R. Parker, Payload Specialist Ronald A. Parise, MS Jeffrey A. Hoffman, and Payload Specialist Samuel T. Durrance.

A "tiger team" of NASA engineers was assigned to resolve the issue. After extensive testing and repairs, including replacing damaged components in the aft compartment, the mission was finally cleared for launch in December 1990, following the completion of STS-38. *Columbia* successfully launched on December 2, 1990, at 1:49 a.m. EST, marking the ninth night launch in the Shuttle program.

The primary goal of STS-35 was to conduct astronomical observations using the ASTRO-1 observatory, which included four telescopes: the Hopkins Ultraviolet Telescope (HUT), the Wisconsin Ultraviolet Photo-Polarimeter Experiment (WUPPE), the Ultraviolet Imaging Telescope (UIT), and the BBXRT, mounted on the Instrument Pointing System (IPS). The crew operated in two shifts, with Gardner, Parker, and Parise forming the Red Team, and Hoffman, Lounge, and Durrance the Blue Team. Commander Brand coordinated mission activities without being assigned to a specific team.

Once in orbit, the telescopes were powered up and raised into position 11 hours into the flight. Observations began 16 hours into the mission after instrument checkouts were completed. Astronomical observations required precise coordination between the Shuttle's crew and ground teams at Marshall Space Flight

Center (MSFC) and Goddard Space Flight Center (GSFC). The flight crew used the Shuttle's thrusters to position the observatory toward celestial targets, with the IPS fine-tuning the alignment based on guide stars.

However, problems arose with the IPS's pointing precision and overheating failures in the data display units, which hindered crew-aiming procedures. Ground teams at MSFC had to step in to assist with telescope aiming, though the BBXRT, operated primarily by GSFC, was not affected by these issues. Despite the challenges, the mission successfully captured 231 observations of 130 celestial objects, achieving 70% of the mission's scientific objectives over a span of 143 hours.

Beyond ASTRO-1, STS-35 supported several other experiments, including the Shuttle Amateur Radio Experiment (SAREX-II), which allowed communication between amateur radio operators on Earth and the Shuttle crew. Payload Specialist Ronald Parise, a licensed operator, managed these communications during his off-duty time. The Space Classroom Program, "Assignment: The Stars," was conducted to inspire student interest in science and technology. During the mission, Hoffman delivered the first classroom lesson broadcast from space on December 7, 1990, teaching about the electromagnetic spectrum and the ASTRO-1 instruments.

The mission was shortened by one day due to bad weather forecasts at Edwards Air Force Base in California, the primary landing site. The Shuttle's Orbital Maneuvering System (OMS) engines were fired over the Indian Ocean on December 10, 1990, to deorbit the spacecraft. *Columbia* touched down on Runway 22 at Edwards Air Force Base at 9:54 p.m. PST after a mission duration of 8 days, 23 hours, 5 minutes, and 8 seconds. This marked the fourth night landing in Shuttle history. *Columbia* was later transported back to Kennedy Space Center on December 20, 1990, aboard the Shuttle Carrier Aircraft.

STS-37: A Milestone in Space Shuttle History

STS-37 marked the thirty-ninth mission of NASA's Space Shuttle program and the eighth flight of *Atlantis*. This six-day mission, launched on April 5, 1991, from Kennedy Space Center, had a primary objective: the deployment of the Compton Gamma Ray Observatory (CGRO). This observatory was the second in NASA's series of Great Observatories, including the visible-spectrum Hubble Space Telescope (HST), the Chandra X-ray Observatory (CXO), and the infrared Spitzer Space Telescope. CGRO's mission aimed to explore high-energy celestial gamma-ray emissions, phenomena that could not be detected from Earth due to atmospheric interference.

Kenneth D. (Ken) Cameron, pilot; Jay Apt, mission specialist; Steven R. Nagel, commander; and Jerry L. Ross and Linda M. Godwin, mission specialists

The crew of STS-37 was led by Commander Steven R. Nagel, who was on his third spaceflight. He was joined by Pilot Kenneth D. Cameron, embarking on his first mission, along with three mission specialists: Linda M. Godwin, also on her first spaceflight, Jerry L. Ross, who served as flight engineer on his third mission, and Jerome Apt, participating in his first mission. Together, they successfully carried out the critical tasks of the mission, including deploying CGRO and conducting the first spacewalks since 1985.

STS-37 was launched at 9:22 a.m. EST on April 5, 1991, after a brief delay caused by concerns over weather conditions that could have impacted the launch. The Shuttle ascended smoothly into an orbit inclined at 28.45 degrees. The primary payload, CGRO, weighing approximately 16,000 kg (35,000 lb), was successfully deployed on the mission's third day. This massive observatory was the heaviest payload ever deployed from the Shuttle into low Earth orbit. Designed to last two years, CGRO was equipped with four scientific instruments: the Burst and Transient Source Experiment (BATSE), the Imaging Compton Telescope (COMPTEL), the Energetic Gamma Ray Experiment Telescope (EGRET), and the Oriented Scintillation Spectrometer Experiment (OSSE). These instruments would help scientists study gamma rays from various cosmic sources.

One of the most notable events of the mission occurred when the CGRO's high-gain antenna failed to deploy. After multiple unsuccessful attempts to resolve the issue remotely, Mission Specialists Jerry Ross and Jerome Apt performed an unscheduled spacewalk on April 7, 1991, to manually free the antenna. This contingency extravehicular activity (EVA) was the first unscheduled spacewalk since 1985. Within 17 minutes, Ross successfully freed the antenna, allowing the mission to proceed as planned.

The following day, Ross and Apt conducted a scheduled EVA, the first of its kind since November 1985. This spacewalk tested techniques for moving astronauts and equipment around the proposed Space Station Freedom. The duo evaluated manual, mechanical, and electrical methods of propelling a cart around a simulated large space structure. Ultimately, they found that hand-over-hand propulsion was the most effective method.

A stainless steel bar punctured Apt's right glove during the second EVA. Fortunately, his silk comfort glove and hand partially sealed the hole, preventing any pressure loss. The incident went unnoticed until postflight inspection. In total, Ross and Apt spent 10 hours and 29 minutes conducting spacewalks during STS-37, laying the groundwork for future space station operations.

Beyond the deployment of CGRO, the mission carried several secondary payloads. The Crew and Equipment Translation Aid (CETA) was part of the EVA tests, while other experiments included the Ascent Particle Monitor (APM), the Shuttle Amateur Radio Experiment (SAREX II), the Protein Crystal Growth experiment, and the Bioserve/Instrumentation Technology Associates Materials Dispersion Apparatus (BIMDA). These experiments explored the potential for manufacturing processes, biomedical research, and fluid sciences in space.

Pilot Kenneth D. Cameron was the primary operator of the SAREX II experiment, allowing the crew to communicate with amateur radio operators on Earth, including receiving live television video from a ham radio club station at Johnson Space Center. The crew also captured striking images of the Kuwaiti oil fires, a stark reminder of the ongoing Gulf War, adding a historical significance to their mission.

After a successful mission, *Atlantis* returned to Earth on April 11, 1991, landing at Edwards Air Force Base in California at 6:55 a.m. PDT. The landing was delayed by a day due to weather conditions at both Edwards and Kennedy Space Center. As *Atlantis* touched down, it rolled to a stop after covering 1,940 meters (6,360 feet) in 56 seconds. The orbiter landed 623 feet short of the runway's threshold marking due to an incorrect wind call, but this presented no issue on the dry lakebed. The orbiter was returned to Kennedy Space Center on April 18, 1991, with a landing weight of 86,227 kg (190,098 lb).

The STS-37 mission was a resounding success, with its key objectives achieved and significant contributions made to gamma-ray astronomy and spacewalk procedures. The Compton Gamma Ray Observatory went on to provide invaluable data for nearly a decade, solidifying its place as a cornerstone of space-based astronomical research.

STS-39: Advancing Space-Based Defense and Scientific Research

Charles L. Veach, mission specialist 5; Michael L. Coats, commander; Gregory J. Harbaugh, mission specialist 2; Donald R. McMonagle, mission specialist 4; L. Blaine Hammond, pilot; Richard J. Hieb, mission specialist 3; and Guion S. Buford, Jr., mission specialist 1.

STS-39, launched on April 28, 1991, from Kennedy Space Center, was a landmark mission that combined military and scientific objectives. Flown by the Space Shuttle Discovery, it was the 40th flight in the program and the 12th flight of Discovery. It was the first unclassified Department of Defense mission since the Challenger disaster, highlighting the Space Shuttle's role in advancing national security and scientific knowledge.

The crew of STS-39 consisted of seven astronauts: Commander Michael L. Coats, Pilot L. Blaine Hammond Jr., Mission Specialist 1 Richard J. Hieb, Mission Specialist 2 Donald R. McMonagle, Mission Specialist 3 Guy S. Gardner, Mission Specialist 4 Charles D. Gemar, and Payload Specialist 1 Gregory J. Harbaugh.

Launch preparations for Space Shuttle Discovery's STS-39 mission began in early 1991, with the initial launch scheduled for March 9. However, during routine processing at Launch Complex 39A, engineers discovered significant structural issues—specifically, cracks on all four lug hinges of the external tank umbilical door drive mechanisms. Recognizing the potential danger, NASA management rolled the shuttle back to the Vehicle Assembly Building (VAB) on March 7 for repairs. Discovery was subsequently transferred to the Orbiter Processing Facility (OPF), where the damaged hinges were replaced with reinforced units taken from the orbiter Columbia.

After the repairs, Discovery was returned to the launch pad on April 1, and the launch was rescheduled for April 23. However, another issue arose during the prelaunch fueling process, when a transducer on the high-pressure oxidizer turbopump of main engine number three produced readings that were out of specification. The faulty transducer and its cable harness were replaced and thoroughly tested before a new launch date was set for April 28.

On April 28, 1991, at 7:33:14 a.m. EDT, Discovery successfully lifted off from Kennedy Space Center, Florida, carrying a launch weight of 112,207 kilograms (247,374 pounds). This marked the beginning of STS-39, a mission dedicated to the U.S. Department of Defense (DoD), with both classified and unclassified objectives.

STS-39 was a pivotal mission, being the first unclassified DoD-dedicated Space Shuttle flight. Previous Shuttle missions for the DoD had been classified, with little to no information released regarding the payloads or experiments. On STS-39, only the payload contained within the Multi-Purpose Experiment Canister (MPEC) remained classified, while the rest of the mission's objectives were openly disclosed. Among the unclassified payloads were the Air Force Program-675 (AFP-675), the Infrared Background Signature Survey (IBSS), and Space Test Payload-1 (STP-1), alongside several advanced scientific experiments including the Critical Ionization Velocity (CIV) and Chemical Release Observation (CRO) investigations.

The crew of STS-39 was divided into two teams, enabling continuous around-the-clock operations. Their mission involved a series of observations related to Earth's atmosphere, gas releases, Discovery's orbital environment, and engine firings, using sensors that ranged from infrared to far-ultraviolet wavelengths. The mission's complexity was further underscored by deploying five spacecraft or satellites from Discovery's payload bay, one of which was retrieved later in the mission.

A key mission component was the deployment of the Shuttle Pallet Satellite-II (SPAS-II), which carried the IBSS instrumentation. SPAS-II, deployed using Discovery's Remote Manipulator System (Canadarm), conducted detailed observations of Discovery's orbital maneuvers. One such maneuver, known as the "Malarkey Milkshake," was carefully monitored by the IBSS. While the SPAS-II and IBSS deployment was initially delayed to prioritize the Cryogenic Infrared Radiance Instrumentation for Shuttle (CIRRIS) experiment, the satellite was eventually deployed on Flight Day Four. CIRRIS had been depleting its liquid helium supply faster than anticipated, and its timely completion was crucial for capturing critical data on auroral and airglow emissions.

Despite the mission's overall success, the crew encountered technical challenges. Two tape recorders crucial for recording data from instruments aboard AFP-675 malfunctioned after just four hours of operation. In a complex repair procedure, the astronauts rerouted wires and connected a splice wire to the Ku-band antenna system, enabling direct data transmission to ground control.

STS-39's high orbital inclination of 57.01 degrees allowed the crew to fly over vast expanses of Earth's landmasses, facilitating critical observations of environmental resources and problem areas. The scientific and defense-related data gathered during the mission were of immense value, both for Earth observation and military applications.

After a successful nine-day mission, Discovery prepared for landing. Initially, Edwards Air Force Base in California was the designated landing site, but high winds there forced NASA to divert the Shuttle to Kennedy Space Center in Florida. Discovery touched down on May 6, 1991, at 2:55:35 p.m. EDT, on Runway 15. The Shuttle's landing weight was 102,755 kilograms (226,536 pounds), and it rolled to a stop after a 2,877-meter (9,439-foot) rollout, lasting 55 seconds.

STS-40: A Pioneering Mission in Space Life Sciences

F. Drew Gaffney, payload specialist 1; Milli-Hughes Fulford, payload specialist 2; M. Rhea Seddon, mission specialist 3; and James P. Bagian, mission specialist 1. Standing in the rear, left to right, are Bryan D. O'Connor, commander; Tamara E. Jernigan, mission specialist 2; and Sidney M. Gutierrez, pilot.

STS-40, launched on June 5, 1991, from Kennedy Space Center, was a groundbreaking mission focused on biomedical research. Flown by the Space Shuttle Columbia, this mission was the 41st flight in the Space Shuttle program and marked the first dedicated life sciences mission in the program's history. Designated as Spacelab Life Sciences-1 (SLS-1), STS-40 was a key step in understanding how living organisms, including humans, adapt to the microgravity space environment.

The crew of STS-40 consisted of seven astronauts: Commander Bryan D. O'Connor, Pilot Sidney M. Gutierrez, Mission Specialist 1 James P. Bagian, Mission Specialist 2 Tamara E. Jernigan, Mission Specialist 3 M. Rhea Seddon, Payload Specialist 1 F. Drew Gaffney, and Payload Specialist 2 Millie Hughes-Fulford. This diverse team of astronauts, including physicians and scientists, was selected for their expertise in medicine, physiology, and space operations, making them ideally suited for the mission's objectives.

The launch of Space Shuttle Columbia for mission STS-40 was originally scheduled for

May 22, 1991. However, the mission was postponed less than 48 hours before liftoff due to a critical issue with the orbiter's main propulsion system. A leaking liquid hydrogen transducer, which had been removed and replaced during leak testing in 1990, was found to have failed a vendor analysis. This raised concerns among engineers that one or more of the nine liquid hydrogen and liquid oxygen transducers, which protruded into the fuel and oxidizer lines, could break off and be ingested by the engine turbopumps, potentially causing catastrophic engine failure. This type of failure posed a significant risk to the mission and the crew, prompting the delay.

Compounding the issue, one of Columbia's five General Purpose Computers (GPCs) experienced a complete failure, along with a multiplexer demultiplexer (MDM) that controlled critical systems, including the orbiter's hydraulics, ordnance, and the Orbital Maneuvering System/Reaction Control System (OMS/RCS) functions housed in the aft compartment. Both the GPC and MDM were promptly replaced and tested to ensure the orbiter's systems were fully operational.

In parallel with these repairs, engineers replaced one liquid hydrogen and two liquid oxygen transducers located upstream in the propellant flow system near the 43 cm (17 in) disconnect area, which was protected by an internal screen. Additionally, three liquid oxygen transducers in the engine manifold area were replaced, while the three liquid hydrogen transducers in this section were removed, and the openings were sealed.

Despite these efforts, further complications arose during preparations for the rescheduled launch. After several attempts to calibrate the Inertial Measurement Unit (IMU) 2 failed, it became necessary to replace the unit and retest the systems. Following this additional setback, the launch was rescheduled for June 5, 1991.

On June 5, 1991, at 9:24:51 a.m. EDT, Columbia successfully launched from Kennedy Space Center. The mission's launch weight was 114,290 kg (251,970 lb), and the liftoff was captured on IMAX cameras, later featured in the 2015 documentary film *Journey to Space*.

STS-40 was the fifth dedicated Spacelab mission and the first focused entirely on life sciences, known as Spacelab Life Sciences-1 (SLS-1). This mission marked a significant milestone in spaceflight research, offering the most comprehensive and interrelated physiological measurements conducted in space since the Skylab missions of the early 1970s. The mission involved human subjects, 30 rodents, and thousands of tiny jellyfish, exploring the effects of spaceflight on six critical body systems.

The primary experiments targeted the cardiovascular and cardiopulmonary systems (heart, lungs, and blood vessels), the renal and endocrine systems (kidneys and hormone-secreting organs), the blood system (blood plasma), the immune system (white blood cells), the musculoskeletal system (muscles and bones), and the neurovestibular system (brain, nerves, eyes, and inner ear). This research provided invaluable insights into how prolonged space travel affects human health and paved the way for future long-duration missions.

In addition to life sciences, other payloads included twelve Getaway Special (GAS) canisters, which housed experiments in materials science, plant biology, and cosmic radiation. The Middeck Zero-Gravity Dynamics Experiment (MODE) and seven Orbiter Experiments (OEX) were also conducted, further contributing to the body of knowledge on space phenomena.

After a successful mission, Columbia returned to Earth, landing at Edwards Air Force Base in California on June 14, 1991, at 8:39:11 a.m. PDT, touching down on Runway 22. The orbiter was later transported back to Kennedy Space Center on June 21, 1991, concluding a mission that significantly advanced the understanding of human physiology in space.

STS-43: Advancing Communications and Earth Observation with Space Shuttle Atlantis

On August 2, 1991, the Space Shuttle Atlantis roared into space on mission STS-43, marking the ninth flight of the Atlantis orbiter. The mission, commanded by John E. Blaha, was primarily focused on deploying the fifth Tracking and Data Relay Satellite (TDRS-5), a crucial component of NASA's TDRS system. The TDRS system revolutionized space communications by providing continuous, high-bandwidth data links between spacecraft and ground stations on Earth.

Joining Commander Blaha were Pilot Michael A. Baker and Mission Specialists Shannon W. Lucid, G. David Low, and James C. Adamson. This experienced crew was tasked with a mission that combined cutting-edge technology with crucial scientific research, reflecting the multifaceted capabilities of the Space Shuttle program.

Mission Specialist (MS) Shannon W. Lucid, MS James C. Adamson, Commander John E. Blaha, MS G. David Low, and Pilot Michael A. Baker.

On August 2, 1991, at 15:02:00 UTC (11:02 a.m. EDT, local time), Space Shuttle Atlantis lifted off from Kennedy Space Center for mission STS-43, marking another milestone in NASA's Space Shuttle program. Originally scheduled for July 23, 1991, the launch faced multiple delays. The first postponement, to July 24, was necessary to replace a faulty integrated electronics assembly responsible for controlling the separation between the orbiter and the external tank. On the rescheduled day, just five hours before liftoff, the mission was delayed again due to a malfunction in the main engine controller of the shuttle's number three engine. After replacing and retesting the controller, the launch was reset

for August 1. Still, poor weather conditions at the return-to-launch site and technical issues involving a cabin pressure vent valve pushed the launch to August 2. Finally, Atlantis launched without further complications, beginning its nine-day mission.

The primary objective of STS-43 was to deploy the fifth Tracking and Data Relay Satellite, TDRS-E, which was attached to an Inertial Upper Stage (IUS) booster. Approximately six hours into the mission, TDRS-E was successfully deployed, and the IUS performed two critical burns that propelled the satellite into its designated geosynchronous orbit, over 22,000 miles above Earth. TDRS-E, later renamed TDRS-5, joined NASA's orbiting TDRS cluster, ensuring near-continuous communication between Earth and low-orbiting spacecraft like the Space Shuttle. By extending coverage from only 15% of each orbit to nearly 100%, depending on spacecraft altitude, the TDRS network revolutionized communications for NASA's space operations. TDRS-5 was activated and began operations in its west location over the Pacific Ocean on October 7, 1991, replacing TDRS-B, which was lost in the 1986 Challenger disaster.

Beyond the deployment of TDRS-E, the STS-43 mission featured an array of secondary experiments to advance space technology and scientific research. One of the key experiments was the Space Station Heat Pipe Advanced Radiator Element (SHARE-II), which tested a cooling system designed for the future Space Station Freedom. Another notable experiment, the Solid Surface Combustion Experiment (SSCE), explored how fire behaves in the microgravity environment of space. The results contributed to understanding combustion processes in low-gravity conditions, critical for ensuring safety in space habitats.

Other onboard experiments included the Shuttle Solar Backscatter Ultraviolet (SSBUV) instrument, which provided important data on Earth's ozone layer, and the Protein Crystal Growth (PCG III) experiment aimed to grow purer protein crystals for medical research. The crew also worked with the Space Acceleration Measurement System (SAMS), which monitored microgravity conditions, and the Air Force's Maui Optical Site (AMOS) experiment, which studied how optical systems perform in space.

The crew's ingenuity was tested when they had to improvise a solution to replace a missing camera part. Working with flight controllers on the ground, they adapted an available component, demonstrating the problem-solving skills required for extended missions in space.

One of the most unique aspects of the STS-43 mission was the sending of the first email from space. On August 9, 1991, astronauts Shannon Lucid and James Adamson used an Apple Macintosh Portable computer to send a message via AppleLink to Earth, specifically to Marsha S. Ivins at Johnson Space Center. The email read:

"Hello Earth! Greetings from the STS-43 crew. This is the first AppleLink from space. Having a GREAT time, wish you were here, ... send cryo and RCS! Hasta la vista, baby, ... we'll be back!"

This began electronic communications between astronauts and mission control via email—a simple but profound leap in space communication technology.

Throughout the mission, the crew experienced only minor technical issues, none of which threatened the success or safety of the flight. One such issue involved the Auxiliary Power Unit (APU 2) cooling system, which failed to activate during an on-orbit test. However, as one of three redundant systems, APU 2 was available for use during reentry and landing, ensuring mission safety.

On August 11, 1991, Atlantis made a smooth return to Earth, touching down on Runway 15 at Kennedy Space Center at 8:23:25 a.m. EDT, concluding its successful mission. STS-43's contributions advanced NASA's communications capabilities and scientific understanding, making it a critical chapter in the Space Shuttle program's history.

STS-48

The STS-48, launched on September 12, 1991, aboard the Space Shuttle Discovery, was a significant mission focused on Earth observation and atmospheric studies. The mission's primary objective was to deploy the Upper Atmosphere Research Satellite (UARS), a key component of NASA's Mission to Planet Earth program, to understand the Earth's environment and the impacts of human activities on the planet.

Commander John O. Creighton led the crew of STS-48, with Pilot Kenneth S. Reightler Jr. and Mission Specialists Charles D. Gemar, James H. Buchli, and Mark N. Brown. Each astronaut played a critical role in the mission's success, particularly in the deployment and operation of the UARS satellite.

Mark N. Brown, mission specialist; John O. Creighton, commander; and Kenneth S. Reightler, pilot. Pictured on the back row (left to right) are mission specialists Charles D. (Sam) Gemar, and

Space Shuttle Discovery was launched from Kennedy Space Center (KSC) at Launch Complex 39A at 7:11 p.m. EDT on September 12, 1991, into a 57.00° inclination orbit. The launch, originally set for 6:57 p.m., was delayed for 14 minutes due to interference in the air-to-ground communication link. The issue resolved itself at the T-5 minute mark, and the countdown proceeded normally, allowing Discovery to lift off smoothly.

The primary mission objective was to deploy the Upper Atmosphere Research Satellite (UARS), a critical component of NASA's Mission to Planet Earth program. On the third day of the mission, UARS was successfully deployed from Discovery's payload bay at an altitude of 650 kilometers (400 miles) above

Earth. This satellite was designed to study the impact of human activities on the planet's atmosphere, with a particular focus on the ozone layer. UARS sought to deepen scientific understanding of energy input into the upper atmosphere, global photochemistry, atmospheric dynamics, and the interplay between the upper and lower atmosphere. It provided valuable data on the structure, chemistry, energy balance, and physical processes within Earth's middle atmosphere, ranging from 16 to 97 kilometers (9.9 to 60 miles) above the surface. This mission marked a significant milestone in NASA's broader effort to monitor Earth as a complex environmental system.

The UARS is grappled by the RMS on STS-48. The Upper Atmosphere Research Satellite (UARS) is in the grasp of the remote manipulator system (RMS) end effector above the payload bay (PLB) of the Earth-orbiting Discovery, Orbiter Vehicle (OV) 103 during STS-48 pre-deployment checkout procedures. UARS solar array (SA) is being deployed.

UARS carried ten sophisticated instruments: the Cryogenic Limb Array Etalon Spectrometer (CLAES), the Improved Stratospheric and Mesospheric Sounder (ISAMS), the Microwave Limb Sounder (MLS), the Halogen Occultation Experiment (HALOE), the High-Resolution Doppler Imager (HRDI), the Wind Imaging Interferometer (WINDII), the Solar Ultraviolet Spectral Irradiance Monitor (SUSIM), the Solar/Stellar Irradiance Comparison Experiment (SOLSTICE), the Particle Environment Monitor (PEM), and the Active Cavity Radiometer Irradiance Monitor (ACRIM II). Though initially planned for an 18-month mission, UARS exceeded expectations, operating for 14 years before being decommissioned.

In addition to UARS, Discovery carried several secondary payloads, including the Ascent Particle Monitor (APM), Middeck 0-Gravity Dynamics Experiment (MODE), Shuttle Activation Monitor (SAM), Cosmic Ray Effects and Activation Monitor (CREAM), Physiological and Anatomical Rodent Experiment (PARE), Protein Crystal Growth (PCG II-2), Investigations into Polymer Membrane Processing (IPMP), and the Air Force Maui Optical Site (AMOS) experiment.

Notably, this flight also marked the first use of an electronic still camera in space. A modified Nikon NASA F4 captured monochrome images with 256 gray levels per pixel. These images were stored on a removable hard disk and could be viewed, enhanced, and transmitted to the ground using a modified laptop aboard Discovery.

STS-48 was also significant as it was the second post-Challenger mission to target Kennedy Space Center as the landing site, and it was the first mission to plan a night landing there. However, due to adverse weather conditions at KSC, Discovery flew an additional orbit before landing at Edwards Air Force Base in California at 3:38 a.m. EDT on September 18, 1991. The orbiter returned to Kennedy Space Center on September 26, 1991.

During the mission, on September 15, video footage captured a flash of light followed by several objects that seemed to move in a controlled manner. NASA explained these objects as ice particles reacting to engine jets, a conclusion supported by astronomer Philip C. Plait in his book *Bad Astronomy*. The event sparked interest and was later featured in an episode of *UFO Hunters*.

STS-44: A Mission of Strategic Significance and Technological Advancement

On November 24, 1991, the Space Shuttle

Atlantis embarked on mission STS-44, a critical mission primarily focused on national security and technological innovation. This mission, the tenth flight of Atlantis and the 44th Space Shuttle mission, was dedicated to deploying a Defense Support Program (DSP) satellite—a key component of the United States early warning defense system against missile launches.

Commander Frederick D. Gregory, an experienced astronaut with a distinguished background as a U.S. Air Force test pilot, led the STS-44 crew. Accompanying him were Pilot Terence T. Henricks, Mission Specialists Story Musgrave, James S. Voss, and Mario Runco Jr., along with Payload Specialist Thomas J. Hennen, a U.S. Army major and the first Army astronaut to fly in space. Hennen's inclusion highlighted the mission's significant military component, as his expertise in surveillance and reconnaissance was crucial for the success of the DSP satellite deployment.

Terence T. Hendricks, pilot; Frederick D. Gregory, commander; and F. Story Musgrave, mission specialist. Standing on the back row (left to right) are James S. Voss, mission specialist; Thomas J. Hennen, payload specialist; and Mario Runco, Jr., mission specialist

The launch of Space Shuttle Atlantis on STS-44 occurred on November 24, 1991, at 23:44 UTC, following a series of delays. Originally scheduled for November 19, 1991, the launch was postponed due to the need to replace and test a malfunctioning redundant inertial measurement unit on the Inertial Upper Stage (IUS) booster, which was attached to the Defense Support Program (DSP) satellite. Once resolved, the launch was reset for November 24 but was delayed by an additional 13 minutes. This was necessary to allow an orbiting spacecraft to pass and to replenish the liquid oxygen in the external tank after minor repairs to a valve in the liquid oxygen replenishment system on the mobile launcher platform. At liftoff, the shuttle's launch weight was 117,766 kilograms (259,630 pounds).

The primary objective of STS-44 was to deploy the DSP-16 satellite, part of the Defense Support Program, a long-standing U.S. initiative aimed at providing missile launch detection and nuclear detonation monitoring through infrared sensors. The DSP-16 satellite, mounted on the IUS-14 booster, was successfully deployed into geostationary orbit on the first day of the mission, further enhancing the United States' early warning capabilities during the tense geopolitical climate of the Cold War. The satellite deployment was a critical success, ensuring continued global coverage for detecting missile launches and other thermal events.

STS-44 also carried a variety of secondary payloads, underscoring the versatility of the Space Shuttle as a platform for scientific and military research. One of the key secondary payloads was Terra Scout, a military experiment designed to demonstrate the feasibility of using space-based platforms for real-time reconnaissance and battlefield management. This experiment represented a significant step forward in integrating space technology into military operations, providing a glimpse into the future of space-based intelligence gathering.

In addition to its military objectives, the mission supported several scientific and biomedical experiments. The crew researched microgravity's effects on the human body, focusing on issues like muscle atrophy, fluid shifts, and cardiovascular changes. These studies were essential for understanding the physiological challenges astronauts would face during long-duration space missions, providing valuable data for future missions requiring extended space stays.

Another important experiment aboard STS-44 was the Space Tissue Loss study, which aimed to explore the effects of microgravity on both human and animal cells. This research was vital for advancing the understanding of cellular processes in space, with potential applications in medical research and biotechnology.

The crew also participated in Earth observation activities throughout the mission, using handheld cameras to capture images of geological, environmental, and meteorological phenomena. These observations contributed to ongoing studies of Earth's environment, offering insights into natural processes and the impact of human activities on the planet.

After nearly seven days in space, Atlantis returned to Earth on December 1, 1991, touching down on Runway 5 at Edwards Air Force Base in California at 22:34 UTC. The rollout lasted 107 seconds, covering 3,411 meters (11,191 feet), with minimal braking to accommodate a test of the shuttle's braking systems. The shuttle's landing weight was 87,918 kilograms (193,826 pounds). Originally, the mission was scheduled to last 10 days, but it was shortened after one of the orbiter's three inertial measurement units failed on November 30. This was the last shuttle landing on a dry lake bed runway, as Atlantis returned to Kennedy Space Center on December 8, 1991.

STS-42: Exploring the Frontiers of Space Science with the International Microgravity Laboratory

STS-42, a NASA Space Shuttle Discovery mission, launched with the Spacelab module to study microgravity's effects on various organisms. Originally scheduled for liftoff at 8:45 EST (13:45 UTC) on January 22, 1992, the launch was delayed due to weather constraints. However, Discovery successfully lifted off an hour later, at 9:52:33 EST (14:52:33 UTC), from Kennedy Space Center, Florida. This marked the first of two flights for Discovery in 1992, with the second mission, STS-53, occurring in December. It was also the last mission of Discovery to carry a seven-member crew until STS-82 in 1997.

The STS-42 mission's primary objective was to investigate the impact of microgravity on living organisms and materials. The crew conducted research aboard the International Microgravity Laboratory-1 (IML-1), a pressurized, crewed Spacelab module. This facility enabled the international crew to perform numerous experiments focused on the nervous system's adaptation to low gravity and microgravity's effects on various life forms, including shrimp eggs, lentil seedlings, fruit fly eggs, and bacteria. Additional experiments involved processing low-gravity materials, such as crystal growth from enzymes, mercury, iodine, and viruses.

Stephen S. Oswald, pilot; Roberta L. Bondar, payload specialist 1; Norman E. Thagard, mission specialist 1; Ronald J. Grabe, commander; David C. Hilmers, mission specialist 2; Ulf D. Merbold, payload specialist 2 and William F. Readdy, mission specialist 3.

The crew, divided into two teams, Red and Blue, ensured round-the-clock monitoring of the experiments. The astronauts on the Blue Team included Commander Ronald J. Grabe, making his third spaceflight; Pilot Stephen S. Oswald, on his first spaceflight; Mission Specialist Norman Thagard, a seasoned astronaut on his fourth flight; and Canadian Payload Specialist Roberta Bondar, making her first and only spaceflight as Canada's first female astronaut. The Red Team consisted of Mission Specialist and Flight Engineer William F. Readdy, on his first spaceflight; Mission Specialist David C. Hilmers, completing his fourth and final mission; and Ulf Merbold, a Payload Specialist from Germany, flying on his second spaceflight and notable as the first West German astronaut.

The mission had a unique international aspect, as it marked the first flight since the Challenger disaster to include non-American astronauts, with Canada and Germany represented by Bondar and Merbold. Originally, astronaut Mary L. Cleave had been selected for the mission but withdrew for personal reasons. She was replaced by Manley Lanier "Sonny" Carter Jr., who tragically died in a plane crash seven months before the launch. David Hilmers then stepped in to fill the role of Mission Specialist 3.

STS-42 carried a variety of payloads in addition to IML-1, including ten Get Away Special (GAS) canisters, middeck payloads such as Gelation of SOLS: Applied Microgravity Research (GOSAMR), Investigations into Polymer Membrane Processing (IPMP), and the Radiation Monitoring Experiment (RME-III). The shuttle also hosted two Shuttle Student Involvement Program (SSIP) experiments and an Australian-developed ultraviolet telescope named Endeavour.

After a one-day mission extension to allow for additional scientific experimentation, Discovery returned to Earth, landing on Runway 22 at Edwards Air Force Base, California, at 8:07:17 PST (16:07:17 UTC) on January 30, 1992. The orbiter's landing weight was 218,016 pounds (98,890 kg), and the rollout distance measured 9,811 feet (2,990 meters). After landing, Discovery was ferried back to Kennedy Space Center on February 16, 1992, where it was prepared for future missions.

STS-45

The STS-45, launched on March 24, 1992, was a milestone in NASA's history. It marked the first mission dedicated entirely to conducting scientific experiments focused on Earth and space sciences. The Space Shuttle Atlantis executed the mission with a crew of seven astronauts who carried out the Atmospheric Laboratory for Applications and Science-1 (ATLAS-1) payload.

This mission was notable for its international collaboration, with instruments onboard from the United States, France, Germany, Belgium, Switzerland, the Netherlands, and Japan. The ATLAS-1 payload was designed to study the Earth's atmosphere, solar radiation, and its interaction with the Earth's magnetic field. These studies were critical for understanding the planet's climate and environmental changes, which were becoming increasingly relevant to global discussions on climate change.

Brian Duffy, pilot (seated on left); and Charles F. Bolden, Jr., commander (seated on right). Standing on the back row (left to right) are Byron K. Lichtenberg, payload specialist 1; C. Michael Foale, mission specialist 3; David C. Leestma, mission specialist 2; Kathryn D. Sullivan, payload commander; and Dirk D. Frimout, payload specialist 2.

The crew, led by Commander Charles D. Gemar, included Pilot Brian Duffy, Mission Specialists Kathryn D. Sullivan, C. Michael Foale, and Byron K. Lichtenberg, Payload Specialist Dirk D. Frimout (the first Belgian astronaut), and Payload Specialist David C. Leestma. Throughout their nine-day mission, the crew operated instruments that measured ozone levels, observed solar radiation, and monitored the Earth's middle atmosphere.

On March 24, 1992, at precisely 8:13 a.m. EST, Space Shuttle *Atlantis* lifted off from Kennedy Space Center, marking the commencement of mission STS-45. Initially slated for launch on March 23, 1992, the mission encountered a 24-hour delay due to technical complications. During the fueling process, engineers detected higher-than-allowable concentrations of liquid hydrogen and liquid oxygen within the orbiter's aft compartment. Despite extensive troubleshooting efforts, the source of the leaks could not be replicated. Engineers hypothesized that the anomalies stemmed from the main propulsion system's plumbing, which had not yet been adequately thermally conditioned to the cryogenic propellants. Once resolved, the launch was rescheduled, and *Atlantis*, weighing 105,982 kg (233,650 lb) at liftoff, successfully launched the following day.

STS-45 carried the first Atmospheric Laboratory for Applications and Science (ATLAS-1), a pivotal scientific payload designed to enhance understanding of Earth's atmosphere and the broader environment of space. ATLAS-1, housed on Spacelab pallets in *Atlantis*' payload bay, consisted of 12 instruments representing contributions from the United States, France, Germany, Belgium, Switzerland, the Netherlands, and Japan. This international collaboration enabled a wide range of scientific investigations across multiple disciplines, including atmospheric chemistry, solar radiation, space plasma physics, and ultraviolet astronomy.

Among the key instruments on ATLAS-1 were the Atmospheric Trace Molecule Spectroscopy (ATMOS) and the Grille Spectrometer, which were essential for analyzing atmospheric composition. The Millimeter Wave Atmospheric Sounder (MAS) and Imaging Spectrometric Observatory (ISO) contributed data on weather and space plasma physics. At the same time, the Atmospheric Lyman-Alpha Emissions (ALAE) and Atmospheric Emissions Photometric Imager (AEPI) offered insights into atmospheric emissions. The Space Experiments with Particle Accelerators (SEPAC), Active Cavity Radiometer (ACR), and the Solar Ultraviolet Spectral Irradiance Monitor (SUSIM) monitored solar activity, with the latter two instruments focused on measuring solar energy and ultraviolet radiation. Additional instruments

supported ultraviolet astronomy and atmospheric studies, including the Far Ultraviolet Space Telescope (FAUST) and the Solar Backscatter Ultraviolet (SSBUV) experiment.

STS-45 also carried a variety of other payloads, including a Get Away Special (GAS) experiment and six mid-deck experiments, further broadening the scope of scientific research conducted during the mission.

Due to the success of the scientific experiments aboard *Atlantis*, NASA extended the mission by an additional day, allowing more time to gather valuable data. After spending more than nine days in space, *Atlantis* returned to Earth on April 2, 1992, at 6:23 a.m. EST. The shuttle touched down smoothly on Runway 33 at the Shuttle Landing Facility at Kennedy Space Center. After a rollout of 2,812 meters (9,226 feet), the orbiter, now weighing 93,005 kg (205,041 lb), completed another successful mission in NASA's Space Shuttle program.

STS-49

Mission Specialist (MS) Kathryn C. Thornton, MS Bruce E. Melnick, MS Pierre J. Thuot, Commander Daniel C. Brandenstein, Pilot Kevin P. Chilton, MS Thomas D. Akers, and MS Richard J. Hieb

STS-49, the maiden flight of the Space Shuttle Endeavour, was a historic and ambitious mission from May 7 to May 16, 1992. This mission was notable not only because it was the first flight of the newly built Endeavour, constructed as a replacement for the Space Shuttle Challenger, but also because it involved one of the most complex and daring satellite repair operations ever attempted in space.

Endeavour, OV-105, was named by students in elementary and secondary schools nationwide after a ship chartered to traverse the South Pacific in 1768. Endeavour was the last space shuttle built and was ordered to replace Challenger. The shuttle immediately imprinted on space history in May 1992 during its first mission, STS-49. Three spacewalking astronauts made the unprecedented effort to grab an orbiting satellite with their gloved hands and pull it into Endeavour's cargo bay to be repaired and re-launched from the shuttle. Endeavour also accomplished the first repair mission to NASA's Hubble Space Telescope, giving the telescope contact lenses to peer to the farthest edges of the universe. Endeavour was the second shuttle to retire after its successful 25th mission, STS-134, which delivered the Alpha Magnetic Spectrometer-2 (AMS) to the ISS in May/June 2011.

The primary objective of STS-49 was the capture, repair, and redeployment of the Intelsat VI (F-3) communications satellite, which had been stranded in an unusable orbit since its launch in March 1990 due to a malfunction of its booster rocket. The mission was critical for restoring the satellite to operational status, and it showcased the Space Shuttle's unique capabilities as a platform for in-orbit satellite servicing and repair.

The STS-49 crew was commanded by Daniel C. Brandenstein, with Kevin P. Chilton serving as the pilot. The mission specialists were Richard J. Hieb, Bruce E. Melnick, Pierre J. Thuot, Kathryn C. Thornton, Thomas D. Akers, and Richard H. Hieb. This seven-member crew was tasked with one of the most challenging missions in the Space Shuttle program's history.

The highlight of STS-49 was the extraordinary spacewalks conducted to capture and repair the Intelsat VI satellite. Initially, two spacewalks were planned, but the mission required an unprecedented three spacewalks to capture the satellite successfully. On May 10, during the first attempt, astronauts Pierre Thuot

and Richard Hieb tried to capture the satellite using a specially designed capture bar. However, the operation proved more difficult than anticipated due to the satellite's large size and the lack of proper handholds.

In an unprecedented move, NASA decided to conduct a third spacewalk with three astronauts working simultaneously—a first in spaceflight history. On May 13, 1992, astronauts Pierre Thuot, Richard Hieb, and Thomas Akers carried out a meticulously coordinated and physically demanding spacewalk. The three astronauts manually grabbed hold of the satellite, securing it so the Endeavour's robotic arm could attach the new rocket motor. This remarkable achievement required precise coordination and teamwork, and it demonstrated the versatility and adaptability of the Shuttle crew in overcoming unexpected challenges.

Once the Intelsat VI satellite was secured, the crew attached a new perigee kick motor, boosting it into its proper geostationary orbit. The satellite was successfully redeployed on May 14, 1992, marking the mission as a success and restoring vital communications capabilities the satellite was designed to provide.

In addition to the primary mission objectives, STS-49 also conducted several secondary experiments, including testing the assembly of structures in space, which provided valuable data for constructing future space stations, including the International Space Station.

STS-49 concluded with Endeavour's return to Earth, landing at Edwards Air Force Base on May 16, 1992. The mission was celebrated as a triumph of human ingenuity and perseverance. It demonstrated the Space Shuttle's unique role in satellite repair and servicing and solidified Endeavor's reputation as a reliable and capable addition to NASA's fleet. The mission's success underscored the importance of having a human presence in space to conduct complex operations that would be impossible to accomplish with robotic systems alone.

The legacy of STS-49 extended far beyond the mission itself. The techniques and lessons learned from the satellite repair operation paved the way for future missions, including the servicing of the Hubble Space Telescope. They reinforced the importance of human spaceflight in maintaining and enhancing space-based assets.

STS-50: A Historic Step in Space Laboratory Research

Space Shuttle Columbia embarked on its fourteenth flight, STS-50, on June 25, 1992, lifting off from Kennedy Space Center's Launch Complex 39A at 12:12 p.m. EDT, carrying a crew of seven astronauts: Commander Richard N. Richards, Pilot Kenneth D. Bowersox, Mission Specialists Bonnie J. Dunbar, Ellen S. Baker, Carl J. Meade, Payload Commander Lawrence J. DeLucas, and Payload Specialist Eugene H. Trinh. The mission was planned for a 13-day duration, making it one of the longest Shuttle missions to that date.

Ellen S. Baker, mission specialist; Kenneth D. Bowersox, pilot; Bonnie J. Dunbar, payload commander; Richard N. Richards, commander; Carl J. Meade, mission specialist; Eugene H. Trinh, payload specialist; and Lawrence J. DeLucas, payload specialist.

The U.S. Microgravity Laboratory-1 (USML-1) mission, flown aboard the Space Shuttle Columbia during STS-50, marked a pivotal step in advancing microgravity research. As the first flight of a Space Shuttle equipped with the Extended Duration Orbiter (EDO) hardware, this mission allowed for longer stays in orbit, pushing the Shuttle's operational limits and setting the stage for future long-term space habitation on Space Station Freedom.

Columbia remained in orbit for almost 14 days—at the time, the longest Shuttle flight in history. The mission's primary goal was to carry the USML-1 Spacelab module into space, enabling scientists to conduct material science, fluid physics, and biotechnology experiments under microgravity conditions. Microgravity refers to the near weightlessness experienced by orbiting spacecraft, where the effects of gravity are significantly reduced due to free fall.

The heart of the mission was the USML-1 Spacelab, housed in Columbia's payload bay. Scientists onboard conducted over 30 experiments across five key areas: fluid dynamics, materials science, combustion science, biotechnology, and technology demonstrations. These investigations were conducted around the clock, maximizing the mission's scientific yield. The experiments on fluid dynamics explored how liquids and gases behave in microgravity, while materials science research focused on crystal growth and solidification. Combustion science examined the behavior of flames and combustion processes in space, and biotechnology experiments studied biological materials and organisms in the unique environment of space.

One of the major milestones of STS-50 was the introduction of three new experiment facilities: the Crystal Growth Furnace (CGF), the Surface Tension Driven Convection Experiment (STDCE) apparatus, and the Drop Physics Module (DPM). The CGF allowed scientists to grow crystals in microgravity, providing valuable data on how crystals form without gravity. This knowledge is essential for improving materials used in semiconductors, lasers, and other advanced technologies. The STDCE was the first experiment to study thermocapillary flows—fluid movements driven by surface tension—without the interference of buoyancy forces present on Earth. Meanwhile, the DPM enabled the study of liquids without the constraints of a container, using sound waves to manipulate and suspend drops in mid-air, offering insights into fluid behavior that could lead to advancements in medical treatments.

Another notable piece of hardware was the Glovebox Facility (GBX), a versatile tool that allowed astronauts to conduct hands-on experiments safely by isolating hazardous materials. The GBX proved indispensable in the growth of protein crystals, allowing crew members to adjust the chemical composition of the growth medium to optimize results, a first for space. Protein crystal growth experiments were particularly important, with around 300 samples grown during the mission. These crystals, including ones from the HIV Reverse Transcriptase Complex and human immune system enzymes, were later studied on Earth using X-ray crystallography. The larger and purer crystals produced in microgravity could lead to advances in drug design and treatment strategies for diseases such as AIDS.

The Generic Bioprocessing Apparatus (GBA) was another critical mission component, processing over 130 experiments involving living cells and microorganisms. These studies have potential applications in ecological waste treatment and biomedical research, particularly in developing drug delivery systems that can target specific tissues in the body.

One of the more fascinating aspects of the mission was the Astroculture-1 experiment, which evaluated a water delivery system for growing plants in microgravity. This research is a critical step toward enabling long-term human habitation in space, where plants could provide food, oxygen, and water purification. Growing plants in space presents unique challenges due to the way fluids behave in microgravity, necessitating innovative solutions like the Astroculture system.

The Space Acceleration Measurement System (SAMS) played an important role in monitoring the microgravity conditions experienced by the experiments. These measurements were vital for scientists to distinguish between true microgravity effects and external disturbances that could affect the results. The SAMS had flown on more than 20 Shuttle missions, providing invaluable data for microgravity research.

In addition to the USML-1 experiments, Columbia's mid-deck hosted further microgravity investigations. These included additional protein crystal growth experiments, the Zeolite Crystal Growth study—focused on materials used in fluid purification and waste cleanup—and Astroculture, which tested methods for plant growth in space. These secondary experiments expanded the mission's scientific scope, contributing to a better understanding of how microgravity affects various biological and material processes.

STS-50 also marked the debut of the Extended Duration Orbiter (EDO) kit, which enabled Columbia to stay in orbit for nearly two weeks. The EDO kit added extra hydrogen and oxygen tanks for power generation, additional nitrogen tanks for maintaining cabin atmosphere, and an improved air regeneration system for removing carbon dioxide. This extended duration was a crucial step toward preparing for long-term missions aboard Space Station Freedom and other future space habitats.

The crew's health and performance during extended missions were a key focus, leading to the development of the Extended Duration Orbiter Medical Project (EDOMP). Crew members monitored their cardiovascular health, took cabin atmosphere samples, and evaluated countermeasures such as the Lower Body Negative Pressure (LBNP) device, which aimed to mitigate the loss of body fluids in microgravity. This research was essential for ensuring astronaut health on long-duration flights and improving their ability to readapt to Earth's gravity upon reentry.

Other payloads aboard Columbia included the Shuttle Amateur Radio Experiment (SAREX), which allowed astronauts to communicate with amateur radio operators and schools around the world, fostering public engagement with space exploration. The Investigations into Polymer Membrane Processing (IPMP) experiment, flown on previous Shuttle missions, studied the formation of polymer membranes, which are used in medical and industrial processes.

Despite its scientific success, Columbia's extended orbital attitude increased its vulnerability to impacts from space debris and micrometeoroids. The orbiter sustained several impacts during the mission, including damage to its windows and carbon-carbon wing leading edges. These impacts highlighted the risks associated with longer space missions and the need for improved shielding and monitoring technologies.

In sum, STS-50 was a landmark mission that expanded the capabilities of the Space Shuttle program and advanced the frontiers of microgravity research. It laid the groundwork for future long-duration spaceflights and provided invaluable insights into how humans and materials behave in the unique environment of space. The scientific achievements of USML-1 not only contributed to space exploration but also paved the way for advancements in materials science, biotechnology, and future space station operations.

STS-46

The STS-46, launched on July 31, 1992, was a significant mission for NASA, highlighting the collaborative efforts between the United States and Europe in space exploration. The Space Shuttle Atlantis conducted the mission, and it focused on deploying two significant payloads: the European Retrievable Carrier (EURECA) and the Tethered Satellite System (TSS-1), both of which were critical for advancing scientific research in space.

The crew of STS-46, commanded by Loren J. Shriver, included Pilot Andrew M. Allen and Mission Specialists Claude Nicollier, Marsha S. Ivins, Jeffrey A. Hoffman, Franklin R. Chang-Díaz, and Payload Specialist Franco Malerba, who became the first Italian in space. Each astronaut played a vital role in the mission's objectives, particularly with the deployment and operation of the Tethered Satellite System.

EURECA, developed by the European Space Agency (ESA), was a free-flying satellite designed to carry out a range of experiments in microgravity over an extended period. Once deployed, EURECA would orbit independently, conducting materials science, solar physics, and biology experiments. The deployment of EURECA demonstrated the Shuttle's ability to deliver and position scientific instruments in orbit and then retrieve them later for analysis and reuse.

Andrew M. Allen, pilot; and Loren J. Shriver, commander. Standing (left to right) are Marsha S. Ivins, mission specialist 4; Claude Nicollier, mission specialist 3; Jeffrey A. Hoffman, mission specialist 1; Franklin R. Chang-Diaz, mission specialist 2; and Franco Malerba, payload specialist 1.

The primary objectives of the mission were the deployment of two significant payloads: the European Space Agency's European Retrievable Carrier (EURECA) and the joint NASA/Italian Space Agency (ASI) Tethered Satellite System (TSS-1). EURECA, a science platform designed for experiments in microgravity, was originally scheduled for deployment early in the mission but was delayed by a day due to a problem with its data handling system. After successfully resolving the issue, EURECA was deployed and, seven and a half hours later, its thrusters were fired to boost the spacecraft to its planned operating altitude of approximately 500 kilometers (310 miles). However, the thruster burn, intended to last 24 minutes, was unexpectedly cut short after only six minutes due to unusual attitude data from the spacecraft. Engineers quickly diagnosed and corrected the issue, allowing EURECA to be successfully boosted to its intended orbit on the sixth day of the mission.

The deployment of the Tethered Satellite System (TSS-1), a joint NASA and ASI project, was also delayed by a day due to the focus on resolving the EURECA problems. TSS-1 was designed to demonstrate electricity generation through the interaction between the tethered satellite and Earth's magnetic field. Unfortunately, during deployment, the satellite only reached a maximum distance of 260 meters (850 feet) from the orbiter, far short of the planned 20 kilometers (12 miles), due to a jammed tether line. Despite repeated attempts over several days to free the tether, the issue could not be resolved. As a result, the TSS-1 operations were curtailed, and the satellite was stowed for its return to Earth. TSS-1 would later be reflown on the STS-75 mission in 1996, with astronauts Franklin Chang-Díaz, Jeffrey Hoffman, Claude Nicollier, and Franklin Allen participating in that mission as well.

The European Retrievable Carrier 1L (EURECA-1L) spacecraft, with solar array panels extended, drifts in space after deployment from the payload bay of Atlantis, Orbiter Vehicle (OV) 104, during STS-46

In addition to the primary payloads, the mission carried several secondary experiments to advance scientific research. These included the Evaluation of Oxygen Integration with Materials/Thermal Management Processes (EOIM-III/TEMP 2A), which investigated how materials perform in space environments, and the Consortium for Materials Development in Space Complex Autonomous Payload (CONCAP II and CONCAP III), which focused on the development of new materials in microgravity. The mission also included the IMAX Cargo Bay Camera (ICBC), which captured high-resolution footage for future space documentaries, and the Limited Duration Space Environment Candidate Materials Exposure (LDCE), which studied the durability of materials in space. Additionally, the

Pituitary Growth Hormone Cell Function (PHCF) experiment explored the effects of microgravity on human cell growth, while the Ultraviolet Plume Instrument (UVPI) was used to observe spacecraft plumes in ultraviolet light.

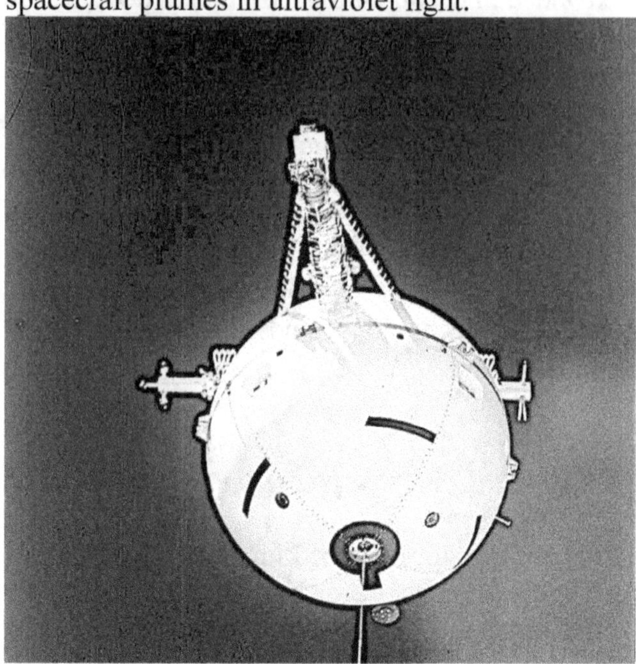

This Space Shuttle Orbiter Atlantis (STS-46) onboard photo is a close-up view of the Tethered Satellite System (TSS-1) in orbit above the Shuttle. A cooperative development effort by the Italian Space Agency (ASI) and NASA, the Tethered Satellite System (TSS) made capable the deployment and retrieval of a satellite which is attached by a wire tether from distances up to 100 km from the Orbiter. These free-flying satellites are used as observation platforms outside of the Orbiter.

Due to the delays encountered with the deployment of EURECA and TSS-1, the mission was extended by an additional day, allowing the crew to complete their scientific objectives and gather valuable data across multiple disciplines.

The mission concluded on August 8, 1992, with Atlantis landing safely at the Kennedy Space Center. STS-46 advanced scientific knowledge and strengthened international cooperation in space exploration.

STS-47

The STS-47, launched on September 12, 1992, aboard the Space Shuttle Endeavour, was a landmark mission for several reasons, mainly its focus on international cooperation and diversity. Designated as Spacelab-J, the mission was a joint venture between NASA and the National Space Development Agency of Japan (NASDA), underscoring the growing collaboration between the United States and Japan in space exploration.

Mission Specialist (MS) Jerome Apt and Pilot Curtis L. Brown, Jr (both holding launch and entry helmets (LEHs)); and (left to right, rear) MS N. Jan Davis, MS and Payload Commander (PLC) Mark C. Lee, Commander Robert L. Gibson, MS Mae C. Jemison, and Japanese Payload Specialist Mamoru Mohri.

STS-47 was notable for its diverse crew, including the first African-American woman in space, Dr. Mae Jemison, and the first married couple to fly together in space, Mark Lee and Jan Davis. Robert L. Gibson commanded the mission, with Curtis L. Brown Jr. serving as the pilot. The mission specialists were Mae Jemison, Mark Lee, Jan Davi, and N. Jan Davis, along with Payload Specialists Mamoru Mohri from Japan and J. David Low from NASA. Mamoru Mohri was the first Japanese astronaut to fly in a Space Shuttle, marking a significant milestone for Japan's participation in manned spaceflight.

The primary objective of STS-47 was to conduct scientific research in the Spacelab-J laboratory housed in the Shuttle's payload bay. The Spacelab-J module contained over 40 experiments across multiple disciplines, including life sciences, materials science, fluid physics, and biological sciences. These experiments were designed to take advantage of the microgravity space environment to gather data that would be impossible to obtain on Earth.

Among the key experiments conducted during STS-47 were studies on the effects of microgravity on plant growth, protein crystal growth, and fluid dynamics. These experiments were crucial for understanding how living organisms and physical processes behave without Earth's gravity, with potential applications for long-duration spaceflight and developing new materials and medicines.

One of the highlights of the mission was Dr. Mae Jemison's participation in a series of life sciences experiments, including studies on bone cell activity in microgravity and the effects of space travel on the human cardiovascular system. Dr. Jemison's presence on the mission was a significant moment in history, as she became a role model and inspiration for women and minorities in STEM fields.

STS-47 also featured the first implementation of the Classroom in Space program. Students on Earth could interact with the crew and observe experiments in real time, fostering a deeper interest in science and technology among young people.

The mission lasted nearly eight days, during which the crew worked around the clock in shifts to ensure the success of the numerous experiments. The data collected from these experiments provided valuable insights that contributed to advancements in multiple scientific fields.

On September 20, 1992, Endeavour returned to Earth, landing at the Kennedy Space Center. The success of STS-47 was a testament to the Space Shuttle's capabilities as a platform for scientific research and international cooperation. The mission's legacy continued to influence future collaborations between NASA and international space agencies, paving the way for a more interconnected approach to space exploration.

STS-52

STS-52 was the 13th flight of the Space Shuttle Columbia and the 51st mission in

NASA's Space Shuttle program. Launched on October 22, 1992, from Kennedy Space Center's Launch Complex 39B, STS-52 was primarily focused on scientific research and deploying a significant satellite that would contribute to Earth's geodetic and geophysical studies.

Michael A. Baker, pilot; James B. Wetherbee, commander; and Steven G. Maclean, payload specialist. On the front row, left to right, are mission specialists Charles (Lacy) Veach, Tamara Jernigan, and William Shepherd.

The crew of STS-52 consisted of six astronauts: Commander James D. Wetherbee, Pilot Michael A. Baker, and Mission Specialists Charles L. Veach, William M. Shepherd, Tamara E. Jernigan, and Payload Specialist Steven G. MacLean, who represented the Canadian Space Agency. This diverse and experienced team was tasked with carrying out a series of scientific experiments in addition to the primary satellite deployment.

One of the primary objectives of STS-52 was deploying the Laser Geodynamic Satellite II (LAGEOS-II). This satellite was part of a joint mission between NASA and the Italian Space Agency (ASI). LAGEOS-II was designed to study the Earth's shape, gravitational field, and rotational dynamics with unprecedented precision. The satellite, a spherical object covered with retroreflectors, was deployed into a medium Earth orbit. By bouncing laser beams off the retroreflectors, scientists on the ground could measure the satellite's position with extreme accuracy, allowing for detailed analysis of tectonic plate movements, sea level changes, and the Earth's gravitational field. The data from LAGEOS-II has been invaluable in understanding our planet's geophysical processes.

In addition to LAGEOS-II, the crew of STS-52 conducted a series of scientific experiments aboard Columbia. One of the key experiments was the United States Microgravity Payload-1 (USMP-1), which consisted of four experiments designed to investigate the effects of microgravity on materials science and fluid physics. These experiments aimed to improve the understanding of how materials behave without gravity, directly affecting manufacturing processes on Earth and in space.

Another important experiment was the Canadian Target Assembly (CTA), which tested the accuracy of the Space Shuttle's remote manipulator system, also known as the Canadarm. This was particularly significant because it demonstrated the ability to maneuver objects in space precisely, a capability critical for future missions involving satellite repair, construction, or retrieval.

STS-52 also carried out the Space Tissue Loss (STL) experiment, designed to study the effects of microgravity on human tissues. This experiment provided valuable data on how cells and tissues react to the space environment, contributing to the broader understanding of human health during long-duration space missions.

Throughout the mission, the crew completed all assigned tasks, demonstrating the versatility and capability of the Space Shuttle as a platform for scientific research. The mission also highlighted the importance of international collaboration, with contributions from NASA and the Canadian Space Agency.

After eight days in orbit, Columbia and its crew returned to Earth, landing at Kennedy Space Center's Shuttle Landing Facility on October 30, 1992. The landing was smooth, and the mission was deemed a success. STS-52 contributed valuable data to geophysics, materials science, and biomedical research, reinforcing the Space Shuttle's role as a key asset in advancing human knowledge through space exploration.

STS-53

STS-53 was the 52nd mission in NASA's Space Shuttle program and the 15th flight of the Space Shuttle Discovery. Launched on December 2, 1992, from Kennedy Space Center's Launch Complex 39A, STS-53 was primarily a Department of Defense (DoD) mission, and much of its content remains classified. However, the mission also included several unclassified experiments and payloads, contributing to national security and scientific research.

The crew of STS-53 consisted of five astronauts: Commander David M. Walker, Pilot Robert D. Cabana, and Mission Specialists Guion S. Bluford, James S. Voss, and Michael R. Clifford. Each astronaut brought significant experience to the mission, particularly in handling sensitive and classified operations.

Guion S. Bluford, and James S. Voss, mission specialists. On the back row, left to right, are David M. Walker, commander; Robert D. Cabana, Pilot; and Michael R. (Rick) Clifford, mission specialist.

In this mission, Space Shuttle Discovery carried a classified primary payload for the United States Department of Defense (DoD), marking the final significant payload for the DoD in the shuttle program. Known as "DoD-1," or USA-89 (1992-086B), the primary payload was a Satellite Data System-2 (SDS 2-3) military communications satellite. This launch followed two earlier deployments of the SDS series—USA-40 on STS-28 and USA-67 on STS-38. These satellites were vital for the DoD, providing secure communications and data relay for military operations. With the deployment of USA-89, the shuttle concluded its role in launching major defense-related payloads, reflecting the evolving relationship between NASA's shuttle program and military objectives in space.

In addition to the classified payload, Discovery carried two unclassified secondary payloads and supported a variety of middeck experiments. One of the secondary payloads consisted of the Orbital Debris Radar Calibration Spheres (ODERACS) satellites, a project designed to help track space debris by using radar to monitor the calibration spheres deployed from the shuttle. The other secondary payload included the combined Shuttle Glow Experiment/Cryogenic Heat Pipe Experiment (GCP), which investigated phenomena such as surface glow on the shuttle in space and the behavior of heat pipes in cryogenic conditions.

Discovery's middeck experiments covered a broad range of scientific and technical objectives. These included *Microcapsules in Space* (MIS-1), an experiment aimed at improving the production of microcapsules used in medical and industrial applications, and the *Space Tissue Loss* (STL) experiment, which studied the effects of microgravity on living tissue, an essential line of research for understanding the biological impacts of long-term spaceflight.

Further experiments such as the *Visual Function Tester* (VFT-2) examined how spaceflight affects astronauts' vision, while the *Cosmic Radiation Effects and Activation Monitor* (CREAM) provided critical data on the radiation environment in space, helping to ensure the safety of crewed missions. The *Radiation Monitoring Equipment* (RME-III) continued the important task of tracking radiation exposure in the shuttle's environment.

Additional scientific endeavors included the *Fluid Acquisition and Resupply Experiment* (FARE), which tested technologies critical for refueling spacecraft in orbit, and *HERCULES* (Hand-held, Earth-oriented, Real-time, Cooperative, User-friendly, Location-targeting, and Environmental System), a system designed to improve the targeting and navigation of satellite-based sensors.

Lastly, two defense-oriented experiments were also onboard. The *Battlefield Laser Acquisition Sensor Test* (BLAST) was designed to evaluate sensors capable of detecting laser emissions on the battlefield, and *CLOUDS* (Cloud Logic to Optimize Use of Defense Systems) sought to improve the understanding of cloud dynamics to optimize military systems operating in atmospheric conditions.

After nearly eight days in orbit, Discovery and its crew returned to Earth, landing at Edwards Air Force Base in California on December 9, 1992. The mission was completed successfully, with all objectives achieved, and Discovery was later flown back to Kennedy Space Center for post-flight processing.

STS-54

STS-54 was the 53rd mission of NASA's Space Shuttle program and the 14th flight of the Space Shuttle Endeavour. Launched on January 13, 1993, from Kennedy Space Center's Launch Complex 39B, the mission was primarily focused on deploying a critical communications satellite and conducting a series of scientific experiments that expanded our understanding of the space environment.

Mario Runco, Jr., mission specialist; John H. Casper, commander; Donald R. McMonagle, pilot; and mission specialists Susan J. Helms, and Gregory J. Harbaugh.

The crew of STS-54 consisted of five astronauts: Commander John H. Casper, Pilot Donald R. McMonagle, and Mission Specialists Mario Runco Jr., Gregory J. Harbaugh, and Susan J. Helms. Each crew member brought a wealth of experience and expertise to the mission, which was essential given the complex tasks they were to undertake.

The primary objective of Space Shuttle mission STS-54, launched aboard *Endeavour*, was the deployment of the fifth Tracking and Data Relay Satellite (TDRS-F), a crucial addition to NASA's Tracking and Data Relay Satellite System (TDRSS). This system provided continuous communication between orbiting spacecraft—such as the Space Shuttle and the Hubble Space Telescope—and ground stations on Earth. TDRSS played a vital role in ensuring that data could be transmitted in real-time, enabling constant contact with astronauts in orbit and greatly enhancing NASA's ability to manage multiple missions simultaneously.

On the first day of the mission, January 13, 1993, the crew successfully deployed TDRS-F from *Endeavour*'s payload bay. The satellite was then transferred to its geostationary orbit using an Inertial Upper Stage (IUS) booster, joining the existing constellation of TDRS satellites. This addition improved the system's capacity and redundancy, supporting NASA's broader space communication network.

In addition to deploying the satellite, the mission included several important scientific experiments, particularly focused on space science and life sciences research. One key payload in *Endeavour*'s bay was the Diffuse X-ray Spectrometer (DXS), a Hitchhiker experiment designed to collect data on X-ray radiation from diffuse sources in deep space. The DXS helped scientists study low-energy X-rays thought to originate from hot gases in our galaxy, contributing to a greater understanding of the interstellar medium and the broader processes that shape the Milky Way.

On the middeck, the mission carried several experiments to test the effects of microgravity on biological and physical systems. These included the Commercial General Bioprocessing Apparatus (CGPA) for life sciences research, the

Chromosome and Plant Cell Division in Space Experiment (CHROMEX), which studied plant growth in microgravity, and the Physiological and Anatomical Rodent Experiment (PARE), which examined the skeletal system and the adaptation of bone to space flight. Additionally, the Space Acceleration Measurement Equipment (SAMS) measured and recorded the microgravity acceleration environment of middeck experiments, while the Solid Surface Combustion Experiment (SSCE) measured the rate of flame spread and temperature of burning filter paper. These experiments provided valuable data for understanding the effects of spaceflight on biological processes and material behavior, both critical for future long-duration missions.

Astronauts Gregory Harbaugh and Mario Runco, Jr. during the EVA on STS-54. Harbaugh uses Runco's EMU mini-workstation as a handhold. The objective of this exercise is to simulate carrying a large object. It will also evaluate the ability of an astronaut to move about it space with a "bulky" object in hand.

On the fifth day of the mission, astronauts Mario Runco Jr. and Gregory J. Harbaugh conducted a nearly five-hour extravehicular activity (EVA) in the open cargo bay. The spacewalk was designed to test tools and techniques for space station assembly, honing NASA's capabilities to construct the upcoming International Space Station (ISS). During the EVA, the astronauts performed various tasks such as moving freely in the cargo bay, climbing into foot restraints without using their hands, and simulating the carrying of large objects in the microgravity environment. This spacewalk, although a late addition to the mission, was crucial in developing the expertise needed for the ISS assembly, demonstrating the feasibility of performing complex construction tasks in space. The EVA, which lasted 4 hours and 28 minutes, provided valuable data for NASA's future missions, particularly those involving long-term space structures like the ISS.

The STS-54 mission also continued NASA's outreach efforts through the American Space Experience (ASE) program, an educational initiative that connected students with live broadcasts from space. The program allowed students to interact with the astronauts, fostering interest in science, technology, engineering, and mathematics (STEM) by providing a unique, real-time learning experience.

After completing its objectives, *Endeavour* and its crew returned to Earth on January 19, 1993, landing safely at Kennedy Space Center's Shuttle Landing Facility. The mission was considered a success, having achieved all its primary and secondary goals, contributing to both scientific understanding and the advancement of human space exploration.

STS-56

STS-56, also known as ATLAS-2 (Atmospheric Laboratory for Applications and Science-2), was a crucial mission that advanced our understanding of Earth's atmosphere and its interaction with solar energy. Launched by the Space Shuttle Discovery, this mission was part of a series of flights dedicated to studying the Earth's middle and upper atmosphere, focusing on the effects of solar energy on our planet's climate and environment.

STS-56 launched on April 8, 1993, from Kennedy Space Center's Launch Complex 39B. Kenneth Cameron commanded the mission, with Stephen S. Oswald serving as the pilot. The crew included three mission specialists: Michael Foale, Kenneth D. Cockrell, and Ellen Ochoa.

Ellen Ochoa, a mission specialist, became the first Hispanic woman to go to space on this mission, marking a significant milestone in the history of NASA's astronaut corps.

Stephen S. Oswald, pilot; and Kenneth D. Cameron, commander. Standing, from the left, are mission specialists Kenneth D. Cockrell, C. Michael Foale, and Ellen Ochoa.

The primary objective of STS-56 was to conduct atmospheric science experiments using the ATLAS-2 payload. This payload was a collection of instruments designed to study the Earth's atmosphere, focusing on the interactions between solar radiation and the middle and upper atmosphere. These studies were essential for understanding the processes that drive weather patterns, climate change, and the depletion of the ozone layer.

Among the key instruments on ATLAS-2 were the Shuttle Solar Backscatter Ultraviolet (SSBUV) experiment, the Atmospheric Trace Molecule Spectroscopy (ATMOS) experiment, and the Solar Ultraviolet Spectral Irradiance Monitor (SUSIM). The SSBUV experiment was designed to calibrate satellite instruments that measure ozone concentrations, while ATMOS measured the concentration of trace gases in the atmosphere, such as methane and chlorofluorocarbons (CFCs). SUSIM focused on measuring solar ultraviolet radiation, critical for understanding how solar energy affects the Earth's atmosphere.

One of the highlights of STS-56 was the successful deployment and retrieval of the Spartan 201 satellite, a free-flying platform equipped with instruments to study the solar corona. The Spartan 201 mission was a key component of the overall objectives of STS-56, providing valuable data on solar activity and its effects on the Earth's upper atmosphere.

Throughout the mission, the crew conducted extensive observations and data collection, contributing to a better understanding of the complex interactions between the Sun and the Earth's atmosphere. The data gathered during the mission were crucial for improving climate models and enhancing the accuracy of weather forecasts.

Ellen Ochoa operated the Remote Manipulator System (RMS) during the Spartan 201 deployment, demonstrating her skills and contributing to the mission's success. The operation of the RMS was a critical aspect of the mission, requiring precision and coordination to ensure the safe deployment and retrieval of the Spartan satellite.

STS-56 concluded successfully on April 17, 1993, when Discovery landed at Kennedy Space Center. The mission was widely regarded as a success, with all primary objectives met and significant scientific data returned for analysis.

The legacy of STS-56 lies in its contributions to atmospheric science and the study of solar-terrestrial interactions. The data collected during the mission provided valuable insights into the processes that govern the Earth's climate and the effects of human activities on the atmosphere. The mission also demonstrated the importance of international cooperation in space-based research, as many of the experiments on ATLAS-2 were developed in collaboration with scientists from around the world.

STS-56 stands out as a mission that significantly advanced our understanding of the Earth's atmosphere and the influence of solar energy on climate and weather patterns. The findings from this mission continue to inform scientific research and contribute to our ability to monitor and protect the Earth's environment.

STS-55

STS-55, also known as Spacelab D-2, was a pivotal mission that showcased the versatility and international cooperation embodied in NASA's Shuttle program. The mission, flown by Space Shuttle Columbia, marked the second dedicated German Spacelab flight and underscored the collaboration between NASA and the German Aerospace Center (DLR).

STS-55 launched on April 26, 1993, from Kennedy Space Center's Launch Complex 39A, following a delay due to technical issues. The mission was commanded by Steven R. Nagel, a veteran astronaut, with Terence T. Henricks serving as the pilot. The crew included five mission specialists: Jerry L. Ross, Charles D. Gemar, Bernard A. Harris Jr., Ulrich Walter, and Hans Schlegel. Notably, Walter and Schlegel were European Space Agency (ESA) astronauts representing Germany, reflecting the international nature of the mission.

Terence (Tom) Henricks, pilot; Steven R. Nagel, commander; and Charles J. Precourt, mission specialist. On the back row, from left to right, are Bernard A. Harris, mission specialist; Hans Schlegel, payload specialist; Jerry L. Ross, mission specialist; and Ulrich Walter, payload specialist.

In April 1993, Space Shuttle *Columbia* embarked on its fourteenth mission, STS-55, carrying the second reusable German Spacelab, known as D-2, into orbit. This mission was a testament to the shuttle program's ability to foster international cooperation and advance scientific space research. The D-2 payload, which included the Spacelab module and an exterior experiment support structure housed in *Columbia's* payload bay, built upon the success of the first German Spacelab flight, D-1, which had flown aboard Shuttle mission STS-61-A in October 1985. Together, these missions solidified the collaboration between the United States and Germany, setting the stage for future space station operations.

The D-2 mission furthered the German microgravity research program initiated by D-1, with oversight by the German Aerospace Center (DLR). The German Space Agency (DARA – Deutsche Agentur für Raumfahrtangelegenheiten) commissioned DLR to conduct the second Spacelab mission, expanding the research agenda from D-1. The mission involved contributions from a diverse array of international partners, including NASA, the European Space Agency (ESA), and space agencies from France and Japan. Eleven nations participated in the scientific investigations, demonstrating the global nature of space exploration. Of the 88 experiments conducted aboard D-2, four were sponsored by NASA, underscoring the mission's significance to both the United States and Germany.

The crew of STS-55 worked in two alternating shifts to ensure continuous operations during the mission. They conducted research in a variety of fields, including fluid physics, materials science, life sciences, biological sciences, technology, Earth observations, atmospheric physics, and astronomy. Many of the experiments were extensions of the research conducted during the D-1 mission, now utilizing advanced technology and upgraded hardware developed in the eight years since 1985. Additionally, several new experiments were introduced, further expanding the scope of scientific inquiry aboard the shuttle.

The D-2 mission achieved several significant milestones. It marked the Space Shuttle program's 365th cumulative day in space and *Columbia's* 100th day of flight time, highlighting the shuttle's role as the workhorse of NASA's fleet. Moreover, the mission set a precedent in space robotics, with the first successful telerobotic capture of a free-floating object by flight controllers based in Germany.

Medical research also featured prominently in the mission, as the crew performed the first intravenous saline solution injection in space. This experiment aimed to study how the human body responds to direct fluid replacement, an important countermeasure for the fluid loss experienced during extended spaceflight. In addition, the crew successfully completed an in-flight maintenance procedure to collect orbiter wastewater, which allowed the mission to proceed without interruption.

The STS-55 mission also included opportunities for public engagement. The crew participated in two amateur radio experiments, SAREX II from the United States and SAFEX from Germany. These experiments allowed students and amateur radio enthusiasts worldwide to communicate directly with the shuttle, fostering interest in space exploration. The crew also conducted a space medicine conference with the Mayo Clinic, sharing insights into the health challenges posed by space travel.

STS-55 concluded successfully on May 6, 1993, when Columbia touched down at Edwards Air Force Base in California. The mission was hailed as a success, with all primary objectives achieved and a wealth of scientific data returned for analysis.

STS-57

STS-57, launched by the Space Shuttle Endeavour, was a significant mission highlighting NASA's commitment to scientific research and developing new technologies in space. The mission, which took place in the summer of 1993, was notable for several key accomplishments, including the retrieval of the European Retrievable Carrier (EURECA) satellite and the first flight of the Spacehab module, a new pressurized laboratory designed to enhance the shuttle's research capabilities.

STS-57 launched on June 21, 1993, from Kennedy Space Center's Launch Complex 39B. Ronald J. Grabe, a veteran astronaut with two prior spaceflights, commanded the mission. Brian

Duffy served as the pilot, and the crew included five mission specialists: G. David Low, Nancy J. Sherlock (Currie), Peter J. K. Wisoff, Janice E. Voss, and Jeffrey N. Williams.

Brian Duffy, pilot; and Ronald J. Grabe, commander. On the back row (left to right) are Peter J. Wisoff, Nancy J. Sherlock, and Janice E. Voss, all mission specialists, and G. David Low, payload commander.

The mission was particularly notable for including Nancy Currie, who operated the Shuttle's Remote Manipulator System (RMS) during the critical EURECA retrieval operation. This mission was also the first spaceflight for Janice Voss and Peter Wisoff, who would have distinguished careers in NASA's astronaut corps.

The primary objective of STS-57 was to retrieve the European Retrievable Carrier (EURECA), a satellite deployed by STS-46 in August 1992. EURECA was an uncrewed platform designed by the European Space Agency (ESA) to conduct multiple scientific experiments in orbit. These experiments covered materials science, biology, and solar physics. After nearly a year in space, EURECA had completed its mission, and STS-57 was tasked with bringing it back to Earth for analysis and reuse.

The retrieval of EURECA was a complex operation that required precise maneuvering by the Shuttle and using the RMS to capture and secure the satellite in the Shuttle's payload bay. Nancy Currie skillfully operated the RMS to capture EURECA, demonstrating the high expertise and training required for such operations.

In addition to the EURECA retrieval, STS-57 marked the first flight of the Spacehab module, a commercially developed pressurized laboratory that expanded the Shuttle's research capabilities. Spacehab provided additional working space for the crew and housed several experiments, including materials science, life sciences, and fluid physics studies. The module's introduction was a significant milestone for NASA, representing a new era of public-private partnerships in space exploration.

One of the most significant highlights of STS-57 was the successful retrieval of EURECA. The satellite had been in orbit for nearly 11 months, conducting valuable research in a microgravity environment. Its retrieval was a testament to the Shuttle's versatility and the crew's skill. Once secured in the payload bay, EURECA was returned to Earth, where its experiments were analyzed, and the satellite itself was refurbished for potential future use.

Another key aspect of the mission was the use of the Spacehab module. This module allowed the crew to conduct multiple scientific experiments in a more controlled environment than the Shuttle's middeck. Experiments included developing new materials, studies on the behavior of fluids in microgravity, and investigations into the effects of spaceflight on biological systems.

STS-57 also included an extravehicular activity (EVA) or spacewalk, performed by astronauts David Low and Peter Wisoff. The EVA, lasting approximately five hours, involved testing tools and techniques that would be used in future space missions, including the assembly of the International Space Station. The spacewalk was a success and provided valuable data for NASA's ongoing development of space operations.

STS-57 concluded successfully on July 1, 1993, with a smooth landing at Kennedy Space Center. The mission was hailed as a success, achieving its objectives, contributing significantly to scientific knowledge, and developing spaceflight technology.

The legacy of STS-57 is multifaceted. The mission demonstrated the Space Shuttle's ability to retrieve and return satellites from orbit, a unique capability of the Shuttle program. The successful use of the Spacehab module marked the beginning of a new era in Shuttle operations, where commercial partnerships played a key role in expanding NASA's research capabilities.

Additionally, the EVA conducted during the mission provided critical insights that would inform the construction of the International Space Station in the coming years. The successful completion of STS-57 reinforced the Shuttle program's role as a vital tool for advancing both science and technology in space.

STS-51

STS-51: A Mission of Milestones and Scientific Achievements

In September 1993, NASA launched the Space Shuttle Discovery on mission STS-51, a significant and multifaceted expedition that contributed to advancing space technology and scientific research. The mission's primary objective was to deploy the Advanced Communications Technology Satellite (ACTS), a project that would set new standards in satellite communications. This mission also marked the deployment and retrieval of the SPAS-ORFEUS satellite, equipped with an IMAX camera that captured breathtaking footage of Discovery in space. Additionally, the mission featured a critical spacewalk to evaluate tools and techniques for the upcoming Hubble Space Telescope (HST) servicing mission later that year.

The STS-51 mission was commanded by Frank L. Culbertson Jr., on his second spaceflight, with William F. Readdy serving as the pilot, also on his second mission. The crew included mission specialists James H. Newman, Daniel W. Bursch, and Carl E. Walz, each embarking on their first journey into space. Their seats were assigned on the flight deck and mid-deck of Discovery, with Culbertson and Readdy positioned at the helm during both launch and landing.

Frank L. Culbertson, commander; Daniel W. Bursch, mission specialist; Carl E. Walz, mission specialist; William F. Readdy, pilot; and James H. Newman, mission specialist

Several delays initially plagued the mission. On July 17, 1993, the launch was scrubbed due to a flaw in the pyrotechnic initiator controller responsible for releasing the solid rocket boosters from the mobile launcher platform. A week later, on July 24, issues with a hydraulic power unit in one of the boosters caused another scrub. The Perseids meteor shower further delayed the launch, pushing the next window to the second week of August. On August 12, the countdown reached T-3 seconds, with the Space Shuttle Main Engines igniting, only to be halted by faulty fuel flow sensors. Finally, after these challenges, STS-51 successfully launched on September 12, 1993.

The Advanced Communications Technology Satellite (ACTS), developed by Lockheed Martin Astro Space for NASA, was a pioneering project to test advanced communications satellite technology. The satellite was deployed on the first day of the mission, but the initial attempt was delayed due to a brief loss of two-way communication between the crew and Mission Control. After resolving the issue, the satellite was successfully deployed, and its Transfer Orbit Stage (TOS) propelled it into geosynchronous orbit.

ACTS represented a significant advancement in satellite communications, utilizing multiple spot beam antennas and sophisticated onboard processing systems. This mission paved the way for future innovations in satellite technology. Despite its initial success, ACTS was eventually retired in 2004 after a decade of service, with the satellite placed in a graveyard orbit to prevent potential collisions with other space assets.

Another key element of the STS-51 mission was the deployment of the Orbiting Retrievable Far and Extreme Ultraviolet Spectrometer (ORFEUS) telescope, mounted on the Shuttle Pallet Satellite (SPAS). ORFEUS was designed to study the birth and death of stars and the composition of interstellar clouds by measuring ultraviolet radiation. The telescope, a collaboration between NASA and the German space agency DARA, was equipped with a 1-meter mirror provided by France's REOSC and an Interstellar Medium Absorption Profile Spectrograph (IMAPS), which allowed for high-resolution observations of hot galactic objects.

Released on September 13, 1993, ORFEUS operated for several days before being retrieved on September 19. The mission provided valuable data for understanding the processes that govern the life cycle of stars and the behavior of interstellar gases. The IMAX camera, also part of the SPAS payload, captured stunning footage of the Shuttle in orbit, some of which was later featured in the films Destiny in Space and Space Station 3D.

On September 16, 1993, mission specialists James H. Newman and Carl E. Walz conducted a spacewalk to test equipment and procedures for the upcoming Hubble Space Telescope servicing mission. The EVA, which lasted over seven hours, was crucial in validating the tools and techniques that would be used to repair and upgrade the Hubble later that year. Despite a minor delay caused by a stubborn toolbox lid, the astronauts successfully completed all their tasks, providing confidence to mission planners and engineers.

STS-51 was also a platform for various secondary experiments. These included the Air Force Maui Optical Site (AMOS) for auroral photography, the Commercial Protein Crystal Growth (CPCG) experiment, and the High Resolution Shuttle Glow Spectroscopy (HRSGS-A), among others. These experiments aimed to enhance understanding of various phenomena, from protein crystallization in microgravity to studying atmospheric gases that cause the Shuttle's surface to glow.

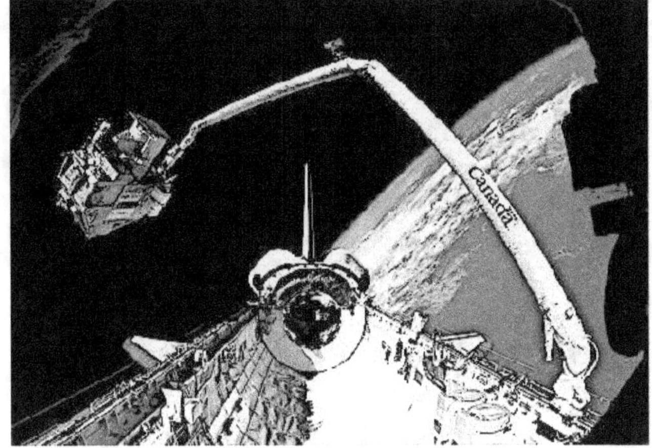

The ORFEUS/SPAS platform is captured by the Canadarm.

Medical and technological evaluations were conducted in-cabin, with crew members participating in studies on the effects of weightlessness on vision and balance and tests on the Shuttle's navigation systems using a Global Positioning System (GPS) receiver. These in-cabin activities contributed to NASA's growing knowledge on human spaceflight and spacecraft operations.

STS-51 was a mission marked by both technological innovation and scientific discovery. From the deployment of ACTS and SPAS-ORFEUS to the successful EVA that paved the way for Hubble's servicing, the mission demonstrated the versatility and capability of the Space Shuttle program. Each element of the mission, from the challenges faced during launch preparations to the successful completion of experiments and spacewalks, underscored the importance of meticulous planning, adaptability, and the pursuit of knowledge in advancing human space exploration.

STS-58

STS-58, flown by the Shuttle Columbia, was a landmark mission in life sciences research. Launched in late 1993, this mission was the most comprehensive study of the effects of spaceflight on living organisms at the time, significantly contributing to our understanding of how space conditions impact human and animal physiology. STS-58 was designated as SLS-2, the second Spacelab Life Sciences mission, and it played a critical role in laying the groundwork for long-term human spaceflight.

STS-58 launched on October 18, 1993, from Kennedy Space Center's Launch Complex 39B. John E. Blaha, an experienced astronaut with several prior spaceflights, commanded the mission. Richard A. Searfoss served as the pilot. The crew comprised six mission specialists: M. Rhea Seddon, William S. McArthur Jr., David A. Wolf, Shannon W. Lucid, Martin J. Fettman, and Margaret R. (Rhea) Seddon. Notably, a veteran astronaut, Shannon Lucid, would later set a record for the longest time in spaceflight by an American.

David A. Wolf, Shannon W. Lucid, and Rhea Seddon, all mission specialists; and Richard A. Searfoss, pilot. Standing in the rear, left to right, are John E. Blaha, commander; William S. McArthur, Jr., mission specialist; and Martin J. Fettmen, payload specialist

The primary objective of STS-58 was to conduct a series of life sciences experiments aboard the Spacelab module, focusing on how spaceflight affects biological processes in humans and animals. The mission was part of NASA's long-term goal to prepare for extended human presence in space, including potential missions to Mars.

STS-58, a 1993 Space Shuttle mission aboard Columbia, was dedicated to life sciences research. It marked a significant step in understanding how the human body adapts to the weightless environment of space. Launched on October 18, 1993, this mission was part of NASA's ongoing efforts to gather detailed physiological data, building upon the knowledge acquired during the Skylab program in the early 1970s and the Spacelab Life Sciences-1 (SLS-1) mission, STS-40, in June 1991.

Columbia's crew, commanded by John Blaha, included Pilot Richard Searfoss, Mission Specialists William McArthur, Shannon Lucid, David Wolf, Payload Specialist Martin Fettman, and Payload Commander Margaret Rhea Seddon. This team was tasked with conducting a series of experiments focused on the cardiovascular, regulatory, DNA, neurovestibular, and musculoskeletal systems. The crew's research extended to human subjects and laboratory animals, specifically 48 rats housed in 24 cages aboard the spacecraft.

The crew conducted neuro-vestibular experiments on the mission's second day to understand space motion sickness and sensory perception changes. Notably, astronauts Lucid and Fettman wore an Accelerometer Recording Unit (ARU), a specialized headset designed to record head movements continuously throughout the day. This contributed valuable data to the study of how spaceflight impacts the human vestibular system.

A minor technical issue arose on October 19, 1993, when a circuit breaker tripped, temporarily cutting power to one of the rodent cages. Flight controllers in Houston quickly addressed the problem, resetting the breaker and restoring power without further incident.

As the mission progressed, the crew employed various devices and procedures to counteract the effects of microgravity. On flight day three, McArthur and Blaha began using the Lower Body Negative Pressure (LBNP) device, designed to mitigate the detrimental effects of prolonged weightlessness on the cardiovascular system. The crew also participated in the Energy Utilization detailed supplementary objective (DSO 612), which involved collecting urine and saliva samples and keeping detailed logs of their exercise, food, and fluid intake. This experiment was crucial in understanding astronauts'

nutritional and energy requirements on long-duration space flights and the relationship between fluid and food consumption.

On October 20, 1993, the crew encountered a minor issue with the space toilet, detecting a slight leak around the filter door. The astronauts promptly addressed the problem, removing the filter and cleaning up the small amount of water that had escaped. They activated a secondary fan separator unit to ensure the air was properly cycled back into the cabin.

The following day, October 21, 1993, the crew continued their rigorous schedule of experiments. Seddon, Lucid, Wolf, and Fettman focused on metabolic studies, collecting additional blood and urine samples. These experiments were part of a broader investigation into calcium absorption and bone metabolism in space, a critical area of research given the known effects of microgravity on bone density. The findings from this mission and data from the SLS-1 mission provided crucial insights into how spaceflight accelerates bone breakdown, a condition not fully compensated by bone formation.

In addition to their scientific duties, the crew engaged with the public through the Shuttle Amateur Radio Experiment (SAREX-2). On October 22, 1993, Blaha and Searfoss contacted students at various schools, including Sycamore Middle School in Pleasant View, Tennessee, Gardendale Elementary in Pasadena, Texas, and Naparima College in Trinidad and Tobago. These communications offered a unique opportunity for students to interact directly with astronauts in orbit, fostering interest in space exploration.

The mission also included tests of the Standard Interface Rack (SIR) and the Portable In-flight Landing Operations Trainer (PILOT). The SIR test, conducted by Searfoss, demonstrated the feasibility of removing and reintegrating equipment into different rack locations during orbital operations. Meanwhile, the PILOT system, a laptop computer simulator controlled by a joystick, was tested to help the mission commander and pilot maintain proficiency for landing during longer-duration Shuttle flights.

On October 23, 1993, the crew focused on

the metabolic studies of the 48 rodents onboard. Seddon, Wolf, Lucid, and Fettman drew blood samples and injecting isotopes to measure plasma volume and red blood cell counts. These experiments aimed to understand how microgravity affects the physiology of living organisms, providing a vital comparison between human and animal responses to spaceflight.

As the mission neared its conclusion, the crew continued collecting scientific data while preparing for their return to Earth. On October 27, 1993, Pilot Rick Searfoss conducted maneuvers as part of the Orbital Acceleration Research Experiment (OARE), which sought to measure the aerodynamic forces acting on the Shuttle in orbit. This data was essential for future Spacelab microgravity research flights, ensuring that experiments could be conducted in the most stable environment possible.

The crew also took time to acknowledge events on Earth, with Searfoss capturing infrared images of wildfires in southern California. The crew expressed their concern for the firefighters and residents affected by the fires, with Searfoss noting that the photographs would be valuable for various scientific disciplines upon their return.

A lighter moment during the mission occurred when Payload Commander Rhea Seddon surpassed her husband, Astronaut Office Chief Robert L. Gibson's total time in space, teasing him in a special message sent down to Earth. Despite surpassing his time in orbit, Seddon acknowledged that her husband still held the record for more launches and landings.

STS-61

STS-61 marked a pivotal moment in the history of NASA's Space Shuttle program as the first mission to service the Hubble Space Telescope (HST). Launched on December 2, 1993, from Kennedy Space Center (KSC) in Florida aboard the Space Shuttle *Endeavour*, the mission aimed to correct a critical flaw in Hubble's optical system. This flaw, known as spherical aberration, had marred the telescope's ability to capture clear images since its

deployment in April 1990 aboard STS-31. The STS-61 mission was not only about rectifying this issue but also upgrading key instruments to enhance Hubble's functionality.

Richard O. Covey, commander; and mission specialists Jeffrey A. Hoffman, and Thomas D. Akers. Seated left to right are Kenneth D. Bowersox, pilot; Kathryn C. Thornton, mission specialist; F. Story Musgrave, payload commander; and Claude Nicollier, mission specialist.

The mission was an intricate and highly demanding endeavor, involving the installation of the Corrective Optics Space Telescope Axial Replacement (COSTAR) system to correct the optical flaw, and the replacement of the Wide Field/Planetary Camera with an improved version. Additionally, the crew replaced the telescope's solar arrays with newer, more efficient ones, further enhancing its capabilities. The complexity of these tasks made STS-61 one of the most challenging missions in Shuttle history, requiring extensive preparation and coordination.

The mission lasted nearly 11 days and included an unprecedented five spacewalks—a record at the time. The demanding schedule allowed for two additional contingency spacewalks, though unnecessary. To manage the workload and prevent fatigue, the spacewalks alternated between two astronauts: Story Musgrave and Jeffrey A. Hoffman, and Kathryn C. Thornton and Thomas Akers. The crew executed these extravehicular activities (EVAs) flawlessly, with each spacewalk contributing significantly to the mission's overall success.

The EVAs began with Musgrave and Hoffman, who performed the first spacewalk on December 5, 1993. Lasting 7 hours and 54 minutes, they successfully installed the corrective optics package. On December 6, Thornton and Akers conducted the second EVA, which lasted 6 hours and 36 minutes, focusing on replacing the solar arrays. Musgrave and Hoffman resumed duties on December 7 for the third spacewalk, which lasted 6 hours and 47 minutes, to continue work on the telescope's instrumentation. Thornton and Akers followed with the fourth EVA on December 8, lasting 6 hours and 50 minutes, and Musgrave and Hoffman concluded the spacewalks on December 9 with a 7-hour, 21-minute EVA, cementing the mission's triumph.

Despite the demanding schedule, the astronauts engaged in some light-hearted activities. Hoffman, in a moment of cultural outreach, spun a dreidel aboard the Shuttle in celebration of Hanukkah, broadcasting this moment to a global audience.

The crew of STS-61 consisted of seven astronauts, each bringing substantial experience to the mission. Commander Richard O. Covey led the flight, marking his fourth and final mission, with Ken Bowersox as Pilot on his second mission. Mission Specialist Kathryn C. Thornton, also on her third mission, played a vital role alongside Claude Nicollier, an ESA astronaut from Switzerland, who served as the flight engineer. Rounding out the crew were veterans Jeffrey A. Hoffman and Story Musgrave, both on their fourth and fifth missions, respectively, and Thomas Akers, on his third spaceflight.

The mission began with some technical challenges. Endeavour was initially scheduled to launch from Pad 39A, but a windstorm on October 30, 1993, caused contamination at the site, prompting a switch to Pad 39B. The contamination did not affect the Hubble payload, which had been securely sealed. Another technical issue emerged on November 18, when a hydraulic actuator on the shuttle's elevon experienced a transducer failure. While replacing the actuator would have delayed the mission, NASA opted to proceed with the launch after depinning the malfunctioning transducer.

After a weather-related scrub on December 1, 1993, the shuttle launched successfully the next day. Weighing 113,541 kg (250,315 lbs), *Endeavour* carried a payload mass of 8,011 kg (17,661 lbs). Upon reaching orbit, the crew began a series of burns to align their trajectory with the Hubble Space Telescope. By Flight Day 3, astronaut Jeffrey Hoffman sighted Hubble, observing that one of the solar arrays was bent at a 90-degree angle, a problem that would be rectified during the mission.

The approach to Hubble was executed with precision. After a series of burns to adjust speed and trajectory, the shuttle rendezvoused with the telescope. At 08:48 UTC on December 4, 1993, astronaut Claude Nicollier used the Canadarm to capture the telescope, bringing it into the shuttle's payload bay for servicing. Once secured, the crew performed visual inspections and prepared for the first spacewalk to begin Hubble's upgrades.

Controllers at NASA's Goddard Space Flight Center had already stowed Hubble's high-gain antennas in preparation for the mission, though minor issues with antenna latches were noted. Despite this, the antennas remained stable, and the team proceeded with the mission without further complications.

On Flight Day 4 of the STS-61 mission, astronauts Story Musgrave and Jeffrey A. Hoffman began their first extravehicular activity (EVA) approximately one hour ahead of schedule, stepping into the cargo bay at 03:46 UTC. Their mission: to conduct critical repairs and upgrades to the Hubble Space Telescope (HST). The first task involved unpacking tools, safety tethers, and work platforms. Hoffman then installed a foot restraint on the end of the shuttle's remote manipulator arm, known as the Canadarm, securing himself for the upcoming work. With astronaut Claude Nicollier controlling the arm from inside the shuttle, Hoffman was maneuvered around the telescope while Musgrave began preparing Hubble for its upgrades.

Musgrave's first task was to install protective covers on Hubble's aft low-gain antenna and on exposed voltage-bearing connector covers, ensuring the safety of both the astronauts and the telescope's systems during the repair work. Next, they opened the Hubble's equipment bay doors to access the interior. Inside, Hoffman installed another foot restraint, securing himself as he began the intricate task of replacing the telescope's Rate Sensing Units (RSUs), which contain gyroscopes essential for keeping Hubble precisely oriented.

The RSUs were critical to the telescope's operation, and Hoffman worked methodically to replace two sets of units. By 17:24 UTC, he had successfully replaced RSU-2, containing gyroscopes 2–3 and 2–4, followed by RSU-3, which housed gyroscopes 3–5 and 3–6. These upgrades were crucial to ensuring that Hubble would continue to operate with pinpoint accuracy. After replacing the RSUs, Musgrave and Hoffman turned their attention to replacing a pair of electrical control units (ECUs) — ECU-3 and ECU-1 — responsible for controlling RSU-3 and RSU-1. In addition, the astronauts replaced eight fuse plugs to protect the telescope's electrical circuits, restoring Hubble to a full set of six fully operational gyroscopes.

However, the operation did not go entirely smoothly. The astronauts encountered difficulty when two of the four bolts on the gyro access doors failed to reset after installing the new gyroscope packages. Engineers on the ground hypothesized that temperature fluctuations caused by the opening and closing of the doors may have caused the metal to expand or contract, preventing the bolts from resetting properly. Musgrave and Hoffman, with assistance from the engineers, used a coordinated approach to solve the issue. An anchored by a payload retention device, Musgrave applied force to the bottom of the door, while Hoffman, secured to the Canadarm, worked at the top. By applying pressure simultaneously at both ends, they could finally secure the latches and complete the installation.

Following the RSU and ECU replacements, Musgrave and Hoffman prepared for the next phase of the mission: the replacement of Hubble's two solar arrays, scheduled for the second spacewalk. They set up the payload bay for their fellow mission specialists, Tom Akers and Kathy Thornton, who would carry out the solar array replacement. To assist with the task, Musgrave and Hoffman prepared the solar array carrier, located at the forward end of the cargo bay, and attached a foot restraint to the telescope to aid in the upcoming operation.

Musgrave and Hoffman's EVA lasted 7 hours and 50 minutes, making it the second longest spacewalk in NASA's history at that time, surpassed only by the 8-hour 29-minute spacewalk during STS-49 in May 1992, during *Endeavour's* maiden flight. Though subsequent spacewalks would surpass these durations, the significance of this EVA was immense, contributing to the success of the mission.

Despite a slight kink in one of the solar arrays — roughly a panel and a half from the end — Hubble program managers and flight controllers decided to proceed with the original plan to stow the arrays. The stowage process involved two steps: first, rolling up the solar arrays, followed by folding them against the telescope for transport. Each solar array, mounted on a four-foot mast, extended 12 meters (39 feet) in length and 2.5 meters (8 feet 2 inches) in width, generating 4.5 kilowatts of power to fuel the telescope's scientific instruments.

On December 6, 1993, Flight Day 5 of the STS-61 mission aboard *Endeavour* began at 15:35 UTC. The astronauts, Thomas D. Akers and Kathryn C. Thornton, commenced the second extravehicular activity (EVA) to replace the Hubble Space Telescope's (HST) aging solar arrays. Thornton's spacesuit had red dashed stripes to distinguish the spacewalkers, while Akers' had diagonal red dashed stripes. The spacewalk started with a minor issue— Thornton's suit had abnormally low pressure in the vent garment, which was quickly resolved once a suspected ice plug melted. Additionally, her communications receiver malfunctioned, preventing direct contact with Mission Control. Instead, the team decided to relay commands through Akers, allowing the EVA to proceed.

Akers initiated the operation by installing a foot restraint on the Canadarm for Thornton, followed by disconnecting the electrical connectors and clamp assembly on the solar array. Despite minor difficulties with the clamp, the process was completed by 04:17 UTC.

Thornton held the large solar array, preventing it from drifting into space. Weighing 160 kg (350 lb) and measuring 5 meters (16 feet) when folded, the array was carefully dismounted at 04:40 UTC while *Endeavour* passed over the Sahara. During the next daylight pass, Thornton jettisoned the array over Somalia at 04:52 UTC. The precise daytime release allowed flight controllers to track its position and speed, with the array gradually moving away from the shuttle at 1.5 m/s (4.9 ft/s), increasing the distance by 18-19 km (11-12 mi) with each orbit.

The astronauts then successfully installed a new solar array, completing the task by 06:40 UTC. After rotating the telescope by 180 degrees, they replaced the second array, stowing the old one for return to the European Space Agency. Following the 6.5-hour EVA, functional tests were performed on four of HST's six gyroscopes, critical to the telescope's orientation control.

On December 7, 1993, at 03:34 UTC, the third spacewalk began while *Endeavour* orbited over Australia. Astronauts Story Musgrave and Jeffrey Hoffman were tasked with installing a new Wide Field and Planetary Camera 2 (WFPC2), a critical upgrade to the Hubble Telescope. Hoffman, assisted by the Canadarm operated by Claude Nicollier, prepared the old camera for removal, while Musgrave worked on opening an access door to monitor the camera's status lights. By 04:48 UTC, the camera was detached and moved into its storage container, while the protective hood on the new WFPC2 was carefully removed. The upgraded WFPC2, weighing 280 kg (620 lb), featured enhanced capabilities, especially in ultraviolet imaging, and its own built-in spherical aberration correction system.

Once installed at 06:05 UTC, ground controllers performed a series of "Aliveness Tests," confirming the new camera was functioning correctly. The team then replaced two magnetometers on HST, responsible for determining the telescope's orientation using Earth's magnetic field. The 6-hour, 47-minute EVA concluded successfully.

The fourth EVA, led by Thornton and Akers, began on December 8, 1993, at 03:13 UTC. Their primary task was to install the Corrective Optics Space Telescope Axial Replacement (COSTAR) system, designed to correct Hubble's spherical aberration caused by its flawed main mirror. After receiving the go-ahead, Akers opened the telescope's aft shroud doors during a night pass to reduce the risk of thermal stress and contamination.

By 04:27 UTC, the High-Speed Photometer (HSP) had been removed, and 05:35 UTC successfully installed COSTAR. The astronauts also upgraded the telescope's onboard computer with additional memory and a co-processor, which passed functional tests. The EVA, lasting 6 hours and 50 minutes, concluded with the telescope being placed in a circular orbit of 596 km (370 mi).

On December 9, 1993, at 03:14 UTC, Musgrave and Hoffman embarked on the mission's final spacewalk. The goal was to replace the Solar Array Drive Electronics (SADE) and ensure the successful deployment of the solar arrays. However, the array failed to rotate upon the first command from ground controllers. Eventually, Musgrave manually cranked the deployment mechanism, and the arrays successfully deployed.

The astronauts then installed covers on the telescope's magnetometers to protect them from debris caused by UV decay. This 7-hour, 21-minute EVA concluded a total of 35 hours and 28 minutes of spacewalking activities for the mission. As a final touch, Hoffman celebrated Hanukkah by spinning a dreidel for viewers back on Earth, bringing a symbolic close to the mission's final spacewalk.

Flight Day 9 began on December 9, 1993, but concerns about one of HST's four onboard Data Interface Units (DIUs) delayed release. Each of the 16 kg (35 lb) DIUs transfer data between the HST's main computer, solar arrays and other critical systems.

A failure on Side A of DIU #2 experienced erratic current fluctuations and some data dropouts. Controllers at the STOCC and mission control developed a troubleshooting procedure to determine the extent of the problem. HST was transferred to internal power and disconnected from its power umbilical at 04:43 UTC.

Controllers then switched channels on the DIU from the A side to the B side and then back to the A side.

They determined HST should be deployed. The drum brakes on the new solar array were applied to prevent them from vibrating during future observations. Claude Nicollier then took hold of the satellite with the Canadarm. The satellite was then lifted and moved away from Endeavour.

The telescope's aperture door was then reopened (a 33-minute procedure) and then released at 10:26 UTC. Commander Dick Covey and pilot Kenneth D. Bowersox fired *Endeavour*'s small maneuvering jets and moved the shuttle slowly away from HST. The Shuttle's landing occurred at Kennedy Space Center on Runway 33 at 05:26 UTC on December 13, 1993.

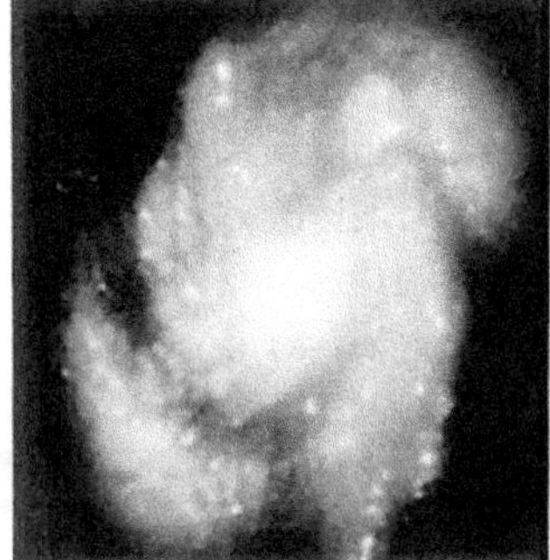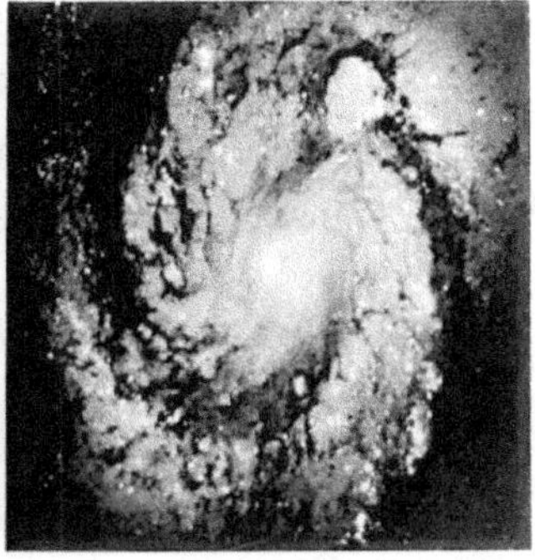

Hubble Telescope images before and after the STS-61 mission

Chapter 5 - International Collaboration: Shuttle-Mir (1994–1997)

STS-60

Astronauts Kenneth S. Reightler Jr., and Charles F. Bolden Jr., pilot and commander, respectively. On middle row are Astronauts Franklin R. Chang-Diaz and N. Jan Davis, mission specialists. On back row are Astronaut Ronald M. Sega (left) and Russia's Sergei K. Krikalev, both mission specialists.

On February 3, 1994, Space Shuttle *Discovery* launched from Kennedy Space Center's Launch Pad 39A, marking the beginning of STS-60, the first mission in the U.S./Russian Shuttle-Mir program. Aboard *Discovery* was Sergei K. Krikalev, the first Russian cosmonaut to fly on a Space Shuttle, underscoring the growing collaboration between the United States and Russia in space exploration. The mission carried several important payloads, including the Wake Shield Facility and the SPACEHAB module, which housed numerous scientific experiments.

The crew, commanded by Charles F. Bolden Jr., included pilot Kenneth S. Reightler Jr. and mission specialists N. Jan Davis, Ronald M. Sega, Franklin Chang-Díaz, and Sergei Krikalev.

Together, they embarked on a mission to conduct a variety of scientific and technological experiments aimed at furthering human knowledge of microgravity and improving future space operations.

After achieving orbit, *Discovery*'s crew opened the payload bay doors and began activating the SPACEHAB laboratory module. Among the experiments was the Organic Separations payload, designed to study cell separation techniques for pharmaceutical purposes, and the Equipment for Controlled Liquid Phase Sintering Experiment (ECLiPSE), which explored the creation of stronger and more durable metals in microgravity. Another critical experiment, Immune-1, investigated the effects of spaceflight on rats' immune systems, while the Commercial Protein Crystal Growth package sought to grow large protein crystals for medical research.

On February 5, a brief incident occurred when *Discovery* passed through a cloud of wastewater ice crystals, though flight controllers determined the cause and no further issues arose. That same day, the crew prepared to deploy the Wake Shield Facility, a free-flying platform designed to create an ultra-high vacuum for growing thin films of semiconductor materials. However, deployment was delayed due to difficulties reading the facility's status lights and interference between its transmitter and the Shuttle's receiver. Despite attempts to resolve the issues, the Wake Shield Facility remained attached to the Shuttle's robotic arm, the Canadarm, for the duration of the mission, and its free-flying operations were ultimately canceled. Nonetheless, it successfully grew five thin films while attached to the arm.

In addition to the WSF, the mission featured the Spacehab module, a pressurized laboratory in the Shuttle's payload bay. Spacehab allowed the crew to conduct multiple scientific experiments in microgravity, ranging from materials science to biological studies. The experiments were designed to leverage the unique space environment to yield results that could not be achieved on Earth.

One of the mission's highlights was testing the Russian-built Shuttle-Mir docking hardware, an essential precursor to future Shuttle-Mir and ISS missions. Although no docking occurred on this flight, the testing provided valuable data to inform the design and operation of the docking mechanisms used in later missions.

STS-60 concluded successfully with Discovery's landing at Kennedy Space Center on February 11, 1994, at 2:19 a.m. EST. The mission lasted eight days, 7 hours, and 9 minutes, orbiting the Earth 130 times and traveling over 3.4 million miles. The successful completion of STS-60 not only achieved its scientific and technical objectives but also symbolized a new era of collaboration between the United States and Russia, setting the stage for more ambitious joint ventures in space exploration and ultimately leading to the construction of the ISS.

"The Wake Shield Facility (WSF) is a free-flying research and development facility that is designed to use the pure vacuum of space to conduct scientific research in the development of new materials. The thin film materials technology developed by the WSF could someday lead to applications such as faster electronics components for computers.

This mission demonstrated the feasibility of international collaboration in space. This theme would become increasingly central to NASA's endeavors as the agency transitioned from the Shuttle era to the era of the International Space Station.

STS-62: A Chronicle of Precision and Progress

STS-62, a pivotal mission in the Space Shuttle program, was flown aboard *Space Shuttle Columbia* and launched on March 4, 1994. This 14-day flight was dedicated to advancing the understanding of microgravity's effects on various physical, biological, and technological processes. The mission was a part of NASA's broader efforts to conduct long-duration research in space, making significant contributions to the growing body of knowledge about the impact of spaceflight on both humans and materials.

mission specialists Charles D. Gemar, Marsha S. Ivins, and Pierre J. Thuot. Seated left to right are Andrew M. Allen, pilot, and John H. Casper, commander.

The primary payload for this mission included the United States Microgravity Payload-02 (USMP-02), a research package designed to investigate the behavior of materials and fluids in a microgravity environment. This set of experiments focused on how weightlessness affects the structure and properties of materials, with implications for manufacturing processes both in space and on Earth. The second major payload was the Office of Aeronautics and Space Technology (OAST-2) package, which carried various engineering and technology experiments to test new technologies for future space missions.

STS-62 also served as a platform for biomedical experiments, as scientists sought to understand better the physiological impacts of long-duration spaceflight on the human body. These experiments, involving both the crew and specialized equipment, focused on cardiovascular health, muscle atrophy, and fluid shifts in the body that occur in the microgravity environment.

A unique aspect of STS-62 was its inclusion of cultural elements. A C.F. Martin backpacker guitar was flown aboard *Columbia*, symbolizing the fusion of science, exploration, and human creativity. Crew members played the guitar during their downtime, reflecting NASA's commitment to maintaining the psychological well-being of astronauts on extended missions.

The crew aboard *Columbia* was a seasoned and diverse group of astronauts. Commander John H. Casper, a distinguished Air Force officer and veteran of the Gulf War, led the mission. Pilot Andrew M. Allen, a former Marine Corps aviator and experienced shuttle pilot, was making his second flight after previously flying on STS-46. Mission Specialist Pierre J. Thuot, an aerospace engineer and naval aviator, brought his expertise in space operations, having previously flown on two shuttle missions. Charles D. Gemar, an Army aviator, was embarking on his third spaceflight, bringing a wealth of experience in shuttle operations. The fifth crew member, Marsha S. Ivins, was an engineer and filmmaker renowned for her precise work on shuttle missions. Ivins was not only an expert in mission logistics but also played a key role in documenting the flight, contributing to the overall understanding of life aboard the Space Shuttle.

On Friday, March 4, 1994, *Space Shuttle Columbia* lifted off from Kennedy Space Center, marking the beginning of STS-62. The primary task of Flight Day One was focused on ascent operations and configuring the orbiter for orbital activities. Shortly after reaching space, the crew performed an Orbital Maneuvering System (OMS) burn, known as OMS-2, to circularize *Columbia's* orbit at approximately 160 by 163 nautical miles (296 by 302 km). The payload bay doors were opened at 10:26 AM EST, allowing the team to begin activating the mission's primary experiments.

The United States Microgravity Payload-02 (USMP-2) was powered up in the orbiter's cargo bay. The Protein Crystal Growth Experiment (PCGE) and the Physiological Systems Experiment (PSE) were initialized, along with other vital systems such as the Active Protein Crystal Growth (APCG) and the Commercial Protein Crystal Growth (CPCG) modules. The Remote Manipulator System (RMS) was also tested, ensuring it was functioning correctly. Additionally, the Cryogenic Gravitational Biology Assembly (CGBA) and other critical payload experiments were activated, marking the start of the mission's scientific operations.

On Saturday, March 5, the crew's second day in orbit began with a focus on exercise and maintaining their physical health, a vital part of any long-duration space mission. Astronauts took turns on the crew cabin's exercise facility, a necessary measure to counter the muscle atrophy resulting from prolonged microgravity exposure. Meanwhile, Pilot Andrew M. Allen and Mission Specialist Charles D. Gemar utilized the Lower Body Negative Pressure (LBNP) device, designed to simulate the effects of gravity on the body's lower extremities, assisting in maintaining cardiovascular health.

Mission Specialists Pierre J. Thuot and Marsha S. Ivins began overseeing the Protein Crystal Growth Experiment (PCGE) and Physiological Systems Experiment (PSE) as these exercises continued. These experiments aimed to provide scientists with valuable data on how proteins and biological systems behave in microgravity. Back on Earth, the Payload Operations Control Center monitored eleven other experiments mounted in the orbiter's payload bay.

The mission team on the ground encountered a minor issue with one of *Columbia's* Auxiliary Power Units (APUs). A pressure sensor in a fuel line indicated higher-than-normal readings, raising concerns about the APU's functionality. The unit was crucial for hydraulic power during landing, although only one of the shuttle's three APUs was necessary for a safe landing. Engineers activated heaters on the unit, which returned the pressure readings to normal. Nevertheless, mission rules dictated that any failure of an APU would lead to a shortened mission, so the team kept a close eye on the system.

Sunday, March 6, brought a mix of scientific research and physical activity. The morning was dedicated to medical studies, with astronauts taking turns on a stationary bicycle mounted in *Columbia's* middeck. This exercise equipment, a staple of shuttle missions, was equipped with shock-absorbing springs to minimize vibrations that could interfere with sensitive experiments aboard the orbiter. Later in the day, Mission Specialist Charles D. Gemar set up a model of a scaffold-like truss structure designed to test the behavior of future space station components in a zero-gravity environment. This model, connected to recorders inside the shuttle, was tested in various configurations to study how similar structures would perform in orbit.

The crew also captured stunning photographs of the "shuttle glow" phenomenon, a visual effect caused by the orbiter's outer skin interacting with atomic oxygen in the upper atmosphere. Simultaneously, they monitored the progress of the protein crystal growth experiments while scientists on the ground collected and analyzed data from the USMP-2 experiments in the payload bay.

One of the day's key experiments was the Critical Fluid Light Scattering Experiment (ZENO), which aimed to locate the critical temperature of xenon, a point at which the fluid exists simultaneously as a gas and a liquid. The team on the ground closely watched data from the experiment and pinpointed the critical temperature. This breakthrough allowed them to study the behavior of xenon at its critical point, providing insights into fluid dynamics that have applications in both physics and material science.

Meanwhile, the Space Acceleration Measurement System (SAMS) continued monitoring the orbiter's microgravity environment. SAMS collected data during activities that could potentially disturb the microgravity environment, such as crew exercises and antenna movements. This data was crucial in ensuring that external disturbances did not affect experiments. A related system, the Orbital Acceleration Research Experiment (OARE), focused on low-frequency accelerations, such as the shuttle's interaction with the rarefied upper atmosphere.

Other notable experiments included the Isothermal Dendritic Growth Experiment (IDGE), which studied how gravity affects the solidification of molten materials, and the Advanced Automated Directional Solidification Furnace (AADSF), which explored the growth of semiconductor crystals in microgravity. The AADSF experiment team closely monitored the solidification of a mercury cadmium telluride crystal. This process benefited from the absence of convective movements that would otherwise cause defects in the crystal structure. These findings were expected to have significant implications for developing advanced electronic devices.

Additionally, the MEPHISTO team conducted studies on the directional solidification of metals, specifically bismuth-tin alloys. Although a minor issue with one of the measurements was encountered, the team was able to troubleshoot the problem and continued gathering data.

Flight controllers in Mission Control reported a relatively quiet day aboard *Columbia*, with no major issues. The earlier concerns regarding high pressure in the APU fuel line had subsided, and confidence grew that the unit would perform as needed. The crew ended their day with eight hours of rest, ensuring they were prepared for the continued scientific investigations ahead.

Flight Day 4 of the STS-62 mission began at 12:53 AM on Monday, March 7, 1994. The day commenced with a medley of armed forces anthems sung by the U.S. Military Academy Glee Club, honoring the crew's service across all four branches of the U.S. military. At the time, Commander John Casper was a colonel in the U.S. Air Force, Pilot Andrew Allen a major in the U.S. Marine Corps, Mission Specialist Charles Gemar a lieutenant colonel in the U.S. Army, and Mission Specialist Pierre Thuot a commander in the U.S. Navy.

After completing their post-sleep routines, the crew focused on the day's scientific tasks. Among the primary activities were checking the Protein Crystal Growth Experiment (PCGE) and monitoring the rodents housed on the middeck for the Physiological Systems Experiment (PSE). These studies were crucial for understanding biological processes in microgravity, especially regarding protein formation and physiological changes in living organisms.

Mission Specialist Charles Gemar continued work on the Middeck 0-Gravity Dynamics Experiment (MODE), which investigated hybrid structures' non-linear, gravity-dependent behavior. This research was significant for designing large structures in space, such as those planned for the International Space Station.

Commander John Casper's presentation on the Space Acceleration Measurement System (SAMS) was a highlight of the day. SAMS, a frequent flyer on shuttle missions, used accelerometers to measure minute vibrations and accelerations that could affect sensitive microgravity experiments. These measurements allowed scientists to fine-tune their experiments to minimize the impact of disturbances and maximize the accuracy of their results.

While much of the crew remained busy with the mission's objectives, Pilot Andrew Allen and Mission Specialist Charles Gemar enjoyed a half-day off from their schedule—a break designed to alleviate the mental and physical strain of the long 14-day mission.

The scientific work continued in earnest, with ongoing research involving the Middeck 0-Gravity Dynamics Experiment and a model of a scaffold-like truss structure, which was analyzed for its behavior in the weightless environment. The truss structure, under consideration for use on future space stations, floated free in the middeck as the crew studied its dynamics.

Meanwhile, the United States Microgravity Payload-2 (USMP-2) experiments continued around the clock, monitored by scientists on the ground. Among the noteworthy ongoing projects was the Space Shuttle Backscatter Ultraviolet (SSBUV) instrument, which had been operating since the first day of the mission. On Day 4, ground controllers planned to use SSBUV to detect sulfur dioxide emissions from Central American volcanoes, testing the instrument's capability to observe atmospheric changes from orbit. More broadly, SSBUV's data was instrumental in calibrating satellites that monitored ozone and other atmospheric gases.

Progress in the Critical Fluid Light Scattering Experiment (ZENO) also made headlines on Flight Day 4. The team reported overnight that they had observed behaviors in the fluid xenon sample that had never been seen on Earth. These findings suggested the experiment had successfully passed through xenon's critical point, where it exists simultaneously as a gas and a liquid. Scientists began preparing for a series of precise measurements at this critical point, using laser light scattering to capture fluctuations in density that could deepen their understanding of fluid dynamics in microgravity.

Meanwhile, the MEPHISTO furnace team achieved breakthroughs in their metal solidification studies. The team successfully completed a Seebeck measurement, a critical process that monitors temperature variations at the solidification front where liquid metal transitions into a solid. The MEPHISTO experiment studied the solidification behavior of a bismuth-tin alloy, providing valuable data on the crystallization process in microgravity, which has important applications for materials science and semiconductor manufacturing.

The Isothermal Dendritic Growth Experiment (IDGE) continued its investigations into the formation of dendrites—branching crystal structures that grow as a material solidifies. The team performed multiple growth cycles, each at different levels of supercooling, with two 35mm cameras capturing detailed images of the process. The experiment's goal was to better understand how dendritic growth is affected by the absence of gravity, and the data collected would contribute to the fundamental understanding of solidification processes.

The Advanced Automated Directional Solidification Furnace (AADSF) remained operational, steadily growing a crystal of mercury cadmium telluride, a material used in infrared detectors. Microgravity allowed the crystal to form without the convective movements caused by gravity on Earth, which can lead to imperfections in the crystal structure. By studying the growth of semiconductor crystals in space, scientists hoped to unlock new methods for producing high-quality materials more efficiently and at lower costs.

As the day concluded, the crew settled into an eight-hour sleep period at 4:53 PM EST, resting in preparation for the upcoming challenges and scientific discoveries of Flight Day 5.

The crew's efforts with USMP-2 operations continued to yield important scientific results. The ZENO experiment remained a focal point, as researchers fine-tuned the temperature manipulations of the xenon sample to capture critical point phenomena. These observations provided valuable insights into the behavior of fluids at extreme conditions, a field of study with broad implications for both basic physics and practical applications in fluid dynamics.

Likewise, the AADSF and MEPHISTO teams continued their work, using the microgravity environment to refine their understanding of solidification processes. The ongoing experiments with semiconductor and metal crystal growth were crucial for advancing materials science, with the data gathered from these experiments informing future technological innovations on Earth.

On Tuesday, March 8, 1994, the crew of the Space Shuttle *Columbia* continued their daily regimen, engaging in exercise, photography, and monitoring experiments related to crystal growth and bioprocessing aboard the orbiter. These activities were critical for understanding how the microgravity environment influences biological and physical processes. Ground-based researchers, remotely controlling experiments housed in Columbia's payload bay, maintained constant observation and data collection.

One key instrument, the Space Shuttle Backscatter Ultraviolet (SSBUV), was used to study Earth's atmosphere. Researchers collected valuable data on tropospheric emissions from Mexican and Central American volcanoes, sulfur dioxide from industrial byproducts above China and Japan, and upper-atmosphere phenomena above Mexico's Colima volcano. These studies enhanced understanding of atmospheric chemistry and global pollution patterns.

Additionally, *Columbia* carried the Office of Aeronautics and Space Technology-2 (OAST-2) payload, which included various experiments to advance materials technology for future spacecraft. The SAMPIE experiment exposed newly designed materials to the harsh environment of space, examining plasma interactions and the performance of advanced solar energy cells. Other significant achievements included ten freeze-thaw cycles of a new cooling technology and airglow spectrometer readings in the upper atmosphere. The SKIRT instrument, which measured atomic oxygen interactions with the orbiter's surface, contributed to understanding how spacecraft materials deteriorate in low Earth orbit.

Despite the intensity of their work, the crew members—Commander John Casper, Mission Specialist Pierre Thuot, and Mission Specialist Marsha Ivins—enjoyed a half-day of rest. The remainder of the mission proceeded smoothly, with *Columbia* maintaining a stable orbit at altitudes between 298 and 302 kilometers. At 2:53 PM CST, the crew began an eight-hour sleep period, awakening later that evening to begin their sixth day in space.

On Wednesday, March 9, the *Columbia* crew dedicated their efforts to secondary experiments conducted on the Shuttle's middeck. Astronaut Tom Gemar continued working with the Middeck 0-Gravity Dynamics Experiment (MODE), which focused on studying how objects behave in a zero-gravity environment. Meanwhile, fellow astronaut Charles D. Gemar took a break to communicate with reporters from Cleveland, Philadelphia, and Knoxville, discussing the medical tests conducted before, during, and after the flight. These tests involved collecting blood and urine samples to monitor physiological changes in space, such as chemical regulatory responses, balance, and exercise capacity.

Throughout the day, the crew also monitored ongoing protein crystal growth experiments and photographed auroral phenomena, checking *Columbia*'s windows for any signs of debris impacts. In the evening, they exercised using the Shuttle's ergometer to counteract the effects of prolonged weightlessness on their bodies.

Meanwhile, back on Earth, Spacelab Mission Operations Control at the Marshall Space Flight Center reported the continued success of the United States Microgravity Payload-2 (USMP-2) experiments. One notable achievement came from the Critical Fluid Light Scattering Experiment, or ZENO, which identified the critical point of xenon. In this state, the element exhibits properties of both a liquid and a gas. Scientists were eager to conduct further optical measurements in this region to deepen their understanding of fluid behavior in microgravity.

The Advanced Automated Directional Solidification Furnace (AADSF) experiment, focused on growing a single crystal of mercury cadmium telluride, continued to yield promising results. Scientists monitoring the growth process reported exceptional progress, adding valuable data to the field of materials science. Likewise, the Isothermal Dendritic Growth Experiment (IDGE), which studied crystal formation in microgravity, received high praise from its research team. Slow-scan video images of dendrites growing within the apparatus were carefully monitored to optimize the experiment's results.

The Space Acceleration Measurement System (SAMS), which continuously recorded vibration levels aboard *Columbia*, ensured that the Shuttle provided a stable platform for these sensitive experiments. The data collected from SAMS helped refine future experiments and spacecraft designs to minimize onboard disturbances.

On Thursday, March 10, the mission marked a personal milestone for Commander Casper, who was informed of his promotion to Lieutenant Colonel in the United States Marine Corps. The crew celebrated this news while continuing their scientific duties.

Later that day, *Columbia* altered its orientation for the first time since launch, maneuvering to a position where the orbiter's tail pointed toward Earth and its payload bay faced forward into the direction of travel. This new orientation allowed for the exposure of materials in the Long Duration Space Environment Candidate Materials Exposure (LDCE) experiment. The LDCE experiment exposed 264 samples of various materials to the space environment under different conditions. One set of materials remained exposed throughout the mission, while others were selectively exposed depending on the Shuttle's orientation, providing valuable insights for future spacecraft material development.

Mission Specialist Marsha Ivins conducted a live interview with students at the Bronx High School of Science, answering questions about the microgravity experiments aboard *Columbia* and providing an insider's view of life in space. Meanwhile, astronauts Tom Gemar and Charles Allen conducted medical tests using the lower body negative pressure unit, designed to study the effects of spaceflight on the human circulatory system.

By Friday, March 11, *Columbia* had reached the midpoint of its mission. Commander Casper initiated routine checks of the Shuttle's environmental control systems, switching several critical systems—such as the humidity separator, cabin pressure, and temperature control—to their backup modes. This procedure was essential for testing the spacecraft's redundancy and ensuring the crew's safety in case of system failures.

Payload bay operations continued as the Shuttle reoriented itself, and crew members monitored protein crystal growth and rodent experiments. Flight controllers on the ground finalized plans to uplink additional digital video to the crew, enhancing real-time communication and data exchange between Mission Control and the Shuttle. The crew concluded their day with a scheduled sleep period, preparing for the final leg of their journey aboard *Columbia*.

On Saturday, March 12, *Columbia's* crew focused on a series of scientific experiments, including the Auroral Photography Experiment, the Commercial Protein Crystal Growth experiment, and the Limited Duration Space Environment Candidate Exposure (LDCE) experiment. These experiments were vital for understanding the effects of space on biological and physical processes, and for furthering space research in areas like protein crystallization, which could benefit medical and pharmaceutical industries on Earth.

Later in the day, the crew employed the Shuttle's Remote Manipulator System (RMS) to troubleshoot a reception issue with the Experimental Investigation of Spacecraft Glow (EISG) instrument located in the payload bay.

The RMS's end effector camera provided a valuable bird's-eye view, allowing the crew to observe and monitor the instrument in operation. The EISG experiment studied the faint glow that spacecraft experience due to interactions with the atomic oxygen and other particles present in low Earth orbit. This phenomenon affected spacecraft surfaces and instruments.

Sunday, March 13, was a relatively light day for *Columbia's* crew, allowing them to enjoy some well-deserved rest. The astronauts spent the first half of the day off before focusing on middeck experiments during the second half. These experiments provided valuable data on microgravity's effects on various materials and biological processes.

During an in-flight news conference, the crew fielded questions on topics ranging from budget constraints and safety concerns to experimentation aboard the Shuttle and the anticipated construction of the International Space Station (ISS). Their responses highlighted the challenges and opportunities of space exploration and the critical role of the Shuttle in laying the groundwork for future space stations.

Meanwhile, activities at Mission Control focused on preparing updated flight plans and uplinking new messages to the crew, outlining changes for Flight Day 11. The crew ended their day with an eight-hour sleep shift, resting in preparation for another busy day in orbit.

On Monday, March 14, *Columbia's* crew awoke to the song "Starship Trooper" by Yes, signaling the start of another day of scientific and operational work. The day's primary focus was a series of two Orbital Maneuvering System (OMS) burns designed to lower the Shuttle's orbit and shift the mission's focus to a new set of scientific goals. The first burn (OMS-3) lowered *Columbia's* orbit to 259 by 291 kilometers, and the second (OMS-4) further refined the orbit to a near-circular 257 by 259 kilometers.

This descent allowed the Shuttle's instruments to engage in more detailed studies of the spacecraft's interaction with atomic oxygen, nitrogen, and other gases in the lower orbit. These gases produce a glow around spacecraft surfaces, a well-documented phenomenon that researchers continued to investigate.

Observations from the Office of Aeronautics and Space Technology-2 (OAST-2) instruments took center stage for the remainder of the mission, including a three-minute release of nitrogen gas from the Shuttle's payload bay to study its effects on the spacecraft's glow. This experiment utilized materials that could one day be used in future satellites, helping to improve material durability in space.

Additionally, the crew executed a 25-minute series of 360-degree spins with the Shuttle's tail pointed toward Earth. This maneuver allowed the Spacecraft Kinetic Infrared Test (SKIRT) instrument to observe the effects of space on different surfaces and materials, providing valuable data for the development of future spacecraft.

Mission Specialists Marsha Ivins and Tom Gemar evaluated the Dexterous End Effector (DEE) experiment, which tested a tracking system for Columbia's RMS. Using a mirror, light-emitting diodes, and a camera mounted in the payload bay, the DEE experiment tested the RMS's accuracy in aligning with specific targets. Additionally, the system measured forces generated by the arm's movements when its magnetic end effector was engaged, with a force torque sensor recording the data.

As part of their daily routine, the astronauts continued their exercise regimen and monitored the remaining experiments onboard. The crew ended their 11th day in space with an eight-hour sleep period, confident that *Columbia* and its systems were in excellent condition as they orbited Earth at 260 kilometers.

Before retiring for the night, the crew sent a special goodnight message to Mission Control in Houston, playing Bette Midler's "From a Distance" as a tribute to the team overseeing their mission. The day marked a transition from the United States Microgravity Payload-2 experiments to a renewed focus on the OAST-2 payload, underscoring the versatile scientific capabilities of the Space Shuttle.

On Tuesday, March 15, the crew of *Columbia* was greeted with the song "View From Above," written and performed by Allison Brown. Inspired by Mission Specialist Marsha Ivins, the song set the tone for a day filled with scientific experiments and technological evaluations. The primary activities of the day centered on the continued operation of the Dexterous End Effector (DEE) and the Experimental Investigation of Spacecraft Glow (EISG) experiments, as well as further use of the Lower Body Negative Pressure (LBNP) device.

Ivins, along with fellow astronauts Pierre Thuot and Tom Gemar, took turns testing the Shuttle's Remote Manipulator System (RMS) as part of the DEE experiment. The crew tested the 50-foot-long arm by inserting pins into sockets with progressively smaller clearances, ranging from 3 millimeters to a tight 0.76 millimeters. They also moved a flat beam within a slot to measure forces detected by a sensor, providing valuable data on the arm's precision and dexterity.

Meanwhile, Gemar and Charles Allen completed sessions in the LBNP device, which simulated the effects of gravity on the body's circulatory system. These medical experiments contributed to understanding how long-duration spaceflight impacts human physiology.

The Office of Aeronautics and Space Technology-2 (OAST-2) payload took precedence in the Shuttle's payload bay, with crew members assisting investigators of the EISG experiment. They positioned the RMS camera to observe the sample plate during operations. Despite a low-light camera failure earlier in the mission, the crew continued gathering data on how nitrogen gas releases impacted the Shuttle's glow. This phenomenon occurs as atomic oxygen interacts with spacecraft surfaces in orbit.

Throughout the day, the Space Shuttle Backscatter Ultraviolet (SSBUV) instrument also continued taking atmospheric readings, helping calibrate satellites monitoring Earth's ozone levels. At 2:08 AM CST the following day, *Columbia* conducted a fifth Orbital Maneuvering System (OMS) burn, lowering its perigee to 194 kilometers, enabling more measurements of the spacecraft glow effect.

On Wednesday, March 16, *Columbia* executed another orbital change with an OMS-5 burn, lowering the orbit to 105 by 138 nautical miles. This adjustment facilitated continued observations of the glowing effect caused by atomic oxygen interactions in the lower orbit. The glowing effect, more pronounced at this altitude, provided invaluable data for understanding the long-term impact of low Earth orbit conditions on spacecraft materials.

As the crew continued their daily tasks, mission focus shifted toward preparing for reentry. Commander John Casper and Pilot Charles Allen performed standard checks on the orbiter's systems, particularly focusing on the flight control systems. The crew used Auxiliary Power Unit (APU) 3 to test the hydraulic systems, which are critical for landing. Despite concerns earlier in the mission over high pressure readings in a fuel line, the APU functioned normally.

The day also included the operation of the Commercial Generic Bioprocessing Apparatus (CGBA) and the Commercial Protein Crystal Growth (CPCG) experiment. Housed in the Shuttle's middeck, these experiments aimed to advance the understanding of protein crystallization and biological processes in space, laying the groundwork for potential medical advancements.

Later, the RMS was employed once again for the DEE experiment, as the crew tested the system's magnetic grapple and alignment capabilities, utilizing force sensors to gather data on arm movements. The crew also activated the Limited Duration Candidate Materials Exposure (LDCE) experiment, exposing materials to the harsh conditions of low Earth orbit for further study.

The day ended with *Columbia* orbiting at a high point of 260 kilometers and a low point of 194 kilometers, while the crew was awakened to the song "Traveling Prayer" by Billy Joel.

Thursday, March 17, marked the final full day of operations aboard *Columbia*. The crew's tasks included a hot firing of the Reaction Control System (RCS) to prepare for reentry, checks of the Shuttle's flight control system, cabin stowage, and deactivation of the SSBUV instrument. The day's wake-up song, "Living in Paradise" by the Brothers Cazimero, accompanied the crew as they began stowing gear and wrapping up their scientific experiments.

Commander Casper and Pilot Allen test-fired *Columbia's* primary steering jets early in the morning, confirming they were in good condition for reentry. The astronauts also spent time practicing landings using a computer simulation designed for the Shuttle, ensuring they were ready for the upcoming descent. Meanwhile, Tom Gemar underwent his final session in the LBNP device, contributing to ongoing research on the effects of spaceflight on the human body.

Marsha Ivins powered down and secured the RMS in its cradle, and Thuot concluded operations of the protein crystal growth experiments, preparing them for landing. Throughout the day, the crew conducted final observations of the Shuttle glow effect, performing a series of spins and nitrogen gas releases to gather more data on how the spacecraft interacted with the gases in low Earth orbit.

By the end of the day, *Columbia* was in an orbit with a high point of 257 kilometers and a low point of 194 kilometers. The crew completed their preparations for the mission's conclusion and readied themselves for reentry.

Friday, March 18, marked the end of *Columbia's* successful mission. The crew began preparations for reentry with a deorbit burn scheduled for 6:18 AM CST. This burn, planned to reduce velocity by 209 feet per second, initiated the Shuttle's descent toward Earth.

At approximately 8:10 AM EST, *Columbia* touched down smoothly on Runway 33 of the Kennedy Space Center's Shuttle Landing Facility. The landing marked the successful conclusion of the STS-62 mission, which had provided significant scientific contributions, from advancing materials science to studying the physiological effects of space on the human body. The data gathered would go on to inform future space missions, including those involving the International Space Station.

STS-59

STS-59, flown by Endeavour, was a pivotal mission dedicated to Earth science. It focused on studying the planet's surface, atmosphere, and climate. The mission, part of NASA's Mission to Planet Earth program, carried the Spaceborne Imaging Radar-C and X-band Synthetic Aperture Radar (SIR-C/X-SAR) payload, marking a significant advancement in radar imaging technology. This mission contributed critical data to understanding Earth's environmental processes, natural hazards, and global change.

Kevin P. Chilton, pilot; and Sidney M. Gutierrez, commander. Seated left to right are Linda M. Godwin, payload commander, and mission specialists Thomas D. Jones, Jay Apt, and Michael R.

STS-59 launched on April 9, 1994, from Kennedy Space Center's Launch Complex 39A. Sidney M. Gutierrez commanded the mission, with Kevin P. Chilton serving as the pilot. The crew consisted of four mission specialists: Linda M. Godwin, Jay Apt, Michael R. Clifford, and Thomas D. Jones. This was the first spaceflight for Thomas Jones, who would go on to participate in three more Shuttle missions.

The shuttle's six astronauts swiftly began activating the sophisticated radar equipment housed in the payload bay. These instruments were integral to NASA's Space Radar Laboratory-1 experiments, part of the agency's "Mission to Planet Earth," designed to study Earth's ecosystems. By 8 p.m., all the instruments were operational, marking the start of the around-the-clock data collection scheduled for the next ten days.

Key instruments on board included the Spaceborne Imaging Radar-C (SIR-C) and the X-

Band Synthetic Aperture Radar (X-SAR), which were tasked with capturing detailed radar images of Earth's surface. However, engineers encountered an initial issue with the X-SAR's power amplifier, which could not be fully activated due to a problem with its internal low-voltage circuit. This issue was traced to an overly sensitive protection circuit, which functioned like a circuit breaker. After bypassing the circuit, the radar functioned without incident, completing its observations as scheduled. Meanwhile, the Measurement of Air Pollution from Satellite (MAPS) instrument successfully began monitoring atmospheric carbon monoxide levels, providing critical data on the distribution of this pollutant in the troposphere.

By April 10, just over a day into the mission, the radar laboratory had already captured data from over 40 observation sites worldwide. Among the notable targets were Howland, Maine; Macquarie Island; the Black Sea; Matera, Italy; and the Strait of Gibraltar. The mission's highest-priority targets, "supersites," were also being monitored. On that Sunday, these included Duke Forest in North Carolina for global carbon and hydrologic cycles, Austria's Ötztal Alps for hydrologic studies, and Lake Chad in the Sahara for geological research.

As the day progressed, further data was collected from sites including Gippsland in Australia, Toronto in Canada, Mammoth Mountain in California, and the Bighorn Basin in Wyoming. Supersite observations continued, focusing on ecological interactions at Raco, Michigan, and hydrologic cycles near Bebedouro, Brazil. The crew also reported excellent Earth observation photography opportunities, capturing images of the frozen lakes around Raco and wildfires in Mexico's Sierra Madre mountains.

The crew operated in two shifts. The Red Team, led by Commander Sidney M. Gutierrez, along with Pilot Kevin P. Chilton and Mission Specialist Linda M. Godwin, began their rest period at 5 p.m. Central time, while the Blue Team, consisting of Mission Specialists Jay Apt, Michael R. Clifford, and Thomas D. Jones, awakened around 4 p.m. to continue the mission's critical scientific work.

By April 11, the mission's third day, the radar lab had downlinked three real-time images of Earth, including a geological study of the Sahara Desert in Algeria. Collaborations with international scientists were in full swing, with radar systems being calibrated over Matera, Italy, and Oberpfaffenhofen, Germany, in conjunction with students from the University of Munich, who gathered data on soil moisture and agricultural biomass.

That same day, the crew made real-time observations of thunderstorms over Taiwan and dust storms off the northwest coast of Australia. The MAPS instrument continued to deliver valuable data on atmospheric pollution, with scientists on the ground confirming strong correlations between satellite and surface measurements.

Throughout April 11, Endeavour's radar systems gathered data from a wide array of sites. Geology observations were made over Japan and Wyoming's Bighorn Basin, while hydrology studies focused on California's Mammoth Mountain. Ecology data was gathered over Canada's Altona, and oceanography passes included the Gulf Stream and Southern Ocean. Linda M. Godwin reported vivid photography of severe thunderstorms over South America and wind patterns near the Galapagos Islands.

In addition to scientific work, the Shuttle Amateur Radio Experiment allowed the crew to communicate with students and Boy Scouts in the United States. However, a minor issue was reported when air bubbles appeared in the shuttle's galley water supply, although this did not interfere with mission operations.

On April 12, 1994, at 3 a.m. Eastern time, Space Shuttle Endeavour's X-SAR radar system downlinked a real-time image of the Andes Mountains in Bolivia, adding to the growing collection of data gathered during the STS-59 mission. The Blue Team, comprising astronauts Jay Apt, Michael R. Clifford, and Thomas D. Jones, continued their work, capturing radar images of key oceanography sites in the South Pacific, East Australian Ocean currents, and the North Atlantic. The team also focused on geological sites like Cerro Laukaru in Chile, snow cover in the Austrian Alps at Otztal, and the arid landscapes of Ha Meshar in Israel. Ecology sites in Howland, Maine, and Duke Forest, North Carolina, were also monitored during this shift, contributing to the mission's comprehensive study of Earth's ecosystems.

Later that morning, the Red Team, led by Commander Sidney M. Gutierrez and Pilot Kevin P. Chilton, resumed their duties around 7 a.m. Eastern time. Both Gutierrez and Chilton had delayed their sleep schedule by an hour due to late-night work on an in-flight maintenance procedure to address air bubbles in the drinking and food preparation water. By bypassing the galley water outlet and connecting the hose directly to the supply tank, they attempted to eliminate the bubbles. However, tests during the Blue Team's shift indicated that some bubbles still entered the drink bags.

During the Red Team's shift, X-SAR transmitted live radar images from Sarobetsu, Japan, one of the mission's high-priority calibration sites. Scientists on the ground worked to measure radar signal strength and optimize radar swath size for mapping purposes. At the same time, researchers developed topographic maps of Japan and investigated how to best utilize the three radar antennas for monitoring rice fields. Additional images of the Bay of Campeche in the Gulf of Mexico and the land surrounding Veracruz, Mexico, provided valuable ecological data, particularly soil and vegetation information during the dry season in the tropical forests.

Mission Specialist Linda M. Godwin reported a successful cloud-free photography opportunity over Chickasha, Oklahoma—one of the mission's designated "supersites." The crew also observed sea ice along the coast of Russia's Kamchatka Peninsula, enhancing their understanding of polar and oceanographic processes.

On April 13, 1994, at 7 a.m. Eastern time, the Blue Team completed their fifth workday and handed over duties to the Red Team. The Blue Team's shift had included live X-SAR images of the Namib Desert in South Africa, helping scientists study radar backscatter and improve models of surface and subsurface structures. Radar imagery also captured the seasonal sea ice melt in the Sea of Okhotsk, off the coast of Siberia, and data from the drought-stricken Sahel region in Sudan.

During this shift, astronaut Jay Apt made a unique radio contact with Russian cosmonauts aboard the Mir space station. As Endeavour passed over Australia, the two spacecraft were within 1,200 nautical miles of each other. Using amateur radio equipment, Apt and the cosmonauts exchanged greetings, a moment shared with amateur radio operators on Earth.

The Blue Team astronauts also continued participating in biomedical studies, exercising on the bicycle ergometer to gather data on cardiovascular deconditioning during space missions. Clifford had the second half of his workday off, while in-flight maintenance procedures to install a makeshift seal on the galley's water dispenser showed promising results in reducing bubbles in the water supply.

At 10:30 a.m. Central time, the Red Team began their fifth mission shift. Tom Snyder interviewed Commander Gutierrez for CNBC, and later that evening, Clifford participated in an interview for Mutual Radio Network's "Jim Bohannon Show," answering questions from listeners. The crew also reported striking photography opportunities over the lakes of Manitoba, Canada, where the water appeared more "bluish" than expected.

On April 14, 1994, at 3:30 a.m. Central time, Rich Clifford continued engaging with the public, answering questions about spaceflight, mission objectives, and life aboard Endeavour during a 20-minute interview on Mutual Radio. That day, the Blue Team documented a significant fire-scarred area in China that had burned in 1987. This site was particularly interesting to the Measurement of Atmospheric Pollution (MAPS) experiment, which studied forest regrowth following fire events.

Jay Apt enjoyed some off-duty time in the early hours of the day, during which he exercised on the bicycle ergometer, recording his heart rate and perceived exertion for the ongoing biomedical studies. Later, he rejoined his teammates as the mission progressed. Meanwhile, X-SAR's quick-look processor provided moving radar images of the Chickasha site in Oklahoma, capturing data from just north of the Kansas border down to the Texas panhandle. Hydrologists analyzed this data to assess how well the radar could detect fluctuations in soil moisture over time.

On April 15, 1994, Space Shuttle Endeavour continued its data collection mission, with astronaut Tom Jones noting a significant reduction in the pollution cloud over Manila Bay in the Philippines compared to earlier in the mission. At 1:50 a.m. Central time, Jones also reported sightings of fires along Burma's west coast and smoke over Tasmania, contributing to the mission's environmental observations. The MAPS instrument, which had been critical in studying atmospheric pollution, exhausted its supply of infrared film, marking the end of one of its key data collection capabilities.

Later that day, at 6 p.m. Central time, Pilot Kevin Chilton explained to the public how a global network of scientists and students stationed at various observation sites worldwide supported the shuttle's radar operations. The mission was a collaboration between space and ground-based scientists, with real-time data being processed by global teams. Mission Specialist Linda Godwin participated in a live Q&A session, responding to questions submitted by CNN viewers, while the crew continued troubleshooting the issue of bubbles in the water supply, a minor but persistent inconvenience.

By April 16, 1994, the mission's ninth day, astronauts Jay Apt and his Blue Team colleagues—Rich Clifford and Tom Jones—used Endeavour's Shuttle Amateur Radio to speak with fellow astronauts Norm Thagard and Bonnie Dunbar, along with two Russian cosmonauts at the Star City training center near Moscow. These cross-continental conversations, which occurred at 11:30 p.m. and 1:15 a.m. Central time, provided a rare and cherished opportunity for communication between astronauts and cosmonauts.

The Blue Team also made several visual observations, including fires burning across Africa and a line of thunderstorms over northeastern Brazil. At the request of payload scientists, they added Rügen Island, located off the northern coast of Germany, to their list of Earth observation sites, capturing vital photography of the Baltic Sea region. Meanwhile, the instruments aboard the shuttle, including the Space Radar Lab-1 equipment, continued to function smoothly, delivering scheduled observations of various global sites, including regions in Japan and Italy. By this point in the mission, all targeted observation sites had been recorded at least once, with the remaining data collected to supplement previous observations. A persistent annoyance—the presence of air bubbles in the crew's water supply—was finally resolved through a successful in-flight maintenance procedure.

On April 17, 1994, the Blue Team began their workday at 3 a.m. Central time, focusing on radar imaging for scientists studying Earth's land surfaces, water resources, and ecosystems. Their work contributed to a greater understanding of how various elements of Earth's environment interact to sustain life. Later that day, at 12:30 p.m. Central time, the shuttle's flight control surfaces and thruster jets were tested to ensure everything was in working order for the upcoming landing at Kennedy Space Center.

In a moment of personal significance, Commander Sidney M. Gutierrez, Pilot Kevin Chilton, and Mission Specialist Tom Jones participated in a Roman Catholic service of Holy Communion, marking a reflective and spiritual moment two weeks after Easter Sunday. The consecrated wafers were distributed among the three astronauts using a golden pyx. Afterward, the entire crew participated in an in-flight news conference, discussing the significance of the mission and answering questions from media representatives. Following the conference, Gutierrez, Chilton, and Clifford conducted routine checks of Endeavour's systems, while Godwin, Apt, and Jones focused on documenting the ongoing scientific activities.

On April 18, 1994, as the mission neared its conclusion, the crew began packing up their equipment. At 2 p.m. Central time, final radar observations were made while the Space Radar Laboratory continued to gather valuable data. Preparations were underway for a landing scheduled for the next day at 10:52 a.m. Central time. The weather forecast for Florida looked favorable, though flight controllers remained watchful of the possibility of low clouds and slight showers.

However, on April 19, 1994, conditions at Kennedy Space Center forced Endeavour to remain in space for an additional day. Clouds and high winds in the area delayed the landing, and the crew quickly reconfigured the orbiter's systems for the extra day in orbit. A portion of the Space Radar Laboratory payload, specifically the Space Imaging Radar (SIR-C), was reactivated to take advantage of the extended mission time. Throughout the STS-59 mission, the radar system had mapped approximately 70 million square kilometers of Earth's surface, covering both land and sea, amounting to roughly 12 percent of the planet's total surface area. Radar images had been captured for about 25 percent of Earth's land surfaces, providing a vast repository of data equivalent to 20,000 encyclopedia volumes.

On April 20, 1994, Endeavour successfully landed at Edwards Air Force Base in California at 9:54 a.m. Central time, concluding its highly successful 11-day mission. The crew's extensive radar observations and environmental monitoring efforts contributed to an unprecedented understanding of Earth's complex ecosystems and environmental processes.

STS-65: Unraveling the Mysteries of Microgravity

STS-65 was launched aboard *Columbia* from Kennedy Space Center, Florida, on July 8, 1994. Commanded by Robert D. Cabana, who would later become the director of Kennedy Space Center. The crew of seven embarked on a 15-day journey focused on microgravity science, carrying the International Microgravity Laboratory (IML-2), a sophisticated suite of experiments designed to study the effects of near-weightlessness on various physical and biological processes.

The seven-member crew aboard included Commander Robert D. Cabana, an experienced naval aviator and veteran of two previous shuttle flights; Pilot James D. Halsell Jr., an Air Force

test pilot making his first journey into space; Mission Specialist Richard J. Hieb, an engineer on his third spaceflight; Mission Specialist Carl E. Walz, an Air Force Colonel on his second mission; Mission Specialist Leroy Chiao, an engineer and flight test engineer on his first shuttle flight; Payload Commander Donald A. Thomas, a physicist also making his first trip to space; and Japanese Payload Specialist Chiaki Mukai, a physician and Japan's first female astronaut.

Richard J. (Rick) Hieb, payload commander; Robert D. (Bob) Cabana, commander; and Donald A. Thomas, mission specialist. Standing, from left to right, are Leroy Chiao, mission specialist; James D. Halsell, pilot; Chiaki Naito-Mukai, payload specialist; and Carl E. Walz, mission specialist.

The International Microgravity Laboratory (IML-2) was the second in a series of Spacelab (SL) flights designed to conduct research in a microgravity environment. The IML concept enabled a scientist to apply results from one mission to the next and to broaden the scope and variety of investigations between missions. Data from the IML missions contributed to the research base for the space station.[2]

As the name implies, IML-2 was an international mission. Scientists from the European Space Agency (ESA), Canada, France, Germany, and Japan collaborated with NASA on the IML-2 mission to provide the worldwide science community with a variety of complementary facilities and experiments mounted in twenty 19" racks in the IML 2

Module.

Research on IML-2 was dedicated to microgravity and life sciences. Microgravity science covers a broad range of activities from understanding the fundamental physics involved in material behavior to using those effects to generate materials that cannot otherwise be made in the gravitational environment of the Earth. In life sciences research, a reduction of gravitation's effect allows certain characteristics of cells and organisms to be studied in isolation. These reduced gravitational effects also pose poorly understood occupational health problems for space crews ranging from space adaptation syndrome to long-term hormonal changes. On IML-2, the microgravity science and life sciences experiments were complementary in their use of SL resources. Microgravity science tends to draw heavily on spacecraft power while life sciences places the greatest demand on crew time.

Life Sciences Experiments and facilities on IML-2 included: Aquatic Animal Experiment Unit (AAEU) in Rack 3, Biorack (BR) in Rack 5, Biostack (BSK) in Rack 9, Extended Duration Orbiter Medical Program (EDOMP) and Spinal Changes in Microgravity (SCM) in the Center Isle, Lower Body Negative Pressure Device (LBNPD), Microbial Air Sampler (MAS), Performance Assessment Workstation (PAWS) in the middeck, Slow Rotating Centrifuge Microscope (NIZEMI) in Rack 7, Real Time Radiation Monitoring Device (RRMD) and the Thermoelectric Incubator (TEI) both in Rack 3.

Microgravity experiments and facilities on IML-2 included: Applied Research on Separation Methods (RAMSES) in Rack 6, Bubble, Drop and Particle Unit (BDPU) in Rack 8, Critical Point Facility (CPF) in Rack 9, Electromagnetic Containerless Processing Facility (TEMPUS) in Rack 10, Free Flow Electrophoresis Unit (FFEU) in Rack 3, Large Isothermal Furnace (LIF) in Rack 7, Quasi Steady Acceleration Measurement (QSAM) in Rack 3, Space Acceleration Measurement System (SAMS) in the Center Isle, and Vibration Isolation Box Experiment System (VIBES) in Rack 3.

Other payloads on this mission were: Advanced Protein Crystallization Facility (APCF), Commercial Protein Crystal Growth (CPCG), Air Force Maui Optical Site (AMOS) Calibration Test, Orbital Acceleration Research Experiment (OARE), Military Application of Ship Tracks (MAST), Shuttle Amateur Radio Experiment-II (SAREX-II). Columbia flew with an Extended Duration Orbiter (ED0) pallet and no RMS Arm was installed. This was also the 1st flight of the payload bay door torque box modification on *Columbia* and the 1st flight of new OI-6 main engine software.

The second mission in the International Microgravity Laboratory (IML) series, IML-2, launched aboard the Space Shuttle *Columbia* on the STS-65 mission on July 8, 1994. Over 15 days, the crew conducted groundbreaking research on the behavior of materials and life in microgravity, advancing knowledge crucial for future space exploration. *Columbia* safely returned to Earth on July 23, 1994, with a seven-member crew that included Chiaki Mukai, the first Japanese woman in space.

This mission was a globally collaborative effort, with contributions from NASA, the European Space Agency (ESA), and the space agencies of Japan (NASDA), Canada (CSA), Germany (DLR), and France (CNES). Investigators from 13 countries participated in a series of over 80 experiments designed to explore microgravity and life sciences, many of which laid the foundation for future space station research. Among these experiments were five life science investigations spearheaded by American researchers, including studies sponsored by NASA's Ames Research Center and Kennedy Space Center (KSC).

Ames Research Center led two key experiments using newts and jellyfish. The newt experiment sought to understand the early development of gravity-sensing organs, such as the utricle and saccule, present in the inner ears of all vertebrate animals. These organs contain otoliths—tiny calcium carbonate stones—that rest on a gelatinous membrane overlaying sensory hair cells. On Earth, gravity pulls on these otoliths, sending the resulting signals to the brain through nerve fibers. In space, the experiment aimed to determine whether the production of otoliths and the development of associated nerve fibers would be affected by microgravity.

The jellyfish experiment focused on studying both behavioral and developmental processes in space. Scientists observed how the jellyfish swam, pulsed, and oriented themselves in microgravity, particularly interested in the development of their gravity-sensing organs. One goal of this research was to determine how much artificial gravity would be required to counteract any negative effects on their physiological development during spaceflight.

The PEMBSIS experiment, sponsored by Kennedy Space Center, studied plant embryogenesis in space. Researchers aimed to assess whether the space environment altered the transition between developmental stages in embryonic daylilies. They also examined how cell division and chromosome behavior might be impacted in microgravity, providing insight into the broader biological effects of space travel.

The choice of organisms for these studies was critical to the experiments' success. For example, the Japanese Red-Bellied Newt (*Cynopus pyrrhogaster*) was selected for its unique developmental timing, which aligned with the mission duration. Female newts could be induced to lay eggs in space, allowing researchers to study the early stages of development in a microgravity environment. The Moon Jellyfish (*Aurelia aurita*) was used in both its sedentary polyp stage and its free-swimming ephyra stage, offering a detailed look at how different life stages of an organism respond to spaceflight. Meanwhile, The PEMBSIS experiment focused on embryogenically competent daylily cells (Hemerocallis cv. Autumn Blaze), which were particularly well-suited to studying plant growth in space.

Sophisticated hardware supported these experiments. Newts were housed in cassette-type water tanks within the Aquatic Animal Experiment Unit (AAEU), developed by NASDA. This life support system kept aquatic animals alive for the entire mission, while individual egg containers within the AAEU allowed for the study of early developmental stages. Jellyfish were housed in the ESA's Biorack facility, with Type I containers accommodating both polyp and ephyra stages.

The slow rotating centrifuge microscope, Nizemi, developed by Germany's DLR, enabled researchers to observe and record the behavior of jellyfish under varying gravitational forces, providing invaluable data on the effects of microgravity on living organisms.

The PEMBSIS experiment relied on the Plant Fixation Chambers (PFCs) provided by NASDA. These sealed containers allowed researchers to chemically fix plant cells in orbit, preserving the samples for later analysis on Earth. Meanwhile, the jellyfish specimens were held in the Refrigerator/Incubator Module (R/IM), a temperature-controlled environment that housed the jellyfish post-experiment, ensuring they were preserved at the correct temperature for subsequent study.

In preparation for the IML-2 mission, the PEMBSIS cell cultures were established approximately one week before the launch. Twelve chambers were filled with a semi-solid medium, six of which were transported to Kennedy Space Center (KSC) and kept in a dark incubator at 22±2°C until being loaded onto the Shuttle. The remaining six chambers were retained as ground controls for comparative analysis.

Approximately 36 hours before launch, the Aquatic Animal Experiment Unit (AAEU) was loaded with 148 prefertilized newt eggs and four adult newts. The newts were distributed across three cassettes: two cassettes contained one adult newt each, while the third held two. Fresh, aerated water at 24°C circulated continuously through the AAEU to sustain the aquatic environment. A duplicate unit was maintained at KSC for ground control.

The jellyfish samples, key to the behavior and development studies, required specific preflight preparation. Twenty-four hours before launch, four groups of six jellyfish polyps were treated with iodine in artificial seawater to induce strobilization, a process by which polyps transform into the free-swimming ephyrae stage. Shortly before flight, the samples were loaded into 10 cuvettes filled with artificial seawater and placed into Type I containers. For the behavior study, a group of normal ephyrae and a group of ephyrae without statoliths were placed in the

Biorack incubator at 22°C, while a third group was placed in the Biorack 1-G centrifuge. Two groups of polyps were similarly prepared, one placed in the incubator and the other in the centrifuge. Ground control equipment at KSC mirrored these setups for comparison.

In orbit, experiments proceeded under carefully controlled conditions. The Ambient Temperature Recorder (ATR-4), a self-contained battery-powered device roughly the size of a deck of cards, was deployed to record temperature data in up to four channels throughout the mission.

On flight days 6, 8, and 11, the crew video-observed the developing newt eggs, documenting their growth. The adult newts were also observed at designated intervals. Unfortunately, on the fifth and ninth days of flight, one adult newt in each of the two cassettes died, contaminating some of the newt eggs. Despite these losses, the remaining two adult newts survived the mission and were retrieved alive after landing.

The jellyfish experiments were similarly thorough. Video recordings of jellyfish ephyrae and polyps were made at regular intervals using the Nizemi rotating microscope/centrifuge, allowing scientists to determine the gravitational thresholds necessary to elicit normal swimming behavior in the ephyrae. On flight day five, the statolith-containing ephyrae hatched on Earth were fixed, while on flight day 13, two of the four polyp groups were preserved for further analysis. The remaining jellyfish were returned to Earth for postflight studies.

The PEMBSIS plant experiment also followed a precise protocol. Shortly before landing, the crew chemically fixed some of the plant cell cultures in a three-percent glutaraldehyde solution to allow for direct comparisons between flight-fixed and ground-fixed groups.

Postflight analysis commenced promptly after landing. The flight cassettes containing the newts were retrieved approximately six hours after touchdown. Some larvae were preserved for detailed study, while others were used to assess how spaceflight affected the otolith-ocular reflex. Researchers measured the otolith volumes and the areas of the associated sensory epithelia to gauge any alterations caused by the microgravity environment.

Similarly, the jellyfish were examined within five hours of landing. Researchers counted, coded, and photographed the living jellyfish, analyzing their pulse rates, the number of arms, rhopalia, and statoliths. Jellyfish with abnormal pulsing behavior were videotaped upon landing and again 24 hours later. Both flight and control jellyfish were allowed to form clones, which were subsequently examined for structural differences, including arm number.

The PEMBSIS plant cell cultures were also analyzed postflight. Within nine hours of landing, the living cells and somatic embryos were photographed, counted, and chemically fixed. Researchers measured and compared chromosomes within and across the cultures to assess any microgravity-induced changes before the first cell division cycle on Earth was complete.

These meticulously executed experiments, both in flight and postflight, provided valuable insights into how organisms and biological processes adapt to the unique conditions of space, furthering scientific understanding crucial for long-term human space exploration.

The newt study aboard IML-2 provided important insights into how microgravity affects the development of gravity-sensing organs. Both the flight and ground-control newt larvae developed at similar rates according to morphological analysis. However, three-dimensional reconstructions revealed significant differences. The larvae reared in microgravity had a larger mean endolymphatic sac (ES) and duct volume than their ground-based counterparts. Moreover, the volume of otoconia—tiny calcium carbonate structures contributing to balance and orientation—within the ES was notably larger in the flight-reared larvae. The development of otoconia in space was also markedly accelerated, suggesting that microgravity might enhance the formation of these structures in developing organisms.

In the jellyfish study, significant abnormalities were observed in the ephyrae (the early life stage of jellyfish) that developed in microgravity. These space-reared ephyrae displayed irregular arm numbers, contrasting the more typical morphology observed in 1-G flight and ground controls. Postflight analysis revealed that fewer ephyrae developed in space were able to swim when tested, indicating that the absence of gravity affected their motor functions. However, polyps that budded in space exhibited advanced development compared to ground controls, producing more buds and progressing faster through their life cycle. Although the jellyfish adapted well to space conditions overall, the irregular arm development suggested that some species are more sensitive to the effects of microgravity than others.

The daylily cell study, or PEMBSIS experiment, revealed striking cytological changes in the cells exposed to microgravity. Both flight-fixed and ground-fixed cells exhibited chromosomal aberrations, but flight samples contained a large number of binucleate cells—cells with two nuclei. In contrast, all ground-control samples were uninucleate. These binucleate cells indicated that the space environment might interfere with normal cell division, leading to abnormalities in the number of nuclei within a cell.

The newt study also experienced some challenges during the mission. At least two of the four adult Japanese Red-bellied Newts (*Cynops pyrrhogaster*) died during the flight. The first death was attributed to stress, a common issue for animals in space. The second death, which occurred on July 17, 1994, was more perplexing. As astronaut Donald A. Thomas monitored the tanks, he discovered the second dead newt, a situation described as "peculiar" by Dr. Michael Wiederhold, a scientist overseeing the study. The weightlessness of space made removing the deceased newt difficult, and there was concern that leaving the body in the tank could contaminate the environment, potentially affecting the surviving newts and their eggs.

Despite these difficulties, the newt and jellyfish studies aboard IML-2 yielded valuable data on how microgravity impacts the development of sensory organs and biological processes, providing insights critical for understanding long-term human and animal adaptation to space.

After completing its objectives, *Columbia* safely returned to Earth, landing at Kennedy Space Center on July 23, 1994. The success of STS-65 advanced microgravity research and strengthened international partnerships in space exploration, setting the stage for future cooperative missions aboard the Space Shuttle and the International Space Station.

STS-64: Pioneering Spacewalking and Advancing Atmospheric Science

STS-64 was a pivotal mission of the Space Shuttle program, undertaken by *Space Shuttle Discovery* and launched from Kennedy Space Center, Florida, on September 9, 1994. Over the course of 11 days, the crew successfully conducted a series of groundbreaking experiments and spacewalks, pushing the boundaries of technology and space science. The mission concluded with a smooth landing at Edwards Air Force Base on September 20, 1994.

Richard N. Richards, commander (center front); L. Blaine Hammond Jr., pilot (front left); and Susan J. Helms, mission specialist (front right). On the back row, from left to right, are Mark C. Lee, Jerry M. Linenger, and Carl J. Meade, all mission specialists.

The STS-64 crew was led by Commander Richard N. Richards, a seasoned astronaut on his fourth and final spaceflight. He was joined by Pilot L. Blaine Hammond, Jr., who completed his second and last spaceflight, and Mission Specialists Jerry M. Linenger, Susan J. Helms,

Carl J. Meade, and Mark C. Lee. Linenger, a flight surgeon and naval aviator, embarked on his first space mission. Helms, flying for the second time, served as the mission's flight engineer. Meade, in his third and final spaceflight, and Lee, on his third mission, were tasked with conducting the mission's Extravehicular Activities (EVAs).

The mission's major highlight was the extravehicular activity (EVA) performed by Mark C. Lee and Carl J. Meade on September 16, 1994. This was the 28th EVA of the Space Shuttle program, lasting 6 hours and 51 minutes. During this EVA, they tested the Simplified Aid for EVA Rescue (SAFER), a new jetpack designed to help astronauts return to safety in case they became untethered during spacewalks. This marked the first untethered U.S. EVA since 1984 and the last of its kind in the Shuttle program. The SAFER system would later prove essential for U.S. and international spacewalks during the construction of the International Space Station (ISS).

One of the most significant aspects of STS-64 was the deployment of the Lidar In-Space Technology Experiment (LITE), marking the first time lidar technology was used in space. Lidar (light detection and ranging) uses laser pulses to study atmospheric conditions. Over 53 hours of operation, LITE gathered more than 43 hours of high-quality data. This data provided unprecedented views of cloud structures, dust clouds, storm systems, and pollutants, contributing significantly to NASA's *Mission to Planet Earth*. Ground-based and airborne teams from 65 groups across 20 countries conducted validation measurements, further enriching the mission's scientific outcomes.

STS-64 also saw the deployment of the Shuttle Pointed Autonomous Research Tool for Astronomy (SPARTAN-201), a free-flying satellite designed to study solar wind acceleration and the Sun's corona. SPARTAN-201 was released from Discovery's cargo bay on the fifth day of the mission, collected data for two days, and was then retrieved for return to Earth. The data gathered would help scientists better understand solar phenomena, which directly impact space weather and satellite communications.

Carl Meade drifts over Discovery's payload bay, STS-64.

The Shuttle Plume Impingement Flight Experiment (SPIFEX) was another critical payload. It was a 33-foot-long extension of the Shuttle's robotic arm, designed to study the effects of the Shuttle's Reaction Control System (RCS) thrusters on large space structures. This data was essential for the planning and operating future space stations, including the Russian Mir and the ISS.

The mission also tested the first U.S. robotics system operated in space, the Robot Operated Processing System (ROMPS). ROMPS was mounted in two Get Away Special (GAS) canisters attached to the cargo bay wall, marking a new step in space robotics. In addition, the Shuttle carried 12 GAS canisters holding self-contained experiments.

The middeck of Discovery hosted various experiments to further scientific knowledge in space. The Biological Research in Canister (BRIC) experiment studied the effects of spaceflight on plant specimens, contributing to our understanding of biology in microgravity. The Military Application of Ship Tracks (MAST) experiment captured high-resolution images of ship tracks to analyze how wakes form and dissipate. The Solid Surface Combustion Experiment (SSCE) provided insights into flame propagation in a weightless environment, while the Radiation Monitoring Equipment III (RME III) measured ionizing radiation aboard the Shuttle.

Additionally, the Shuttle Amateur Radio Experiment II (SAREX II) demonstrated the ability to establish short-wave radio communications between the orbiter and amateur radio operators on the ground, strengthening public engagement with space missions. The Air Force Maui Optical Station (AMOS) test, which did not require any onboard hardware, was conducted to gather data for military applications.

STS-64 was also notable for introducing the Advanced Crew Escape Suit, a full-pressure suit that replaced the earlier partial-pressure Launch Entry Suit. This new suit became a staple of Shuttle missions, enhancing astronaut safety during launch and reentry.

STS-68

Standing are, left to right, Michael A. Baker, mission commander; and Terrence W. Wilcutt, pilot. On the front row are, left to right, Thomas D. Jones, payload commander; and Peter J. K. (Jeff) Wisoff, Steven L. Smith and Daniel W. Bursch, all mission specialists.

STS-68, the seventh flight of the Space Shuttle Endeavour, launched on September 30, 1994, from Kennedy Space Center. This mission was the second flight of the Space Radar Laboratory (SRL-2), a significant project to advance Earth sciences through radar imaging. The mission underscored NASA's commitment to using space-based assets to understand Earth's environment and natural processes better.

The primary objective of STS-68 was to continue the work begun on STS-59, the first SRL mission, by using advanced radar systems to gather detailed data on the Earth's surface. The SRL-2 payload consisted of three major instruments: the Spaceborne Imaging Radar-C (SIR-C), the X-band Synthetic Aperture Radar (X-SAR), and the Measurement of Air Pollution from Satellites (MAPS) experiment.

SIR-C and X-SAR were designed to operate in multiple radar bands, allowing them to penetrate multiple types of terrain, from dense forests to deserts, and capture high-resolution images of the Earth's surface. These instruments gave scientists critical data on environmental changes, deforestation, geological features, and hydrological processes. Observing the Earth in different radar frequencies enabled the mapping of soil moisture levels, vegetation cover, and even subsurface geological formations.

On the other hand, the MAPS experiment focused on measuring the global distribution of carbon monoxide in the Earth's atmosphere. This data was vital for understanding the impact of human activities, such as fossil fuel combustion and biomass burning, on atmospheric composition and air quality. MAPS also contributed to global climate models by providing a clearer picture of how pollutants are transported across continents and oceans.

STS-68's mission profile included 183 Earth orbits over 11 days, allowing the crew to collect an extensive dataset covering multiple regions of the planet. The crew consisted of seven astronauts: Commander Michael A. Baker, Pilot Terrence W. Wilcutt, Mission Specialists Thomas D. Jones, Steven L. Smith, Daniel W. Bursch, and Peter J.K. Wisoff, along with Payload Specialist Thomas D. "Tom" Jones. The crew's diverse expertise was crucial in managing the complex operations of the SRL-2 instruments and ensuring the mission's success.

One of STS-68's notable achievements was the extensive use of real-time data downlinks, which allowed scientists on the ground to direct the Shuttle's observations and adjust the radar systems as needed. This capability enhanced the mission's scientific return by enabling targeted observations of areas of particular interest, such as regions experiencing rapid environmental changes.

Throughout the mission, the crew worked in shifts to ensure continuous operation of the SRL-2 payload. They encountered and overcame several technical challenges, including issues with the radar calibration and data storage systems, demonstrating the resilience and problem-solving capabilities hallmarks of Space Shuttle missions.

The data collected during STS-68 significantly advanced the understanding of Earth's surface processes and contributed to developing new remote sensing techniques. The mission's findings were used in multiple scientific fields, including geology, ecology, hydrology, and atmospheric science. The high-resolution radar images provided by SIR-C and X-SAR offered new insights into Earth's complex systems, from glaciers and rivers' dynamics to deforestation and urbanization's impacts.

Endeavour safely returned to Earth on October 11, 1994, landing at Edwards Air Force Base in California. The success of STS-68 demonstrated the value of the Space Shuttle as a versatile platform for Earth science research.

STS-66: Probing Earth's Atmosphere and Advancing Atmospheric Science

On November 3, 1994, the Space Shuttle Atlantis was poised on Launch Pad 39B at Kennedy Space Center. It was ready to embark on a mission to understand Earth's atmosphere and its interactions with solar energy. Designated STS-66, this mission was part of NASA's Atmospheric Laboratory for Applications and Science (ATLAS) program, a series of missions to study the delicate balance between Earth's atmosphere and the Sun.

At 11:59 a.m. EST, Atlantis's engines ignited, and the shuttle soared into the sky, beginning its journey into orbit. The crew of STS-66 consisted of six astronauts, each bringing a unique set of

skills and experience to the mission: Commander Donald R. McMonagle, a seasoned Air Force Colonel on his third shuttle flight; Pilot Curtis L. Brown Jr., an Air Force Lieutenant Colonel on his second mission; Payload Commander Ellen Ochoa, an electrical engineer and the first Hispanic woman to travel to space, on her second flight; Mission Specialist Joseph R. Tanner, a naval aviator and engineer making his first journey into space; Mission Specialist Jean-François Clervoy, a French aerospace engineer and the third European Space Agency (ESA) astronaut to fly on the shuttle; and Mission Specialist Scott E. Parazynski, a physician and biomedical engineer on his first shuttle mission.

Jean-Francois Clervoy, mission specialist; Scott E. Parazynski, mission specialist; Curtis L. Brown, pilot; Joseph R. Tanner, mission specialist; Donald R. McMonagle, commander; and Ellen S. Ochoa, payload commander.

The Atmospheric Laboratory for Applications and Sciences – 3 (ATLAS-03) was the primary payload aboard the Space Shuttle Atlantis on mission STS-66, continuing a series of Spacelab flights aimed at studying the sun's energy and its impact on Earth's climate and environment. Launched in November 1994, this mission provided the first detailed measurements of the Northern Hemisphere's middle atmosphere during late fall. The mission's timing was significant, as it coincided with the diminishing of the Antarctic ozone hole, enabling scientists to investigate its effects on mid-latitudes, how Antarctic air recovered, and how the northern atmosphere responded as winter approached.

In addition to the ATLAS-03 investigations, STS-66 deployed and later retrieved the Cryogenic Infrared Spectrometer Telescope for Atmosphere (CRISTA), mounted on the Shuttle Pallet Satellite (SPAS). This payload, a collaborative effort between NASA and the German Space Agency (DARA, now the DLR), was designed to explore atmospheric variability. CRISTA provided measurements that complemented those collected by the Upper Atmosphere Research Satellite (UARS), launched aboard Discovery in 1991. The data from CRISTA-SPAS helped scientists better understand the dynamic processes in Earth's middle atmosphere, offering insights into disturbances in trace gases that affect the planet's energy balance.

The ATLAS-03 mission was part of a broader effort to study the sun's energy output and the composition of Earth's atmosphere, focusing particularly on global ozone levels. ATLAS-03 carried seven instruments, many of which had flown on previous ATLAS missions. These instruments collected invaluable atmospheric measurements, contributing to a more comprehensive understanding of Earth's environment.

One key instrument aboard ATLAS-03 was the Atmospheric Trace Molecule Spectroscopy (ATMOS), which gathered more data on atmospheric trace gases than on its three previous flights combined. The Shuttle Solar Backscatter Ultraviolet Spectrometer (SSBUV), another instrument, measured ozone levels and calibrated data from the aging NOAA-9 satellite, while also working in tandem with other ATLAS-03 instruments. The Active Cavity Radiometer Irradiance Monitor (ACRIM) measured solar radiation with remarkable precision, acting as a calibration tool for its counterpart on UARS. Other solar instruments, including the Measurement of the Solar Constant (SOLCON), the Solar Spectrum Measurement (SOLSPEC), and the Solar Ultraviolet Spectral Irradiance Monitor (SUSIM), contributed to tracking changes in solar radiation over time. These measurements were critical for understanding the sun's influence on Earth's atmosphere.

Additionally, the Millimeter Wave Atmospheric Sounder (MAS), though operational for only nine hours before a computer malfunction halted its functions, collected valuable data on water vapor, chlorine monoxide, and ozone distribution at altitudes between 12 and 60 miles (20 to 100 kilometers). These observations helped scientists refine their models of the atmosphere.

On the second day of the mission, CRISTA-SPAS was released from the Shuttle's Remote Manipulator System (RMS) arm. For over eight days, CRISTA-SPAS flew at a distance of 25 to 44 miles (40 to 70 kilometers) behind Atlantis, gathering its first-ever global data on medium- and small-scale disturbances in trace gases in the middle atmosphere. This data would ultimately improve atmospheric models and enhance understanding of Earth's energy balance. Another instrument on CRISTA-SPAS, the Middle Atmosphere High Resolution Spectrograph Investigation (MAHRSI), measured ozone-destroying hydroxyl and nitric oxide levels. MAHRSI delivered the first complete global maps of hydroxyl in the atmosphere, providing key information on the chemistry of ozone depletion.

For the retrieval of CRISTA-SPAS, the crew tested a new R-Bar approach method, which was designed to minimize fuel usage and reduce the risk of contaminating Mir's systems with Shuttle thruster emissions. This technique would later be used during U.S. Shuttle missions to the Russian space station Mir.

STS-66 marked the last solo flight for Atlantis for over 14 years, as its subsequent missions were dedicated to the Mir and International Space Station (ISS) programs. Atlantis would not fly another solo mission until STS-125, the final mission to service the Hubble Space Telescope.

STS-63: A Mission of Firsts and a Historic Rendezvous

STS-63 marked a significant milestone in the US/Russian Shuttle-Mir Program, as it conducted the first rendezvous between an American Space Shuttle and Russia's Mir space station. Known as the 'Near-Mir' mission, Space Shuttle *Discovery* launched from Kennedy Space Center's launch pad 39B on February 3, 1995, at night, marking *Discovery*'s 20th flight. This mission was notable for several historic firsts: Eileen Collins became the first female pilot of a Space Shuttle, and astronauts Michael Foale, born in the UK, and Bernard A. Harris Jr., the first African-American astronaut to perform an extravehicular activity (EVA), took part in the mission's spacewalks. Additionally, the mission carried out a successful rendezvous and flyaround of Mir, which served as crucial preparation for the future STS-71 mission that would achieve the first docking between Shuttle and Mir.

Janice E. Voss, mission specialist; Eileen M. Collins, pilot; (the first woman to pilot a Space Shuttle), James D. Wetherbee, commander; and Vladmir G. Titov (Cosmonaut). Standing in the rear are mission specialists Bernard A. Harris (the first Afro-American to walk in space), and C. Michael Foale.

The *Discovery* crew was composed of six members. Commander James D. Wetherbee led the mission, while Eileen Collins piloted the shuttle on her first spaceflight. Bernard A. Harris Jr. and Michael Foale, both on their third spaceflights, served as mission specialists, alongside Janice E. Voss on her second flight and Russian cosmonaut Vladimir G. Titov, who brought valuable experience from his long-duration stay on Mir.

The primary mission objective was to perform a rendezvous with Mir to test proximity operations, verifying flight techniques, communications, navigation systems, and other interfaces required for future joint missions. The mission also had secondary objectives, including deploying and retrieving the Spartan-204 platform, a retrievable satellite designed to gather data from ultraviolet light sources in space. The mission also supported various scientific experiments housed in SPACEHAB-3, a commercial module that carried 20 experiments in biotechnology, materials science, and technology demonstrations.

Throughout the mission, *Discovery* performed a series of thruster burns to close the distance to Mir. Although initial plans were to bring *Discovery* within 10 meters of the station, leaks in three of the orbiter's Reaction Control System thrusters posed a potential issue for the rendezvous. After extensive discussions between U.S. and Russian mission control teams, it was decided to proceed with the rendezvous using backup systems. Radio contact was made with the Mir crew ahead of schedule, and excitement built as *Discovery* approached within 11 meters of the Russian station on February 6, 1995. Commander Wetherbee's words as *Discovery* neared Mir—"As we bring our spaceships closer together, we are bringing our nations closer together"—underscored the broader diplomatic significance of the mission.

Following the successful rendezvous, *Discovery* conducted a full flyaround of Mir while filming and photographing the station. The Mir crew, consisting of Commander Alexander Viktorenko, Flight Engineer Yelena Kondakova, and Valery Polyakov, observed the flyaround. Polyakov, who was on a record-setting mission for the longest time in space, watched from the Mir Core Module. No disruptions to Mir's operations, including its solar arrays, were reported during the approach.

Among the scientific payloads aboard *Discovery* was the Spartan-204 satellite, which was deployed and retrieved during the mission. Spartan-204 gathered data on ultraviolet light sources, while other experiments onboard the shuttle contributed to research in various scientific fields. For instance, the Cryo System Experiment (CSE) studied cryogenic fluids, and the Shuttle Glow experiment (GLO-2) focused on understanding the phenomenon of shuttle glow. Another significant experiment was ODERACS-2, which deployed six small calibration targets into orbit to help improve radar tracking of space debris.

The Mir Space Station as viewed from the Space Shuttle Discovery during STS-63.

On February 9, 1995, astronauts Harris and Foale conducted an EVA, lasting 4 hours and 38 minutes. Suspended on *Discovery*'s robotic arm, the astronauts tested modifications to their spacesuits designed to enhance thermal protection during spacewalks. They also rehearsed handling the Spartan-204 satellite, an essential practice for future space station assembly tasks. The EVA marked the first for both Harris and Foale, and it was particularly notable as the first spacewalk conducted by an astronaut of African heritage.

The SPACEHAB-3 module carried a wide array of experiments, including Astroculture, which focused on plant growth in space as part of ongoing research to develop life support systems for long-duration space missions. Astroculture also had potential Earth applications, such as energy-efficient lighting and air purification technologies. Additionally, the Immune experiment examined the effects of spaceflight on the human immune system, research with potential benefits for immunocompromised patients on Earth.

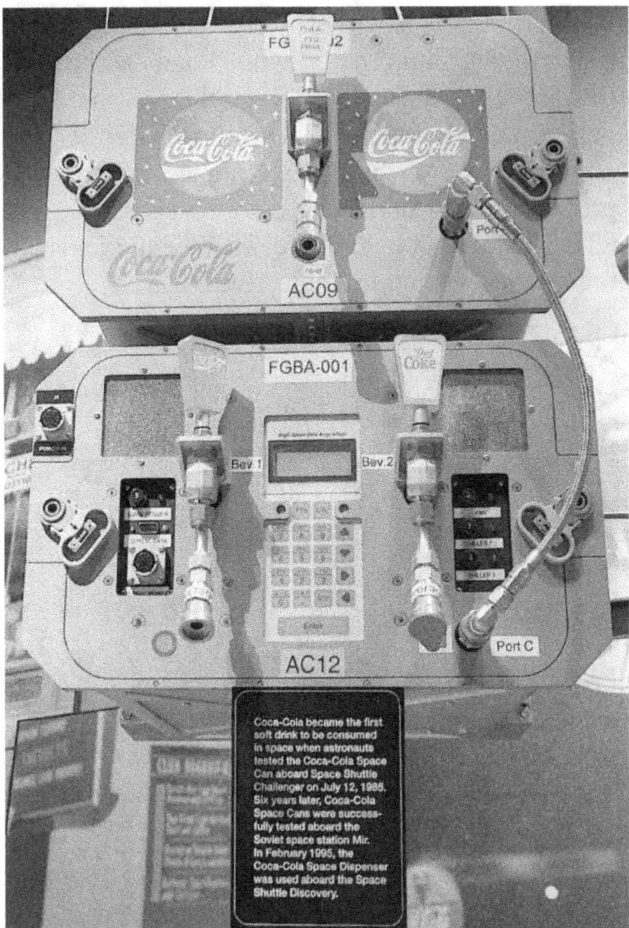

One of the mission's lighter moments was the testing of a Coca-Cola dispenser aboard *Discovery*. Developed by BioServe Space Technologies, the dispenser allowed astronauts to sample pre-mixed soda in microgravity, offering insights into how taste and fluid dynamics are altered in space.

STS-67

STS-67, the eighth flight of the Space Shuttle Endeavour and the second mission dedicated to the Spacelab program's ASTRO observatory, launched from Kennedy Space Center on March 2, 1995.

The STS-67 mission was originally scheduled for 16 days but was extended to nearly 17 days, making it one of the longest Shuttle missions. This extension allowed the crew to maximize their observations, taking full advantage of the unique conditions provided by the Shuttle's orbit.

The crew of STS-67 consisted of seven astronauts: Commander Stephen S. Oswald, Pilot

William G. Gregory, Mission Specialists Tamara E. Jernigan, John M. Lounge, and Wendy B. Lawrence, and Payload Specialists Ronald A. Parise and Samuel T. Durrance. Parise and Durrance were both astronomers who had previously flown on the first ASTRO mission, STS-35, and brought a wealth of experience to the mission's scientific objectives.

Stephen S. Oswald, commander; Tamara E. Jernigan, payload commander; and William G. (Bill) Gregory, pilot. On the back row (left to right) are Ronald A. Parise, payload specialist; Wendy B. Lawrence, mission specialist; John M. Grunsfeld, mission specialist; and Samual T. Durrance, payload

Astro-2, the second dedicated Spacelab mission for astronomical observations in the ultraviolet (UV) spectral regions, built upon the foundation set by its predecessor, Astro-1, which had flown on STS-35. Astro-2, launched aboard Space Shuttle Endeavour during STS-67, carried a sophisticated suite of three instruments: the Hopkins Ultraviolet Telescope (HUT), the Ultraviolet Imaging Telescope (UIT), and the Wisconsin Ultraviolet Photo-Polarimeter Experiment (WUPPE). Together, these instruments enabled scientists to gather data on objects within the Solar System and beyond, including stars, nebulae, supernova remnants, galaxies, and active extragalactic sources. This wealth of information complemented the findings from Astro-1, significantly enhancing our understanding of the universe in the ultraviolet spectrum.

The Ultraviolet Imaging Telescope (UIT) was a crucial component of both missions, as it allowed observations of UV radiation from space—wavelengths that are absorbed by Earth's atmosphere and cannot be detected from ground-based observatories. During Astro-1, the UIT provided the first precise measurements of ultraviolet radiation across various celestial bodies. However, the UIT flown on Astro-2 was significantly more capable, capturing nearly double the ultraviolet spectrum compared to its predecessor. Additionally, since STS-67 launched at a different time of year than STS-35, Astro-2 was able to observe sections of the sky that had been inaccessible to Astro-1, offering a broader and more detailed view of the universe in ultraviolet light.

Aboard the Middeck, several critical science experiments were conducted, including the Protein Crystal Growth Thermal Enclosure System Vapor Diffusion Apparatus-03 (PCG-TES-03), the Protein Crystal Growth Single Thermal Enclosure System-02 (PCG-STES-02), the Shuttle Amateur Radio Experiment-II (SAREX-II), the Commercial Materials Dispersion Apparatus Instrumentation Technology Associates Experiments-03 (CMIX-03), and the Middeck Active Control Experiment (MACE). Among these, MACE stood out as a key engineering research payload, focusing on spacecraft motion control technology.

The MACE experiment was designed to evaluate a closed-loop control system that could compensate for motion disturbances in spacecraft. The system was composed of several components, including rate gyros, reaction wheels, and a precision pointing payload, all of which worked together to simulate and mitigate motion disturbances. Over 45 hours of testing, Commander Stephen S. Oswald and Pilot William G. Gregory conducted around 200 different motion disturbance scenarios, assessing the ability of the control system to adapt and correct. The data collected from MACE was instrumental in the future development of more advanced spacecraft control systems, enabling better stability and precision during space missions.

STS-67 also carried two Get Away Special (GAS) payloads, designated G-387 and G-388, sponsored by the Australian Space Office and

AUSPACE Ltd. These experiments focused on ultraviolet observations of deep space and nearby galaxies, aiming to study various astrophysical phenomena. The payloads were designed to investigate the structure of galactic supernova remnants, the distribution of hot gas in the Magellanic Clouds, and the hot galactic halo emissions. Additionally, they explored emission patterns associated with galactic cooling flows and jets.

The two GAS canisters were interconnected with a cable and worked in tandem to collect data. Canister 1 housed a UV telescope equipped with a motorized door assembly that opened to expose the telescope to space, while UV reflective filters determined the specific ultraviolet bandpass. Canister 2 contained two video recorders for data storage and batteries to power the experiment. These observations contributed to the growing body of knowledge about the universe's ultraviolet emissions. Notably, STS-67 became the longest flight of Space Shuttle Endeavour, underscoring the significance and breadth of the mission's scientific contributions.

STS-71: A Historic First Docking with Mir

On June 27, 1995, Space Shuttle Atlantis launched from Kennedy Space Center's Launchpad 39A for STS-71, a historic mission that would become the first U.S. Space Shuttle to dock with the Russian Mir space station. This mission marked a significant milestone in the U.S.-Russian Shuttle-Mir Program, as it represented a new era of international collaboration in space exploration.

The crew of STS-71 consisted of seven astronauts: Commander Robert "Hoot" Gibson, Pilot Charles Precourt, and Mission Specialists Ellen Baker, Gregory Harbaugh, Bonnie Dunbar, along with two cosmonauts, Anatoly Solovyev and Nikolai Budarin. These two cosmonauts were part of the Mir-19 crew and were scheduled to stay on Mir following the shuttle's departure.

The crew assigned to the STS-71 mission included (front left to right) Vladimir N. Dezhurov, Mir 18 crew download; Robert L. Gibson, commander; and Anatoly Y. Solovyev, Mir 19 crew upload. On the back row, left to right, are Norman E. Thagard, Mir 18 crew download; Gennadiy Strekalov, Mir 18 crew download; Gregory J. Harbaugh, mission specialist; Ellen S. Baker, mission specialist; Charles J. Precourt, pilot; Bonnie J. Dunbar, mission specialist; and Nikolai Budarin, Mir 19 crew upload

After its smooth ascent into orbit, Atlantis began its rendezvous sequence with Mir, leading to the first docking of a Space Shuttle with a space station on June 29, 1995. Using the R-bar approach, a docking technique that minimized the need for Shuttle jet firings, Atlantis and Mir successfully docked while orbiting 216 nautical miles (400 kilometers) above the Lake Baikal region of Russia. The Orbiter Docking System (ODS), located in Atlantis's payload bay, flawlessly connected to the Kristall module on Mir, symbolizing the technological achievements of both space programs.

With the docking complete, the combined mass of Atlantis and Mir, at nearly 225 metric tons, became the largest spacecraft ever in orbit at that time. The event was also notable for being the 100th crewed space launch by the United States, and it demonstrated the Space Shuttle's unique ability to act as both a transporter and a temporary extension of a space station.

The STS-71 crew delivered two cosmonauts, Anatoly Solovyev and Nikolai Budarin, to replace the Mir-18 crew, which included U.S. astronaut Norman Thagard. Thagard, along with

Vladimir Dezhurov and Gennady Strekalov, had spent over 100 days aboard Mir. This crew exchange was the first time a Shuttle mission involved switching members with a space station crew, further cementing the growing partnership between the U.S. and Russia.

Vladimir Dezhurov and "Hoot" Gibson shake hands in orbit, a homage to the Apollo-Soyuz Test Project (ASTP). Later that day, President Bill Clinton announced that this handshake was a major breakthrough towards the end of the Cold War.

The primary objective of the STS-71 mission, launched on June 29, 1995, was to achieve a historic rendezvous and docking between the Space Shuttle Atlantis and the Russian Space Station Mir. This mission marked the first U.S.-Russian docking in two decades, a significant milestone in space collaboration following the end of the Cold War. Atlantis successfully delivered a relief crew of two cosmonauts, Anatoly Solovyev and Nikolai Budarin, to Mir, facilitating the exchange of personnel aboard the space station.

In addition to the primary docking mission, STS-71 focused on conducting joint U.S.-Russian life sciences investigations aboard the Spacelab module, which was docked to Mir. These investigations were pivotal in advancing our understanding of the effects of long-term spaceflight on the human body. The mission also included the logistical resupply of Mir and the recovery of U.S. astronaut Norman E. Thagard, who had spent over 100 days on the space station.

Secondary objectives of the mission included filming with the IMAX camera, capturing the grandeur of space and the intricacies of the docking, and conducting the Shuttle Amateur Radio Experiment-II (SAREX-II), which allowed students and amateur radio operators on Earth to communicate directly with the crew.

STS-71 was notable for several historic achievements. It was the 100th U.S. human space launch conducted from Cape Canaveral and the first docking between a U.S. Space Shuttle and a Russian Space Station. When Atlantis and Mir were docked, they formed the largest spacecraft ever assembled in orbit, with a combined mass of approximately 225 metric tons. This mission also saw the first on-orbit changeout of a Shuttle crew.

The rendezvous sequence began with a lift-off at 15:32:19 EDT, timed precisely to align with Mir's orbit. The ascent was smooth, with no need for the OMS 1 burn, and the OMS 2 burn followed, adjusting the orbit to a 160 x 85.3 nautical mile trajectory—the lowest perigee altitude ever flown by an orbiter. This low altitude allowed for a rapid catch-up rate with Mir, which was critical for the docking procedure. About three hours later, the orbit was raised to 210 x 159 nautical miles, providing a more stable trajectory for the final approach.

Docking occurred at 9:00 a.m. EDT on June 29, 1995, using the R-Bar, or Earth radius vector, approach. This technique involved Atlantis closing in on Mir from directly below, leveraging natural gravitational forces to slow the orbiter's approach and reducing the need for jet firings. Commander Hoot Gibson manually piloted the final phase of the docking, bringing Atlantis to a near-perfect alignment with Mir. The docking took place approximately 216 nautical miles above the Lake Baikal region of the Russian Federation, with the Orbiter Docking System (ODS) connecting flawlessly to Mir's Kristall module.

Once docked, Atlantis and Mir formed a massive orbiting complex. After opening hatches, the STS-71 crew entered Mir for a welcoming ceremony. On the same day, the Mir 18 crew officially handed over control of the station to the Mir 19 crew, marking a seamless transition of responsibilities. A symbolic handshake between

Vladimir Dezhurov and Hoot Gibson paid homage to the Apollo-Soyuz Test Project (ASTP) of 1975. President Bill Clinton later remarked that this gesture symbolized a major breakthrough in U.S.-Russian relations.

The joint crew conducted extensive biomedical investigations for the next five days and transferred essential supplies between the spacecraft. The Spacelab module aboard Atlantis hosted 15 biomedical and scientific investigations spanning seven disciplines, including cardiovascular and pulmonary functions, human metabolism, neuroscience, hygiene, sanitation, radiation, behavioral performance, biology, and microgravity research. The Mir 18 crew served as primary test subjects, and their physiological responses were meticulously documented to understand the impact of prolonged spaceflight better.

The mission also involved transferring various materials and samples between the spacecraft. Atlantis brought back more than 100 urine and saliva samples, about 30 blood samples, and numerous air, water, and surface samples from Mir. Additionally, the Shuttle delivered over 450 kilograms of water, specially designed spacewalking tools for the Mir 19 crew, and oxygen and nitrogen supplies to increase the air pressure aboard Mir.

Throughout the mission, the Shuttle Amateur Radio Experiment-II (SAREX-II) provided an opportunity for public outreach. During this portion of the flight, the crew communicated with several schools, including Redlands High School in California. Students and technicians had the rare chance to ask the astronauts questions, one of which was light-hearted: "What would happen if an astronaut sneezed inside their helmet?" Pilot Charlie Precourt responded with humor, explaining that a sneeze would likely "spray the face shield," adding a relatable, human touch to the mission.

After five days of joint operations, Atlantis undocked from Mir on July 4, following a farewell ceremony. After undocking from Mir, Atlantis performed a flyaround of the space station, allowing the crew to capture breathtaking images of both spacecraft orbiting above Earth. The undocking sequence, likened to a "cosmic ballet" by Commander Gibson, was monitored by both the Shuttle crew and the Mir 19 crew aboard their Soyuz spacecraft. During the undocking, Mir experienced a computer malfunction, causing the station to drift in attitude. The Mir 19 crew promptly re-docked to the station and restored attitude control by replacing the faulty computer hardware.

Atlantis returned to Earth with a crew of eight, equaling the largest crew in Shuttle history. The returning Mir 18 crew members were seated in custom-made recumbent seats to ease their re-entry into the gravity environment after more than 100 days in space. Despite a minor glitch with one of the Shuttle's General Purpose Computers, which was quickly resolved, the mission was a resounding success.

STS-71 concluded with a successful landing at Kennedy Space Center on July 7, 1995, at 10:54 a.m. EDT. The mission lasted nearly 10 days and was widely celebrated as a landmark achievement in space exploration. This mission demonstrated the Shuttle-Mir program's technical capabilities and symbolized the strengthening cooperation between the United States and Russia. It set the foundation for future Shuttle-Mir missions, which would continue to advance international collaboration in space and ultimately contribute to developing and operating the International Space Station (ISS).

STS-70

STS-70 marked the 21st flight of Space Shuttle *Discovery* and the final mission to deploy a Tracking and Data Relay Satellite (TDRS). This historic mission was the first to be managed from the new mission control center at Johnson Space Center in Houston, showcasing NASA's evolving infrastructure. The shuttle carried the TDRS-G satellite, marking the end of an era for shuttle-based TDRS deployments.

Launched from Kennedy Space Center on July 13, 1995, at 9:41 a.m. EDT, *STS-70* followed only six days after the landing of *Atlantis*, setting a record for the fastest turnaround between two shuttle flights. This mission also introduced the new Block 1 orbiter main engine, a significant advancement to improve performance and safety.

The five-member crew of *STS-70* included Commander Terence T. Henricks, Pilot Kevin R. Kregel, and Mission Specialists Donald A. Thomas, Nancy J. Currie-Gregg, and Mary Ellen Weber. The mission was notable for featuring Kregel and Weber's first spaceflights, while the rest of the crew had prior spaceflight experience. This diverse team worked together to accomplish the mission's primary objectives, including the successful deployment of TDRS-G.

Astronauts Brian Duffy (right front) and Brent W. Jett (left front) are mission commander and pilot, respectively. Mission specialists (back row, left to right) are Winston E. Scott, Leroy Chiao, Koichi Wakata, and Daniel T. Barry. Wakata is an international mission specialist representing Japan's National Space Development Agency (NASDA) based at the Johnson Space Center (JSC).

Originally slated to fly ahead of *STS-71*, *STS-70* was delayed due to an unusual issue—damage to the external tank caused by nesting flicker woodpeckers. On May 31, 1995, NASA discovered 71 holes in the tank's thermal insulation, some as large as four inches in diameter. This required immediate attention. Technicians installed safeguards against further damage, and on June 2, NASA made the decision to delay *STS-70* to conduct repairs, moving *STS-71* ahead. On June 8, *Discovery* was rolled back to the Vehicle Assembly Building (VAB) for repair and returned to the launch pad by June 15.

After a smooth countdown, the launch on July 13 proceeded with only a minor hold of 55 seconds at T-31 due to a fluctuation in the external tank's range safety system. Once resolved, *Discovery* successfully lifted off, carrying the new Block 1 engine, which featured several design improvements, including a high-pressure liquid oxygen turbopump and a new powerhead configuration.

Sunburst over the payload bay

The primary goal of *STS-70* was the deployment of TDRS-G, the seventh satellite in the Tracking and Data Relay Satellite System. Weighing approximately 2,200 kilograms, the satellite was released from *Discovery's* payload bay six hours into the mission. The deployment was a complex operation overseen by Mission Specialists Donald Thomas and Mary Ellen Weber, and the shuttle maneuvered away from the satellite to allow the Inertial Upper Stage (IUS) booster to carry TDRS-G into its final geostationary orbit 22,000 miles above the central Pacific.

Once in orbit, the TDRS-G satellite joined a network of operational satellites responsible for maintaining constant communication with various NASA spacecraft, including the Space Shuttle, the International Space Station, and the Hubble Space Telescope. At this time, the on-orbit network included operational spacecraft in both the TDRS East and West positions, as well as a functional spare. However, the network was aging, with the first satellite, TDRS-1, far exceeding its planned seven-year lifespan.

Secondary Payloads and Experiments

In addition to the primary mission, *STS-70* carried a wide array of scientific payloads. Among these was the Bioreactor Demonstration System (BDS), designed to simulate the growth of tissues in microgravity using a rotating cylinder filled with growth medium. This experiment aimed to grow organized tissue structures, offering insights into cellular processes in space that could lead to advancements in medical research.

Other experiments included the Commercial Protein Crystal Growth (CPCG) experiment, designed to improve protein crystallization techniques in space, and the Space Tissue Loss/National Institutes of Health Cells (STL/NIH-C) experiment, which focused on understanding how cells behave in the absence of gravity. Additional experiments included the Shuttle Amateur Radio Experiment-II (SAREX-II), and the Military Applications of Ship Tracks (MAST), which studied how ship exhausts affected cloud formations.

STS-70 was initially scheduled to land at Kennedy Space Center on July 21, 1995. However, heavy fog over the Shuttle Landing Facility led to two consecutive landing delays. Flight Director Rich Jackson decided to wave off the landing, and the crew remained in orbit for an additional day. Weather reconnaissance by astronaut Steve Oswald confirmed that the poor visibility made it impossible to land safely.

On July 22, *Discovery* successfully touched down at 8:02 a.m. EDT on Runway 33 at Kennedy Space Center. The mission lasted eight days, 22 hours, and 20 minutes, marking a triumphant conclusion to a complex and historic flight.

STS-69

STS-69, the 11th flight of the Space Shuttle Endeavour, launched on September 7, 1995, from Kennedy Space Center. The mission was a multifaceted one, with objectives that included deploying and retrieving satellites, conducting scientific experiments, and testing new spaceflight technologies. This mission exemplified the versatility of the Space Shuttle program, as it managed a complex array of tasks in a single flight.

David M. Walker (right front) mission commander; with Kenneth D. Cockrell (left front) pilot. On the back row are (left to right) Michael L. Gernhardt and James H. Newman, both mission specialists; and James S. Voss, payload commander.

The STS-69 crew consisted of five astronauts: Commander David M. Walker, Pilot Kenneth D. Cockrell, and Mission Specialists James S. Voss, James H. Newman, and Michael L. Gernhardt.

The 11-day mission marked the second flight of the Wake Shield Facility (WSF), a saucer-shaped satellite designed to operate independently of the Space Shuttle for several days. The WSF's primary objective was to grow thin films in a near-perfect vacuum, which was created by the satellite's wake as it traveled through space. These thin films, essential for advanced electronics, could only be produced in the unique conditions of space. During this mission, the Shuttle crew successfully deployed and later retrieved the Spartan 201 astronomy satellite, conducted a six-hour spacewalk to test construction techniques for the International Space Station (ISS), and assessed improvements made to spacesuits, focusing on thermal protection during spacewalks.

Spartan 201, a free-flyer satellite, was

making its third flight aboard the Space Shuttle. This satellite's mission was to study the Sun's outer atmosphere and its interaction with the solar wind—a constant stream of charged particles that flows through the solar system. By examining the solar wind's origin and its effects on Earth, Spartan 201 contributed valuable data to ongoing research on solar-terrestrial interactions, which have significant implications for satellite operations and communications systems on Earth.

The mission also saw the first flight of the International Extreme Ultraviolet Hitchhiker (IEH-1), part of a series of five planned flights. IEH-1 aimed to measure and monitor long-term variations in extreme ultraviolet (EUV) emissions from the Sun and to study the plasma torus surrounding Jupiter, a region influenced by volcanic activity on its moon Io. The data gathered from this payload provided critical insights into the dynamics of solar and planetary systems.

Additional payloads, including the combined Capillary Pumped Loop-2/Gas Bridge Assembly (CAPL-2/GBA) experiment were aboard the Shuttle Endeavor. This payload demonstrated a cooling system in microgravity, a key technology for future Earth-observing systems. Also part of this suite of experiments was the Thermal Energy Storage-2 (TES-2) payload, which explored the behavior of thermal energy storage salts in space, providing valuable data for future solar power systems. The TES-2 payload specifically studied lithium fluoride-calcium fluoride eutectic salts, designed to absorb and store heat, which would play a role in advanced energy systems for long-duration space missions.

Among the mission's scientific endeavors was the Electrolysis Performance Improvement Concept Study (EPICS), which focused on generating oxygen and hydrogen by electrolyzing water in space. This technology was crucial for long-term missions, such as those aboard the ISS, as it could significantly reduce the need for resupplying oxygen from Earth, lowering logistical costs and enhancing mission sustainability.

Other notable payloads included the National Institutes of Health Cells-4 (NIH-C4) experiment, which investigated bone loss in microgravity—an important study for understanding the effects of long-duration space travel on the human body. The Biological Research in Canister-6 (BRIC-6) experiment examined the gravity-sensing mechanisms in mammalian cells, furthering scientific understanding of cellular responses to microgravity.

In addition, two commercial experiments were flown on this mission. The CMIX-4 experiment studied cell changes in microgravity, with a particular focus on neuromuscular disorders, while the Commercial Generic Bioprocessing Apparatus-7 (CGBA-7) experiment supported research in pharmaceuticals, biomedicine, agriculture, and environmental science.

Each of these experiments, from cutting-edge solar system research to crucial biomedical studies, showcased the Shuttle's unique role in advancing scientific knowledge and supporting future space exploration.

STS-73: A Journey of Scientific Discovery

Albert Sacco Jr., payload specialist; Kent V. Rominger, pilot; Michael E. Lopez-Alegria, mission specialist. On the back row are, left to right, Catherine G. Coleman, mission specialist; Kenneth D. Bowersox, commander; Fred W. Leslie, payload specialist; and Kathryn C. Thornton, payload

On October 20, 1995, Space Shuttle Columbia launched from Kennedy Space Center's Launch Complex 39B at 9:53 a.m. EDT, embarking on STS-73, a mission dedicated to scientific research and exploration. This mission, Columbia's 18th flight and the 73rd flight of the Space Shuttle program, was primarily focused on the United States Microgravity Laboratory-2 (USML-2), a sophisticated laboratory designed to conduct experiments in the unique microgravity space environment.

The STS-73 crew consisted of seven astronauts: Commander Kenneth D. Bowersox, Pilot Kent V. Rominger, Mission Specialists Catherine G. Coleman, Michael E. Lopez-Alegria, Fred W. Leslie, Albert Sacco Jr., and Payload Commander Kathryn C. Thornton. Each astronaut played a critical role in the mission's success, contributing their expertise to the diverse array of scientific investigations conducted aboard Columbia.

The second United States Microgravity Laboratory (USML-2) mission was the primary payload aboard Space Shuttle Columbia's STS-73 flight, which launched on October 20, 1995. This 16-day mission represented a significant milestone in the ongoing cooperative efforts of the U.S. government, universities, and industry to advance scientific knowledge in the near-weightless environment of space. The USML-2 mission built upon the successes of its predecessor, USML-1, which had flown aboard Columbia during the STS-50 mission in 1992. Both missions aimed to explore the fundamental science of fluid dynamics, combustion, crystal growth, and other physical processes, seeking to understand how these phenomena are altered in microgravity.

USML-2 was not merely a repetition of the earlier mission but an evolution, incorporating new insights gained from USML-1. For example, the first mission's data significantly expanded scientific models of fluid physics, revealing how gravity influences the behavior of fluids and the propagation of flames. Additionally, it advanced understanding of semiconductor crystal formation in space. The experiments on USML-2 were carefully refined based on these findings, allowing scientists to push their research further.

Some of the most critical experiments involved the growth of protein crystals, which had demonstrated promise during USML-1. These protein crystal structures, vital for drug development and medical research, were further studied to unlock new molecular insights.

During the mission, flight controllers and experiment scientists directed operations from NASA's Spacelab Mission Operations Control facility at the Marshall Space Flight Center. Teams stationed at various NASA centers and universities across the U.S. monitored and supported the mission's scientific objectives, ensuring seamless coordination of the experiments. The payload aboard USML-2 included a variety of instruments designed to measure and analyze the microgravity environment. Among these were the Orbital Acceleration Research Experiment (OARE), Space Acceleration Measurement System (SAMS), and Three-Dimensional Microgravity Accelerometer (3DMA). These instruments helped quantify even the smallest forces acting on the spacecraft and the experiments within, improving the accuracy and reliability of the scientific results. Other payloads included the Suppression of Transient Accelerations By Levitation Evaluation (STABLE) system, designed to minimize the effects of vibrations, and the High-Packed Digital Television Technical Demonstration system, which aimed to test new technologies for video transmissions from space.

A unique cultural highlight of STS-73 occurred on October 26, 1995, when Mission Commander Ken Bowersox threw the ceremonial first pitch for Game 5 of the World Series between the Cleveland Indians and Atlanta Braves. Bowersox's pitch, pre-recorded from space, symbolized the growing integration of space exploration into popular culture, showcasing how the Shuttle program captivated the public imagination even during routine scientific missions.

USML-2 experienced several challenges before it could lift off. Originally scheduled for September 25, 1995, the launch endured six scrubbed attempts, ultimately taking off on its seventh try. This tied STS-73 with STS-61C for the record of the most scrubbed launch attempts. However, once aloft, the mission proceeded smoothly, advancing scientific knowledge in microgravity and laying the groundwork for future research aboard the International Space Station (ISS).

Upon its return to Earth, the data and insights gathered during USML-2 contributed to a deeper understanding of basic physical processes for space exploration and applications on Earth. The mission helped prepare for more advanced operations aboard the ISS and other future space programs, marking another significant step forward in humanity's exploration of the final frontier.

STS-74: Strengthening Ties with Russia in Space

James D. Halsell, pilot and Kenneth D. Cameron, commander. Standing, left to right, are mission specialists William S. McArthur, Jerry L. Ross, and Chris A. Hadfield.

STS-74 was the fourth mission of the US/Russian Shuttle-Mir program and the second docking of the Space Shuttle with Mir. Space Shuttle Atlantis lifted off from Kennedy Space Center launch pad 39A on 12 November 1995. This mission, the 15th flight of Atlantis and the 74th Space Shuttle mission overall was a significant step in the ongoing collaboration between the United States and Russia in space exploration. The primary objective of STS-74

was to deliver and install the Russian-built Docking Module to the Russian Space Station Mir, a crucial component that would facilitate future Shuttle-Mir docking missions.

The crew of STS-74 consisted of five astronauts: Commander Kenneth D. Cameron, Pilot James D. Halsell Jr., and Mission Specialists Jerry L. Ross, William S. McArthur Jr., and Chris A. Hadfield. Notably, Chris Hadfield, a Canadian Space Agency astronaut, became the first Canadian to visit the Russian Space Station Mir during this mission.

The crew of Space Shuttle Atlantis began preparing for the STS-74 mission in 1994, thirteen months before launch. Their training encompassed the operation of the shuttle, the intricate procedures required for mating and docking with the Russian space station Mir, and the management of various scientific experiments aboard the orbiter.

Atlantis itself was readied for the mission with meticulous care. On August 25, 1995, three thrusters in its right-hand Orbital Maneuvering System pod were replaced at the Orbiter Processing Facility. By September 5, 1995, Atlantis's three Space Shuttle Main Engines (SSMEs) was installed, and preparations for the Russian-built docking module were finalized.

As the launch date approached, engineers conducted detailed inspections to ensure the Solid Rocket Boosters (SRBs) readiness. On November 7, 1995, engineers determined that no additional work was needed, despite earlier concerns about small cracks in the hold-down posts on boosters used in previous flights. A thorough inspection confirmed that the STS-74 boosters were free of defects.

On November 9, the crew and engineers prepared to load cryogenic oxygen and hydrogen into the onboard tanks. These reactants powered the shuttle's fuel cells, providing electricity and water for the mission.

The initial launch attempt was scheduled for November 11, 1995, at 7:56 a.m. EST, but it was postponed just five minutes before liftoff due to poor weather conditions at the Transatlantic Abort Landing (TAL) site. The crew was already aboard the shuttle when the decision was made to scrub the launch. The countdown resumed the following day, November 12, and Atlantis successfully launched at 7:30:43 a.m. EST. After a flawless countdown, the shuttle ascended into the sky within a 10-minute launch window. Approximately 43 minutes later, a two-minute engine burn adjusted the shuttle's trajectory to a circular orbit at 162 nautical miles.

Once in orbit, the five-member crew began configuring Atlantis for its on-orbit operations. About 90 minutes after liftoff, the payload bay doors were opened, and the orbiter was given the "go" for operations. Commander Ken Cameron and Pilot Jim Halsell conducted the first of several thruster burns to adjust Atlantis's path towards Mir. Meanwhile, Canadian astronaut Chris Hadfield activated the Russian docking module stored in the shuttle's payload bay, preparing it for the docking operations scheduled for flight day two.

On November 13, 1995, the crew spent their first full day in space preparing for the docking module's transfer and the upcoming link-up with Mir. Mission specialists Jerry Ross and Bill McArthur inspected spacesuits in case a spacewalk became necessary. At the same time, Hadfield powered up the shuttle's robotic arm, which would be used to move the docking module.

Hadfield also tested the Advanced Space Vision System (ASVS), a precise alignment tool consisting of large dots placed on the docking module and system. This system would ensure accurate alignment during the module's transfer to the Orbiter Docking System.

Commander Cameron oversaw the installation and alignment of the centerline camera, mounted on the Orbiter Docking System, to assist in the final stages of docking. By early morning, Atlantis was 4,000 miles behind Mir, closing the gap at a rate of 380 miles per orbit. The day ended with the crew settling down for an eight-hour sleep period.

On November 14, the crew successfully mated the Russian docking module to the shuttle's Orbiter Docking System. Hadfield operated the robotic arm to lift the docking module from the payload bay, rotate it, and align it with the docking system. The shuttle's steering jets then brought the two spacecraft together, securing the docking module with hooks and latches to form an airtight seal.

The successful mating occurred at 1:17 a.m. CST, with Atlantis soaring over Eastern Europe. By 3:00 a.m., the robotic arm was disengaged, and the crew raised the cabin pressure to standard levels after completing the task. Preparations then began for the crucial docking with Mir the following day.

On November 15, Atlantis completed a historic docking with the Russian space station Mir. Tension ran high as Cameron skillfully maneuvered the shuttle into position using its thrusters. At 6:27:38 UTC, the two spacecraft docked flawlessly, marking the second Shuttle-Mir docking. After a series of checks, the hatches were opened, and the five Atlantis crew members joined the Mir crew—Russian cosmonauts Yuri Gidzenko, Sergei Avdeyev, and ESA astronaut Thomas Reiter—for three days of combined operations. The crews exchanged gifts, including flowers and chocolates, and prepared for joint activities.

The Atlantis crew worked alongside the Mir crew for the next three days to transfer vital supplies, equipment, and water to the space station. Two new solar arrays, one Russian and one jointly developed, were among the items transferred, critical upgrades for Mir's power supply. The crews also exchanged scientific samples and equipment, including the University of California Berkeley Trek Experiment, which had been aboard Mir for four years.

The mission also included scientific research aboard Atlantis. The shuttle carried the Goddard Payload Package (GPP), which used broadband spectroscopy to study Earth's thermosphere, ionosphere, and mesosphere. The crew also conducted the Photogrammetric Appendage Structural Dynamics Experiment (PASDE), which used cameras to analyze the structural dynamics of Mir's solar arrays during the docked phase. This data would prove valuable for future space missions, including the International Space Station.

On November 18, at 8:15:44 UTC, Atlantis undocked from Mir, leaving the docking module permanently attached to the Kristall module. The docking module would ensure future space shuttles had sufficient clearance from Mir's solar arrays.

The Mir Docking Module, positioned in Atlantis's payload bay, ready to be docked to the Kristall module of space station Mir

On November 20, 1995, Atlantis prepared for its return to Earth. The deorbit burn occurred at around 11:00 a.m. EST on orbit 128, initiating the shuttle's descent. Atlantis touched down at Kennedy Space Center's Shuttle Landing Facility at 12:01:27 p.m. EST, completing a mission elapsed time of 8 days, 4 hours, 30 minutes, and 44 seconds. With a smooth landing, the STS-74 mission came to a successful conclusion.

STS-72: Capturing Satellites and Exploring Space

On January 11, 1996, Space Shuttle Endeavour embarked on its 10th mission, STS-72, from Kennedy Space Center, Florida. This mission's primary objective was capturing and returning the Japanese Space Flyer Unit (SFU), a microgravity research spacecraft. Launched at 4:41 AM EST, STS-72 marked the 74th flight in the Space Shuttle program and showcased the Shuttle's versatility and capability as a platform for scientific research and technological demonstrations.

Under the command of astronaut Brian Duffy, the six-member crew included Pilot Brent W. Jett, Mission Specialists Leroy Chiao, Winston E. Scott, Koichi Wakata, and Daniel T. Barry. This mission marked their first journey into space for Jett, Scott, Wakata, and Barry, while Duffy and Chiao brought prior experience to this critical mission.

The crew's seating arrangements on both launch and landing reflected their roles, with Duffy and Jett occupying the forward flight deck positions and the rest seated across the flight and mid-decks.

Astronauts Brian Duffy (right front) and Brent W. Jett (left front) are mission commander and pilot, respectively. Mission specialists (back row, left to right) are Winston E. Scott, Leroy Chiao, Koichi Wakata, and Daniel T. Barry. Wakata is an international mission specialist representing Japan's National Space Development Agency (NASDA) based at the Johnson Space Center (JSC).

The Space Flyer Unit (SFU) had been launched by Japan's National Space Development Agency (NASDA) from Tanegashima Space Center on March 18, 1995. Weighing approximately 3,577 kilograms (7,886 pounds), the SFU conducted automated research in various fields, including materials science, biology, engineering, and astronomy, for nearly ten months. The task of retrieving this spacecraft fell to Mission Specialist Koichi Wakata, who operated Endeavour's remote manipulator system (RMS) arm. On January 11, 1996, Space Shuttle *Endeavour* lifted off from Kennedy Space Center on its 10th mission, STS-72. The early morning launch at 4:41 AM EST marked the 74th flight in the Space Shuttle program and demonstrated the orbiter's versatility in supporting scientific research and technological advancements. Commanded by astronaut Brian Duffy, the six-member crew also included Pilot Brent W. Jett and Mission Specialists Leroy Chiao, Winston E. Scott, Koichi Wakata, and Daniel T. Barry. For Jett, Scott, Wakata, and Barry, STS-72 was their first venture into space, while Duffy and Chiao brought prior mission experience.

The mission's primary objective was retrieving the Japanese Space Flyer Unit (SFU), a microgravity research satellite launched by Japan's National Space Development Agency (NASDA) on March 18, 1995, from Tanegashima Space Center. Weighing approximately 3,577 kilograms (7,886 pounds), the SFU conducted automated experiments in fields such as materials science, biology, and astronomy during its 10 months in orbit. On flight day three, Mission Specialist Koichi Wakata operated *Endeavour*'s Remote Manipulator System (RMS) to capture the SFU. The satellite's solar arrays, which had failed to latch properly during retraction, were jettisoned as a contingency procedure before the satellite was safely stowed in *Endeavour*'s payload bay.

A key feature of STS-72 was its two extravehicular activities (EVAs), or spacewalks, aimed at testing tools and techniques for assembling the International Space Station (ISS), which was set to begin construction in 1998. On January 15, astronauts Leroy Chiao and Daniel Barry conducted the first EVA, lasting 6 hours and 9 minutes. Their tasks included attaching a portable work platform to the Shuttle's robotic arm, mimicking the movement of large ISS components, and testing a rigid umbilical cable tray designed to connect space station modules. The second EVA, on January 17, featured Chiao and Winston Scott. Lasting 6 hours and 53 minutes, this spacewalk focused on gathering additional data on ISS construction techniques, including utility box installations and the thermal evaluation of spacesuits in extreme cold temperatures. Scott tested his suit's performance by positioning himself on the *OAST-Flyer* platform, with *Endeavour* maneuvered into deep space to expose the suit to temperatures as low as minus 104 degrees Fahrenheit.

In addition to the retrieval of the SFU, STS-72 deployed and later recaptured the *OAST-Flyer*, a Spartan carrier spacecraft. This reusable platform carried four experiments, including the Return Flux Experiment (REFLEX), which tested spacecraft contamination models, and the Global Positioning System Attitude Determination and Control Experiment (GADACS), which demonstrated the use of GPS technology in space. Wakata once again operated the RMS for both the deployment on flight day four and the retrieval on flight day six.

The mission also carried a diverse array of scientific payloads, including the Shuttle Solar Backscatter Ultraviolet Experiment (SSBUV-8), which measured atmospheric ozone levels, and the Space Tissue Loss Experiment, which studied the effects of microgravity on biological tissues. The National Institutes of Health's NIH-R3 experiment explored muscle atrophy in space, while the Thermal Energy Storage (TES-2) experiment examined innovative methods of heat storage. Other payloads, such as the Pool Boiling Experiment (PBE), contributed to the study of phase transitions in microgravity, further broadening the scientific scope of the mission.

STS-72 was documented extensively by a PBS film crew, which followed the crew and their families through training and the mission itself. The resulting 90-minute documentary, *Astronauts*, narrated by Bill Nye, aired in July 1997, offering the public an intimate view of life aboard a Space Shuttle mission. In December 2020, unseen photographs from the mission's preparation were released by photographer John Angerson, offering new insights into the meticulous process of readying astronauts and spacecraft for flight.

STS-75

STS-75, launched on February 22, 1996, was a significant and ambitious mission to advance scientific research in microgravity. The mission, flown by Space Shuttle Columbia, was the 75th shuttle mission and Columbia's 19th flight. The primary objective of STS-75 was to deploy the Tethered Satellite System (TSS-1R) and conduct experiments in the unique environment of low Earth orbit.

During this period, the crew conducted a

wide range of scientific experiments and technological demonstrations. Andrew M. Allen commanded the mission, with Scott J. Horowitz serving as the pilot. The mission specialists were Franklin R. Chang-Díaz, Jeffrey A. Hoffman, Maurizio Cheli, and Claude Nicollier, with payload specialist Umberto Guidoni representing the Italian Space Agency.

Scott J. Horowiz, pilot; Andrew M. Allen, commander; and Franklin R. Chang-Diaz, mission specialist. Standing, left to right, are Maurizio Cheli, mission specialist of the European Space Agency (ESA); Umberto Guidoni, payload specialist (Italy); Jeffrey A. Hoffman, mission specialist; and Claude Nicollier, mission specialist (ESA).

The Tethered Satellite System was a collaborative project between NASA and the Italian Space Agency (ASI). The TSS-1R was designed to investigate the electrodynamics of a tethered system in space. The concept involved deploying a satellite connected to the shuttle by a 12.8-mile (20.7-kilometer) conductive tether. The satellite, when deployed, would generate electricity by moving through Earth's magnetic field, producing a high voltage along the tether that could be harnessed for scientific purposes.

On February 25, 1996, three days after launch, the crew began the deployment of the Tethered Satellite. The deployment initially proceeded as planned, and the satellite extended to over 12 miles (approximately 19 kilometers) from the orbiter. However, the tether unexpectedly broke shortly after reaching full deployment, and the satellite drifted away from the shuttle. The break occurred due to excessive stress on the tether, caused by a combination of factors including higher-than-expected electrical currents and mechanical wear on the tether's insulation.

Despite the satellite's loss, the mission provided valuable data on the dynamics of tethered systems in space. The breakage of the tether allowed scientists to study the behavior of a free-flying satellite in low Earth orbit and the effects of rapid deployment and unexpected tether failure. These insights would prove crucial for the future design and deployment of tethered satellite systems.

In addition to the TSS-1R experiment, the STS-75 mission carried several other scientific payloads. One of the key experiments was the United States Microgravity Payload (USMP-3), which focused on materials science and biotechnology. These experiments were designed to use the microgravity environment to study fundamental processes such as crystal growth, fluid dynamics, and biological cell development.

The STS-75 mission lasted 15 days, 17 hours, and 41 minutes, making it one of the longest shuttle missions.

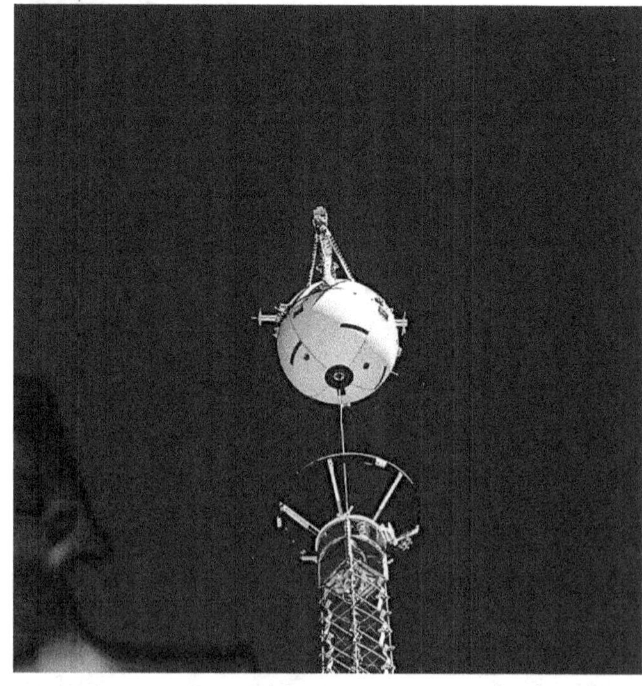

The Tethered Satellite System (TSS) and part of its supportive boom device prior to deployment operations.

Despite the setback with the Tethered Satellite, STS-75 was a success in terms of the data collected and the experiments conducted. The mission's findings contributed to a better understanding of the challenges associated with tethered systems in space and provided insights that would influence future missions. The scientific experiments aboard Columbia during STS-75 also added to the growing knowledge about microgravity's effects on materials and biological systems, furthering NASA's long-term goals of space exploration and scientific discovery.

The Space Shuttle Columbia and its crew safely returned to Earth on March 9, 1996, landing at Kennedy Space Center in Florida. The lessons learned from STS-75 will continue to inform NASA's endeavors, ensuring that each mission built upon the successes and challenges of the past and pushed the boundaries of what was possible in space exploration.

STS-76

STS-76, NASA's 76th Space Shuttle mission and the 16th for *Atlantis*, launched from Kennedy Space Center's launch pad 39B at 08:13:04 UTC on March 22, 1996. This mission, which lasted just over nine days, traveled approximately 6.1 million kilometers (3.8 million miles) while orbiting Earth 145 times. *Atlantis* returned to Earth on March 31, 1996, landing on runway 22 at Edwards Air Force Base at 13:28:57 UTC.

The mission marked a significant milestone in the Shuttle–Mir program as the third docking of a U.S. Space Shuttle with the Russian Space Station Mir. One of the mission's primary objectives was to transport astronaut Shannon Lucid to Mir, where she would replace NASA astronaut Norman Thagard, becoming the first American woman to live aboard the station. Lucid's four-and-a-half-month stay set a new record for the longest-duration U.S. spaceflight, surpassing Thagard's previous record. Astronaut John E. Blaha later replaced Lucid during the STS-79 mission in August 1996, and her time on Mir became part of a continuous U.S. presence in space for the next two years.

Aboard *Atlantis*, the crew of STS-76 consisted of six astronauts. Commander Kevin P.

Chilton was on his third and final spaceflight, while Pilot Richard A. Searfoss flew his second mission. Mission specialists included Ronald M. Sega (second and final flight), Michael R. Clifford (third and final flight), and Linda M. Godwin (third spaceflight). Shannon Lucid was on her fifth and final spaceflight, joining the Expedition 21 crew on Mir.

astronauts Ronald M. Sega, mission specialist; Kevin P. Chilton, mission commander; and Richard A. Searfoss, pilot. On the back row, left to right, are mission specialists Michael R. (Rich) Clifford, Shannon W. Lucid, and Linda M. Godwin.

One of the highlights of the mission occurred on flight day six when Godwin and Clifford conducted an extravehicular activity (EVA). They attached four Mir Environmental Effects Payload (MEEP) experiments to Mir's docking module during this six-hour and two-minute spacewalk. These experiments aimed to monitor the environment surrounding the space station over an 18-month period. This EVA was notable for being the first U.S. spacewalk around two docked spacecraft since the Skylab program in 1974. Godwin and Clifford wore Simplified Aid for EVA Rescue (SAFER) units, a propulsive device first tested during the STS-64 mission.

The payload bay configuration for STS-76 included the Orbiter Docking System at the forward end and a SPACEHAB single module in the aft. This was the first time a pressurized SPACEHAB module was used to support Shuttle-Mir dockings, primarily serving as a stowage area for supplies bound for the space station. In addition to the delivery of equipment, water, and resupply items totaling about 2,500 kilograms (5,500 pounds), the mission carried the European Space Agency's Biorack experiment for scientific research. The Biorack facility enabled 11 experiments, investigating the effects of microgravity and cosmic radiation on a variety of biological samples, including plants, cells, and bacteria.

The docking with Mir occurred on flight day three using the R-bar approach, a technique previously employed during STS-74. The actual connection between *Atlantis* and Mir's Kristall module docking port occurred at 02:50 UTC on March 24, 1996. After docking, the hatches between the two spacecraft opened less than two hours later, allowing the crew of *Atlantis* to be welcomed by Mir 21 Commander Yury Onufriyenko and Flight Engineer Yuri Usachov, who had been on Mir since February 1996.

During the five days of docked operations, the crew transferred water, scientific equipment, and supplies between the Shuttle and Mir. They also delivered the Mir Glovebox Stowage (MGBX) to replenish the glovebox onboard the station and carried out the Queen's University Experiment in Liquid Diffusion (QUELD) and the High-Temperature Liquid Phase Sintering (LPS) experiment. Additionally, *Atlantis* returned with experiment samples and equipment, furthering research in microgravity's effects on materials and biological systems.

Among other notable payloads were the Shuttle Amateur Radio Experiment (SAREX), which allowed students and amateur radio operators to communicate with the Shuttle crew, and KidSat, a project designed to allow middle school students to participate in space exploration by selecting and analyzing Earth observation images. Another payload, the Trapped Ions in Space (TRIS) experiment, developed by the Naval Research Laboratory, was housed in a Getaway Special canister in the payload bay.

After a mission lasting 9 days, 5 hours, and 16 minutes, Atlantis and her crew safely returned to Earth, landing at Edwards Air Force Base in California on March 31, 1996. The success of STS-76 further solidified the partnership between the United States and Russia in space exploration. It set the stage for the construction and operation of the International Space Station, a symbol of international collaboration in space.

STS-77

STS-77, launched on May 19, 1996, was the 77th flight of NASA's Space Shuttle program and the 11th flight of Endeavour. The mission focused on deploying and operating commercial and technological payloads, showcasing NASA's commitment to fostering the development of space-based technologies and scientific research in a microgravity environment.

The STS-77 mission was commanded by John H. Casper, with Curtis L. Brown Jr. serving as the pilot. The mission specialists were Daniel W. Bursch, Mario Runco Jr., Marc Garneau, and Andrew S. W. Thomas. Marc Garneau, a Canadian astronaut, was the first Canadian to fly in space and was on his second spaceflight, further strengthening the collaboration between NASA and the Canadian Space Agency (CSA).

Curtis L. Brown, pilot; and John H. Casper, commander. Standing, left to right, are mission specialists Daniel W. Bursch, Mario Runco, Marc Garneau (CSA), and Andrew S. W. Thomas.

NASA's flight of Space Shuttle *Endeavour* on STS-77 was a critical mission dedicated to advancing the commercial space frontier. This mission not only highlighted NASA's commitment to fostering space-based technologies but also showcased the increasing role of international collaboration and commercial partnerships in space exploration.

A key mission component was the *SPACEHAB* module, a commercially owned and operated pressurized laboratory housed in the Shuttle's payload bay. The *SPACEHAB* module carried nearly 1,400 kilograms (3,100 pounds) of experiments and support equipment, aimed at commercial space product development in diverse fields, including biotechnology, materials science, and agriculture. One notable experiment was the *Commercial Float Zone Facility (CFZF)*, a collaboration between the U.S., Canada, and Germany, which used the float-zone technique to heat and process electronic and semiconductor materials. Another important experiment in the module, the *Space Experiment Facility (SEF)*, focused on growing crystals through vapor diffusion, offering potential advances in producing high-quality materials.

Among the mission's significant achievements was the deployment of the *Spartan-207* satellite, which carried the *Inflatable Antenna Experiment (IAE)*, a pioneering demonstration of inflatable space structures. The IAE tested the feasibility of deploying large, lightweight antennas in space, a technology that could revolutionize future spacecraft design by minimizing mass and volume. On May 20, 1996, *Spartan-207* was released from *Endeavour's* payload bay, and the antenna inflated to a diameter of 50 feet (15 meters). Over a ninety-minute period, the crew and ground controllers gathered valuable data on the antenna's performance in the vacuum of space. After the experiment concluded, the antenna was jettisoned, and the Spartan-207 satellite was successfully recovered, proving the viability of inflatable structures for future space communications and remote sensing technologies.

The mission also carried four technology experiments as part of the *Technology Experiments for Advancing Missions in Space (TEAMS)* program. These included the *Global Positioning System (GPS) Attitude and Navigation Experiment (GANE)*, which tested the accuracy of GPS for providing attitude information to spacecraft; the *Vented Tank Resupply Experiment (VTRE)*, aimed at improving in-space refueling methods; the *Liquid Metal Thermal Experiment (LMTE)*, which evaluated the performance of liquid metal heat pipes in microgravity; and the *Passive Aerodynamically Stabilized Magnetically Damped Satellite (PAMS)*, designed to demonstrate aerodynamic stabilization principles in the upper atmosphere. The PAMS satellite was deployed from the Shuttle and its movements were tracked using onboard cameras, further expanding the knowledge base for future satellite stabilization techniques.

Several secondary experiments also flew aboard *Endeavour*. These included the *Brilliant Eyes Ten Kelvin Sorption Cryocooler Experiment (BETSCE)*, which focused on cooling technologies for space-based observation systems, and the *Biological Research in a Canister (BRIC)* experiment, which studied the effects of microgravity on plant and animal cells. Another key biological experiment was the *Plant-Generic Bioprocessing Apparatus (P-GBA)*, which tested plant growth in microgravity, contributing to research on the potential for agriculture in space.

A unique addition to the mission was a Coca-Cola fountain dispenser, officially known as *Fluids Generic Bioprocessing Apparatus-2 (FGBA-2)*. This apparatus tested the feasibility of producing carbonated beverages in space by mixing water, carbon dioxide, and flavored syrups. The successful operation of this experiment demonstrated the potential for human comfort and lifestyle enhancements during long-term space missions.

The mission concluded successfully on May 29, 1996, with *Endeavour* and its crew safely landing at Kennedy Space Center after a mission duration of ten days, 0 hours, and 40 minutes. STS-77 was pivotal in advancing NASA's efforts to develop space-based technologies, with its wide array of commercial, scientific, and technological experiments contributing significantly to future missions. The successful deployment of the IAE, along with the extensive research conducted aboard the *SPACEHAB* module, underscored the potential of space as a platform for both commercial ventures and scientific discoveries.

STS-78

STS-78, launched on June 20, 1996, was the 78th flight of NASA's Space Shuttle program and the 20th flight of Columbia. This mission, also known as the Life and Microgravity Spacelab (LMS) mission, was dedicated to advancing scientific understanding in two critical areas: life sciences and microgravity research. The mission's success added to NASA's growing knowledge on how humans and materials behave in the unique space environment.

The STS-78 mission was commanded by Terence T. Henricks, with Kevin R. Kregel serving as the pilot. The mission specialists included Susan J. Helms, Richard M. Linnehan, Charles E. Brady Jr., Jean-Jacques Favier, and Robert Brent Thirsk. Jean-Jacques Favier, a French physicist, and Robert Brent Thirsk, a Canadian physician, were international astronauts, reflecting the global collaboration in space research. This mission marked the first time a Canadian astronaut conducted research aboard the Space Shuttle on behalf of the Canadian Space Agency.

Terrence T. (Tom) Henricks, commander; and Kevin R. Kregel, pilot. Standing, left to right, are Jean-Jacques Favier (CNES), payload specialist; Richard M. Linneham, mission specialist; Susan J. Helms, payload commander; Charles E. Brady, mission specialist; and Robert Brent Thirsk (CSA).

The mission objectives of this Space Shuttle flight were pivotal in preparing for long-duration spaceflights, specifically in support of future International Space Station (ISS) expeditions. Central to the mission was research into the effects of extended space travel on human physiology. This was conducted through 22 life science and microgravity experiments using the Orbiter's pressurized Life and Microgravity Spacelab module (LM2). Additionally, the mission aimed to test the use of the Orbiter's Reaction Control System (RCS) jets to raise the altitude of orbiting satellites—a procedure crucial for future ISS operations.

After launch, Columbia ascended to an altitude of 278 kilometers (173 miles) with an orbital inclination of 39 degrees to the Earth's equator. This altitude was carefully chosen to allow the seven-member crew to maintain their Earth-based sleep rhythms while reducing vibrations and directional forces that could interfere with the delicate microgravity experiments onboard.

Once in orbit, the crew entered the 40-foot (12-meter) long Spacelab module to begin a rigorous schedule of over 40 science experiments. These experiments utilized not only the Spacelab's facilities but also lockers in the orbiter's middeck section. Thirteen of the experiments focused on the effects of microgravity on the human body, examining crucial issues such as bone density loss and muscle atrophy—conditions that would be vital to understand for long-term human habitation aboard the ISS. Another six experiments investigated fluid dynamics and metallurgy in space, studying the behavior of fluids and metals in near-weightlessness and the production of metallic alloys and protein crystals.

One of the mission's most notable experiments was the first comprehensive study of sleep patterns in microgravity. The crew also performed in-flight repairs to the Bubble, Drop, and Particle Unit (BDPU), an apparatus designed to explore fluid physics. These experiments laid the groundwork for future scientific research aboard the ISS.

The mission also featured a critical test of a technique later used to boost the altitude of the

Hubble Space Telescope. During the mission, Columbia's vernier Reaction Control System jets were pulsed gently to raise the Shuttle's altitude without damaging the payloads onboard. This procedure was later used during the second Hubble servicing mission (STS-82) and became a standard method for boosting the ISS's orbit when docked with a Shuttle.

However, despite the mission's scientific successes, a post-flight analysis revealed a significant anomaly with the solid rocket boosters (SRBs). After their disassembly in June, engineers discovered damage to the SRB field joints, likely caused by hot gases. Although this damage did not compromise astronaut safety, it raised concerns due to the similarities to the O-ring failure that had caused the catastrophic loss of Challenger in 1986. In this instance, the gas path had traveled through the SRB field joints but had not penetrated the critical capture joint where the O-rings were located.

This issue was attributed to the use of new Environmental Protection Agency (EPA)-mandated adhesive and cleaning fluids. As a result, the following mission, STS-79, which was scheduled to dock with the Russian Space Station Mir and return astronaut Shannon Lucid, was delayed. Although there were discussions about using a Soyuz spacecraft to return Lucid, the Shuttle was deemed safe, and the mission proceeded as planned, ensuring her safe return aboard the Shuttle.

STS-79

STS-79, launched on September 16, 1996, was the 79th flight of NASA's Space Shuttle program and the 17th flight of Atlantis. This mission was pivotal in the Shuttle-Mir Program, representing the fourth docking of a Space Shuttle with the Russian space station Mir and the first to perform a crew exchange. The mission's objectives included the delivery of supplies and scientific equipment, the transfer of U.S. astronaut John Blaha to Mir, and Shannon Lucid's return to Earth after her record-setting stay aboard the station.

Jerome (Jay) Apt, mission specialist; Terrence W. Wilcutt, pilot; William F. Readdy, commander; Thomas D. Akers, and Carl E. Walz, both mission specialists. On the back row (left to right) are mission specialists Shannon W. Lucid, and John E. Blaha.

The STS-79 mission was commanded by William F. Readdy, with Terrence W. Wilcutt serving as the pilot. The mission specialists were Thomas D. Akers, Jay Apt, Carl E. Walz, and John E. Blaha. Blaha was a crucial member of the mission, as he was scheduled to replace Shannon Lucid on Mir, allowing her to return to Earth after an unprecedented 188 days in space— the longest single spaceflight by an American and the longest by a woman at that time.

STS-79 marked a pivotal moment in the Shuttle-Mir program as the first Space Shuttle mission to dock with a fully completed Mir space station, following the arrival of its Priroda module. Launched aboard Atlantis, this mission was highlighted by several significant milestones, most notably the return of American astronaut Shannon Lucid after an unprecedented 188 days in space. Lucid's long-duration spaceflight set a new American record and also established a worldwide record for the longest spaceflight by a woman at that time.

The mission carried the Orbiter Docking System, weighing 1,821 kilograms (4,015 pounds), and was the fourth Shuttle-Mir docking overall. Lucid, who had traveled to Mir aboard STS-76 on March 22, 1996, was replaced by astronaut John Blaha, who would remain on Mir for approximately four months. This marked the first American crew exchange on the Russian space station. Blaha joined the Mir 22 crew, led by commander Valery Korzun and flight engineer Alexander Kaleri. In January 1997, Blaha returned to Earth aboard STS-81 and was succeeded on Mir by astronaut Jerry Linenger.

STS-79 also featured the second flight of the SPACEHAB module in support of Shuttle-Mir operations and the debut of the SPACEHAB Double Module configuration. The forward section of the double module was equipped with various experiments, while the aft section housed essential logistics, including food, clothing, scientific supplies, and spare equipment for Mir. The total mass of the SPACEHAB Double Module was 4,774 kilograms (10,525 pounds).

Atlantis performed a flawless R-bar approach, linking up with Mir on September 18, 1996, at 15:13 UTC. The hatches between the two spacecraft opened early the next day, at 05:40 UTC, allowing for the exchange of crew members and the transfer of vital supplies. In total, over 1,814 kilograms (3,999 pounds) of logistics, food, water, and experiments were transferred to Mir, while 907 kilograms (2,000 pounds) of experiment samples and equipment were brought back to Atlantis. This extensive exchange resulted in a total logistical transfer of more than 2,722 kilograms (6,001 pounds), the most extensive of any Shuttle-Mir mission to date.

Shannon Lucid's time aboard Mir was marked by significant scientific research. She conducted experiments in advanced technology, Earth sciences, biology, human life sciences, microgravity, and space sciences. Among the experiments were Environmental Radiation Measurements to assess ionizing radiation aboard Mir, and the Greenhouse-Integrated Plant Experiments, which studied the effects of microgravity on plant growth, specifically focusing on dwarf wheat. Additionally, Lucid collected data on the immune system's response to long-duration spaceflight through the Assessment of Humoral Immune Function, which involved gathering blood and saliva samples. Much of this research took place in the Priroda module, the last component of the Mir space station, which had arrived during Lucid's mission.

A view of the newly-completed Soviet/Russian space station Mir, shown over the limb of the Earth, as seen from the Space Shuttle Atlantis following undocking during STS-79.

Atlantis itself hosted several notable experiments during the flight. These included the Extreme Temperature Translation Furnace (ETTF), capable of processing materials in space at temperatures exceeding 871 degrees Celsius (1,600 degrees Fahrenheit), and the Commercial Protein Crystal Growth (CPCG) experiment, which studied the crystallization of proteins in microgravity. The Mechanics of Granular Materials experiment was designed to investigate how cohesionless materials behave, which could provide insights into natural phenomena such as earthquakes and landslides.

The mission also tested new technology as part of ongoing risk-reduction efforts for the future International Space Station (ISS). Notably, the Active Rack Isolation System (ARIS), designed to reduce vibrations and disturbances for scientific experiments, flew for the first time on STS-79. This technology would later play a crucial role in enhancing the stability of experiments aboard the ISS.

Towards the end of the mission, a test was conducted using Atlantis's small vernier jets to lower its orbit, similar to a maneuver performed during STS-82 to reboost the Hubble Space

Telescope while it remained in the orbiter's payload bay.

STS-79 was a highly successful mission that advanced international cooperation between the U.S. and Russia and contributed significantly to scientific research and technology development, laying the groundwork for future operations aboard the ISS.

STS-80

STS-80, flown by Space Shuttle *Columbia*, was a milestone in the history of space exploration. Initially scheduled for launch on October 31, 1996, the mission experienced several delays, primarily due to technical issues and weather concerns, pushing the launch to November 19, 1996. The planned 16-day mission ended up lasting 17 days, 15 hours, and 53 minutes, making it the longest shuttle mission in history at the time.

Kent V. Rominger, pilot; and Kenneth D. Cockrell, commander. Standing (left to right) are mission specialists Tamara E. Jernigan, F. Story Musgrave, and Thomas D. Jones.

Commander Kenneth D. Cockrell led a crew of five, including Pilot Kent V. Rominger and Mission Specialists F. Story Musgrave, Thomas D. Jones, and Tamara E. Jernigan. Musgrave, on his sixth and final spaceflight, became the only astronaut to fly on all five Space Shuttles— *Challenger*, *Atlantis*, *Discovery*, *Endeavour*, and *Columbia*. He also set a record for being the oldest person to fly in space at age 61, a distinction that would later be surpassed.

The mission had ambitious goals, including deploying and retrieving two satellites and two planned spacewalks to test International Space Station (ISS) construction equipment. Unfortunately, both spacewalks were canceled due to a mechanical issue with the airlock hatch, believed to be caused by a loose screw, preventing Jones and Jernigan from exiting the orbiter.

Despite the setback with the spacewalks, the mission succeeded in deploying and retrieving the *Orbiting and Retrievable Far and Extreme Ultraviolet Spectrometer-Shuttle Pallet Satellite II* (ORFEUS-SPAS II) and the *Wake Shield Facility-3* (WSF-3). ORFEUS-SPAS II was released on the first day of the mission and retrieved on the sixteenth day. The satellite collected valuable data, with its two spectrographs (far and extreme ultraviolet) taking over 400 observations of celestial bodies. This second flight of ORFEUS-SPAS II improved significantly upon its predecessor by providing more than twice the data.

The WSF-3, designed by the University of Houston's Space Vacuum Epitaxy Center, aimed to generate ultra-thin semiconductor films in space using the vacuum of space. This was its third mission, and despite technical issues on previous flights, WSF-3 completed seven successful film creations before Columbia recaptured it.

Several other scientific experiments were also part of STS-80's payload. These included biological experiments such as the NIH-R4, which investigated vascular constriction in space for the National Institute of Health. Additionally, the *Biological Research in Canister* (BRIC) experiment studied the growth of tobacco and tomato seedlings under microgravity, while the *Visualization in an Experimental Water Capillary Pumped Loop* (VIEW-CPL) tested new methods for thermal management in spacecraft. These diverse experiments contributed to understanding the effects of spaceflight on both biological and material systems, laying the groundwork for future space missions.

The mission concluded with a successful landing on December 7, 1996, after weather delays postponed the planned landing for two days. During re-entry, astronaut Tom Jones had planned to capture footage of the plasma trailing behind the shuttle using a video camera, but instead, he ended up assisting Story Musgrave, who remained on the flight deck during re-entry. Despite the challenges faced, STS-80 remains a landmark mission for its scientific achievements and record-setting duration in space.

STS-81

STS-81, launched aboard the Space Shuttle Atlantis in January 1997, marked the fifth of nine planned missions to the Russian space station Mir. As part of NASA's ongoing collaboration with the Russian space program, this mission played a key role in advancing joint space exploration efforts. STS-81 was significant for its astronaut exchange and extensive logistical transfer between the two spacecraft, setting new records in cooperation and scientific achievement.

The crew of STS-81 was led by Commander Michael A. Baker, who was on his fourth and final spaceflight. He was joined by Pilot Brent W. Jett, Jr., on his second spaceflight, and Mission Specialists Peter J.K. Wisoff (third spaceflight), John M. Grunsfeld (second spaceflight), and Marsha Ivins (fourth spaceflight). Also aboard was astronaut Jerry M. Linenger, who was embarking on his second and final space mission as he prepared to replace John E. Blaha on Mir. Blaha had been part of the

Mir 22 crew since September 19, 1996, as part of a long-duration stay.

The mission began with the docking of Atlantis to Mir on January 14, 1997, at 22:55 EST. The crew opened the hatch between the two spacecraft at 00:57 EST on January 15, allowing the exchange of astronauts to proceed. Linenger officially replaced Blaha at 04:45 EST, marking the second NASA-Russian astronaut exchange. Blaha had spent 118 days on Mir, contributing to joint NASA-Russian experiments during his time aboard the station. His total time in orbit, including the shuttle mission, reached 128 days by the time he returned to Earth.

One of the primary objectives of STS-81 was to transfer vital supplies between Atlantis and Mir. This included an unprecedented exchange of nearly 6,000 pounds (2,722 kilograms) of logistics. Among the materials transferred to Mir were 1,400 pounds (635 kilograms) of water, 1,138 pounds (516 kilograms) of U.S. scientific equipment, and over 2,200 pounds (1,001 kilograms) of Russian logistical supplies. The return payload to Earth included 1,256 pounds (570 kilograms) of scientific materials, Russian logistics, and miscellaneous items. This logistical exchange was the largest in the history of the shuttle-Mir missions up to that point.

In addition to logistical operations, STS-81 also carried the SPACEHAB double module, which provided the crew with additional space for scientific experiments. One notable experiment was the successful growth of wheat plants aboard Mir, marking the first time a crop had completed a full life cycle—from seed to seed—in space. This experiment highlighted the potential for long-term space habitation and the future of growing food in space environments.

Another critical experiment aboard Atlantis was the testing of the Treadmill Vibration Isolation and Stabilization System (TVIS), designed for use in the Russian Service Module of the International Space Station (ISS). The TVIS helped assess how exercise in space could be isolated from spacecraft vibrations, a necessary advancement for maintaining astronaut health during extended missions. Atlantis also fired its small vernier jet thrusters during mated operations with Mir to gather engineering data for future International Space Station construction and maintenance.

After five days of successful docked operations, Atlantis undocked from Mir on January 19, 1997, at 09:15 EST. The shuttle performed a fly-around of the station to capture photographic and video documentation of Mir's condition before returning to Earth. No significant in-flight anomalies were reported during the mission, showcasing the shuttle's reliability and the smooth collaboration between the U.S. and Russian space programs.

Atlantis touched down at Kennedy Space Center, concluding a mission that strengthened international ties and advanced scientific research and logistical capabilities in space. Linenger remained on Mir for more than four months, contributing to NASA-Russian joint experiments until his return aboard STS-84 in May 1997. Blaha's fifth and final spaceflight added a remarkable chapter to his distinguished career.

STS-82

Space Shuttle Discovery launched on February 11, 1997, for mission STS-82, the 82nd flight of the Space Shuttle program and the second mission dedicated to servicing the Hubble Space Telescope (HST). This mission was a crucial part of NASA's commitment to maintaining and upgrading Hubble, ensuring the continuation of its groundbreaking contributions to astronomy and our understanding of the universe.

The crew of STS-82 consisted of seven astronauts: Commander Kenneth D. Bowersox, Pilot Scott J. Horowitz, and Mission Specialists Mark C. Lee, Steven A. Hawley, Gregory J. Harbaugh, Steven L. Smith, and Joseph R. Tanner. This team was highly experienced, with several members having previous spaceflight experience, critical given the complexity and precision required for the mission.

The STS-82 mission, flown by the Space Shuttle *Discovery*, marked the second of five planned servicing missions for the Hubble Space Telescope (HST), which had been placed into orbit on April 24, 1990, by the Shuttle *Discovery* during STS-31. This mission was a crucial step in maintaining and enhancing the capabilities of the Hubble, an observatory that has revolutionized astronomy. The first servicing mission, completed by *Endeavour* during STS-61 in December 1993, corrected Hubble's flawed optics and ensured its continued operational success.

Launched in February 1997, the primary objectives of STS-82 were to upgrade and replace aging components on the Hubble and to install new scientific instruments to boost the telescope's capabilities significantly. On the third day of the mission, *Discovery*'s seven-member crew initiated the first of four planned spacewalks—referred to as Extra-Vehicular Activities (EVAs)—to begin their work on the telescope. Using *Discovery*'s robotic arm, astronaut Steve Hawley, who had deployed the telescope seven years prior, retrieved Hubble from orbit and secured it in the shuttle's payload bay.

The crew's responsibilities were extensive. They replaced two of Hubble's original instruments, the Goddard High Resolution Spectrograph (GHRS) and the Faint Object Spectrograph (FOS), with the Space Telescope Imaging Spectrograph (STIS) and the Near Infrared Camera and Multi-Object Spectrometer (NICMOS), respectively. These new instruments extended Hubble's reach into unseen realms of the universe, with STIS designed to explore supermassive black holes and NICMOS offering a clear view of the universe in the near-infrared spectrum. The mission also installed a refurbished Fine Guidance Sensor (FGS), used to orient the telescope and conduct astrometric science precisely. It replaced one of Hubble's reel-to-reel tape recorders with a modern Solid State Recorder (SSR), capable of storing ten times more data than its predecessor.

During EVA 1, conducted by astronauts Mark Lee and Steven Smith on February 13, the team worked for six hours and 42 minutes to begin the instrument exchange and perform key upgrades. Despite a minor disturbance when a gust of air from the shuttle's airlock unexpectedly affected one of Hubble's solar arrays, the crew remained focused, seamlessly removing and replacing the aged instruments.

EVA 2 held the following day, lasted seven hours and 27 minutes. Astronauts Greg Harbaugh and Joe Tanner installed a new Fine Guidance Sensor and upgraded a failed Engineering and Science Tape Recorder. During this EVA, the astronauts discovered signs of wear on the telescope's thermal insulation, particularly on the sun-facing side, which prompted additional repairs.

On February 15, EVA 3 saw Lee and Smith replace a Data Interface Unit, install the new SSR, and change out one of Hubble's four Reaction Wheel Assemblies (RWA), responsible for pointing the telescope and maintaining its stability. Given the mission's complexity and the growing list of repairs, mission controllers approved a fifth EVA to address the degraded insulation.

EVA 4, completed on February 16, involved Harbaugh and Tanner replacing Hubble's Solar Array Drive Electronics, responsible for controlling the positioning of the solar arrays. The astronauts also installed multi-layer thermal blankets to cover areas where the insulation had deteriorated. Meanwhile, inside *Discovery*, astronauts worked to fabricate additional insulation materials for the telescope.

The final spacewalk, EVA 5, held on February 17, lasted five hours and 17 minutes. Lee and Smith attached thermal insulation blankets to key equipment compartments at the top of Hubble's Support Systems Module, safeguarding its critical electronics and data processing units. By the end of the mission, the crew had completed all planned upgrades and repairs, amassing 33 hours and 11 minutes of EVA time.

Discovery's maneuvering jets were fired periodically throughout the mission to boost Hubble's orbit by eight nautical miles, positioning the telescope in a 335-nautical-mile by 321-nautical-mile orbit. Hubble was redeployed on February 19, 1997, at 1:41 a.m. EST. Initial checks of the newly installed instruments indicated they were functioning as expected. Over the subsequent weeks, NASA conducted extensive calibrations of the new equipment, with the first images and data expected eight to ten weeks later.

STS-82 demonstrated the remarkable versatility of the Space Shuttle as a platform for servicing orbiting satellites. The mission's success in upgrading Hubble's scientific capabilities ensured the telescope would continue to provide groundbreaking discoveries, setting the stage for future missions, including STS-103 in 1999 and STS-109 in 2002.

STS-83

On April 4, 1997, Space Shuttle *Columbia* launched for mission STS-83, the 83rd flight in NASA's Space Shuttle program. This mission, designated for the Microgravity Science Laboratory (MSL-1), aimed to conduct a series of scientific experiments under microgravity conditions. The MSL-1 was housed within the Spacelab module, a proven platform for in-orbit research. The mission held high expectations, set to last 15 days, during which scientists hoped to gain insights into combustion, fluid physics, and material science—fields critical to both space exploration and terrestrial applications.

The crew of STS-83 was composed of seven highly skilled astronauts: Commander James D. Halsell Jr., Pilot Susan L. Still, and Mission Specialists Janice E. Voss, Donald A. Thomas, Michael L. Gernhardt, Roger K. Crouch, and Gregory T. Linteris. Each member played a pivotal role in ensuring the operation of the Spacelab module and managing the complex array of scientific experiments aboard the Shuttle. These experiments included studying microgravity's effects on combustion processes, solidification of metals, and other key materials science investigations that could potentially enhance manufacturing processes on Earth.

However, just three days into the mission, a critical issue arose. Flight controllers on the ground had been closely monitoring Fuel Cell #2, one of the Shuttle's three electrical power-generating units. A persistent anomaly indicated that oxygen and hydrogen were potentially mixing within the cell, a dangerous scenario that risked detonation—a situation reminiscent of the Apollo 13 incident. Despite extensive troubleshooting, the anomaly worsened, and flight rules dictated that the cell be shut down to prevent further escalation. With only two of three fuel cells operational, NASA decided to terminate the mission early, as the Shuttle required all fuel cells to function optimally for a full mission duration. After just 3 days and 23 hours, STS-83 returned to Earth, landing safely on April 8, 1997.

Reflight as STS-94

Despite its abbreviated nature, STS-83 was not considered a failure. After recognizing the significance of the experiments on board, NASA quickly decided to reattempt the mission. In an unprecedented move, *Columbia* was reprocessed for a rapid reflight, bypassing the typical post-mission maintenance cycle. This swift turnaround highlighted NASA's adaptability and commitment to scientific progress. The same crew returned to space on July 1, 1997, under the designation STS-94, successfully completing the mission's original objectives.

The scientific payload of STS-83 and STS-94 was centered around the Microgravity Science Laboratory, a sophisticated platform with facilities for cutting-edge experiments. The Large Isothermal Furnace (LIF), originally developed by the Japanese Space Agency (NASDA), was one of the major facilities aboard. It played a key role in studying the diffusion of liquid metals and alloys, offering valuable insights into the behavior of these materials in the absence of gravity. Another crucial experiment was the Electromagnetic Containerless Processing Facility (TEMPUS), which examined nucleation and surface tension in metallic alloys. TEMPUS enabled scientists to observe the thermophysical properties of undercooled materials, providing data that could improve material stability and performance both in space and on Earth.

Combustion research was also a focus of the mission, with the Combustion Module-1 (CM-1) supporting a series of experiments, including the Laminar Soot Processes (LSP) and the Structure of Flame Balls at Low Lewis-number (SOFBALL). These experiments aimed to better understand how fire behaves in a microgravity environment—a critical area of research for improving fire safety on Earth and future space missions.

The STS-94 mission, which lasted the full 15 days as originally planned, also featured the Droplet Combustion Experiment (DCE). This study explored the fundamental combustion properties of single droplets under various pressure and oxygen levels, which could lead to improvements in combustion engines and environmental control systems.

Advanced measurement tools, such as the Space Acceleration Measurement System (SAMS) and the Orbital Acceleration Research Experiment (OARE), supported these experiments. These instruments monitored the spacecraft's environment, ensuring that the experiments were conducted under optimal microgravity conditions.

Astronaut Chris Hadfield, who served as

CAPCOM during both missions, reflected on the importance of adhering to Flight Rules, noting that they provide clear guidelines for critical decision-making.

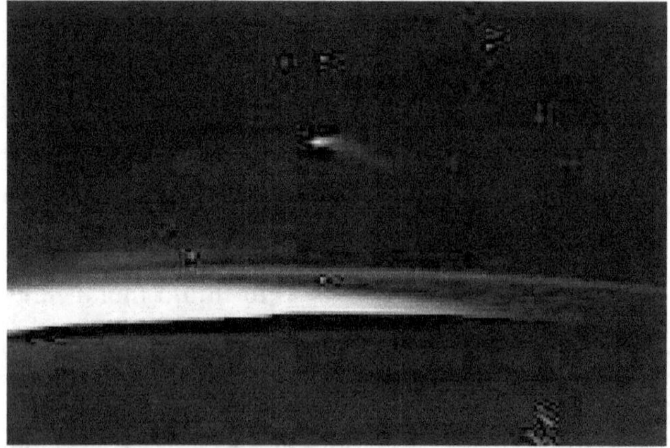

Comet Hale-Bopp as seen from the shuttle

Following the success of STS-94, the mission was lauded as a major achievement in microgravity research. The data collected from the experiments aboard *Columbia* during both missions has since advanced scientific understanding in fields ranging from material science to combustion processes. The swift reflight also demonstrated NASA's commitment to ensuring that scientific objectives are met, even in the face of technical setbacks.

On July 17, 1997, *Columbia* and her crew returned to Earth, successfully completing one of NASA's most unique mission sequences. The legacy of STS-83 and STS-94 lives on through the invaluable data and insights gained from the Microgravity Science Laboratory, which continue to inform scientific advancements on Earth and in space.

STS-84

STS-84, launched on May 15, 1997, marked the sixth Shuttle-Mir docking mission and the ninth overall mission of the Atlantis orbiter. The mission was a significant milestone in the collaborative space program between the United States and Russia, aiming to enhance international cooperation in space exploration and prepare for the construction of the International Space Station (ISS).

Atlantis lifted off from Launch Complex 39A at Kennedy Space Center at 4:07 a.m. EDT, carrying a crew of seven astronauts. The primary objective of STS-84 was to deliver supplies, conduct experiments, and exchange crew members aboard the Russian space station Mir. The mission was commanded by Charles J. Precourt, with Eileen M. Collins serving as the pilot, marking her third spaceflight. The mission specialists included Carlos I. Noriega, Edward T. Lu, Jean-François Clervoy, and Elena V. Kondakova, the latter the first Russian woman to fly aboard the Space Shuttle. Michael Foale, the second British-born astronaut, was also on board, set to replace Jerry Linenger, who had been on Mir since January 1997.

The STS-84 mission, part of NASA's Shuttle-Mir program, marked the sixth docking between the Space Shuttle and the Russian space station Mir. This series of joint missions, which consisted of nine Shuttle-Mir dockings and seven long-duration U.S. astronaut stays aboard Mir, was crucial in expanding international cooperation in space and advancing scientific research. Prior to STS-84, the Shuttle missions involved in the program were STS-71, STS-74, STS-76, STS-79, and STS-81. The program allowed U.S. astronauts to launch and land aboard the Space Shuttle while serving extended stays on Mir, where they contributed to ongoing scientific experiments. Russian cosmonauts utilized their Soyuz spacecraft for launch and return.

During STS-84, which launched on May 15, 1997, Atlantis delivered 3,318 kilograms (7,315 pounds) of supplies, including water, scientific equipment, and logistical materials, to the Mir space station. In total, 465 kilograms (1,025 pounds) of water, 383.2 kilograms (845 pounds) of U.S. science equipment, 1,168.6 kilograms (2,576 pounds) of Russian logistics, and 178.1 kilograms (393 pounds) of miscellaneous materials were transferred to the station. After a successful stay, Atlantis returned to Earth with 407.1 kilograms (898 pounds) of U.S. science materials, 531.2 kilograms (1,171 pounds) of Russian logistics, 14 kilograms (31 pounds) of European Space Agency (ESA) materials, and 170.7 kilograms (376 pounds) of miscellaneous items.

A significant milestone of STS-84 was the exchange of U.S. astronaut C. Michael Foale with Jerry Linenger, marking the fourth successive U.S. astronaut to live aboard Mir. Linenger, who had arrived on Mir aboard STS-81 on January 15, 1997, spent a total of 123 days aboard the station and 132 days in space, placing him second only to Shannon Lucid for the most time spent in orbit by an American astronaut at that point. Linenger's mission also marked the first anniversary of continuous U.S. presence in space, which had begun with Lucid's arrival on March 22, 1996.

Linenger's stay on Mir was eventful. One of the most notable achievements was the first U.S.-Russian spacewalk on April 29, 1997. During this five-hour extravehicular activity (EVA), Linenger and Mir 23 Commander Vasily Tsibliyev installed the Optical Properties Monitor (OPM) on the station's exterior. This experiment was designed to study the effects of space on optical materials, particularly those used in telescopes, over a nine-month period. However, not all events were planned. On February 23, a fire broke out aboard the aging space station. While the fire caused minimal damage, the crew was required to wear protective masks for 36 hours until the station's air was fully cleaned. At the time of the incident, Linenger was aboard Mir with two Russian cosmonauts from Mir 22, two cosmonauts from Mir 23, and a German cosmonaut.

The actual docking of STS-84 with Mir occurred on May 17, 1997, at 02:33 UTC, over the Adriatic Sea. The hatches between the two spacecraft were opened just before dawn, at 04:25, and the crews—comprising the Shuttle and Mir 23 members—exchanged greetings and conducted a safety briefing. By 14:15 UTC, Linenger and Foale officially exchanged places, continuing the U.S. presence aboard Mir.

The transfer of supplies and equipment between the Shuttle and Mir was a significant part of the mission, with the transfer completed ahead of schedule. One of the most critical items moved to the station was an Elektron oxygen-generating unit. In total, 249 items, including nearly 450 kilograms (990 pounds) of water, were transferred between the two spacecraft. The total weight of water, experiment samples, supplies, and hardware exchanged during STS-84 was around 3,400 kilograms (7,500 pounds).

Foale's research program aboard Mir included 35 investigations across six scientific disciplines, including advanced technology, Earth observations, biology, human life sciences, space station risk mitigation, and microgravity sciences. Of these, 28 were continued from previous missions, while seven were newly initiated, particularly in the areas of biological and crystal growth studies, and materials processing. One notable observation during his mission was the imaging of Comet Hale-Bopp, a celestial event visible from space.

On May 22, 1997, at 01:04 UTC, Atlantis undocked from Mir. Unlike previous Shuttle-Mir missions, there was no flyaround of the station. However, Atlantis paused three times during its departure to allow data collection from a European sensor device, intended to assist future rendezvous operations of the proposed European Space Agency (ESA) Automated Transfer Vehicle (ATV) for resupplying the International Space Station (ISS).

During the mission, the crew also conducted various experiments using the Biorack facility, housed in Atlantis's SpaceHab Double Module, and performed environmental air samplings, radiation monitoring, and a detailed photographic survey of Mir during the docked phase. Throughout the mission, Atlantis performed exceptionally, equipped with the SpaceHab Double Module weighing 4,187 kilograms (9,231 pounds) and the Orbiter Docking System weighing 1,922 kilograms (4,237 pounds).

STS-84 concluded on May 24, 1997, when Atlantis touched down, marking another successful milestone in the Shuttle-Mir program and contributing valuable experience and data to the cooperative effort between NASA and the Russian space agency, laying the groundwork for the future International Space Station.

STS-85

STS-85, launched on August 7, 1997, was a vital mission in exploring and understanding Earth's atmosphere and climate. The mission, the 23rd flight of the orbiter Discovery, emphasized atmospheric research, deploying and operating several scientific payloads designed to study the Earth's environment and advance knowledge in atmospheric sciences.

The mission was commanded by Curtis L. Brown, Jr., with Kent V. Rominger serving as the pilot. The crew included four mission specialists: N. Jan Davis, Robert L. Curbeam, Jr., Stephen K. Robinson, and Canadian Space Agency astronaut Bjarni Tryggvason. This diverse and experienced crew conducted multiple experiments and deployed critical instruments designed to observe and measure atmospheric phenomena.

One of the primary objectives of STS-85 was the deployment of the CRISTA-SPAS (Cryogenic Infrared Spectrometers and Telescopes for the Atmosphere – Shuttle Pallet Satellite) payload. CRISTA-SPAS was a sophisticated German-built satellite to study the Earth's middle atmosphere. The satellite was equipped with spectrometers and telescopes to measure the atmosphere's composition, explicitly focusing on water vapor, ozone, and other trace gases. These measurements were crucial for understanding the dynamics of the atmosphere and the processes that influence climate change.

After its deployment on the mission's second day, CRISTA-SPAS operated independently, collecting data on the Earth's atmosphere. At the same time, Discovery performed a series of maneuvers to maintain its relative position for later retrieval. The satellite's data contributed to understanding atmospheric circulation patterns and the distribution of trace gases, providing insights essential for climate modeling and environmental science.

In addition to CRISTA-SPAS, STS-85 carried the Japanese Manipulator Flight Demonstration (MFD), an experiment to test the performance of a small robotic arm in space. This experiment was significant for future space missions, as it demonstrated the potential for using robotic systems to perform tasks that would otherwise require human intervention, thus enhancing the safety and efficiency of space operations.

The mission also included the deployment of the Technology Applications and Science-1 (TAS-1) payload, which housed several smaller experiments. The Southwest Ultraviolet Imaging System (SWUIS) observed the Earth's atmosphere and nearby celestial bodies in ultraviolet light. The data collected by SWUIS would help scientists understand more about the ozone layer and other atmospheric components that are difficult to study from the ground.

During the mission, the crew conducted a series of detailed Earth observations, taking advantage of the Shuttle's unique vantage point in low Earth orbit. These observations included studies of land use patterns, ocean dynamics, and atmospheric phenomena. The crew used handheld cameras and specialized instruments to capture images and data valuable for environmental monitoring and scientific research.

On the final day of the mission, the crew successfully retrieved the CRISTA-SPAS satellite, which had completed its atmospheric observations. The retrieval was a delicate operation, requiring precise coordination between the Shuttle and the satellite. The successful completion of this task underscored the versatility and capability of the Shuttle as a platform for deploying and retrieving payloads in space.

STS-85 concluded with a smooth landing at Kennedy Space Center on August 19, 1997, at 7:07 a.m. EDT. The mission lasted 11 days, 20 hours, and 27 minutes, during which the crew orbited the Earth 185 times and traveled over 4.7 million miles. The mission's achievements in atmospheric science and technology demonstrations significantly contributed to the broader goals of Earth observation and space exploration, providing valuable data and experience for future missions.

STS-86

STS-86, launched on September 25, 1997, was a significant chapter in the Shuttle-Mir program. It marked the seventh Shuttle-Mir docking mission and the tenth flight of the orbiter Atlantis. The mission underscored the growing cooperation between the United States and Russia in space exploration, with objectives that included crew exchange, scientific research, and the delivery of supplies to the Mir space station.

The crew of STS-86 was led by Commander James D. Wetherbee, with Pilot Michael J. Bloomfield. The mission specialists were Vladimir G. Titov, Scott E. Parazynski, Jean-Loup Chrétien, Wendy B. Lawrence, and David A. Wolf. Notably, Vladimir Titov, a Russian cosmonaut, brought valuable experience to the mission, having previously flown on both the Soyuz and the Space Shuttle. Jean-Loup Chrétien, a French astronaut, also added an international dimension to the crew, reflecting the global collaboration the Shuttle-Mir program symbolized.

The seventh Mir docking mission, STS-86, carried a SPACEHAB double module to support cargo transfer, an astronaut exchange, and critical logistical operations between the Space Shuttle *Atlantis* and the Russian space station Mir. This mission marked the seventh Shuttle-Mir docking, following STS-71, STS-74, STS-76, STS-79, STS-81, and STS-84, and played a pivotal role in advancing the NASA-Russian Space Agency Phase 1B cooperative effort.

Atlantis spent 10 days in space, with five days of docked operations between the Shuttle and Mir. One of the primary objectives was to exchange astronauts Michael Foale and David Wolf, continuing the presence of an American astronaut aboard the Russian space station. This crew transfer extended the collaboration between the U.S. and Russia in human spaceflight, with Wolf becoming the sixth consecutive U.S. astronaut to live aboard Mir. However, the decision to proceed with the transfer was not without controversy. Due to safety concerns following incidents aboard Mir during Foale's and Jerry Linenger's stays, NASA Administrator Daniel Goldin made the final call to proceed with Wolf's mission the night before launch.

Foale returned to Earth aboard *Atlantis* after spending 145 days in space, including 134 aboard Mir. A critical incident marked his time on Mir on June 25, when a Progress resupply vehicle collided with the Spektr module during manual docking by Mir 23 Commander Vasili Tsibliev. The impact damaged a solar array and depressurized the module, forcing the crew to seal the hatch to Spektr, leaving behind Foale's personal effects and several NASA science experiments. Tsibliev and Flight Engineer Aleksandr Lazutkin had planned a spacewalk to reconnect power to Spektr's undamaged solar arrays, but Tsibliev's health complications, discovered on July 13, shifted these duties to the incoming Mir 24 crew.

NASA also made a critical crew change when Wendy Lawrence, initially scheduled to replace Foale, was replaced by David Wolf due to her inability to perform spacewalks in the Russian Orlan suit. Wolf's experience made him better suited to assist with the planned repairs to Spektr. Upon their arrival on August 7, Mir 24 Commander Anatoly Solovyev and Flight Engineer Pavel Vinogradov conducted an internal spacewalk inside the depressurized Spektr on August 22, reconnecting power cables and restoring partial functionality to the module.

On September 5, Foale and Solovyev performed a six-hour external spacewalk to assess the damage to Spektr. During this EVA, two solar arrays were repositioned, and a radiation experiment from Jerry Linenger's mission was retrieved. This work was crucial for future repairs, although the breach in Spektr's hull remained elusive.

The historic docking of *Atlantis* and Mir occurred at 3:58 pm EDT on September 27, with hatches opened at 5:45 pm. The next day, David Wolf officially became a member of Mir 24, and Foale transitioned to the STS-86 crew. Wolf's assignment marked the continuation of NASA's commitment to maintaining a presence aboard Mir, concluding with Andrew Thomas's transfer during the STS-89 mission in January 1998.

A key highlight of STS-86 was the first joint U.S.-Russian extravehicular activity (EVA) conducted during a Shuttle mission, marking the 39th spacewalk of the Shuttle program. Scott Parazynski and Vladimir Titov performed a five-hour, one-minute spacewalk on October 1, attaching a solar array cap to assist future repairs to Spektr and retrieving four Mir Environmental Effects Payloads (MEEPs). This spacewalk also tested components of the Simplified Aid for EVA Rescue (SAFER) jet packs.

Throughout the docked phase, the combined Mir 24 and STS-86 crews transferred over four tons of cargo from the SPACEHAB module to Mir. This included water, experiment hardware, attitude control computers, and various logistical items critical to the station's continued operations. In exchange, experiment samples and hardware, including an outdated Elektron oxygen generator, were moved to *Atlantis* for return to Earth. The undocking took place at 1:28 pm EDT on October 3, followed by a 46-minute flyaround of Mir to conduct a visual inspection.

During the mission, *Atlantis* fired its small jet thrusters to gather data for the Mir Structural Dynamics Experiment (MISDE), which measured vibrations affecting Mir's structure and solar arrays. Several scientific experiments were also conducted aboard the Shuttle, including the Commercial Protein Crystal Growth experiment and various radiation monitoring and atmospheric experiments.

STS-86 also contributed to NASA's educational outreach efforts, with the "Seeds in Space-II" program and the KidSat project designed to engage students in space science. The mission concluded with *Atlantis* landing at Kennedy Space Center on Runway 15, successfully ending the continued Shuttle-Mir collaboration.

STS-87

STS-87, launched on November 19, 1997, from Launch Complex 39B at the Kennedy Space Center, was the 88th flight of the Space Shuttle program and the 24th flight of Space Shuttle Columbia. This mission had several significant goals, including conducting scientific experiments using the United States Microgravity Payload (USMP-4), deploying and retrieving the SPARTAN-201 satellite, and performing two extravehicular activities (EVAs).

The mission was particularly notable for being the first time that an EVA was conducted from Columbia. Although EVAs were initially planned for STS-5 in 1982 and STS-80 in 1996, they were canceled due to various technical issues, including spacesuit malfunctions and airlock problems. On STS-87, however, the crew successfully carried out two spacewalks, with astronaut Takao Doi making history as the first Japanese astronaut to perform an EVA.

The EVAs, critical to the mission's objectives, involved testing hardware for future space station assembly and retrieving the Spartan satellite after it encountered deployment issues. These tasks advanced both the International Space Station's (ISS) development and ongoing microgravity research, marking STS-87 as a pivotal mission in the Space Shuttle program's legacy.

The mission carried a crew of six astronauts, led by Commander Kevin R. Kregel, with Steven W. Lindsey as the pilot. The mission specialists were Winston E. Scott, Kalpana Chawla, Takao Doi from Japan, and Leonid K. Kadenyuk, the first astronaut from an independent Ukraine. Takao Doi became the first Japanese astronaut to perform an EVA during the mission, a historic moment for Japan's involvement in human spaceflight.

On November 25, 1997, at 00:02 UTC, astronauts Scott and Doi embarked on their first spacewalk (EVA 1) during the STS-87 mission aboard Space Shuttle Columbia. This spacewalk lasted for 7 hours and 43 minutes, concluding at 07:45 UTC. Their second spacewalk (EVA 2) took place on December 3, 1997, starting at 09:09 UTC and lasting for 4 hours and 59 minutes, ending at 14:09 UTC. These spacewalks were part of a critical mission that included various scientific experiments and payloads designed to advance our understanding of microgravity and space physics.

STS-87 carried the United States Microgravity Payload (USMP-4), an advanced set of experiments to study microgravity's effects on various materials and physical processes. Managed by the Marshall Space Flight Center in Huntsville, Alabama, USMP-4 consisted of experiments that were housed in two Mission-Peculiar Experiment Support Structures (MPESS) located in the shuttle's payload bay. These experiments benefitted from the extended mission capabilities provided by the Extended Duration Orbiter (EDO) kit, allowing for more extensive data collection during the mission.

A key experiment during STS-87 was the deployment and retrieval of the Spartan-201 spacecraft, which was designed to study the Sun's outer atmosphere, known as the solar corona. Deployed on the 18th orbit and retrieved on the 52nd, Spartan-201 carried the Ultraviolet Coronal Spectrometer and the White Light Coronograph. These instruments aimed to investigate the solar corona's heating mechanisms and solar wind's acceleration, contributing vital data to solar physics.

In addition to its primary solar physics experiments, Spartan-201 carried several secondary experiments, including the Technology Experiment Augmenting Spartan (TEXAS). This real-time communications link made crucial adjustments to the Spartan's White Light Coronograph based on solar images. The Video Guidance Sensor (VGS) Flight Experiment, another secondary payload, tested a laser guidance system that would later become an integral part of NASA's Automated Rendezvous and Capture (AR&C) system.

Several scientific experiments were conducted in the payload bay, including the Advanced Automated Directional Solidification Furnace (AADSF), which was used to study the solidification of semiconductor crystals. By moving liquid samples through temperature zones, scientists were able to observe the solidification process under microgravity conditions, gathering critical data on the composition and structure of these materials.

Another important experiment was the Confined Helium Experiment (CHeX), which tested the behavior of helium in a confined, two-dimensional space. This experiment provided insights into how boundaries influence material properties, contributing to the understanding of quantum physics and fluid dynamics.

Materials science also played a prominent role in STS-87. The Isothermal Dendritic Growth Experiment (IDGE) investigated dendritic solidification, a process in which tree-like crystal structures form as metals solidify. By studying dendrite size, shape, and interaction, scientists aimed to improve the manufacturing and processing of metallic alloys.

The Material pour l'Étude des Phénomènes Intéressant la Solidification sur Terre et en Orbite (MEPHISTO) experiment, also focused on solidification, processed three identical bismuth and tin alloy samples. Each sample was subjected to different electrical techniques to measure the solidification front's position, temperature, and velocity. Data collected from MEPHISTO were correlated with readings from the Space Acceleration Measurement System (SAMS), providing a clearer understanding of how microgravity affected the materials' behavior.

The SAMS system itself was a key part of the mission, monitoring the microgravity environment during the experiments. The system's sensors detected minute accelerations, capturing data on how vibrations and movements affected the shuttle's cargo bay experiments. This information was crucial for refining experimental conditions and understanding how microgravity influences physical processes.

The Orbital Acceleration Research Experiment (OARE), an additional payload, measured the aerodynamic forces acting on the shuttle at orbital altitudes. This sensitive instrument collected data on low-level aerodynamic acceleration, which supported both space materials processing and the prediction of orbital drag, enhancing NASA's understanding of how spacecraft interact with the Earth's upper atmosphere.

The Shuttle Ozone Limb Sounding Experiment (SOLSE) was designed to study the altitude distribution of ozone in the Earth's atmosphere, providing critical data to predict changes in atmospheric composition. Using advanced Charged Coupled Device (CCD) technology, SOLSE was a simplified, low-cost instrument intended to map ozone distribution. Housed in a Hitchhiker (HH/GAS) canister equipped with a Motorized Door Assembly (HMDA), the instrument included a UV spectrograph, CCD array detector, and visible light cameras. Once in orbit, a crew member activated SOLSE to perform limb and Earth viewing observations. Limb observations targeted altitudes between 20 and 50 kilometers (12 to 31 miles) above the Earth's surface, while Earth viewing observations enabled data correlation with other ozone-mapping instruments.

The Loop Heat Pipe (LHP) experiment tested advanced thermal energy management technology, crucial for future spacecraft. Using anhydrous ammonia as the working fluid, the LHP transported thermal energy with high conductivity in zero gravity. This passive two-phase flow device could transfer up to 400 watts over 5 meters through semiflexible tubes, relying on capillary forces to circulate the working fluid. Heat applied to the evaporator vaporized part of the fluid, which then condensed to release heat and returned via capillary action, making the system self-sustaining.

The Sodium Sulfur Battery Experiment (NaSBE) marked the first space test of sodium-sulfur battery technology. Each of the four battery cells, consisting of a sodium anode, sulfur cathode, and solid ceramic electrolyte, operated at 350°C to liquefy the sodium and sulfur, generating electrical power. Once in orbit, the experiment was activated by the crew, and its performance was monitored from the Payload Operations Control Center (POCC) on Earth. NaSBE's success demonstrated the potential of sodium-sulfur batteries for future space missions.

The Turbulent Gas Jet Diffusion Flames (TGDF) experiment aimed to understand the behavior of gas jet diffusion flames under microgravity conditions. The experiment helped predict how such flames would behave in space by imposing large-scale disturbances on well-defined laminar flames. Controlled disturbances were applied at frequencies of 2.5 Hz, 5 Hz, and 7.5 Hz, providing valuable data for future spacecraft fire safety measures.

The Get Away Special (GAS G-036) payload included four separate experiments that exposed various materials to the space environment. The experiments included the Cement Mixing Experiment (CME), which hydrated cement samples in microgravity, and the Configuration Stability of Fluid Experiment (CSFE), which analyzed fluid behavior. Additionally, the Computer Compact Disc Evaluation Experiment (CDEE) tested the durability of computer discs, and the Asphalt Evaluation Experiment (AEE) examined the effects of the space environment on asphalt samples.

The Extended Duration Orbiter (EDO) Pallet, a 15-foot (4.6-meter) diameter cryo-kit structure, enabled the shuttle to support missions lasting up to 18 days. Weighing 352 kilograms (776 lbs), the pallet housed liquid hydrogen and liquid oxygen tanks, which supplied the shuttle's fuel cells. These cells generated enough electrical power to support the equivalent of four households for six months and produced over 1,360 kilograms (3,000 lbs) of drinking water. This extended mission duration allowed for enhanced scientific research in microgravity, life sciences, and Earth observation.

The Middeck Glovebox (MGBX) was a critical facility for handling materials science and biological experiments in a controlled environment. On this mission, it hosted three key experiments:

Wetting Characteristics of Immiscibles (WCI), which investigated how wetting characteristics influenced the segregation of immiscible liquids during microgravity processing.

Enclosed Laminar Flames (ELF), which aimed to validate models of flame stabilization in zero gravity and study the interaction between diffusion flames and vortex flow.

Particle Engulfment and Pushing by Solidifying Interfaces (PEP), which explored the behavior of particles at liquid/solid interfaces, enhancing understanding of metal-ceramic mixtures and solidification physics in space.

The Collaborative Ukraine Experiment (CUE) was a significant mid-deck payload aboard the Space Shuttle, designed to investigate the effects of microgravity on plant growth. As a model for international scientific collaboration between Ukraine and the United States, the CUE involved researchers from Kansas State University, Louisiana State University, and Ukrainian institutions. The experiment was conducted using the Plant Growth Facility (PGF) and the Biological Research in Canisters (BRIC), with additional support from a Gaseous Nitrogen (GN2) Freezer and fixation hardware.

The PGF was a sophisticated system that supported plant growth for up to 30 days in space, maintaining optimal environmental conditions for normal plant development. It consisted of several subsystems:

Control and Data Management Subsystems (CDMS) for monitoring and controlling the experiment.

Fluorescent Light Module (FLM), which provided the necessary light for photosynthesis.

Atmospheric Control Module (ACM) to regulate air quality.

Plant Growth Chambers (PGCs), where the plants were housed.

Support Structure Assembly (SSA) and the Generic External Shell (GES) to provide the structural framework.

This entire system was compact enough to fit within one mid-deck locker and operated on a 28 V direct current (DC) power supply. The plant chosen for this study was Brassica rapa, commonly known as turnip. By studying plant growth in the unique environment of space, the CUE aimed to advance our understanding of how microgravity impacts biological processes, with potential applications for long-term space missions.

Another key component of the mission was the Extravehicular Activity Development Flight Test – 05 (EDFT-05). This test focused on validating hardware and procedures for future Extravehicular Activity (EVA) missions on the International Space Station (ISS). As part of Detailed Test Objective (DTO) 671, the EDFT-05 evaluated end-to-end EVA assembly and maintenance operations, which are critical for ISS construction and upkeep. Additional DTOs included testing the Extravehicular Mobility Unit (EMU) Electrical Cuff Checklist (DTO 672) and the EMU Thermal Comfort and EVA Worksite Thermal Environment (DTO 833). Two EVAs were performed during this mission to accomplish these objectives, further expanding the experience base for both flight and ground crews involved in EVA operations.

The Autonomous EVA Robotic Camera (AERCam/Sprint) was a groundbreaking addition to EVA technology on this mission. AERCam/Sprint was a small, free-flying, spherical camera platform designed for use outside the spacecraft. Equipped with a self-contained cold gas propulsion system, the free-flyer could be controlled with six degrees of freedom. Rate sensors provided data to maintain automatic attitude control, allowing the AERCam/Sprint to hover steadily while recording.

Designed with safety in mind, the free-flyer was covered in soft cushioning material to prevent damage in the event of a collision. It moved slowly, minimizing potential impact forces, while its lightweight construction ensured that any energy from an impact would be absorbed. The device was controlled from inside the shuttle via a small control station, where an astronaut used a Simplified Aid For EVA Rescue (SAFER) device controller to input commands. These commands were transmitted to AERCam/Sprint through an ultra-high-frequency (UHF) radio modem link.

The AERCam/Sprint provided a new method for astronauts to inspect the exterior of the spacecraft without the need for a human to perform an EVA, offering a safer and more efficient means of capturing visual data during space missions. This innovation represented a step forward in autonomous systems for space exploration, with potential applications in future missions to the ISS and beyond.

Columbia carried a payload mass of 4,451 kilograms (9,813 pounds) and returned with a landing mass of 102,717 kilograms (226,452 pounds). The mission ended successfully on December 5, 1997, at 12:20 UTC, with a landing at Kennedy Space Center's Shuttle Landing Facility (SLF) on Runway 33.

STS-89

STS-89, launched on January 22, 1998, from Kennedy Space Center in Florida, was the eighth mission to the Russian Space Station Mir under the Shuttle-Mir Program, a collaborative effort between NASA and the Russian space agency, Roscosmos. The mission, flown by the Space

Shuttle Endeavour, was a vital component of the joint program to foster international cooperation in space, exchange scientific knowledge, and prepare for future long-duration missions aboard the International Space Station (ISS).

Commanded by Terrence W. Wilcutt, the STS-89 crew included Pilot Joe F. Edwards Jr. and Mission Specialists Bonnie J. Dunbar, Michael P. Anderson, James F. Reilly, Salizhan S. Sharipov, and Andrew S. W. Thomas. The mission was especially significant for Thomas, who was set to replace astronaut David A. Wolf as the resident NASA astronaut aboard Mir, continuing the tradition of U.S. astronauts living and working aboard the Russian station.

On January 22, 1998, Space Shuttle *Endeavour* launched on STS-89, marking the eighth of nine planned missions to the Russian space station Mir. This mission was notable for being the fifth involving the exchange of U.S. astronauts aboard Mir. Astronaut David Wolf, who had been residing on the station since late September 1997, was replaced by Astronaut Andrew Thomas. Wolf's time on Mir contributed to crucial U.S.-Russian collaboration in space, particularly in preparation for long-duration missions aboard the International Space Station (ISS).

Thomas spent approximately four months aboard Mir, where he conducted a variety of scientific experiments in microgravity. His stay symbolized the deepening partnership between NASA and the Russian space agency as they built operational experience for the future ISS. Thomas returned to Earth in late May 1998 during the final Shuttle-Mir docking mission, STS-91, when Space Shuttle *Discovery* arrived to bring him home.

STS-89, however, was significant not just for the astronaut exchange but also for the scientific cargo it carried. More than 3,175 kilograms (7,000 pounds) of experiments, supplies, and hardware were transferred between the Shuttle and Mir, underscoring the importance of these missions for advancing both American and Russian research in space.

A critical component of the mission was the SPACEHAB module, which housed several important experiments and payloads. The Advanced X-Ray Detector (ADV XDT) and Advanced Commercial Generic Bioprocessing Apparatus (ADV CGBA) were among the key scientific instruments aboard, designed to study biological processes in microgravity and develop commercial applications for space-based research. Other notable payloads included the Mechanics of Granular Materials (MGM) Experiment, which sought to understand how granular materials behave in weightless environments, a finding with potential implications for construction on Earth and in space.

Additionally, the mission included experiments crucial for the development of future space exploration technologies. These included the Intra-Vehicular Radiation Environment Measurements (RME-1312), which monitored radiation levels inside the spacecraft, and the Volatile Removal Assembly prototype, a vital component for the ISS Water Recovery System that would help sustain astronauts in long-term missions by recycling water aboard the station.

In-cabin experiments further contributed to NASA's understanding of life sciences and physical phenomena in space. The Microgravity Plant Nutrient Experiment (MPNE) explored how plants absorb nutrients in a microgravity environment, information that could influence future space agriculture. The TeleMedicine Instrumentation Pack (TMIP) tested equipment that would allow astronauts to perform remote medical diagnostics, a necessary technology for long-duration space missions far from Earth.

STS-89 also carried numerous Getaway Special experiments, which provided opportunities for universities and international partners to conduct space-based research. Notable among these were the University of Michigan's Vortex Ring Transit Experiment (VORTEX) and the German Aerospace Center's experiments on Marangoni convection and glass fining, both investigating fluid dynamics in microgravity. Additionally, the Chinese Academy of Sciences contributed experiments on crystal growth and material sciences, reflecting the global interest in space-based research.

The mission concluded successfully with *Endeavour*'s landing at Kennedy Space Center on January 31, 1998.

STS-90

STS-90, launched on April 17, 1998, from Kennedy Space Center in Florida, was a mission dedicated to advancing our understanding of the human nervous system and the effects of spaceflight on the brain and body. Flown by the Space Shuttle Columbia, this mission was the final flight of the Spacelab module. This reusable laboratory had been a cornerstone of scientific research in space for over a decade.

Commanded by Richard A. Searfoss, the crew of STS-90 included Pilot Scott D. Altman, mission Specialists Richard M. Linnehan, Dafydd (Dave) Williams, Kathryn P. Hire, and Specialists in payload research Dr. Jay C. Buckey and Dr. James A. Pawelczyk. The mission's primary objective was to conduct a series of experiments under the Neurolab program, a collaborative effort between NASA, the National Institutes of Health (NIH), and international partners.

Neurolab was a significant mission within NASA's Space Shuttle program, designed to explore the effects of microgravity on the nervous system. Conducted aboard Space Shuttle *Columbia* during the STS-90 mission, Neurolab focused on advancing the understanding of how the nervous system adapts to the unique environment of space. The primary goals of the mission were to investigate how microgravity influences neurological and behavioral changes, specifically targeting the adaptation of the vestibular system, the central nervous system, and the pathways that control spatial awareness in the absence of gravity. It also sought to explore the impact of microgravity on the development of the nervous system in both humans and animals. Leading the scientific efforts was Mary Anne Frey, whose work played a crucial role in shaping the experiments conducted during the mission.

Neurolab was a collaborative international effort, bringing together six space agencies and seven U.S. research institutions. Scientists and researchers from nine countries participated, conducting 31 different studies in the microgravity environment of space. Key partners included the National Institutes of Health (NIH), the National Science Foundation (NSF), the Office of Naval Research, and space agencies from Canada (CSA), France (CNES), Germany (DLR), Japan (NASDA), and the European Space Agency (ESA).

Of the 31 studies, 26 focused specifically on the nervous system, one of the most intricate and least understood systems in the human body. The experiments involved a variety of test subjects, including rats, mice, crickets, snails, fish, and even the astronauts themselves. These tests were designed to provide insights into how the nervous system develops and functions in space, contributing to our broader understanding of neuroscience and space physiology. The majority of these experiments took place in the pressurized Spacelab long module housed within *Columbia*'s payload bay. Notably, STS-90 marked the 16th and final scheduled flight of the ESA-developed Spacelab module, although Spacelab pallets would continue to be utilized on the International Space Station (ISS) for future research.

While most of the planned research was conducted successfully, the Mammalian Development Team faced unexpected challenges. Due to the high mortality rate of neonatal rats onboard, the team had to adjust its priorities and refocus its science activities. Despite this setback, the mission made substantial contributions to space neuroscience.

In addition to Neurolab's primary experiments, STS-90 also carried several other payloads, including the Shuttle Vibration Forces experiment, the Bioreactor Demonstration System-04, and three Get-Away Special (GAS) canister investigations. These supplementary experiments contributed to the growing body of knowledge about how spaceflight affects both biological and mechanical systems.

STS-90 was notable for being the first mission to perform an Orbital Maneuvering System (OMS) assist burn during ascent, demonstrating the shuttle's increasing versatility in orbital operations. Furthermore, the crew faced an onboard challenge when a valve in the Regenerative Carbon Dioxide Removal System malfunctioned, threatening to cut the mission short. However, with the assistance of engineers on the ground, the crew ingeniously used aluminum tape to bypass the faulty valve, allowing the mission to continue as planned.

The crew, consisting of seven astronauts, included Mission Specialist Kathryn Hire, who was the first employee from Kennedy Space Center to be selected as an astronaut candidate. Three of the astronauts, including mission specialists Jay Buckey, Richard Linnehan, and Payload Specialist James Pawelczyk, made notable public appearances on the Canadian television series *Popular Mechanics for Kids* after the mission, further expanding the reach of NASA's educational initiatives.

Although the Mission Management Team considered extending the mission by one day, they ultimately decided against it, as the science community determined that the additional time was unnecessary. Additionally, weather conditions were predicted to deteriorate after the scheduled landing on May 3, 1998, so the mission concluded as planned.

A curious occurrence during the STS-90 mission was the observation of a bat landing on the shuttle's External Tank before liftoff. While the bat flew away after main engine start, it marked the second such incident in the shuttle program, the first occurring on STS-72. Another similar event would later occur during STS-119, where the bat remained attached to the tank throughout liftoff.

The STS-90 mission and its Neurolab experiments played a pivotal role in advancing our understanding of how microgravity affects the nervous system and contributed significantly to both basic neuroscience research and spaceflight medicine. The findings from this mission continue to inform our knowledge of human physiology in space, particularly in preparation for long-duration missions such as those planned for the Moon and Mars.

STS-91

STS-91, launched on June 2, 1998, from Kennedy Space Center in Florida, was a significant mission in the history of space exploration. It marked the final Shuttle-Mir docking mission. Flown by the Space Shuttle Discovery, this mission was the last in a series of joint missions between NASA and the Russian space agency, Roscosmos, under the Shuttle-Mir Program, a precursor to the International Space Station (ISS).

Commanded by Charles J. Precourt, the crew of STS-91 included Pilot Dominic L. Pudwill Gorie, and Mission Specialists Wendy B. Lawrence, Franklin R. Chang-Díaz, Janet L. Kavandi, and Russian cosmonaut Valery V. Ryumin, who served as a representative of the Russian Space Agency. The mission's primary objectives were to conduct scientific research, deliver supplies to the Mir space station, and bring back NASA astronaut Andrew S. W. Thomas, who had been living on Mir since January 1998 as part of his long-duration

mission.

STS-91 marked a significant milestone in the Space Shuttle program, serving as the final Shuttle-Mir docking mission and the only docking involving Space Shuttle *Discovery*. This mission was a key element of the Phase 1 Program, which laid the groundwork for the International Space Station (ISS) by maintaining a continuous American presence in space. It also played a crucial role in developing the procedures and hardware necessary for international collaboration in space exploration.

A notable feature of STS-91 was the use of the new super lightweight external tank (SLWT) for the first time. Although it maintained the same dimensions as the previous external tanks—154 feet (47 meters) long and 27 feet (8.2 meters) in diameter—the SLWT was 7,500 pounds (3,400 kilograms) lighter. This weight reduction was achieved by using an aluminum-lithium alloy and enhancing the tank's structural design. The redesigned hydrogen tank featured a machined, orthogonal waffle-like pattern, which increased strength and stability. These improvements would later contribute to boosting the payload capacity for missions to the International Space Station, making the Shuttle more versatile for its future role in constructing the ISS.

On June 4, 1998, *Discovery* docked with the Russian space station *Mir* at 16:01 UTC, at an altitude of 208 miles. The hatches between the two spacecraft were opened at 14:34 the same day, marking the official transition of astronaut Andy Thomas from *Mir*'s crew to *Discovery*'s.

Thomas had spent 130 days aboard *Mir*, contributing to a cumulative total of 907 days spent by seven U.S. astronauts aboard the Russian station as long-duration crew members. Over the next four days, the crews of Mir 25 and STS-91 transferred more than 1,100 pounds (500 kilograms) of water and exchanged nearly 4,700 pounds (2,130 kilograms) of cargo, including experiments and supplies. Several long-term U.S. experiments from *Mir* were moved into *Discovery*'s middeck and the SPACEHAB module in the payload bay, including the Space Acceleration Measurement System (SAMS), tissue engineering co-culture (COCULT) investigations, and two crystal growth experiments. Additionally, the crews conducted Risk Mitigation Experiments (RMEs) and Human Life Sciences (HLS) investigations.

When the hatches closed at 09:07 on June 8, and *Discovery* undocked from *Mir* at 12:01 the same day, it brought an end to the historic Shuttle-Mir program, concluding Phase 1 of the International Space Station (ISS) program. The sight of *Mir* drifting away, as seen from *Discovery*, was symbolic of the transition from the cooperative Shuttle-Mir missions to the next era of international partnership in space: the assembly and operation of the ISS.

In addition to the docking, STS-91 carried the prototype of the Alpha Magnetic Spectrometer (AMS), an ambitious experiment designed to search for dark matter and missing matter in the universe. The AMS was powered up on Flight Day 1. However, a malfunction in *Discovery*'s Ku-band communications system prevented the transmission of high-rate data, including television signals, to Earth. The crew managed to bypass this issue by using S-band/FM communications when within range of ground stations, ensuring that AMS data could still be sent. Some data was also recorded onboard the shuttle for analysis post-mission.

The Ku-band system failure was traced to a component inaccessible to the crew, leading to the loss of television transmission for the duration of the mission. Communications with *Mir* were similarly affected, as television signals from the Russian space station were blocked due to a problem between a Russian ground station and mission control near Moscow, reducing real-time imagery to audio-only coverage.

During the mission, the *Discovery* crew also evaluated new electronics and software for the Shuttle's robotic arm, which would be crucial for ISS assembly missions. Additional experiments included eight Get Away Special payloads in the cargo bay, as well as combustion, crystal growth, and radiation monitoring studies conducted within *Discovery*'s mid-deck crew cabin area.

STS-95

STS-95, launched on October 29, 1998, from Kennedy Space Center, Florida, was one of the most iconic missions of the Space Shuttle program. Using the orbiter *Discovery*, this flight marked the 25th mission for *Discovery* and the 92nd overall mission since the Space Shuttle program began in April 1981. The mission was highly publicized, largely due to the participation of John H. Glenn Jr., the former Project Mercury astronaut and U.S. Senator, who returned to space at 77. Glenn had previously made history as the first American to orbit the Earth in 1962, and his return to space 36 years later was a significant moment in both space exploration and American history. As the oldest person to reach Earth orbit, Glenn set a record that remained for 23 years. Though Wally Funk and William Shatner surpassed Glenn's age on suborbital flights in 2021, Glenn remains the oldest person to orbit Earth.

Another remarkable aspect of the mission

was its contribution to television broadcasting technology. The launch of STS-95 was the first event to be broadcast live in the U.S. using the new ATSC HDTV system, providing high-definition, coast-to-coast coverage of the launch. This technological milestone coincided with another first: Pedro Duque of Spain became the first Spaniard to travel to space, adding international significance to the mission.

The primary objectives of STS-95 were centered around life sciences and solar research. The crew conducted numerous experiments aboard the *SpaceHab* module, many of which focused on Senator Glenn's physical condition, investigating how spaceflight affects the aging process. This mission was a continuation of NASA's long-term research on human health in space, contributing to understanding how the human body responds to the unique microgravity conditions.

STS-95 also expanded scientific knowledge of the Sun's influence on Earth. The crew deployed the Spartan 201 spacecraft, which operated independently of the Shuttle to study the acceleration of the solar wind originating from the Sun's outermost layer, the solar corona. The Spartan 201 mission helped scientists better understand solar wind dynamics and its potential impact on Earth's magnetosphere and space weather.

The mission, which lasted just under ten days, was a success. *Discovery* landed at Kennedy Space Center's Shuttle Landing Facility, bringing an end to this historic flight. Interestingly, the official weather forecast, provided by the 45th Weather Squadron, had predicted 100 percent favorable weather for both the launch and landing—a rare occurrence in spaceflight.

STS-95 was also notable for the attendance of then-sitting U.S. President Bill Clinton, who became the second president to witness a crewed space launch, following President Richard Nixon's attendance at the Apollo 12 launch in 1969. Clinton, joined by First Lady Hillary Clinton, observed the launch from the roof of the Launch Control Center.

The crew of STS-95 was composed of astronauts with varied experience levels. Curtis L. Brown Jr. served as Commander, flying his fifth space mission. Steven W. Lindsey, the mission's Pilot, was on his second spaceflight, while Mission Specialists Stephen K. Robinson and Scott E. Parazynski were on their second and third flights, respectively. Chiaki Mukai from Japan's National Space Development Agency (NASDA) flew as a Payload Specialist for her second and final spaceflight. Pedro Duque, flying for the European Space Agency (ESA), made his debut in space on this mission, marking his first flight. Finally, John Glenn, the Payload Specialist, completed his second and final spaceflight, making him the last astronaut from the pre-Shuttle era to fly aboard the Shuttle.

The primary objectives of STS-95 included conducting a series of science experiments within the pressurized *Spacehab* module, deploying and retrieving the Spartan free-flyer payload, and operating the Hubble Space Telescope Orbital Systems Test (HOST) and the International Extreme Ultraviolet Hitchhiker (IEH) payloads. These experiments represented a significant leap in understanding the effects of space on the human body and advancements in space technology.

The *Spacehab* module, provided by the private company Spacehab, Inc., was a key component of the STS-95 mission. It provided additional pressurized workspace for experiments, cargo, and crew activities. The *Spacehab* system had been utilized on various Shuttle science missions and joint Shuttle-Mir flights, offering a flexible environment for conducting a wide array of experiments in space.

For this mission, a single-module *Spacehab* was flown in the forward section of *Discovery's* payload bay. Crew members accessed the module via the airlock tunnel system, and experiments sponsored by NASA, the Japanese Space Agency (NASDA), and the European Space Agency (ESA) focused on life sciences, microgravity sciences, and advanced technology. These experiments were crucial in deepening scientific understanding of how microgravity affects biological and physical processes, with significant implications for future long-duration missions.

The Spartan 201-5 free-flyer payload was a

pivotal element of STS-95, deployed and retrieved using the Shuttle's mechanical arm. Its mission was to study the hot outer layers of the Sun's atmosphere, known as the solar corona. While free-flying from the Shuttle, Spartan gathered critical measurements of the solar wind, an intense stream of charged particles that flows from the Sun. This data was vital for improving understanding of how solar winds influence satellites in orbit and affect weather conditions on Earth, which, in turn, impact global communications systems.

This marked the fifth flight of Spartan, originally flown on STS-56 in 1993. On its previous mission, STS-87, the payload experienced problems after deployment and had to be manually retrieved via spacewalk. For STS-95, the attitude control system that had caused earlier issues was fully repaired, and Spartan successfully completed its experiments.

The HOST platform, another significant payload on STS-95, carried experiments to validate technologies planned for future upgrades to the Hubble Space Telescope (HST). Four key experiments on the platform tested new equipment in Earth's orbit. The NICMOS Cooling System experiment demonstrated a reverse turbocharged Brayton cycle cooler, intended to provide longer operational life than the existing dewar system on Hubble. The HST 486 computer experiment tested for any radiation-susceptible parts in the replacement computer slated for Hubble's next servicing mission. Additionally, the solid-state recorder experiment compared the flight-spare solid-state recorder with the one installed on Hubble. Lastly, a fiber-optic line test collected data for post-flight analysis, ensuring the continued performance and reliability of the Hubble systems.

The IEH payload on STS-95 carried a half-dozen experiments designed to study ultraviolet light and its effects. One of the most prominent experiments, the Solar Extreme Ultraviolet Hitchhiker (SEH), measured ultraviolet and far-ultraviolet fluxes to better understand Earth's upper atmosphere. Other experiments, such as the Ultraviolet Spectrograph Telescope for Astronomical Research (UVSTAR), helped form images of extended plasma sources like Jupiter and hot stars. The IEH suite also included experiments like CONCAP-IV, which focused on growing thin films via physical vapor transport, and the Petite Amateur Navy Satellite (PANSAT), a Department of Defense-managed payload that involved the deployment of a small satellite for digital communication research. Together, these experiments contributed valuable data to fields ranging from astrophysics to materials science.

A highlight of STS-95 was the participation of John Glenn, who, at 77, became the oldest person to travel into space. His role aboard *Discovery* was not merely symbolic—Glenn's participation was pivotal for NASA's aging and geriatric studies. The physiological similarities between the aging process and the effects of spaceflight, such as bone density loss, muscle atrophy, and balance disorders, made Glenn an ideal subject. Sponsored by NASA and the National Institute on Aging, these experiments sought to collect data that could advance scientific understanding of aging processes on Earth.

Glenn's participation in two key experiments involving sleep monitoring and protein use provided researchers with valuable comparative data from his 1962 orbital flight on *Friendship 7*. Although he was disqualified from participating in a melatonin study due to a medical condition, the data gathered from Glenn during STS-95 would aid in developing models for studying aging.

John Glenn's return to space was monumental, making him the third sitting member of Congress to fly into space, following Utah Senator Jake Garn on STS-51-D and Florida Representative (later Senator) Bill Nelson on STS-61-C. Glenn, Ohio's senior Senator at the time, rejoined the pantheon of astronauts who had transitioned into political roles, a group that also included Apollo 17 astronaut Harrison Schmitt and Apollo 13 astronaut Jack Swigert.

In a touching reprise of his *Friendship 7* flight, the citizens of Perth and Rockingham, Australia, once again turned on their lights as *Discovery* passed overhead, welcoming Glenn back to space as they had in 1962.

In recognition of their contributions to space

exploration and public engagement, the crew of STS-95 was awarded the prestigious Douglas S. Morrow Public Outreach Award in 1999 by the Space Foundation. This annual award honors individuals or organizations significantly raising public awareness and understanding of space programs. The inclusion of astronaut John Glenn, a national hero, and the scientific achievements of the mission captured the public's imagination, making STS-95 a landmark mission in terms of outreach.

Another anomaly occurred in orbit when a Reaction Control System (RCS) thruster on the left-hand Orbital Maneuvering System (OMS) pod developed a leak. Engineers observed the gas venting from the thruster and decided to disable the affected jet using an isolation valve. Despite the loss of one thruster, attitude control was maintained due to the redundancy of the system, which featured 44 jets distributed around the orbiter.

STS-95 also made history as the inaugural high-definition television (HDTV) broadcast in the United States. The Advanced Television Systems Committee (ATSC) provided live coast-to-coast coverage of the launch, marking a significant milestone in broadcast technology. This HDTV signal was transmitted to various public theaters, science centers, and other locations equipped to receive the new broadcast format. The Harris Corporation provided the necessary transmission and reception equipment for this groundbreaking event.

The broadcast was hosted by two iconic figures in space and media history: Walter Cronkite, the legendary CBS News anchor, and Pete Conrad, a former astronaut from the Gemini and Apollo programs. The involvement of such well-known personalities added an extra layer of significance to the event, blending the worlds of technology, media, and space exploration in an unprecedented way.

Although the mission was largely successful, a few anomalies occurred during STS-95. One significant issue involved the drag chute door, which detached from the orbiter during main engine ignition at liftoff. This raised concerns about the drag chute potentially deploying prematurely upon re-entry, which could have jeopardized the landing. As a result, the decision was made not to use the drag chute during landing rollout. However, *Discovery* landed safely using only wheel brakes and speedbrakes on Runway 33 at Kennedy Space Center's Shuttle Landing Facility.

Chapter 6 - Constructing a New Future: Building the ISS (1998–2003)

STS-88

STS-88: The First Space Shuttle Mission to the International Space Station (ISS)

STS-88 marked a significant milestone in space exploration as it became the first Space Shuttle mission to visit the International Space Station (ISS). Flown by Space Shuttle Endeavour, the mission's primary objective was to deliver the first American module, the Unity Node, to the station, commencing the assembly of human history's largest orbiting space laboratory.

Launched on December 4, 1998, from Kennedy Space Center's Launch Pad 39A, Endeavour carried a six-member crew led by Commander Robert D. Cabana. The mission's highlight was the mating of the U.S.-built Unity Node to the Russian-built Functional Cargo Block (Zarya), already in orbit. Zarya, developed by Boeing and the Russian Space Agency, had been launched on a Proton rocket from the Baikonur Cosmodrome in Kazakhstan a month prior, on November 20, 1998.

The primary mission task was to connect Unity to Zarya using three planned spacewalks. Astronauts Jerry L. Ross and James H. Newman performed the extravehicular activities (EVAs) to connect power and data cables between Unity and Zarya, ensuring that the two modules could function cohesively. Unity, known as Node 1, was designed with two Pressurized Mating Adapters (PMAs) to facilitate future connections. One PMA was permanently attached to Zarya, while the other would be used for docking with visiting spacecraft.

Besides its primary objective, STS-88 carried several other payloads. These included the IMAX Cargo Bay Camera (ICBC), which captured high-resolution mission footage, the Argentine Scientific Applications Satellite (SAC-A), and the MightySat 1 Hitchhiker payload. The mission also featured the Space Experiment Module (SEM-07) and the University of Michigan-sponsored Getaway Special (G-093).

The crew of the STS-88 mission consisted of six astronauts, three from the United States and three from Russia. Each crew member played a vital role in the mission's success. Commander Robert D. Cabana, who was making his fourth and final spaceflight, piloted the shuttle. Frederick W. Sturckow, who was making his first spaceflight, served as the pilot. Jerry L. Ross, who was making his sixth spaceflight, was the mission specialist 1. Nancy J. Currie, who was making her third spaceflight, served as the mission specialist 2 and flight engineer. James H. Newman, who was also making his third spaceflight, was the mission specialist 3. Sergei K. Krikalev, who was making his fourth spaceflight and representing Russia, was the mission specialist 4. Each crew member contributed to the mission's success in their own unique way.

The mission's complex operations began immediately following the launch. On Flight Day 2, Currie used Endeavour's robotic arm, Canadarm, to thoroughly inspect Unity and the shuttle's payload bay. Preparations for the first spacewalk also took place, with Ross and Newman checking their spacesuits and SAFER units, essential for their upcoming EVAs.

By Flight Day 3, the crew pressurized PMA-2 and initiated the critical task of docking Unity with Zarya. Currie, using the Canadarm, successfully maneuvered Unity into position atop the Orbiter Docking System. On December 6, Flight Day 4, Endeavour executed a series of precise maneuvers to bring the shuttle within reach of Zarya, allowing Currie to capture the module and dock it with Unity.

Three spacewalks were conducted to install and activate Unity. On December 7, Ross and Newman began their first EVA, a 7-hour, 21-minute spacewalk to connect 40 electrical and data cables between Unity and Zarya. These cables were vital for transferring power and data between the two modules, marking the first physical connection between American and Russian segments of the station. The successful connection of Unity's systems, confirmed by mission control in Houston, was a significant achievement.

A second EVA on December 9 saw Ross and Newman installing additional hardware, including antennas for Unity's communication systems. They also removed restraint pins from Unity's common berthing mechanisms and installed protective covers on critical components. Their final spacewalk on December 12 lasted 6 hours and 59 minutes, during which they completed Unity's installation and tested the SAFER jet packs designed for astronaut safety during spacewalks.

Flight Day 8, December 10, 1998, was a historic day as Endeavour's crew became the first humans to enter the newly-constructed ISS. Commander Cabana and Russian cosmonaut Sergei Krikalev opened the hatch to Unity at 2:54 p.m. EST, signifying the beginning of human habitation of the station. Joined by the rest of the crew, they explored the new outpost, setting up lights, unstowing essential gear, and installing an early S-band communications system. This system would allow flight controllers in Houston to send commands to Unity and monitor its systems, greatly enhancing the station's operational capability.

Before departing, the astronauts outfitted the station with equipment and logistical supplies, stowing items for future Shuttle missions. Krikalev and Currie replaced a faulty battery discharging unit in Zarya, restoring full functionality to the module's power system. With Unity and Zarya successfully docked and operational, Endeavour undocked from the station on December 13, leaving the newly assembled complex to orbit Earth unpiloted.

The STS-88 mission concluded with Endeavour's landing at Kennedy Space Center on December 15, 1998. The mission, lasting 12 days, covered 4.7 million miles and represented the first crucial step in building the ISS. The successful docking of Unity and Zarya, along with the mission's spacewalks and cargo deployments, laid the foundation for the ISS's growth into the largest human-made structure in space, an enduring symbol of international cooperation and scientific achievement.

STS-96

STS-96, launched aboard Space Shuttle Discovery on May 27, 1999, was a significant mission in the history of the Space Shuttle program. It marked the first docking of a Space Shuttle with the International Space Station (ISS). This mission, officially designated as ISS Flight 2A.1, was a logistics and resupply flight crucial in preparing the nascent space station for future habitation and operations.

Commanded by Kent V. Rominger, with pilot Rick D. Husband, the crew of STS-96 consisted of seven astronauts, including Mission Specialists Ellen Ochoa, Tamara E. Jernigan, Daniel T. Barry, Julie Payette from the Canadian

Space Agency (CSA), and Valery I. Tokarev from the Russian Space Agency (RKA). The crew's international composition underscored the ISS project's collaborative nature, which brought together space agencies worldwide.

STS-96, launched on May 27, 1999, was a pivotal logistics and resupply mission for the International Space Station (ISS), marking the first docking of Space Shuttle Discovery with the orbiting station. This mission was critical for delivering essential supplies and equipment to the ISS, supporting its continued construction and operation.

The primary payload was the Spacehab Double Module (DM), making its 13th flight overall and its 6th in dual-module configuration. This module, designed to increase the Shuttle's cargo capacity, carried various equipment and supplies for the ISS. Accompanying the Spacehab was the Integrated Cargo Carrier (ICC), which housed crucial components for the ISS's Russian and American segments. Among these were the STRELA cargo crane, mounted externally on the Russian segment, and the "ORU Transfer Device" (OTD), a U.S.-built crane designed to facilitate equipment movement around the station. The ICC also carried the SPACEHAB Oceaneering Space System Box (SHOSS), which provided additional storage and transportation for station hardware.

An important scientific payload aboard STS-96 was the Student Tracked Atmospheric Research Satellite for Heuristic International Networking Equipment (STARSHINE). This satellite, a 19-inch diameter hollow sphere covered in 1,000 small, flat mirrors, was designed for educational purposes. Students around the world were tasked with visually tracking the satellite as it orbited Earth, using its reflective surface to monitor its position during morning and evening twilight. Students could derive data on atmospheric density over time by calculating its orbit and measuring atmospheric drag. The satellite was launched using the Hitchhiker Ejection System (HES), which was integrated into the Shuttle's payload bay along with several auxiliary components like the Lightweight Avionics Plate (LAP) and the Adapter Beam Assembly (ABA).

Another key experiment on this mission was the Shuttle Vibration Forces (SVF) experiment, which aimed to measure the vibratory forces between an aerospace payload and its mounting structure during launch. Using custom force transducers mounted on four brackets attached to the Shuttle's canister, the experiment collected data for about 100 seconds following liftoff. This was the second flight of the SVF experiment, designed to improve understanding of the forces acting on payloads during space missions.

Additionally, the mission included the Orbiter Integrated Vehicle Health Monitoring (IVHM) – HEDS Technology Demonstration (HTD). This experiment was part of an effort to modernize the Space Shuttle fleet by testing off-the-shelf sensor technologies in space. The goal of IVHM was to reduce the time and complexity of ground-based inspections, streamline troubleshooting, and enhance the overall safety of the Shuttle by providing real-time insights into the vehicle's systems during flight. This data was intended to inform future upgrades to the Shuttle's health monitoring systems, ultimately improving mission reliability and safety.

After nine days in orbit, Discovery and its crew returned to Earth, landing at Kennedy Space Center on June 6, 1999. The mission was widely regarded as a success, accomplishing its objectives and further developing the ISS as a key element in human space exploration.

STS-93

STS-93, launched on July 23, 1999, from Kennedy Space Center in Florida, was a historic mission for several reasons, most notably for being the first Space Shuttle mission commanded by a woman, Colonel Eileen M. Collins. This mission, flown by the Space Shuttle Columbia, carried and deployed the Chandra X-ray Observatory, one of NASA's Great Observatories, into orbit, marking a significant advancement in our ability to study the universe.

The crew of STS-93 included Pilot Jeffrey S. Ashby and Mission Specialists Steven A. Hawley, Catherine G. "Cady" Coleman, and Michel Tognini, a French astronaut representing the European Space Agency (ESA). The mission's primary objective was the successful deployment of the Chandra X-ray Observatory, a powerful telescope designed to observe X-rays from high-energy regions of the universe, such as the remnants of exploded stars, black holes, and distant galaxies.

Chandra was the most sophisticated X-ray observatory ever built at the time and was a key part of NASA's Great Observatories program, including the Hubble Space Telescope, the Compton Gamma Ray Observatory, and the Spitzer Space Telescope. The observatory was designed to provide unprecedented resolution and sensitivity, allowing astronomers to explore the universe in ways that had never been possible.

Shortly after launch, however, STS-93 encountered a technical issue that could have jeopardized the mission. During ascent, two of the Shuttle's three main engines experienced a drop in performance due to a hydrogen leak in the propulsion system. The onboard computers compensated by extending the burn time of the engines, ensuring that Columbia reached its intended orbit, but the incident highlighted the complexities and risks inherent in human spaceflight. The mission controllers and crew managed the situation skillfully, and the mission continued as planned.

Once in orbit, the crew's primary task was to deploy the Chandra X-ray Observatory. On July 23, just a few hours after launch, Columbia's payload bay doors opened, and the observatory was released into space. Chandra was equipped with its propulsion system to boost the observatory into its final, highly elliptical orbit, extending from approximately 16,000 kilometers (10,000 miles) to 140,000 kilometers (87,000 miles) above Earth. This orbit allowed Chandra to avoid most of the interference from Earth's X-ray-emitting atmosphere, providing a clear view of the X-ray universe.

The deployment was a success, and Chandra soon began transmitting data back to Earth, offering astronomers new insights into some of the universe's most energetic and enigmatic phenomena. The observatory's first images included breathtaking views of the remnants of supernova explosions, the hot gas surrounding black holes, and the collision of galaxy clusters. These observations have since contributed to numerous scientific discoveries and have significantly advanced our understanding of the cosmos.

In addition to deploying Chandra, the STS-93 crew conducted a series of secondary experiments and technology demonstrations. These included the Southwest Ultraviolet Imaging System (SWUIS), which captured ultraviolet images of the Moon and other celestial objects, and the Midcourse Space Experiment (MSX), which monitored the Earth's atmosphere and space environment.

After completing the mission's objectives, Columbia re-entered Earth's atmosphere and landed safely at Kennedy Space Center on July 27, 1999. The mission lasted 4 days, 22 hours, and 50 minutes, and was considered a resounding success despite the challenges encountered during launch.

STS-93 remains a landmark mission in the history of space exploration, not only for its scientific achievements but also for its role in breaking new ground for women in space. Eileen Collins' leadership as the first female Shuttle commander was a significant milestone, inspiring a new generation of women in science and engineering. Meanwhile, the Chandra X-ray Observatory's successful deployment has had a lasting impact on astronomy, providing invaluable data that continues to shape our understanding of the universe.

STS-103: Revitalizing the Hubble Space Telescope

The 96th Space Shuttle launch, STS-103, marked the 27th flight of *Space Shuttle Discovery* and was a critical Hubble Space Telescope servicing mission. The launch of STS-103 faced numerous delays. Initially scheduled for December 6, 1999, the launch was scrubbed due to technical issues. Further delays followed, with weather conditions postponing the launch on multiple occasions. Finally, on December 19, 1999, after four previous attempts, *Discovery* successfully lifted off, beginning a vital mission to extend the life of one of NASA's most important scientific instruments.

It launched from Kennedy Space Center, Florida, on December 19, 1999, and safely returned on December 27, 1999, making it the last Shuttle mission of the decade. Uniquely, this mission spanned through Christmas, following a series of delays due to technical issues and adverse weather conditions.

The primary objective of STS-103 was to restore the functionality of the Hubble Space Telescope. NASA accelerated the mission after three of the telescope's six gyroscopes failed, threatening its precision pointing capabilities. Hubble required at least three operational gyroscopes to continue its scientific observations, prompting a "call-up" mission to prevent the telescope from entering safe mode, where it would remain in orbit but unable to conduct science.

Commanded by Curtis L. Brown Jr., with Scott J. Kelly as Pilot, the STS-103 crew included Mission Specialists Steven L. Smith, John M. Grunsfeld, C. Michael Foale, Claude Nicollier from the European Space Agency (ESA), and Jean-François Clervoy, also from ESA. The crew was highly experienced, particularly in spacewalks and spacecraft maintenance, making them well-suited for the complex tasks ahead.

The primary objective of STS-103, flown by the Space Shuttle *Discovery*, was the Hubble Space Telescope Servicing Mission 3A, which focused on critical repairs and upgrades to the orbiting observatory. The mission was hastily rescheduled for December 1999 after three of Hubble's six gyroscopes failed, severely limiting the telescope's ability to point precisely at distant astronomical targets. NASA's mission rules required at least three functional gyroscopes for science operations, so a fourth failure would have forced the telescope into safe mode, halting observations entirely.

The crew of STS-103, composed of seven astronauts, was tasked with performing four days of Extravehicular Activities (EVAs) to replace and upgrade key components of the telescope. Working in pairs, the astronauts alternated EVA shifts, efficiently coordinating their efforts to restore the Hubble to full operational capacity.

The gyroscopes, essential to Hubble's precise pointing capability, spin at 19,200 revolutions per minute, using gas bearings in a sealed cylinder that floats in a thick fluid. Over time, thin wires supplying electricity to the motor had corroded and broken, a failure attributed to the oxygen used during assembly. NASA engineers believed they had solved the issue by using nitrogen in the assembly process for the new gyroscopes. During the mission, the astronauts replaced all six gyroscopes, ensuring the telescope's continued functionality. Each gyroscope is part of a Rate Sensor Unit (RSU), which weighs approximately 24 pounds and fits within a compact assembly measuring about 13 by 11 by 9 inches.

In addition to the gyroscopes, the crew replaced one of Hubble's Fine Guidance Sensors (FGS), a vital instrument for accurate pointing.

The new FGS, a refurbished unit from a previous mission, weighed over 470 pounds and stood over 5 feet tall. The crew also installed a new onboard computer, which was 20 times faster and held six times more memory than the old one. This upgrade significantly reduced software maintenance costs and improved overall performance. The new computer, weighing around 70 pounds, was a critical enhancement ensuring smoother operations for future missions.

Another major improvement was the installation of a Voltage/Temperature Improvement Kit (VIK) to safeguard Hubble's batteries from potential overcharging and overheating when in safe mode. This simple but crucial device weighed just over three pounds but was essential in prolonging the telescope's life.

During the mission, a new S-Band Single Access Transmitter (SSAT) was also installed. The SSATs sent Hubble's data to Earth through NASA's Tracking Data Relay Satellite System (TDRSS). One of the transmitters had failed in 1998, so the astronauts replaced it with a new unit weighing about 8.6 pounds. The updated transmitter ensured that Hubble could continue transmitting its invaluable data.

A spare solid-state recorder was also added, replacing an outdated reel-to-reel tape recorder. This new recorder, capable of storing 10 times more data than its predecessor, significantly improved Hubble's data handling capacity, allowing scientists to gather more information from the telescope's observations. The solid-state recorder weighed approximately 25 pounds and became one of the most critical upgrades for scientific research.

In addition to these technical enhancements, the astronauts addressed the deterioration of Hubble's thermal insulation, which had degraded after years in space. They replaced the telescope's outer insulation with a New Outer Blanket Layer (NOBL) and Shell/Shield Replacement Fabric (SSRF). These protective layers ensured that Hubble could withstand the extreme temperature fluctuations experienced during its 90-minute orbits around Earth, safeguarding the telescope's sensitive instruments from the harsh conditions of space.

STS-103 also carried a special payload for the Student Signatures in Space (S3) program. Hundreds of thousands of elementary school students' signatures, collected on posters from schools across the United States, were scanned onto disks and flown aboard *Discovery*. This unique project was part of NASA's effort to inspire the next generation of scientists and engineers.

This mission marked *Discovery*'s final solo flight, as all subsequent missions would be in support of the International Space Station (ISS). One of the astronauts on board, John Grunsfeld, a key figure in NASA's astronomical research, brought along a "Planet Mars" flag as a personal item, symbolizing humanity's ongoing exploration of space.

STS-103 was the final solo mission for *Space Shuttle Discovery*, as all its subsequent missions were dedicated to the International Space Station. This mission underscored NASA's ability to respond swiftly to technical challenges, ensuring that the Hubble Space Telescope remained one of the most productive astronomical observatories in history. The mission also held symbolic value, with astronaut John Grunsfeld bringing a "Planet Mars Flag" aboard, symbolizing humanity's ongoing ambition to explore the cosmos.

The mission concluded with a smooth re-entry and landing at Kennedy Space Center on December 27, 1999.

STS-99

STS-99, a notable mission in the Space

Shuttle program, was launched aboard Space Shuttle Endeavour on February 11, 2000. This mission, officially designated as the Shuttle Radar Topography Mission (SRTM), had a unique and ambitious objective: to map the Earth's surface in unprecedented detail. The mission was a collaboration between NASA, the National Geospatial-Intelligence Agency (NGA), and the German and Italian space agencies, making it a truly international effort.

Commanded by Kevin R. Kregel, with Dominic L. Pudwill Gorie as pilot, the STS-99 crew included five mission specialists: Janice E. Voss, Mamoru Mohri from Japan's NASDA (now JAXA), Gerhard P.J. Thiele from the European Space Agency (ESA), Janet L. Kavandi, and U.S. Air Force Colonel Clifford L. E. Anderson, who served as the payload commander. The diverse and highly skilled crew was tasked with conducting one of the most complex and technically demanding missions of the Space Shuttle era.

The Shuttle Radar Topography Mission (SRTM), launched aboard *Space Shuttle Endeavour* during STS-99, was an ambitious international effort led by NASA and the National Imagery and Mapping Agency (now the National Geospatial-Intelligence Agency) in collaboration with the German Aerospace Center (DLR). The mission's primary goal was to create the most complete and accurate high-resolution digital topographic database of Earth, covering a vast portion of the planet's surface between 60° north and 56° south latitudes. To achieve this, the crew aboard *Endeavour* utilized a highly specialized radar system, capturing around 8 terabytes of data during the 11-day mission, culminating in a near-global 3D map of Earth's surface.

The radar system onboard used interferometric synthetic aperture radar (IFSAR) technology, incorporating both C-band and X-band radar. These instruments allowed the team to obtain digital topographic maps meeting rigorous Interferometric Terrain Height Data (ITHD)-2 specifications. The mission's outcomes included a spatial resolution of 30 meters by 30 meters, with vertical height accuracies of 16 meters absolute and 10 meters relative, alongside

horizontal circular accuracy of 20 meters. These unprecedented measurements—close to 1 trillion data points—enabled scientists to produce more accurate maps, improve water drainage models, enhance the accuracy of flight simulators, and optimize the placement of communication towers, among other practical applications.

The mission began smoothly. After *Endeavour* launched on February 11, 2000, the radar mast was successfully deployed to its full 60-meter length, and the antenna system was activated for mapping. Within 12 hours of launch, the crew initiated data collection, working alternating shifts to ensure round-the-clock mapping operations. The radar system's data was transmitted to NASA's Jet Propulsion Laboratory, where early analysis indicated high-quality results.

Challenges arose on the second day when the cold-gas thrust system designed to counteract the mast's gravity gradient torque failed. This malfunction caused an increase in *Endeavour*'s propellant consumption, threatening the mission's duration. To counter this issue, engineers devised a strategy to conserve enough propellant to complete the mission. Additionally, a series of "flycast" maneuvers were employed to reduce stress on the extended mast during attitude adjustments, helping to maintain the mission's operational integrity.

Mapping operations continued over the next several days, and by the tenth day of the mission, radar data collection concluded after a final sweep across Australia. Over the course of 222 hours and 23 minutes, the radar system had recorded 332 high-density tapes, covering 99.98% of the planned mapping area. Remarkably, only 80,000 square miles—primarily in North America—remained unmapped, most of which had already been extensively charted by other methods. The data collected during the mission was enough to fill the equivalent of 20,000 CDs, providing an invaluable resource for Earth scientists and engineers.

In addition to SRTM, the *Endeavour* crew conducted the EarthKAM project, a student-driven experiment that allowed middle school students to select and capture images of Earth from space. The students' images, totaling 2,715 digital photos, were used in classrooms around the world for projects in geography, mathematics, and space science. The program set a new record for participation, with over 75 middle schools involved.

One noteworthy aspect of the mission was using the Spacelab pallet system. Despite the Spacelab program being officially discontinued nearly two years earlier, the system was recommissioned for STS-99. This marked a rare instance of Spacelab hardware being utilized after its formal retirement, further highlighting the adaptability and resourcefulness of the Shuttle program.

On February 22, 2000, after completing its objectives, *Endeavour* safely returned to Earth, landing at the Shuttle Landing Facility. The success of STS-99 and SRTM significantly advanced Earth observation capabilities, leaving a lasting impact on geospatial mapping and topographic studies.

Additionally, the mission received popular attention through the 2007 documentary *Oasis Earth*, which showcased high-definition footage from *Endeavour*, including spectacular views of Earth, the Moon, and Mount Fuji, captured by Japanese astronaut Mamoru Mohri. This mission also marked a transition for the Space Shuttle program, as STS-99 was the last flight to feature the original cockpit layout. Subsequent flights, beginning with STS-101 aboard *Atlantis*, would utilize the modernized "glass cockpit," representing a new era of Shuttle operations.

After completing its mission, Endeavour returned to Earth, landing safely at Kennedy Space Center on February 22, 2000. The mission was hailed as a resounding success, with the data collected providing an invaluable resource for scientists, engineers, and decision-makers worldwide.

STS-101: Preparing the ISS for a New Era of Habitation

STS-101 was a pivotal mission for the Space Shuttle Atlantis, conducted between May 19, 2000, and May 29, 2000. Designated as mission 2A.2a, it served as a critical resupply mission to the International Space Station (ISS) at a time when the station was still uncrewed. This 10-day mission played a key role in preparing the ISS for future manned expeditions and ensuring its operational viability.

The Space Shuttle Atlantis conducted the mission, carrying a crew of seven astronauts led by Mission Commander James D. Halsell and Pilot Scott J. Horowitz. The mission specialists included Mary E. Weber, Jeffrey N. Williams, James S. Voss, Susan J. Helms, and Russian cosmonaut Yuri I. Malenchenko. The diverse and experienced crew was tasked with a complex set of objectives, reflecting the intricate nature of ISS operations.

Originally planned as mission 2A.2, STS-101 was intended to follow the delivery of the Russian Service Module Zvezda. However, delays in Zvezda's launch necessitated splitting the mission into two parts: 2A.2a, which arrived before Zvezda, and 2A.2b, planned for after the module's arrival. The changes in scheduling also affected the mission's objectives, particularly the planned spacewalk to connect cables to Zvezda. Instead, astronauts Edward Lu, Jeffrey Williams, and Yuri Malenchenko, who were initially scheduled to perform this spacewalk, were reassigned to the STS-106 mission. To fill the crew requirements for STS-101, three members of Expedition 2—Susan Helms, James Voss, and Yuri Usachov—were added to the roster for this short but vital mission to their future home aboard the ISS.

The launch of STS-101 faced multiple delays

in April 2000 due to high winds, but once in orbit, Atlantis traveled approximately 4.1 million miles, completing 155 orbits around Earth. The shuttle safely landed on Runway 15 at Kennedy Space Center, marking the mission's successful completion.

One of the standout features of STS-101 was its use of a newly upgraded "glass cockpit," a significant advancement in Space Shuttle technology that modernized the vehicle's avionics and increased its operational efficiency. This was the first mission to fly with this technology, setting a precedent for future missions.

The mission's primary objective was to deliver crucial supplies to the ISS, utilizing a Spacehab double module and an Integrated Cargo Carrier pallet. The crew also conducted a spacewalk to install and assemble key components of the station. In addition, they performed a reboost, raising the station's altitude from 230 miles (370 km) to 250 miles (400 km) to optimize its orbital position.

Critical tasks during the mission included ingress procedures to ensure safety aboard the ISS. The crew took air samples, monitored carbon dioxide levels, deployed portable personal fans, measured air flow, and reworked and modified the station's ducting system. They also replaced air filters, fire extinguishers, and smoke detectors on the Zarya module. Additionally, four suspect batteries on Zarya were replaced, along with failed or suspect electronics associated with those batteries. The crew also replaced the Radio Telemetry System memory unit, port early communications antenna, Radio Frequency Power Distribution Box, and cleared the Space Vision System target.

In terms of assembly and upgrades, the crew installed the Strela crane, additional exterior handrails, center-line camera cable, and "Komparus" cable inserts. They also reseated the U.S. crane and transferred assembly parts, tools, and equipment for future use on the ISS.

The resupply effort was extensive, providing the station with water, film and video tape for documentation, office supplies, and personal items. Medical support supplies were delivered, along with exercise equipment, a formaldehyde monitoring kit, and a passive dosimetry system to monitor radiation exposure.

One of the most critical incidents during the mission involved damage to a tile seam on Atlantis. This damage caused a breach during reentry, allowing superheated gas to enter the left wing. Fortunately, the breach did not penetrate deeply, and the shuttle could return safely. Had the damage been more severe, it could have led to a catastrophic failure similar to the Columbia disaster in 2003. The issue was identified and repaired before the next flight.

After completing their objectives, the crew of Atlantis undocked from the ISS on May 25, 2000, and began their return journey to Earth. The mission concluded with a smooth landing at Kennedy Space Center on May 29, 2000. The success of STS-101 ensured the ISS was well-prepared for the next phase of its development: continuous human occupancy and the further expansion of the station's capabilities.

STS-106: Preparing for the Arrival of the First Expedition Crew

Space Shuttle Atlantis launched on September 8, 2000, for mission STS-106, marking another critical step in the assembly and preparation of the International Space Station (ISS). This mission was pivotal as it set the stage for the arrival of the first permanent crew to the ISS, Expedition 1. The seven-member crew aboard Atlantis was tasked with preparing the station for long-term human habitation, a milestone in the ISS's ongoing development.

The crew of STS-106 included Commander Terrence Wilcutt, Pilot Scott Altman, and Mission Specialists Daniel Burbank, Edward Lu, Yuri Malenchenko, Richard Mastracchio, and Boris Morukov.

The presence of Malenchenko and Morukov, both Russian cosmonauts, underscored the international collaboration essential to the ISS program. This mission exemplified the partnership between NASA and Roscosmos, as the crew worked together to ensure the ISS was ready for continuous human presence.

The Space Shuttle Atlantis launched on September 8, 2000, marking the beginning of mission STS-106. This mission, known as Space Station assembly flight ISS-2A.2b, was critical in preparing the International Space Station (ISS) for its first long-term crew. The mission featured the SPACEHAB Double Module and the Integrated Cargo Carrier (ICC), which were utilized to transport supplies to the station, along with a scheduled spacewalk. Veteran astronaut Terrence Wilcutt, commanding his second Shuttle flight, led a seven-member crew on an 11-day mission to further equip the ISS for permanent human occupation.

The primary objective of STS-106 was to prepare the newly arrived Zvezda Service Module for the station's first Expedition crew, scheduled to arrive in late 2000. Zvezda, which had been launched and docked to the ISS in July, served as the early living quarters and the cornerstone of Russia's contributions to the station. The mission also involved offloading supplies from both the SPACEHAB cargo module and the Russian Progress M-1 resupply craft, docked to the rear of Zvezda.

STS-106 was added to the Shuttle manifest following delays in launching Zvezda. Initially, these tasks were part of mission STS-101, but the delay caused them to be split, leading to separate missions for the remaining tasks. STS-106 was tasked with unloading cargo and conducting one spacewalk to connect Zvezda fully to the ISS.

On flight day three, Mission Specialists Dr. Ed Lu and Russian cosmonaut Yuri Malenchenko performed a spacewalk lasting 6 hours and 14 minutes. Their primary task was to connect nine power, data, and communication cables between the Zvezda and Zarya modules, linking essential systems. Additionally, they installed a six-foot-long magnetometer to provide orientation data to the ISS's control systems, ensuring precise positioning relative to Earth and reducing the need for fuel-intensive maneuvers.

This spacewalk set a new distance record for tethered spacewalkers, as Lu and Malenchenko ventured over 100 feet above Atlantis's cargo bay, supported by crewmates Dan Burbank and Rick Mastracchio, who controlled the Shuttle's robotic arm. The spacewalk was significant not only for its technical achievements but also for being the 50th conducted in Space Shuttle history and only the second joint U.S.-Russian EVA outside a Shuttle.

On flight day four, the crew entered the ISS through Pressurized Mating Adapter-2 (PMA-2), beginning the process of transferring over three tons of supplies and equipment to the station. The Atlantis crew were the first humans to see the interior of Zvezda since its launch. They transferred critical supplies, including food, water, batteries, and environmental control systems, for the upcoming Expedition 1 crew, set to establish the first long-term human presence aboard the station.

As the crew worked, they also executed a reboost of the ISS, using Atlantis's Reaction Control System (RCS) to raise the station's altitude, ensuring it remained in a stable orbit. Transfer and maintenance tasks continued through flight day five, as Lu and Malenchenko focused on installing key voltage stabilizers and components of the Elektron oxygen-generation system within Zvezda. This system, essential for splitting water into oxygen and hydrogen, would be activated after the arrival of the first permanent crew.

The crew's efforts successfully transferred over 6,000 pounds of materials to the ISS, including water, food, office supplies, and crucial life-support systems. In total, the crew spent 5 days, 9 hours, and 21 minutes inside the ISS, ensuring the station was well-equipped for its upcoming residents.

Atlantis undocked from the ISS on flight day eight, following a series of altitude boosts to place the station in a higher orbit. Commander Wilcutt and Pilot Scott Altman expertly piloted the Shuttle in a double-loop flyaround, providing a comprehensive view of the station before returning home. After traveling over 4.9 million miles, Atlantis touched down at the Kennedy Space Center on September 20, 2000, marking the 23rd consecutive Shuttle landing at the Florida spaceport.

In addition to the mission's primary objectives, STS-106 also delivered the first amateur radio station to the ISS, which the Expedition 1 crew would later use. The mission was a resounding success, bringing the ISS one step closer to permanent human habitation, while also demonstrating the ongoing collaboration between the United States and Russia in space exploration.

STS-92

STS-92 was a Space Shuttle mission to the International Space Station (ISS) flown by Space Shuttle Discovery. It marked the space shuttle's 100th mission and was launched from Kennedy Space Center, Florida, on 11 October 2000.

Commanded by Brian Duffy, the crew of STS-92 included Pilot Pamela Ann Melroy and Mission Specialists Leroy Chiao, William S. McArthur Jr., Peter J. K. Wisoff, Michael E. Lopez-Alegria, and Koichi Wakata, a Japanese astronaut representing the Japan Aerospace Exploration Agency (JAXA).

On October 15, 2000, the Space Shuttle Discovery, as part of mission STS-92, embarked on a critical phase in the assembly of the International Space Station (ISS). This mission, which lasted nearly 13 days, marked a significant step in building the station's backbone, enabling future expansions and preparing the ISS for scientific research and crew habitation. Central to the mission were four spacewalks (extravehicular activities, or EVAs) that involved two teams of astronauts meticulously working to install key components of the station.

The first spacewalk, EVA 1, was conducted by astronauts Leroy Chiao and Bill McArthur on October 15. Beginning at 14:27 UTC and lasting for 6 hours and 28 minutes, the duo worked with precision to install the Z1 truss, a vital exterior framework on the ISS. The Z1 truss, a zenith-port structure, served as the first permanent lattice-work for the station, designed to support future additions, including the U.S. solar arrays. The Z1 would also house essential communications and control systems, enabling the ISS to maintain its proper orientation in orbit using the Control Moment Gyroscopes (CMGs).

EVA 2 took place the following day on October 16. Michael López-Alegría and Jeff Wisoff exited Discovery at 14:15 UTC and spent 7 hours and 7 minutes outside the spacecraft. Their task focused on continuing the installation of the Pressurized Mating Adapter-3 (PMA-3). This docking port would later be used to attach other station modules and provide a secure location for future shuttle dockings.

On October 17, Chiao and McArthur undertook EVA 3, a 6-hour and 48-minute spacewalk. Their primary objective was to complete the installation of the Z1 truss and secure the Control Moment Gyroscopes. Weighing 60 pounds each, these gyroscopes were crucial for maintaining the station's orientation without fuel-consuming thrusters, marking a major step in the station's ability to stay in orbit for long durations.

The final spacewalk, EVA 4, was carried out on October 18 by López-Alegría and Wisoff. Lasting 6 hours and 56 minutes, this spacewalk completed the Z1 truss installation and further secured the PMA-3. During this EVA, the astronauts also tested the Simplified Aid for EVA Rescue (SAFER) jet backpack, a personal propulsion unit designed to assist astronauts in returning to the station should they become untethered. In a controlled test, the astronauts flew up to 50 feet from the spacecraft, demonstrating the capabilities of the SAFER system while ensuring they remained safely tethered.

These spacewalks were essential in preparing the ISS for future missions and scientific research. The installation of the Z1 truss allowed for the attachment of the first U.S. solar arrays during the next shuttle flight, STS-97. Additionally, integrating the Ku-band communication system enhanced the station's ability to support early scientific operations and relay television broadcasts back to Earth. The mission's success set the stage for the continued expansion of the ISS, paving the way for future

laboratories, solar arrays, and modules.

The STS-92 mission also marked an advancement in the station's operational capabilities, as it featured seven days of docked operations with the ISS, two ingress opportunities for crew members, and the delivery of critical components that would further expand the station's structure and functionality. Installing the Control Moment Gyroscopes represented a major leap in the ISS's ability to maintain its orientation in space, making it a more autonomous platform for scientific research and international collaboration.

After completing all mission objectives, Discovery undocked from the ISS on October 20, 2000, and returned to Earth, landing at Edwards Air Force Base in California on October 24, 2000. The mission's achievements were a significant step forward in the ongoing construction of the ISS, laying the groundwork for the station's future as a hub for international scientific research and exploration.

STS-97

STS-97, launched aboard Space Shuttle *Endeavour* on December 1, 2000, played a pivotal role in the ongoing assembly of the International Space Station (ISS). As the last human spaceflight of the 20th century, this mission marked a significant transition in space exploration, bridging past accomplishments with the advancements yet to come. It was *Endeavour's* 15th flight and the 101st mission of the Space Shuttle program, making it a landmark in NASA's efforts to establish a permanent human presence in space.

The primary objective of STS-97 was to install the first set of U.S.-provided solar arrays on the ISS. These arrays, part of the P6 Truss, were critical to expanding the station's energy capacity, essential for its long-term habitation and future scientific endeavors. When fully deployed, the solar arrays measured an impressive 240 feet in length and significantly increased the station's ability to generate power. Their installation required careful coordination, precision, and the expertise of mission specialists Joseph Tanner and Carlos Noriega, who conducted three demanding spacewalks (Extravehicular Activities, or EVAs) to attach and deploy the arrays.

The first EVA, conducted on December 3, 2000, saw Tanner and Noriega begin the task of positioning and securing the solar arrays. Over 7 hours and 33 minutes, the astronauts worked in the vacuum of space to ensure the arrays were properly aligned and ready for deployment. During the second EVA on December 5, lasting 6 hours and 37 minutes, the pair continued their efforts, making critical adjustments and securing additional components. The third and final EVA, on December 7, which lasted 5 hours and 10 minutes, concluded the installation process, leaving the solar arrays fully functional and operational.

In addition to installing the solar arrays, the Endeavour crew prepared the Pressurized Mating Adapter-3 (PMA-3) for the arrival of the Destiny Laboratory Module, one of the most important scientific modules on the ISS. PMA-3 would serve as a docking port for future missions, particularly the Destiny module, which would greatly expand the ISS's capacity for long-term scientific research in microgravity.

The mission also delivered essential supplies to the Expedition 1 crew, who had recently begun the first long-duration stay aboard the ISS. These supplies included equipment, tools, and consumables necessary for the station's continued operation and expansion. The transfer of these materials ensured that the ISS remained fully functional as its infrastructure evolved.

Commanded by Brent Jett and piloted by Michael J. Bloomfield, the STS-97 crew consisted of five astronauts, including Mission Specialists Joseph R. Tanner, Carlos I. Noriega, and Marc Garneau, the first Canadian astronaut to fly aboard the ISS. Garneau played a crucial role in operating the Canadarm, the shuttle's robotic arm, to maneuver the massive P6 Truss into place during the spacewalks.

On Flight Day 3, Jett skillfully guided *Endeavour* to dock with the ISS while orbiting 370 kilometers (230 miles) above northeastern Kazakhstan. The docking allowed for the seamless crew and cargo transfer, marking the first official exchange between the shuttle and the station's Expedition 1 crew, led by Commander William Shepherd, with Pilot Yuri Gidzenko and Flight Engineer Sergei Krikalev.

As the mission unfolded, the crews worked closely together, completing the solar array installation and conducting various structural tests on the station. These tests verified the integrity of the station's new components and ensured that its systems were functioning as intended. The astronauts also exchanged equipment and supplies and conducted checks on the systems that would support future missions, including those involving the Destiny module.

On December 9, after nearly seven days of docked operations, *Endeavour* undocked from the ISS, leaving behind a significantly enhanced station, ready for future additions and scientific endeavors. The undocking took place 235 miles above the border of Kazakhstan and China, with Pilot Michael Bloomfield performing an hour-long flyaround of the station to capture detailed photographs of the newly installed solar arrays and other ISS components.

The successful completion of STS-97 not only advanced the construction of the ISS but also marked a turning point in the history of human spaceflight. The mission's achievements—installing the ISS's first set of solar arrays, supporting the first long-term crew, and preparing the station for future expansions— were critical in transforming the ISS from a small outpost into a more capable and self-sufficient space station. As the final human spaceflight of the 20th century, STS-97 set the stage for continued exploration in the new millennium, symbolizing NASA's enduring commitment to advancing human presence in space.

The first spacewalk began on December 3, 2000, at 18:35 UTC, with mission specialists Joseph Tanner and Carlos Noriega stepping outside the shuttle to begin the process of installing the P6 solar arrays. The astronauts attached the massive arrays over 7 hours and 33 minutes and positioned them for deployment. This intricate task involved precise alignment and handling of the arrays, which spanned 240 feet when fully deployed. EVA 1 concluded on December 4, 2000, at 02:08 UTC, marking the successful initial steps of the mission.

On December 5, 2000, the second spacewalk commenced at 17:21 UTC. Tanner and Noriega continued their work, focusing on configuring the truss for long-term operation and ensuring that the solar arrays were properly secured and functional. Over the next 6 hours and 37 minutes, the duo installed Floating Potential Probes to measure the electrical potential surrounding the ISS, ensuring the station's electrical systems would remain stable. They also installed a camera cable outside the Unity Module to facilitate future missions. The spacewalk concluded at 23:58 UTC, with key tasks completed ahead of schedule.

The third and final spacewalk began on December 7, 2000, at 16:13 UTC. During this 5-hour, 10-minute EVA, Tanner and Noriega completed final preparations for the station's upcoming Destiny Laboratory Module, ensuring the docking port was ready for its installation. They also transferred supplies and equipment between *Endeavour* and the ISS, supporting both ongoing and future operations aboard the station. EVA 3 ended at 21:23 UTC, bringing the spacewalking activities of STS-97 to a successful close.

Beyond the spacewalks, STS-97 involved critical mission operations inside and outside *Endeavour*. On Flight Day 3, Commander Brent Jett expertly piloted *Endeavour* to dock with the ISS at an altitude of 370 kilometers (230 miles) above northeastern Kazakhstan. This docking connected the shuttle to the station's Pressurized Mating Adapter, allowing for seamless crew and cargo transfers between the two spacecraft.

Inside *Endeavour*, Canadian Mission Specialist Marc Garneau operated the Canadarm, the shuttle's robotic arm, to carefully extract the 8-ton P6 Truss from the payload bay. After positioning the truss overnight to allow its components to warm, the crew worked to complete its installation during the spacewalks. While the solar arrays were being deployed and checked out, Tanner and Noriega moved supplies and computer hardware into the station for the Expedition 1 crew, the first long-duration residents of the ISS.

On December 4, 2000, Expedition 1 Commander William Shepherd, Pilot Yuri Gidzenko, and Flight Engineer Sergei Krikalev officially entered the Unity Module for the first time since the shuttle's arrival, retrieving the items left for them by *Endeavour's* crew. The mission provided a critical resupply for the Expedition 1 team, allowing them to continue their work aboard the ISS.

On December 8, 2000, the crew of *Endeavour* paid their first visit to the Expedition 1 crew. Up until that point, the two spacecraft had kept one hatch closed to maintain atmospheric pressure differences, ensuring safe conditions for both spacewalks and daily activities. After a welcome ceremony and briefing, the crews conducted structural tests of the station's new solar arrays, transferred equipment and supplies, and shared a momentous exchange aboard the growing station.

The next day, on December 9, the crews completed final transfers of supplies and other items. At 10:51 EST, the *Endeavour* crew bid farewell to the Expedition 1 team, closing the hatches between the spacecraft. After nearly seven days docked together, *Endeavour* undocked from the ISS at 14:13 EST, with Pilot Michael Bloomfield guiding the shuttle through an hour-long, tail-first flyaround of the station, capturing detailed photographs of the newly installed components. The undocking took place 235 miles above the border of Kazakhstan and China, and the final separation burn occurred near the northeast coast of South America.

The mission duration of STS-97 spanned 10 days, 19 hours, 58 minutes, and 20 seconds, during which the Space Shuttle *Endeavour* and its crew traveled a total distance of 7,203,000 kilometers (4,476,000 miles) while in orbit around the Earth. The mission concluded with a successful landing on December 11, 2000, at 23:04 UTC. *Endeavour* touched down smoothly at Kennedy Space Center's Shuttle Landing Facility (SLF), using Runway 15 to complete its historic journey. This landing marked the culmination of a critical mission in the assembly of the International Space Station and the final human spaceflight of the 20th century, setting the stage for future advances in space exploration.

STS-98

On February 7, 2001, Space Shuttle *Atlantis* launched on STS-98, a mission critical to the ongoing construction of the International Space Station (ISS). Designated ISS Flight 5A, this mission was pivotal in expanding the station's scientific capabilities by delivering and installing the Destiny laboratory module, the first major U.S. science lab on the ISS. The installation of Destiny transformed the station into a world-class research facility, furthering the United States' role in low Earth orbit science.

Astronauts Kenneth D. Cockrell (right front), mission commander; and Mark L. Polansky (left front), pilot; along with astronauts Marsha S. Ivins, Robert L. Curbeam, Jr., (left rear) and Thomas D. Jones (right rear), all mission specialists.

Kenneth D. Cockrell commanded STS -98,

with Mark L. Polansky serving as pilot. The crew also included Mission Specialists Robert L. Curbeam Jr., Marsha S. Ivins, and Thomas D. Jones. Astronaut Mark C. Lee was originally scheduled to be part of the crew, but for undisclosed reasons, he was replaced by Curbeam prior to the mission.

A series of technical challenges marked the preparation for STS-98. The first launch attempt on January 19, 2001, was scrubbed due to the need for booster separation cable inspections. After a 19-day turnaround, *Atlantis* successfully lifted off on February 7, 2001, at 6:11 PM EST from Kennedy Space Center's Launch Pad 39-A.

The Destiny laboratory, delivered aboard *Atlantis*, was the first NASA laboratory module to be permanently used in space since Skylab, nearly three decades earlier. Constructed by Boeing at the Michoud Assembly Facility and the Marshall Space Flight Center in 1997, the module was later equipped with scientific racks, equipment, and cables at Kennedy Space Center's Space Station Processing Facility.

Destiny is a state-of-the-art aluminum module, measuring 28 feet (8.5 meters) in length and 14 feet (4.3 meters) in width. It consists of three cylindrical sections and two end-cones with hatch openings. The blue and white color scheme of the end-cones aids astronauts in navigating the module. A key feature of Destiny is its 20-inch (510 mm) window, designed to provide astronauts with an observation point for Earth and space-based experiments.

Atlantis docked to the ISS's Pressurized Mating Adapter 3 (PMA-3) on the nadir of Node 1. The crew temporarily moved PMA-2 to the Z1 truss using the shuttle's robotic arm, clearing the way for the installation of Destiny. With precision, they lifted the 14.5-ton module from *Atlantis*' payload bay and permanently berthed it to the forward hatch of Node 1.

Once installed, astronauts Thomas Jones and Robert Curbeam conducted a series of three spacewalks to complete the module's integration with the ISS. During these spacewalks, they connected power and data cables, inspected Destiny's nadir window, and repositioned PMA-2 to its final location on the module's forward hatch.

The first spacewalk on February 10, 2001, lasted 7 hours and 34 minutes. Jones and Curbeam began by disconnecting cables and removing protective covers from Destiny. After ensuring the module was securely in place, they connected power and data cables to integrate the lab with the station.

On February 12, Jones and Curbeam assisted with the relocation of PMA-2, detaching it from the Z1 truss and installing it on the forward end of Destiny. They also installed a Power Data Grapple Fixture and a video signal converter to support the future operation of the station's robotic arm, Canadarm2. This spacewalk lasted 6 hours and 50 minutes.

The final spacewalk occurred on February 14, lasting 5 hours and 25 minutes. During this EVA, the astronauts attached a spare communications antenna to the station's exterior, inspected solar array connections, and released a cooling radiator. They also conducted a safety test to evaluate the ability of a spacewalker to transport an immobile crewmember back to the shuttle's airlock.

STS-98 marked the 100th spacewalk in U.S. spaceflight history and represented a major leap forward in the development of the ISS. The mission was conducted while the station's first long-duration crew, Expedition 1, was aboard, underscoring the station's growing capacity as a research outpost. The successful delivery and installation of the Destiny laboratory solidified the ISS's role as a cornerstone of international space collaboration, with long-term implications for scientific research in microgravity

After nearly 13 days in space, Atlantis and its crew returned to Earth, landing at Edwards Air Force Base in California on February 20, 2001. The mission was widely regarded as a success, achieving its objectives and significantly enhancing the ISS's capabilities.

STS-102: Paving the Way for Long-Duration Space Missions

STS-102, launched on March 8, 2001, by the Space Shuttle Discovery, was a pivotal mission that marked the first crew exchange on the International Space Station (ISS), laying the groundwork for long-duration human spaceflight aboard the orbiting laboratory. The mission's objectives centered around rotating the ISS crew, delivering essential supplies, and continuing the assembly and outfitting of the station, underscoring the Space Shuttle's role as an essential tool in the station's development.

The mission was led by Commander James D. Wetherbee and Pilot James M. Kelly, with a crew that included Mission Specialists Andrew S. W. Thomas and Paul W. Richards, along with Expedition 2 astronauts Yury V. Usachov, Susan J. Helms, and James S. Voss. The latter three were set to replace the Expedition 1 crew, who had been living on the ISS since November 2000, marking the beginning of long-duration stays on the station.

For Space Shuttle mission STS-102, crew seat assignments were carefully arranged to accommodate both the mission's operational needs and the unique physical demands of astronauts returning from long-term spaceflight. Seats 1 to 4 were on the flight deck, with 5 to 8 positioned on the mid-deck. Seat 8 was located to the starboard side of Seat 7.

Commander Jim Wetherbee occupied Seat 1, with Pilot James Kelly in Seat 2 on launch. Mission Specialist Andrew Thomas was positioned in Seat 3 for launch but did not require a seat for landing, as he would later transfer to the International Space Station (ISS). Flight Engineer Paul Richards was in Seat 4 for both launch and landing. On the mid-deck, Susan Helms and Yuri Usachyov occupied Seats 5 and

7 respectively during launch, while Helms' landing partner was Sergey Krikalev. Vladimir Voss shared the mid-deck with Helms and Usachyov, seated in position 5. Notably, William Shepherd, Krikalev, and Usachyov all returned to Earth in recumbent couches, a rare seating configuration designed to minimize the effects of re-entry on their deconditioned bodies after spending four months in microgravity aboard the ISS. These couches allowed them to recline, which helped reduce the stress of re-acclimating to Earth's gravity. The recumbent couches were a unique feature of this mission and were not used in subsequent shuttle flights.

Additionally, Andrew Thomas was moved to Seat 5 on the mid-deck to operate the side hatch as necessary, given the special seating arrangements.

The STS-102 mission included two crucial spacewalks (Extravehicular Activities or EVAs) as part of the ongoing assembly of the ISS:

EVA 1 took place on March 11, 2001. Susan Helms and James Voss spent 8 hours and 56 minutes outside the ISS, beginning at 05:12 UTC and concluding at 14:08 UTC. Their mission focused on installing components brought to the station, including tasks related to hardware relocation.

EVA 2, performed by Andrew Thomas and Paul Richards on March 13, 2001, lasted for 6 hours and 21 minutes, starting at 05:23 UTC and ending at 11:44 UTC. During this EVA, they continued installing hardware and carried out additional assembly tasks on the exterior of the ISS.

STS-102 was designated Space Station Assembly Flight ISS-5A.1 and marked the first use of the Multi-Purpose Logistics Module (MPLM), specifically the Leonardo module, to transport supplies to the ISS. The module was a critical component for future logistics operations, capable of carrying up to 16 International Standard Payload Racks (ISPRs) designed for installation in the U.S. Destiny Laboratory on the station.

The shuttle also transported an Integrated Cargo Carrier (ICC), which was equipped with the External Stowage Platform-1 (ESP-1) mounted on its underside. The ESP-1 was later placed on the port side of the Destiny module, serving as a storage platform for Orbital Replacement Units (ORUs), such as spare parts and equipment. The mission's two spacewalks played a vital role in transferring and installing these components to the station's exterior, advancing the construction and functionality of the growing ISS.

STS-102 was a key mission in continuing the assembly of the ISS, facilitating the arrival of Expedition 2 while allowing the Expedition 1 crew to return to Earth.

After nine days in space, Discovery undocked from the ISS on March 18, 2001, carrying the Expedition 1 crew back to Earth. The mission concluded with a safe landing at Kennedy Space Center on March 21, 2001. STS-102 successfully demonstrated the Shuttle's capability to support long-duration missions on the ISS, making it a key mission in the overall ISS program.

STS-100: A Mission of International Collaboration and Technological Achievement

In April 2001, the Space Shuttle Endeavour embarked on a critical mission that highlighted the collaborative spirit and technological advancements of the International Space Station (ISS) program. Designated STS-100, this mission played a pivotal role in the ongoing construction and operation of the ISS, underscoring the shuttle's role as a workhorse for space assembly and logistics.

Launched on April 19, 2001, from Kennedy Space Center, Endeavour carried a crew of seven astronauts led by Mission Commander Kent V. Rominger and Pilot Jeffrey S. Ashby. The crew consisted of experts from multiple spacefaring nations, reflecting the international partnership the ISS embodies. The mission specialists included Chris A. Hadfield, the first Canadian to perform a spacewalk, along with John L. Phillips, Scott E. Parazynski, Umberto Guidoni from the European Space Agency (ESA), and Yuri V. Lonchakov from the Russian Space Agency.

The primary objective of STS-100 was the delivery and installation of the Canadarm2, a sophisticated robotic arm that became an essential tool for the ISS's construction and maintenance. Formally known as the Space Station Remote Manipulator System (SSRMS), Canadarm2 represented the cutting edge of space robotics, with the capability to move along the station's exterior, assemble modules, and assist with docking operations. This 17.6-meter-long robotic arm was a significant upgrade from the original Canadarm, used on the shuttle since the early 1980s.

Endeavour's payload bay was filled with essential equipment, including the Italian-built Raffaello Multi-Purpose Logistics Module (MPLM). The Raffaello module was packed with supplies and experiments crucial to the ISS's operations, including food, clothing, and scientific equipment, demonstrating the shuttle's role in supporting the station's long-term habitability.

Endeavour docked with the ISS upon reaching orbit, and the crew began installing Canadarm2. This task involved two spacewalks, or Extravehicular Activities (EVAs), performed by Hadfield and Parazynski. The first EVA on April 22, 2001, saw the astronauts attaching Canadarm2 to its base on the Destiny laboratory module. Over seven hours, they carefully connected the arm's electrical and data cables, bringing it to life. During the second EVA, conducted on April 24, the astronauts tested the arm's functionality, ensuring it was fully operational for its critical role in future ISS assembly tasks.

One of the mission's most challenging aspects was activating Canadarm2. The crew worked closely with ground controllers to troubleshoot issues, ensuring the arm could perform its tasks precisely. The successful installation and activation of Canadarm2 marked a significant milestone in the ISS program, greatly enhancing the station's ability to expand and adapt to future needs.

The STS-100 mission also demonstrated the importance of international cooperation in space exploration. The presence of astronauts from NASA, ESA, and the Russian Space Agency aboard Endeavour symbolized the global effort to build and operate the ISS. This collaboration was further highlighted by the delivery of Raffaello, a product of Italy's contribution to the program.

After nine days in space, Endeavour undocked from the ISS on April 29, 2001, and began its journey back to Earth. The mission concluded with a smooth landing at Kennedy Space Center on May 1, 2001. STS-100 advanced the ISS's construction and reinforced the shuttle's role as a critical element in the international partnership that made the space station possible.

STS-104: Expanding the International Space Station's Capabilities

STS-104, launched on July 12, 2001, by the Space Shuttle Atlantis, was a significant mission focused on expanding the International Space Station (ISS) capabilities. The mission's primary objective was to deliver and install the Quest Joint Airlock Module. This crucial addition enhanced the station's ability to support spacewalks, or Extravehicular Activities (EVAs), using both American and Russian spacesuits. This mission marked a key milestone in the ISS assembly, furthering its development as a fully functional and self-sustaining orbiting laboratory.

Steven W. Lindsey commanded the mission, with Charles O. Hobaugh serving as Pilot. The crew included Mission Specialists Michael L. Gernhardt, Janet L. Kavandi, and James F. Reilly. Each crew member brought experience, particularly in spacewalks and station assembly, essential for successfully completing STS-104's objectives.

The primary objective of Space Shuttle Atlantis' STS-104 mission was the delivery and installation of the Quest airlock to the International Space Station (ISS). The Quest airlock, also called the Joint Airlock, is a pressurized module composed of two cylindrical chambers connected by a bulkhead and hatch. This airlock was designed to facilitate extravehicular activities (EVAs), serving as the primary entry and exit point for astronauts conducting spacewalks. Once installed and activated, Quest became the primary airlock for U.S. spacesuits, known as Extravehicular Mobility Units (EMUs), while also supporting the use of the Russian Orlan spacesuit for spacewalks, enhancing the versatility of the ISS.

The airlock itself measures 20 feet (6.1 meters) in length, 13 feet (4.0 meters) in diameter, and weighs approximately 6.5 short tons (5.9 metric tons). Constructed from a combination of steel and aluminum, it was manufactured at the Marshall Space Flight Center (MSFC) by Boeing, the primary contractor for the ISS. Quest consists of two primary components: a crew airlock, where astronauts prepare for spacewalks, and an equipment airlock, which stores EVA gear and facilitates pre-EVA preparations. In addition to the airlock, Atlantis carried a Spacelab pallet with four High Pressure Gas Assembly (HPGA) containers, which were later attached to the exterior of the airlock to provide critical support for EVAs.

Mission specialists Michael Gernhardt and James Reilly conducted three spacewalks during STS-104, accumulating a total of 16 hours and 30 minutes outside the station. The first EVA, conducted on July 15, 2001, lasted 5 hours and 59 minutes and focused on the installation of the Quest airlock. Gernhardt and Reilly worked methodically, assisting in securing the airlock to its designated position on the ISS. The second spacewalk, on July 18, lasted 6 hours and 29 minutes and involved the external outfitting of the airlock, including the installation of the four HPGA tanks, handrails, and other essential equipment. The third and final spacewalk took place on July 21, lasting 4 hours and 2 minutes. This EVA marked a significant milestone as it was conducted from the newly installed Quest airlock itself, signaling its full operational status.

STS-104 was notable for being the last Space Shuttle mission with a five-member crew, as all subsequent missions would carry six or seven astronauts, except for the final mission, STS-135, which had a crew of four. The mission also marked the debut of the "Block II" Space Shuttle Main Engine (SSME), a significant upgrade in engine performance. However, post-launch analysis detected an anomaly during engine shutdown, which was later addressed and mitigated on STS-108 in November 2001.

By delivering and installing the Quest Joint Airlock, the Atlantis crew enabled the ISS to support spacewalks using American and Russian spacesuits, a crucial advancement for the station's operational flexibility.

STS-104 concluded with Atlantis undocking from the ISS on July 21, 2001, and returning to Earth. The mission ended with a successful landing at Kennedy Space Center on July 24, 2001.

STS-105: Continuing the Legacy of International Space Station Assembly and Crew Rotation

STS-105 was a mission of the Space Shuttle *Discovery* to the International Space Station

(ISS), launched from Kennedy Space Center, Florida, on August 10, 2001. Commanded by Scott J. Horowitz, with Frederick W. "Rick" Sturckow serving as pilot, the mission included mission specialists Daniel T. Barry and Patrick G. Forrester. Also aboard were the three members of the Expedition 3 crew—Frank L. Culbertson Jr., Mikhail Tyurin, and Vladimir N. Dezhurov—who were set to replace the departing Expedition 2 crew for an extended stay aboard the ISS.

STS-105's launch was initially scheduled for August 9, 2001, but was delayed due to weather concerns. It successfully launched the following day, on August 10, at 5:10:14 PM EDT. Unusually, the launch occurred early in the 10-minute window, a rare event for a shuttle mission, to avoid an approaching storm system that threatened to violate launch criteria.

This mission marked *Discovery*'s last flight before undergoing a significant refit, including an upgrade to the glass cockpit layout, a modernization already implemented on *Atlantis* and *Columbia*. Following the completion of STS-105, *Discovery* would not fly again until STS-114 in 2005 due to the fleet's grounding after the tragic *Columbia* disaster in 2003.

The primary objective of STS-105 was the rotation of the ISS crew. The Expedition 2 crew returned to Earth while Expedition 3 began their extended stay aboard the station. Additionally, the mission delivered essential supplies and scientific equipment using the Italian-built Multi-Purpose Logistics Module (MPLM) *Leonardo*. This was *Leonardo*'s second flight, having previously flown on STS-102. The MPLM, measuring 6.4 meters in length and 4.6 meters in diameter, housed resupply stowage racks, platforms, and two new scientific racks for the U.S. laboratory module, *Destiny*.

Leonardo contained six Resupply Stowage Racks and four Resupply Stowage Platforms, loaded with nearly 3,073 kilograms (6,775 pounds) of equipment and supplies for the ISS. This included two new EXPRESS (Expedite the Processing of Experiments to the Space Station) racks—EXPRESS Racks 4 and 5—which added significant scientific capability to *Destiny*. Each EXPRESS rack weighed over 1,000 pounds when fully loaded, enhancing the station's capacity to conduct scientific research.

Additionally, the mission carried an Integrated Cargo Carrier (ICC) loaded with the Early Ammonia Servicer and the Materials International Space Station Experiment (MISSE) Payload Experiment Containers (PECs 1 & 2). MISSE marked the first set of external experiments on the ISS, designed to test the durability of materials exposed to the harsh space environment for future spacecraft applications. These PECs were originally developed during the Shuttle-Mir Program and had previously flown on Mir in the 1990s.

Mission specialists Daniel Barry and Patrick Forrester conducted two extravehicular activities (EVAs) to install new equipment and perform maintenance on the ISS.

EVA 1, conducted on August 16, 2001, spanned 6 hours and 16 minutes. The astronauts diligently prepared the station for future assembly operations during this EVA.

EVA 2, held on August 18, 2001, lasted 5 hours and 29 minutes. This EVA was dedicated to installing additional scientific hardware and performing routine station maintenance.

STS-105 carried various other payloads, including experiments from NASA's Wallops Flight Facility as part of the Shuttle Small Payloads Project (SSPP). Among these were the Hitchhiker payload *Simplesat*, the Cell Growth in Microgravity experiment, and the Microgravity Smoldering Combustion experiment. These payloads provided valuable data for advancing space technologies and understanding the effects of microgravity on various physical and biological processes.

The mission lasted nearly 12 days, successfully concluding with *Discovery*'s landing at Kennedy Space Center on August 22, 2001. STS-105's success in rotating the ISS crew, delivering essential supplies, and conducting critical scientific experiments further solidified the partnership between the Shuttle and the International Space Station during the early 2000s.

STS-108: A Mission of Unity and Progress

STS-108, a Space Shuttle mission aboard *Endeavour*, launched on December 5, 2001, and marked the 12th shuttle flight to visit the International Space Station (ISS). The primary objective of this mission was to deliver crucial supplies to the ISS and to facilitate crew rotation between Expedition 3 and Expedition 4. It was the first shuttle mission to the station since the installation of the Russian airlock module, Pirs, which had expanded the station's capabilities for spacewalks and docking.

Endeavour carried the Expedition 4 crew to the ISS, while the Expedition 3 crew, who had spent four months aboard the station, returned to Earth. The new crew included Commander Yuri Onufrienko from the Russian Space Agency (RKA), as well as NASA astronauts Carl E. Walz and Daniel W. Bursch. The returning crew comprised Commander Frank L. Culbertson, Jr.,

along with cosmonauts Mikhail Turin and Vladimir Dezhurov, marking the end of their successful Expedition 3 mission.

One of the key highlights of the mission was the attachment of the Raffaello Multi-Purpose Logistics Module (MPLM) to the ISS. The module carried approximately 2.7 metric tons (3 tons) of equipment, scientific experiments, and supplies vital for the station's ongoing operations. After unloading the cargo, the crew detached Raffaello and stowed it back in *Endeavour*'s payload bay for the return journey to Earth.

In addition to cargo delivery, the crew performed a single spacewalk, contributing to the ongoing maintenance and expansion of the ISS. The mission also featured scientific investigations, including educational experiments that involved students from over 25 countries, highlighting NASA's commitment to international collaboration and outreach.

The launch of STS-108 faced several delays. Initially scheduled for November 29, 2001, it was postponed to December 4 to allow the Expedition 3 crew to complete a spacewalk, which cleared an obstruction on the Russian Progress supply vehicle's latching mechanism. However, unfavorable weather conditions at Kennedy Space Center (KSC) caused another delay, as clouds and precipitation moved into the area during the countdown. Finally, on December 5, at 17:12 EST, *Endeavour* successfully lifted off, delivering fresh supplies and crew to the ISS.

The mission also deployed a small satellite, Starshine-2, which was part of an educational initiative involving more than 25,000 students worldwide. This project was designed to teach students about satellite tracking and orbital mechanics, further fostering international scientific engagement.

The crew of STS-108 was led by Commander Dominic L. Gorie, on his third spaceflight, and Pilot Mark Kelly, who was embarking on his first. Joining them were Mission Specialists Linda M. Godwin and Daniel M. Tani, both of whom played key roles in the mission's success. Godwin, making her fourth and final spaceflight, oversaw various scientific operations, while Tani, on his first space mission, contributed to logistics and station activities.

On December 7, 2001, the hatches between *Endeavour* and the International Space Station's (ISS) Destiny Laboratory opened at 22:42 UTC, allowing the ten astronauts and cosmonauts from both spacecraft to meet and exchange greetings. This marked the beginning of an important crew transition. On December 8, the Expedition 3 crew officially ended their 117-day residency aboard the ISS as their custom-fitted Soyuz seat-liners were transferred to *Endeavour* for the return trip to Earth. Simultaneously, the Expedition 4 crew's seat-liners were installed in the Soyuz return vehicle, marking the formal exchange of station crews.

In a critical operation, *Endeavour*'s Pilot, Mark Kelly, and Mission Specialist Linda Godwin used the shuttle's robotic arm to extract the Raffaello Multi-Purpose Logistics Module (MPLM) from the shuttle's payload bay. They secured it to a designated berth on the ISS's Unity node, enabling the crews to begin unloading vital supplies the same day. This module delivered approximately 5,000 pounds of equipment, including food, clothing, experiments, medical supplies, and tools for upcoming spacewalks.

On December 10, 2001, astronauts Linda Godwin and Daniel Tani performed a spacewalk that lasted 4 hours and 12 minutes. Their primary task was to install insulation on mechanisms that rotated the station's main solar arrays, ensuring their continued efficiency. Additionally, the astronauts retrieved a cover from a storage bin that had previously been removed from a station antenna. This cover, upon its return to Earth, could potentially be reused. As part of a "get-ahead" task, they also positioned two switches on the station's exterior, which were scheduled to be installed during a future shuttle mission, STS-110. This EVA concluded a record year for NASA, marking the 18th spacewalk of 2001, with 12 conducted from the shuttle and six from the station itself.

Mission managers extended *Endeavour*'s stay at the ISS to 12 days, allowing additional time for maintenance tasks. These included repairs to the station's treadmill and the replacement of a failed compressor in the Zvezda Service Module's air conditioning system. The astronauts successfully transferred over 5,000 pounds of supplies from both *Endeavour*'s mid-deck and the Raffaello MPLM, including 850 pounds of food, 1,000 pounds of clothing, 300 pounds of experiments, 800 pounds of spacewalking equipment, and 600 pounds of medical supplies. On the return journey, the Raffaello module was packed with items bound for Earth, completing its mission.

On December 12, a minor issue was detected with one of the shuttle's three Inertial Measurement Units (IMUs), which are crucial for navigation. IMU 2 was taken offline as a precaution after exhibiting transient behavior, and IMU 3 was brought online in its place. Although IMU 2 continued to function normally afterward, mission controllers kept it offline for the remainder of the flight. This minor issue did not affect the mission, as the remaining IMUs operated without issue.

On December 15, 2001, *Endeavour* performed a brief engine firing to adjust the ISS's orbit, ensuring that the station would pass at a safe distance from a piece of debris — a spent Russian rocket upper stage launched in the 1970s. Without this reboost, the debris would have passed within three miles of the ISS. Instead, the maneuver pushed the station's trajectory far enough to guarantee a safe passage, about 40 miles from the debris. Due to the additional fuel used for this operation, *Endeavour*'s crew did not perform the planned full-circle flyaround of the station. Instead, the shuttle undocked and completed a quarter-circle flyaround before firing its engines for a final separation burn at 12:20 AM EST, initiating their departure from the orbital outpost.

STS-108 carried home numerous scientific results from experiments conducted during Expedition 3. These included the Advanced Protein Crystallization Facility and the Dynamically Controlled Protein Crystal Growth experiment, which provided new insights into crystallization in microgravity. The shuttle also brought back cells from the Cellular Biotechnology Operations Support System (CBOSS), used for cancer and kidney cell research in space. CBOSS remained active on the

ISS during Expedition 4, continuing experiments related to ovarian and colon cancer cells, as well as kidney cells.

In *Endeavour*'s payload bay, a variety of experiments from around the world were returned to Earth. The Multiple Application Customized Hitchhiker-1 (MACH-1) experiment array included the Prototype Synchrotron Radiation Detector, the Collisions into Dust Experiment-2, and the Capillary Pump Loop. Additionally, the Space Experiment Module (SEM) carried scientific experiments from Argentina, Portugal, Morocco, Australia, and U.S. schoolchildren.

As part of the mission's educational outreach component, *Endeavour* deployed the small STARSHINE 2 satellite, a project involving over 30,000 students from 660 schools in 26 countries. The students helped polish the satellite's 845 mirrors and tracked its orbit for eight months to calculate the density of Earth's upper atmosphere.

STS-108 was the first Space Shuttle mission to launch after the tragic September 11 attacks, and its crew carried symbols of remembrance aboard *Endeavour*. Among these were an American flag recovered from the rubble at the World Trade Center, a U.S. Marine Corps flag from the Pentagon, and 6,000 small American flags to honor the victims and those who served in the recovery efforts. These flags were later given to the families of those who lost their lives on that day. Additionally, the shields of 23 fallen New York Police Department officers and patches from the Fire Department of New York and the Port Authority were flown aboard *Endeavour* as a tribute to the bravery of first responders.

The Office of the New York City Commissioner of Records maintains the American flag from the World Trade Center as a symbol of resilience and unity.

EVA 1, conducted by Linda Godwin and Daniel Tani, commenced on December 10, 2001, at 17:52 UTC and concluded on December 10, 2001, at 22:04 UTC. The duration of this spacewalk was 4 hours and 12 minutes.

The spacewalk successfully installed insulation on the station's solar array mechanisms, retrieved a stowed cover from a previous mission, and prepared switches for future installation during STS-110, completing a busy and productive mission.

STS-108 lasted for 11 days, 19 hours, 36 minutes, and 45 seconds, covering a total distance of approximately 7.7 million kilometers (4.8 million miles) in orbit. After completing its mission to resupply the International Space Station (ISS) and facilitate crew exchanges, *Endeavour* returned to Earth on December 17, 2001. The shuttle touched down at Kennedy Space Center's Shuttle Landing Facility (SLF) on Runway 15 at 17:56:13 UTC, concluding the final Space Shuttle mission of 2001.

STS-109: A Mission of Restoration and Advancement

STS-109, the 108th flight of the Space Shuttle program and the 27th flight of the Space Shuttle Columbia, launched on March 1, 2002, from Kennedy Space Center at 6:22 AM EST. This mission was critical for the continued success of one of NASA's most celebrated and scientifically productive missions: the Hubble Space Telescope. STS-109, also known as the Hubble Space Telescope Servicing Mission 3B, was a testament to human ingenuity and the ability to maintain and enhance our technological tools in space.

After an initial scrub on April 4, 2002, due to a hydrogen leak, Space Shuttle *Atlantis* successfully launched on April 8, 2002, from Launch Complex 39B at the Kennedy Space Center. A minor delay occurred at the T-5 minute mark of the countdown due to a data issue in the backup Launch Processing System, but the problem was quickly resolved, and liftoff occurred with just 11 seconds remaining in the launch window. This mission marked a significant milestone for the shuttle program, as it was the first to utilize the upgraded Block II main engines, designed with an improved fuel pump, a stronger integral shaft and disk, and more robust bearings. These enhancements aimed to increase the reliability and safety of the shuttle's propulsion system while extending the operational life of the engines.

Mission specialist Jerry Ross also made history during STS-110, becoming the first human to travel to space seven times.

The primary objective of STS-109 was to upgrade and extend the life of the Hubble Space Telescope, which had been orbiting Earth since 1990. Columbia carried a crew of seven astronauts: Commander Scott Altman, Pilot Duane Carey, and Mission Specialists John Grunsfeld, Nancy Currie, Richard Linnehan, James Newman, and Michael Massimino. Each astronaut played a crucial role in the mission's success, particularly during the five spacewalks (EVAs) planned to perform intricate repairs and upgrades to Hubble.

The primary objective of STS-110, launched aboard Space Shuttle *Atlantis*, was to attach the S0 Truss segment to the International Space Station (ISS). This stainless steel segment formed the central backbone of the station, allowing future truss segments, such as the S1 and P1, to be connected during subsequent missions, STS-112 and STS-113 respectively. The S0 Truss was attached to the Destiny Laboratory Module, an essential hub for scientific research aboard the ISS.

In addition to the S0 Truss, the mission also delivered the Mobile Transporter (MT), an 885-kilogram (1,951-pound) assembly designed to glide along the station's integrated truss structure. Developed by Astro Aerospace in Carpinteria, California, the MT plays a critical role in station operations. During the following mission, STS-

111, the Mobile Base System (MBS) would be mounted onto the MT, forming part of the Mobile Servicing System (MSS). This system enables the Canadarm2 robotic arm to travel the length of the truss, allowing for repairs, assembly tasks, and other essential functions across the station.

STS-110, a mission that involved four intricate spacewalks, was pivotal in installing and configuring the S0 Truss on the International Space Station (ISS).

The first spacewalk, conducted by astronauts Steven Smith and Rex Walheim on April 11, 2002, lasted 7 hours and 48 minutes. During this EVA, the astronauts successfully installed the S0 Truss onto the Destiny Laboratory Module, a crucial step in expanding the station's truss structure.

Astronauts Jerry Ross and Lee Morin led the second spacewalk, which took place on April 13, 2002. This 7-hour and 30-minute EVA focused on further securing the truss to the station and preparing for future installations.

The third spacewalk, carried out on April 14, 2002, was primarily aimed at reconfiguring the Canadarm2 robotic arm to interact with the newly installed truss. This 6-hour and 27-minute EVA ensured the arm's continued functionality as the station's structure expanded.

The final spacewalk of the mission, conducted by Jerry Ross and Lee Morin on April 16, 2002, lasted 6 hours and 37 minutes. During this EVA, the astronauts installed hardware to support future spacewalks and prepared the station for upcoming assembly tasks.

The crew of STS-110 consisted of seven astronauts, each assigned critical roles during both the launch and landing phases. The commander, Michael Bloomfield, along with pilot Stephen Frick, were seated on the flight deck, while mission specialists such as Jerry Ross, Rex Walheim, and Lee Morin carried out vital tasks in space. Ellen Ochoa, another mission specialist, played a key role in supporting the astronauts during their spacewalks from inside the shuttle.

STS-110 was a technical success and a critical milestone in the construction and expansion of the International Space Station. With the successful delivery and installation of the S0 Truss and Mobile Transporter, the mission laid the groundwork for future assembly tasks and the continued growth of the station's structural and operational capabilities.

Columbia's return to Earth on March 12, 2002, at 4:32 AM EST at Kennedy Space Center marked the successful completion of a mission that had revitalized one of humanity's most important scientific instruments.

STS-110: Advancing the International Space Station

STS-110, the 109th flight of the Space Shuttle program and the 27th mission for the Space Shuttle Atlantis, launched on April 8, 2002, from Kennedy Space Center at 4:44 PM EDT. This mission was a key step in the ongoing assembly of the International Space Station (ISS), as it delivered and installed the S0 (S-Zero) Truss. This critical structural component would be the backbone of the station's expanding truss system.

The crew of STS-110, commanded by Michael J. Bloomfield, included Pilot Stephen N. Frick and Mission Specialists Jerry L. Ross, Steven L. Smith, Ellen Ochoa, Rex J. Walheim, and Lee M.E. Morin. Each astronaut had a crucial role in ensuring the mission's success, particularly during the four spacewalks (EVAs) required to install and activate the S0 Truss.

The S0 Truss, 44 feet long and weighing nearly 27,000 pounds, was the first segment of the station's integrated truss structure, which would eventually span over 300 feet and support the station's solar arrays, thermal radiators, and other critical systems. Installing the S0 Truss was a complex and demanding task, requiring precise coordination between the crew aboard Atlantis and the Expedition 4 crew already aboard the ISS.

First EVA (April 11, 2002): Astronauts Jerry L. Ross and Steven L. Smith conducted the initial EVA, which lasted over 7 hours. They secured the S0 Truss to the Destiny laboratory module using a combination of bolts and latches. The spacewalkers also began the process of connecting power and data cables, which would be completed in subsequent EVAs.

Second EVA (April 13, 2002): During the second spacewalk, Ross and Walheim connected electrical cables between the S0 Truss and the rest of the station. This EVA was particularly challenging due to the complex layout of the cables and the need to ensure that all connections were secure to enable power distribution through the new truss.

Third EVA (April 14, 2002): Ochoa and Morin focused on activating the Mobile Transporter, a railcar system attached to the S0 Truss that would allow the station's robotic arm, Canadarm2, to move along the truss and support future assembly tasks. The spacewalkers also completed additional connections and checks to ensure the system's functionality.

Fourth EVA (April 16, 2002): Walheim and Smith conducted the final spacewalk, which involved installing additional equipment on the S0 Truss, including a series of structural components and a second set of power cables. This EVA concluded the installation process, ensuring the S0 Truss was fully integrated into the ISS.

The mission's success was a major milestone in the assembly of the ISS, as the S0 Truss provided the foundation for the station's extensive truss system, which would later support additional truss segments, solar arrays, and radiators. This truss structure was crucial for the station's long-term sustainability and its ability to support a permanent human presence in space.

STS-110 also marked a significant achievement for astronaut Jerry L. Ross, who became the first person to complete seven spaceflights, setting a record that underscored his extraordinary contributions to human spaceflight and space station assembly.

Atlantis returned to Earth on April 19, 2002, landing at Kennedy Space Center at 12:27 PM EDT. The mission's success reinforced the importance of the Space Shuttle in constructing and maintaining the ISS, highlighting the shuttle's unique capability to deliver and install large, complex components in orbit.

STS-111

STS-111, also known as Endeavour's mission to the International Space Station (ISS), was launched on June 5, 2002. The mission faced several delays due to weather conditions. The first launch attempt on May 30, 2002, was scrubbed due to thunderstorms and electrical activity in the area. A second attempt on May 31 was also canceled because of continued concerns about bad weather, including hail. Finally, after addressing technical issues, including a nitrogen valve problem, Endeavour successfully lifted off on June 5, 2002.

This mission played a critical role in continuing the assembly and operation of the ISS, providing both supplies and new crew members. The shuttle delivered the Expedition 5 crew to the station, consisting of two Russian cosmonauts and one American astronaut, while safely returning the Expedition 4 crew of one Russian and two American astronauts to Earth. This routine rotation ensured that ISS operations continued smoothly with fresh personnel to oversee scientific experiments and station maintenance.

A key payload of STS-111 was the Multi-Purpose Logistics Module (MPLM), which carried vital experiment racks along with three stowage and resupply racks. These supplies were critical to maintaining the station's scientific capabilities and operational needs. Additionally, STS-111 installed a crucial ISS robotic system component: the Mobile Base System (MBS). This component was affixed to the Mobile Transporter (MT), which had been installed earlier during the STS-110 mission. Together, this formed part of the Canadian Mobile Servicing System (MSS), enhancing the capabilities of the station's Canadarm2 robotic arm. The MSS allowed the arm to "inchworm" its way across the station, moving from the U.S. Lab module to the MBS along the station's truss structure. This enhanced the arm's flexibility in accessing various work sites along the station's exterior, a crucial ability for ongoing construction and maintenance tasks.

This mission also marked a significant milestone in European spaceflight history, as it was the last flight for a French astronaut under the banner of CNES (the French space agency). CNES had disbanded its astronaut group and transferred its remaining personnel to the European Space Agency (ESA), concluding France's national manned spaceflight program.

The crew of STS-111 included Commander Kenneth Cockrell, Pilot Paul Lockhart, and Mission Specialists Franklin Chang-Díaz and Philippe Perrin. Chang-Díaz, a veteran astronaut, and Perrin, a French astronaut making his first flight, conducted three spacewalks during the mission. These extravehicular activities (EVAs) were essential for installing new components and maintaining the station's operational capabilities.

The first spacewalk took place on June 9, 2002, lasting 7 hours and 14 minutes. During this EVA, Chang-Díaz and Perrin successfully attached the Power and Data Grapple Fixture to the P6 truss, preparing the station for future expansion. The second spacewalk, on June 11, 2002, lasted 5 hours, during which the two astronauts attached the Mobile Base System to the Mobile Transporter, enabling the Canadarm2 to move more freely along the station's exterior. The third and final spacewalk occurred on June 13, 2002, lasting 7 hours and 17 minutes. During this EVA, the astronauts replaced the wrist joint of Canadarm2, ensuring that the arm remained fully functional for future operations.

STS-111 concluded on June 19, 2002, when Endeavour touched down at Edwards Air Force Base in California after adverse weather conditions at the primary landing site, Kennedy Space Center, forced the shuttle to divert.

STS-112

STS-112 (ISS Assembly Flight 9A) was a pivotal 11-day Space Shuttle mission flown by *Atlantis* to the International Space Station (ISS). Launched from Kennedy Space Center's Launch Pad 39B on 7 October 2002 at 19:45 UTC, *Atlantis* carried the Starboard 1 (S1) truss segment, weighing 28,000 pounds, as its primary payload. Space Shuttle *Atlantis* lifted off under clear skies, with the ascent proceeding according to plan. For the first time in Shuttle history, a rocket-mounted camera transmitted live video of the launch to mission controllers, providing unprecedented visual data until the separation of the solid rocket boosters obscured the camera's view.

This mission marked another critical step in the ISS assembly, contributing to the structural framework of the station by delivering the S1 truss, which supports vital radiator systems. After a successful mission, *Atlantis* concluded its 4.5-million-mile journey with a smooth landing at 15:44 UTC on 18 October 2002 at Kennedy Space Center's Shuttle Landing Facility.

Though the launch went smoothly, it was not without incident. A chunk of foam, approximately 4 x 5 x 12 inches, shed from the External Tank (ET) bipod ramp during liftoff. This debris struck the lower part of the left Solid Rocket Booster (SRB), causing a dent about 4 inches wide and 3 inches deep in the SRB-ET Attach Ring. The damage raised concerns, yet, despite the anomaly, NASA decided to proceed with the next scheduled mission, STS-113. Unfortunately, the decision to overlook these foam strikes contributed to the Space Shuttle *Columbia* disaster on STS-107 in early 2003.

Initially, *Atlantis* had been slated for the STS-114 mission in March 2003, but following the loss of *Columbia*, the Shuttle fleet was grounded for safety evaluations. Consequently, *Atlantis* did not fly again until STS-115 in September 2006, marking the resumption of construction work on the ISS.

The STS-112 crew was led by Commander Jeffrey S. Ashby, making his third and final spaceflight. He was joined by Pilot Pamela A. Melroy, Mission Specialists David A. Wolf, Sandra H. Magnus, Piers Sellers, and Fyodor Yurchikhin of Russia. This mission was the first spaceflight for Magnus, Sellers, and Yurchikhin, while Melroy and Wolf were both veterans of previous spaceflights.

A key mission objective was the delivery and installation of the S1 truss segment, which provides critical support for the space station's radiators. Boeing and Lockheed Martin manufactured the truss, with work commencing in May 1998 at Michoud Assembly Facility and completed in March 1999. By June 2002, the truss was handed over to NASA for final preparations, ready to be launched on *Atlantis*.

Another important payload was the Crew Equipment Translation Aid (CETA) cart, a mobility aid for astronauts working on the station's exterior. It was attached to the Mobile Transporter system, allowing future assembly crews to carry out tasks more efficiently.

STS-112 also carried several science experiments aboard, focusing on biological and crystallization research. The Plant Generic Bioprocessing Apparatus (PGBA) and Commercial Generic Bioprocessing Apparatus (CGBA) were included, alongside the Protein Crystal Growth Single-locker Thermal Enclosure System, which housed the Protein Crystallization Apparatus for Microgravity (PCG-STES-PCAM). Additionally, the mission carried samples for the Zeolite Crystal Growth Furnace (ZCG) experiment, continuing research into material science in microgravity environments.

Preparations for the STS-112 mission experienced a slight delay when tiny cracks were discovered in *Atlantis'* propulsion system. The cracks were found within the metal flow liners of the liquid hydrogen fuel lines feeding the shuttle's main engines. Though the cracks posed no immediate threat to the fuel pipes, concerns arose that metal debris could potentially break off and damage the engines, risking a catastrophic engine shutdown during ascent. NASA addressed the issue before clearing the shuttle for flight.

October 8, Flight Day 2 saw the crew preparing for their rendezvous with the ISS. Pilot Pamela Melroy and Mission Specialists David Wolf and Piers Sellers checked out the spacesuits and EVA equipment, while Commander Ashby and Mission Specialist Sandra Magnus operated *Atlantis'* robotic arm to inspect the shuttle's payload bay. The crew also completed several orbital maneuvers, aligning *Atlantis* with the ISS's orbit.

On Flight Day 3, *Atlantis* docked with the ISS at 15:17 UTC, connecting to the Pressurized Mating Adapter-2 on the Destiny Laboratory. This docking initiated a week of joint operations between the STS-112 and Expedition 5 crews. Following successful pressure checks, the crew of *Atlantis* joined the ISS crew inside the space station, where they immediately began preparations for the first of three planned spacewalks. The crew's work also included configuring the spacesuits and reviewing procedures for installing the new S1 truss segment using the station's robotic arm.

The core objective of STS-112 was the installation of the S1 truss. Over the next several days, astronauts Sellers and Wolf conducted spacewalks to secure the truss to the station and connect its electrical and cooling systems. The installation of the S1 truss added critical infrastructure, extending the station's ability to dissipate heat from its systems.

10 October (Flight Day 4 – EVA 1) began at 3:00 AM CDT with a musical wake-up call to *Atlantis'* crew from Mission Control in Houston. The focus of Flight Day 4 was the installation of the Starboard 1 (S1) truss, a key component of the International Space Station's (ISS) structural framework. Astronauts Peggy Whitson and Sandra Magnus operated the ISS's Canadarm2 robotic arm to remove the S1 truss from *Atlantis'* payload bay and move it into position on the starboard end of the S0 truss. At 8:36 AM CDT, the two trusses were secured together using four motorized bolts, marking a major milestone in the mission.

While this was happening, astronauts David Wolf and Piers Sellers prepared for the mission's first extravehicular activity (EVA). This spacewalk, the 44th conducted in support of ISS assembly and maintenance, began at 11:21 AM EDT. Wolf, wearing a spacesuit with red stripes for identification, and Sellers, in an all-white suit, exited the Quest airlock to begin their tasks. Wolf's primary mission was to connect power, data, and fluid lines between the S0 and S1 trusses, while Sellers focused on unlocking three folded radiators on the S1 truss, allowing them to deploy and optimize the station's cooling system.

The spacewalk lasted seven hours and one minute, extending 31 minutes longer than planned due to a malfunction with the Canadarm2. This glitch required Wolf to manually install a television camera system on the far end of the truss without the robotic arm's assistance. Additionally, Wolf encountered a minor issue with his helmet's earphones, which began losing power near the end of the EVA. Despite these challenges, the crew completed their tasks. Shuttle Commander Jeff Ashby operated *Atlantis'* robotic arm during the spacewalk, providing camera views to document the activities. After a successful day of assembly, the astronauts returned to the airlock, which was re-pressurized at 5:22 PM CDT, bringing EVA 1 to a close.

11 October (Flight Day 5), following the intense work of EVA 1, Flight Day 5 provided a few hours of off-duty time for the combined shuttle and ISS crews. They shared a meal in the Zvezda Service Module, taking a moment to relax before resuming transfer operations. The crew began moving scientific experiments between *Atlantis* and the ISS, including experiments focused on protein crystal growth and liver cell function in microgravity. One of the notable transfers included a set of liver cell tissue samples, which were returned to Earth for analysis.

The crew also delivered fresh supplies to the ISS, including seven water containers and about 15 pounds of nitrogen gas, which Commander Ashby transferred from *Atlantis* to the station. Meanwhile, Wolf and Sellers, with assistance from Pilot Pamela Melroy, prepared for the mission's second spacewalk by recharging the spacesuit water systems and configuring their tools. Throughout the day, the astronauts participated in several live media interviews, discussing their experiences and the challenges of working in space. Wolf and Sellers recounted their manual installation of the S1 truss camera during EVA 1, describing how their heart rates exceeded 170 beats per minute during the strenuous activity.

On 12 October (Flight Day 6 – EVA 2), Wolf and Sellers conducted the second spacewalk of the mission. Beginning at 14:31 UTC, EVA 2 focused on preparing hardware for future extravehicular activities (EVAs) and continuing work on the S1 truss. The spacewalk lasted six hours and four minutes and concluded at 20:35 UTC.

On 13 October (Flight Day 7), the crew spent Flight Day 7 conducting system checks and preparing for the mission's final spacewalk. No major issues were reported, and the crew continued to monitor the performance of the newly installed S1 truss.

14 October (Flight Day 8 – EVA 3), the third and final spacewalk began at 14:08 UTC and lasted six hours and 36 minutes. Wolf and Sellers completed the installation of the S1 truss, ensuring all systems were connected and operational. This marked the successful conclusion of the truss assembly tasks for STS-112.

After the completion of the spacewalks on 15 October (Flight Day 9), the crew focused on finalizing transfer operations and preparing *Atlantis* for its undocking from the ISS. They conducted additional experiments and ensured all systems aboard the ISS functioned correctly after the truss installation.

16 October Flight Day 10, *Atlantis* undocked from the ISS, concluding the joint operations between the shuttle and the station. The crew performed a standard flyaround of the ISS to capture images and document the progress of the station's assembly.

17 October, the crew spent Flight Day 11 preparing for re-entry and landing. Systems aboard *Atlantis* were checked, and the shuttle's payload was secured for its return to Earth.

During EVA 1, held on October 10, 2002, the spacewalk spanned 7 hours and 1 minute. The crew achieved several notable objectives during this EVA. They successfully installed power, data, and fluid lines connecting the S0 and S1 trusses. Additionally, they deployed radiators and installed a television camera system.

EVA 2, conducted on October 12, 2002, lasted for 6 hours and 4 minutes. The primary focus of this EVA was preparing for future EVAs by installing essential hardware for those missions.

EVA 3, held on October 14, 2002, marked a significant milestone as the crew successfully installed the S1 truss. This EVA lasted for 6 hours and 36 minutes.

STS-112 marked the first time a camera was mounted to the shuttle's external tank to capture launch footage. While the footage provided critical insight, the camera became fogged with propellant after the separation of the solid rocket boosters, limiting its usefulness. Following this mission, NASA adjusted the camera placement for future flights. After the *Columbia* disaster in 2003, this camera became a vital tool for detecting falling debris from the external tank during shuttle ascents.

The mission lasted nearly 11 days, concluding on October 18, 2002, when Atlantis landed safely at Kennedy Space Center. The success of STS-112 was celebrated as a critical step in the ISS assembly sequence, with the S1 Truss providing the necessary infrastructure for future modules and systems that would be added to the station in subsequent missions.

STS-113

STS-113, also known as Endeavour's 19th flight, played a crucial role in the ongoing assembly and expansion of the International Space Station (ISS). Launched on November 23, 2002, from Kennedy Space Center in Florida, STS-113 was a milestone mission that advanced the ISS's construction and marked the final shuttle flight of 2002, capping a year of significant achievements in space exploration.

The primary objective of STS-113 was to deliver and install the P1 Truss, a key structural element of the ISS's Integrated Truss Structure (ITS). The P1 Truss, weighing approximately 28,000 pounds and measuring 45 feet in length, was designed to be the mirror image of the S1 Truss, installed during the previous shuttle mission, STS-112. The installation of the P1 Truss was vital for expanding the station's cooling and power systems, as it housed essential radiators and supported critical electrical connections.

Commanded by James Wetherbee, the STS-113 crew included Pilot Paul Lockhart and Mission Specialists Michael Lopez-Alegria and John Herrington. Joining them were the returning members of the ISS Expedition 5 crew—Commander Valery Korzun, Flight Engineer

Peggy Whitson, and Flight Engineer Sergei Treschev—who were replaced by the incoming Expedition 6 crew, consisting of Commander Ken Bowersox, Flight Engineer Nikolai Budarin, and NASA ISS Science Officer Donald Pettit.

Lopez-Alegria and Herrington conducted three Extravehicular Activities (EVAs) to install the P1 truss. These spacewalks, each lasting over six hours, involved complex tasks such as attaching the truss to the station's existing structure, connecting power and data cables, and deploying the truss's radiator panels. The EVAs demonstrated the astronauts' skill and the importance of careful planning and execution in the unforgiving space environment.

STS-113 also marked a historic moment in NASA's space program, with Mission Specialist John Herrington becoming the first enrolled member of a Native American tribe (Chickasaw Nation) to fly in space. Herrington's participation brought additional attention to the mission and highlighted the diverse backgrounds of the individuals contributing to space exploration.

In addition to its construction tasks, STS-113 was vital in rotating the ISS crew, ensuring the continued human presence aboard the station. The handover between Expedition 5 and 6 was well-coordinated, reflecting the growing experience of NASA and its international partners in managing long-duration space missions.

The mission was extended due to unfavorable weather conditions at the primary landing site, leading to 13 days in space. Endeavour ultimately touched down at Kennedy Space

Center on December 7, 2002, successfully concluding STS-113 and bringing home the returning ISS crew members.

STS-113's success was a testament to the capabilities of the Space Shuttle program and the ongoing international collaboration that made the ISS a reality. The installation of the P1 Truss was a critical step in the station's assembly, laying the groundwork for future expansion and enhancing the ISS's ability to support scientific research and exploration activities.

STS-107: The Final Voyage of Space Shuttle Columbia

STS-107 marked the 113th flight of NASA's Space Shuttle program and was the 28th and final mission of *Columbia*. The mission launched on January 16, 2003, from Kennedy Space Center, Florida, and tragically ended on February 1, 2003, when *Columbia* disintegrated during re-entry, claiming the lives of all seven astronauts aboard. This disaster, which occurred just 16 minutes before the scheduled landing at Kennedy Space Center, was the 88th mission following the *Challenger* tragedy in 1986.

The crew of STS-107 consisted of highly skilled astronauts from the United States and Israel. Commander Rick D. Husband, on his second and final spaceflight, led the mission alongside Pilot William C. McCool, also on his first and only spaceflight. The mission specialists included David M. Brown, Kalpana Chawla, Michael P. Anderson, and Laurel B. Clark—all of whom were embarking on either their first or final missions. Completing the crew was Ilan Ramon, Israel's first astronaut, flying as a payload specialist.

During the 16-day mission, *Columbia* orbited Earth conducting an extensive series of scientific experiments. These experiments were part of various international collaborations, utilizing the SPACEHAB Research Double Module (RDM), Freestar payload, and other research platforms. The crew spent 15 days, 22 hours, and 20 minutes in space, conducting investigations in fields such as biology, materials science, and atmospheric studies. One notable experiment captured a possible new atmospheric phenomenon, dubbed the "Transient Ionospheric Glow Emission in Red" (TIGER).

Among the items carried aboard *Columbia* was a copy of a drawing by Petr Ginz, a 14-year-old Holocaust victim, imagined from the Moon's surface. The drawing was carried by Ilan Ramon and tragically lost in the disaster, along with other personal mementos symbolizing humanity's shared aspirations for peace and scientific discovery.

Despite the disaster, significant scientific data was salvaged. Approximately 30% of the mission's experimental data was transmitted back to ground stations before re-entry, and NASA recovered additional samples and hard drives from the debris field in Texas. Experiments such as the Critical Viscosity of Xenon-2 (CVX-2) were among the major recoveries, with scientists able to retrieve critical data even after the catastrophic failure. Another payload, the Commercial Instrumentation Technology Associates Biomedical Experiments-2 (CIBX-2), achieved a remarkable recovery rate, with 90% of its data salvaged, contributing valuable insights into cancer treatment research.

The cause of the *Columbia* disaster was traced back to the mission's launch, when a piece of insulating foam broke off from the external fuel tank and struck the orbiter's left wing. This impact compromised the orbiter's thermal protection system, a series of reinforced carbon-carbon panels and thermal tiles designed to shield the spacecraft during re-entry. As *Columbia* re-entered Earth's atmosphere, superheated air penetrated the damaged wing, leading to the vehicle's loss of control and disintegration over Texas.

This damage echoed concerns from an earlier flight, STS-27, when *Atlantis* sustained similar but less severe damage in 1988. However, Atlantis survived because the impact occurred on a stronger, more protected area of the orbiter. Unfortunately, the weakened section of *Columbia*'s left wing could not withstand the extreme conditions of re-entry.

The aftermath of the *Columbia* disaster led to the creation of the Columbia Accident Investigation Board (CAIB), which conducted a comprehensive investigation into the accident. The CAIB's final report identified organizational and technical failures, including the foam strike, as root causes of the tragedy, leading to significant reforms in NASA's management and engineering practices. The investigation spurred a seven-month search for debris, collecting over 85,000 pieces of the orbiter—approximately 38% of its total structure.

STS-107 remains a solemn chapter in space exploration history, a mission remembered for the scientific discoveries it enabled and the lessons it taught about safety, perseverance, and the unyielding pursuit of knowledge. A memorial inside the Space Shuttle *Atlantis* Pavilion at Kennedy Space Center now honors the legacy of *Columbia* and her crew, including artifacts like the cockpit window frame that survived the disaster, reminding future generations of the risks and triumphs inherent in human spaceflight.

STS-114

STS-114, flown by Space Shuttle *Discovery*, marked NASA's first "Return to Flight" mission following the tragic loss of *Columbia* during STS-107. After 907 days, on July 26, 2005, at 10:39 a.m. EDT (14:39 UTC), *Discovery* launched from Kennedy Space Center, Florida, into clear skies. This highly anticipated flight came after several delays, including an earlier attempt on July 13, which was scrubbed due to unresolved fuel sensor anomalies in the external tank. Despite these lingering concerns, NASA approved the launch to resume shuttle operations.

The STS-114 mission concluded on August 9, 2005, when *Discovery* safely landed at Edwards Air Force Base in California. Originally scheduled to land at Kennedy Space Center, poor weather in Florida forced mission controllers to select the backup landing site. Although the mission itself was a success, the return to Earth was overshadowed by concerns raised during ascent, when footage revealed debris separating from the shuttle's external tank—a key issue that had contributed to the *Columbia* disaster. As a precaution, NASA temporarily suspended future shuttle flights, delaying subsequent missions until additional modifications could be made to enhance the shuttle's safety. Flights resumed nearly a year later with STS-121 on July 4, 2006.

Commanded by veteran astronaut Eileen Collins, STS-114's crew was diverse and highly experienced. Collins, leading her fourth and final spaceflight, became the second female commander of a Space Shuttle mission, her first command having been STS-93. The mission's pilot, James M. Kelly, was on his second and final flight, while mission specialists included Soichi Noguchi of Japan, on his first spaceflight, and Stephen K. Robinson, an experienced astronaut serving as the flight engineer. Other crew members included Wendy Lawrence and Andrew Thomas, both on their fourth and final missions, and Charles Camarda, making his first and only spaceflight.

The original crew for STS-114 had been planned to carry the Expedition 7 crew to the

International Space Station (ISS) and return the Expedition 6 crew to Earth. However, the tragic events of *Columbia* led to changes in mission objectives and personnel.

The primary goal of STS-114 was not just the delivery of supplies to the ISS but also the testing of new safety and inspection techniques for the Shuttle program. This mission was critical in proving that NASA had addressed the vulnerabilities exposed by the *Columbia* disaster. One of the key safety tools introduced was the Orbiter Boom Sensor System (OBSS), a 50-foot (15-meter) extension attached to the Shuttle's robotic arm. Equipped with high-resolution visual cameras and a Laser Dynamic Range Imager (LDRI), the OBSS allowed the crew to perform detailed inspections of the Shuttle's Thermal Protection System (TPS), including the leading edges of the wings, nose cap, and other critical areas.

The crew conducted three spacewalks, all of which contributed to the mission's dual focus on repairs and resupply. During the first spacewalk, the astronauts tested repair techniques on the Shuttle's heat shield. The second spacewalk involved the replacement of a failed Control Moment Gyroscope (CMG) on the ISS, which helps control the station's orientation in space. This gyroscope had been delivered in the Shuttle's cargo bay on the Lightweight Multi-Purpose Experiment Support Structure Carrier (LMC), alongside the TPS Repair Box. The final spacewalk, conducted on August 3, was historic—Stephen Robinson successfully removed two protruding gap fillers from the underside of the Shuttle, marking the first time repairs were performed on the exterior of a spacecraft in orbit. Additionally, concerns over a thermal blanket near the commander's window were addressed after NASA determined it would not affect re-entry.

The mission also carried the Raffaello Multi-Purpose Logistics Module, built by the Italian Space Agency, which was attached to the ISS and used to transfer additional supplies. Another significant payload was the External Stowage Platform-2, mounted to the port side of the Quest Airlock, and the MISSE 5 experiment, which was deployed on the station's exterior to study the effects of the space environment on various materials.

One often overlooked achievement of STS-114 was its role in boosting the ISS's orbit. The Shuttle's arrival increased the ISS's altitude by approximately 4,000 feet (1,200 meters), as the station loses about 100 feet (30 meters) of altitude each day due to atmospheric drag. As uncertainty loomed over when the next Shuttle would visit the station, the crew used the extended mission time to transfer additional items from *Discovery* to the ISS.

During the launch of *Discovery* on STS-114, several notable anomalies occurred that drew attention and later prompted NASA to ground future Shuttle flights until further investigations could be conducted. Just 2.5 seconds after liftoff, a large bird was seen colliding with the top of the external fuel tank and sliding down its surface. Fortunately, this impact did not affect the mission, as it did not strike the orbiter, and the Shuttle was traveling at a relatively low speed at the time.

Shortly after liftoff, a small piece of thermal tile, approximately 1.5 inches (38 mm) in size, detached from the front landing gear door. This was captured on video, showing a small white area where the shard had come off. The detachment occurred before the separation of the solid rocket boosters (SRB), and although the cause of the damage was unclear, engineers flagged the tile for inspection. The Orbiter Boom Sensor System (OBSS) was scheduled to examine the area in detail on July 29, 2005, to ensure that no critical damage had occurred that might compromise reentry.

At 127.1 seconds into the ascent, just 5.3 seconds after the SRB separation, a more significant event occurred. A large piece of debris, thought to be foam, broke away from the Protuberance Air Load (PAL) ramp of the external tank. Measuring roughly 36.3 by 11 by 6.7 inches (922 by 279 by 170 mm) and weighing about 0.45 kilograms (0.99 lb), this piece of foam was half the size of the debris that had caused the *Columbia* disaster. Fortunately, the foam did not strike the orbiter, and the mission continued, though concerns about the integrity of the foam insulation persisted.

In the following moments, a smaller piece of foam struck the right wing of the orbiter. NASA calculated that the impact force exerted only a fraction of the energy required to cause serious damage. Subsequent laser scans and imaging of the wing by the OBSS confirmed that no critical damage had been sustained. However, in light of these anomalies, NASA suspended future Shuttle flights on July 27, 2005, until the foam loss issue could be fully resolved.

Initially, NASA suspected that improper installation and handling of the external tanks at the Michoud Assembly Facility in Louisiana had contributed to the foam loss. NASA Administrator Michael Griffin indicated that the earliest the next Shuttle could launch was September 22, 2005, pending further investigation. However, the aftermath of Hurricane Katrina, which devastated the Gulf Coast, including the Michoud facility and NASA's Stennis Space Center, delayed the next Shuttle mission until July 4, 2006, when STS-121 was successfully launched.

In December 2005, X-ray photographs revealed that the cracks causing foam loss were the result of thermal expansion and contraction during fuel tank filling, rather than human error. NASA officials, including Wayne Hale, formally apologized to the Michoud workers, who had been unfairly blamed for the foam loss that contributed to the *Columbia* disaster.

STS-114 made history not only as the "Return to Flight" mission but also as the first Shuttle mission to conduct in-flight repairs. During the mission, concerns arose about two protruding gap fillers on the underside of the Shuttle. Gap fillers serve different purposes, such as preventing "chattering" of the tiles during ascent and reducing heat transfer between the tiles. While NASA determined that the gap fillers were not necessary for reentry, they opted to remove them due to the risk of disrupting the smooth, laminar air flow over the Shuttle's surface during reentry. If left unaddressed, the fillers could cause turbulence, leading to potentially dangerous temperature variations on the orbiter's heat shield.

Astronaut Stephen K. Robinson, during the third extravehicular activity (EVA), was tasked with removing the gap fillers. Riding the robotic arm of the International Space Station, Robinson ventured underneath *Discovery*. With gentle effort, he successfully removed the gap fillers using only his hands, describing the process as straightforward: "I'm grasping it, and I'm pulling it, and it's coming out very easily." NASA had prepared alternative plans, including the possibility of cutting the gap fillers if manual removal failed, but no tools were needed. This marked the first time in history that repairs were made on the exterior of an orbiting spacecraft.

Another concern during the mission involved a damaged thermal blanket located beneath the commander's window on the orbiter's port side. Although NASA considered a fourth spacewalk to repair or remove the damaged section, wind tunnel testing confirmed that the blanket would not pose a risk during reentry. As a result, plans for an additional EVA were canceled.

Chronology of Key Events During STS-114

On July 13, 2005, NASA prepared for the highly anticipated launch of the Space Shuttle Discovery, marking the first "Return to Flight" mission after the loss of Columbia in 2003. The countdown began with optimism, but events unfolded differently as the launch attempt was scrubbed due to technical issues.

July 13, 2005: Scrubbed Launch Attempt At 11:55 EDT, after a programmed hold, the countdown resumed. By 12:01 EDT, the crew of STS-114 boarded the traditional Astrovan to travel to Launch Pad 39B, where they arrived at 12:30 EDT. They entered the White Room, the final staging area before boarding the shuttle. Excitement filled the air, but at 13:32 EDT, a problem was detected with a liquid hydrogen (LH2) fuel level sensor, leading to the decision to scrub the launch. At 13:34 EDT, the crew began to exit the shuttle, and by 13:59 EDT, they had safely completed egress, allowing the technical team to begin troubleshooting the issue.

July 14, 2005: Technical Meeting and Press Conference
Following the previous night's draining of the external tank, the Mission Management Team convened at 14:00 EDT to assess the situation. NASA confirmed during a press conference at 14:45 EDT that the earliest possible launch

window would now be July 17, and that the technical challenges with Discovery had not affected preparations for Atlantis' upcoming mission (STS-121).

July 26, 2005: Successful Launch of STS-114
At 08:08 EDT on July 26, the shuttle crew completed boarding. By 09:00 EDT, the shuttle's hatch was securely closed. A brief hold occurred at T-minus 20 minutes at 09:24 EDT, and a second hold at T-minus 9 minutes at 09:45 EDT. At 10:27 EDT, launch control cleared the way for launch, and the final countdown resumed at 10:30 EDT. The auxiliary power units (APUs) were activated at T-minus 4 minutes, and at 10:39 EDT, Space Shuttle Discovery lifted off, clearing the tower and soaring into the sky. Discovery's main engines shut down eight minutes after liftoff, and the external fuel tank separated as planned, marking the shuttle's successful entry into orbit.

July 28, 2005: Docking with the International Space Station (ISS)
Two days later, on July 28 at 07:18 EDT, Discovery docked with the ISS. This docking was historic as it involved the first-ever Rendezvous Pitch Maneuver, allowing the ISS crew to photograph Discovery's heat shield for inspection, ensuring the shuttle was safe for reentry.

July 30, 2005: First Spacewalk (EVA-1)
As astronauts Soichi Noguchi and Stephen Robinson stepped into space, the first extravehicular activity (EVA-1) commenced at 05:46 EDT on July 30. Their primary objective was to test new techniques for repairing the shuttle's thermal protection system, crucial in preventing disasters like Columbia's. The spacewalk lasted for 6 hours and 50 minutes, concluding successfully at 12:36 EDT.

August 1, 2005: Second Spacewalk (EVA-2)
On August 1, at 04:44 EDT, Noguchi and Robinson began their second spacewalk (EVA-2), tasked with replacing a failed Control Moment Gyroscope (CMG) on the ISS. The gyroscope played a key role in maintaining the station's orientation. After 6 hours and 30 minutes, the EVA was successfully completed at 11:14 EDT.

August 3, 2005: Third Spacewalk (EVA-3)
The final spacewalk of the mission (EVA-3) occurred on August 3. Beginning at 04:48 EDT, Robinson carefully removed two protruding gap fillers between the shuttle's thermal insulation tiles—an unprecedented repair performed during flight. Meanwhile, Noguchi installed the Materials International Space Station Experiment (MISSE 5) to test solar cells and deployed an amateur radio satellite (PCSat2). The spacewalk, lasting 6 hours and 1 minute, concluded at 10:49 EDT, successfully addressing critical objectives.

August 6, 2005: Farewell to the ISS
On August 6, at 01:14 EDT, the crew of Discovery bid farewell to their ISS counterparts, sealing the hatches between the orbiter and the station. Discovery undocked from the ISS at 03:24 EDT, beginning its journey home.

August 8, 2005: Delayed Landing
Due to weather concerns over Kennedy Space Center, Mission Control waved off the first landing attempt at 03:20 EDT on August 8. A second opportunity at 05:04 EDT was also postponed. NASA tentatively scheduled the next landing attempt for August 9, with Edwards Air Force Base in California as a backup.

August 9, 2005: Final Landing at Edwards Air Force Base.
The first landing attempt on August 9 was again waved off due to bad weather at Kennedy Space Center. At 05:03 EDT, thunderstorms within the safety zone forced NASA to redirect Discovery to Edwards Air Force Base. At 07:06 EDT, Discovery executed its deorbit burn, slowing the shuttle for reentry. The APUs were activated at 07:28 EDT, controlling the descent as Discovery began to experience the effects of the Earth's atmosphere. At 08:08 EDT, Commander Eileen Collins took manual control for the final approach to Runway 22 at Edwards. At 08:11 EDT, Discovery touched down safely, completing the mission. With the call of "wheel stop" at 08:12 EDT, STS-114 came to a successful conclusion, marking NASA's triumphant return to human spaceflight operations following the Columbia disaster.

The crew exited Discovery at 10:13 EDT, having completed a mission that was critical not only for the Shuttle program's future but for the continued advancement of human space exploration.

Conclusion

The STS-114 mission, though beset by technical challenges and weather delays, achieved its objectives of testing shuttle safety protocols and delivering essential supplies to the ISS. After completing its objectives, *Discovery* undocked from the ISS and performed a fly-around of the station, taking detailed photographs before beginning its return journey. Reentry was originally scheduled for August 8, 2005, at Kennedy Space Center, but unsuitable weather conditions forced a delay. The Shuttle touched down the following day at Edwards Air Force Base, California, at 8:11 a.m. EDT (5:11 a.m. PDT, 12:11 UTC), concluding the historic mission.

STS-121: A Critical Return to Space

STS-121 was a pivotal NASA Space Shuttle mission to the International Space Station (ISS), flown by *Space Shuttle Discovery* in 2006. The mission served dual purposes: to test new safety and repair techniques developed after the tragic loss of *Space Shuttle Columbia* in February 2003, and to deliver essential supplies, equipment, and a new crew member, German European Space Agency (ESA) astronaut Thomas Reiter, to the ISS.

After two delays due to weather conditions, *Discovery* successfully launched on July 4, 2006, at 14:37:55 EDT. This marked a historic moment, as it became the only Space Shuttle launch to occur on the United States' Independence Day. The mission lasted 13 days, culminating in a successful landing at Kennedy Space Center on July 17, 2006, at 09:14:43 EDT.

STS-121 was also designated as ISS Assembly Mission ULF 1.1, an important part of the ISS construction effort. Following on from the objectives of the earlier STS-114 mission, STS-121 continued to address the safety recommendations made by the Columbia Accident Investigation Board. As part of NASA's "Return to Flight" test program, this mission demonstrated the Shuttle program's readiness to resume regular launches. Its success paved the way for the continued construction of the ISS.

The crew of STS-121 was led by Commander Steven Lindsey, who was embarking on his fourth spaceflight. He was accompanied by Pilot Mark Kelly, on his second mission, and five mission specialists: Michael E. Fossum, making his first spaceflight; Lisa Nowak, a flight engineer on her only spaceflight; Stephanie Wilson, also on her first spaceflight; Piers Sellers, a veteran of one prior mission; and Thomas Reiter, who was embarking on his second and final spaceflight. Reiter's transfer to the ISS brought the station's crew complement back up to three, restoring its staffing level after it had been reduced following the *Columbia* disaster and the subsequent grounding of the Shuttle fleet.

Reiter was originally scheduled to be replaced by Russian cosmonaut Sergey Volkov before STS-121 was delayed to July 2006. Additionally, Piers Sellers, a British-born astronaut, was a late replacement for Carlos Noriega, who was removed from the mission due to a temporary medical condition.

The mission utilized key Shuttle hardware, including External Tank ET-119, Solid Rocket Boosters BI-126 and RSRM-93, and three Space Shuttle Main Engines (SSMEs), serial numbers 2045, 2051, and 2056. The Orbital Maneuvering System (OMS) engines LP-01/35 and RP-03/33 were also integral to the Shuttle's operations.

During the STS-121 mission to the International Space Station (ISS), the crew aboard *Space Shuttle Discovery* carried out crucial tests of new equipment and procedures designed to enhance the safety of future Space Shuttle flights. These measures focused on inspecting and repairing the thermal protection system, a key component for safeguarding the Shuttle during re-entry. The mission also delivered vital supplies and cargo to the ISS, contributing to its future expansion.

The need for these safety tests arose from the tragic *Columbia* accident in 2003, prompting NASA to require two test flights before resuming regular Shuttle operations. Initially, many activities were planned for STS-114, but NASA divided these tasks across two missions due to the additional post-*Columbia* safety protocols. Before the accident, *Columbia* had been assigned to fly both STS-118 and STS-121. The STS-118 mission, also an ISS-related flight, was reassigned to *Discovery* before later being transferred to *Endeavour*. The designation STS-121, originally intended for *Columbia* to service the Hubble Space Telescope, became available again. With STS-115 through STS-120 already assigned, NASA chose STS-121 for the second test flight following STS-114.

The STS-121 mission, originally planned for *Atlantis* in September 2005, was shifted to *Discovery* following a landing gear issue on *Atlantis*. However, *Discovery* was already scheduled to fly STS-114, and after completing that mission and returning to California, NASA reshuffled the schedules once more. *Atlantis* was assigned to fly STS-115 in August 2006, while *Discovery* was to proceed with STS-121. A series of delays pushed the launch to July 2006 due to concerns about foam debris and the Engine Cut-Off (ECO) sensor issues identified during STS-114.

On May 12, 2006, *Discovery* was moved from the Orbiter Processing Facility to the Vehicle Assembly Building (VAB), where it was mated to its External Tank (ET) and Solid Rocket Boosters (SRBs). The spacecraft rolled out to Pad 39B on May 19, 2006, in preparation for the July launch window, which existed for approximately ten minutes each day between July 1 and July 19.

The mission's cargo included over two tons of supplies, notably the Multi-Purpose Logistics Module (MPLM) *Leonardo*, making its fourth flight to the ISS. Among the significant equipment delivered was the Minus Eighty Degree Laboratory Freezer for ISS (MELFI), a French-built unit with four independent drawers capable of maintaining different temperatures. The MELFI freezer was designed to store reagents and samples at temperatures of -80°C, -26°C, and 4°C. It would not only store samples on the ISS but also transport them to and from the station in a temperature-controlled environment.

Additionally, STS-121 carried the European Modular Cultivation System (EMCS), a gas-tight incubator equipped with centrifuges to conduct biological experiments under various gravity conditions. Ground control units operated simultaneously in Europe and at NASA's Ames Research Center to compare results.

One of the mission's key technological advancements was delivering a new oxygen generation system. Still in the testing phase, this system was seen as a potential solution for long-duration missions to the Moon and Mars. Although it would initially operate below its maximum capacity, it was designed to support a crew of six on the ISS in the future, supplementing the Russian-built Elektron system in the Zvezda module.

Other significant hardware included a Danish-built cycling machine for crew exercise, the Cycle Ergometer with Vibration Isolation System (CEVIS). A replacement heat exchanger for the ISS's common cabin air assembly was also delivered to control the internal air temperature aboard the station.

Discovery's payload bay housed the Integrated Cargo Carrier (ICC) with the Trailing Umbilical System (TUS) for the Mobile Transporter. The mission also brought back an old pump module and installed two Fixed Grapple Bars to facilitate future Extra-Vehicular Activities (EVAs). Additionally, the mission carried the TPS (Thermal Protection System) Repair Box, an essential tool in ensuring the safety of future Shuttle flights.

The *Discovery* crew performed three planned spacewalks, with a fourth contingent on available resources. These EVAs allowed astronauts to perform critical tasks, including testing repair techniques on the Shuttle's thermal protection system. One of the "Get Ahead" tasks, planned to maximize available crew time, was successfully completed during EVA 2. The crew also transferred equipment, supplies, and experiments between the Shuttle and ISS using the *Leonardo* module.

On the day of the second launch attempt for STS-121, at 13:14, NASA's launch director made the decision to scrub the launch due to unfavorable weather conditions. The next launch was rescheduled for July 4, 2006, around 14:38 EDT. In the meantime, the shuttle's fuel cells were replenished to ensure there would be enough electrical power to support the mission's third spacewalk, which was contingent on the availability of resources.

During pre-launch inspections, a small piece of foam, weighing 2.6 grams (0.092 oz), was discovered to have fallen from the insulation on the upper part of the external fuel tank. This foam fragment had come off a bracket holding the oxygen line in place. NASA determined that the foam piece was too small to pose any significant risk to the shuttle, even if it detached during flight. The thermal stress caused by repeatedly filling and emptying the tank with cryogenic fuel was known to cause such damage to the insulation.

Further complicating matters, at 08:35, NASA identified a failure in the backup circuit breaker controlling the primary heaters on the solid rocket boosters' segment joints. However, repairs were deemed unnecessary, as the heaters were only required in colder weather, and the primary heater remained operational.

For the third launch attempt, the weather forecast was much more favorable, with only a 20% chance of conditions preventing launch. Finally, on July 4, 2006, at 14:37:55 EDT, *Space Shuttle Discovery* successfully launched from Kennedy Space Center in Cape Canaveral, Florida. This marked the third attempt and became the first U.S. Space Shuttle launch to take place on Independence Day.

Much of the focus during the launch was on monitoring the external tank for insulation foam loss, a critical safety issue since the *Columbia* disaster. New cameras had been installed on the shuttle to provide better monitoring, with three cameras placed on each solid rocket booster— one to observe separation and two focused on the leading edge. Additionally, a camera was mounted on the external tank to broadcast live images on NASA TV during launch. As soon as the shuttle's main engines cut off, two crew members from the mid-deck immediately left their seats to photograph and record the external tank.

The timing of the launch was influenced by lighting conditions, as NASA aimed to ensure optimal visibility for external tank imaging. However, it was determined that day-to-day differences in lighting were not critical since the sunlight's angle, combined with the unpredictable tumbling of the external tank during separation, would play a more significant role in determining image clarity.

NASA's live broadcast showed no visible signs of large foam pieces detaching from the external tank. However, close inspection revealed that several small pieces of debris had detached and floated away, though they appeared after the critical period of concern. Approximately 23 minutes into the flight, Mission Specialist Michael Fossum reported observing debris floating beside the orbiter. He described it as a 4 to 5-foot-long piece with straps attached, which resembled a thermal protection system blanket. Such blankets, which protect parts of the vehicle that do not reach extreme temperatures, had been observed flapping on the previous mission, STS-114, without causing concern. Further analysis revealed that the debris was actually strips of ice formed on the outside of an engine nozzle, which sublimated and disintegrated during observation—an occurrence seen on earlier missions.

In addition, a tile shim was observed to have come loose during main engine start. Despite this, the mission continued without incident. The orbital maneuvering system thruster, which had experienced a heater failure prior to launch, was warmed by pointing it at the sun, allowing it to function properly during ISS docking operations.

En route to the ISS, the crew used the Orbiter Boom Sensor System (OBSS), a 50-foot (15 m) extension fitted with two types of lasers and a high-resolution camera, to inspect the shuttle's underside for damage, with special attention paid to the leading edges of the wings.

During a post-flight briefing on day two, it was revealed that a gap filler was protruding from the port side lower wing, though it was not in a location of particular concern. This gap filler had been part of the shuttle since 1982, predating any modifications made after the STS-114 mission. In a more unusual finding, bird droppings were discovered on the leading edge of the right wing, a remnant from before the launch. Upon landing, the crew humorously remarked during a press conference that the droppings, though charred, were still visible on the orbiter.

Following the rendezvous with the International Space Station (ISS), *Space Shuttle Discovery* performed the standard rendezvous pitch maneuver, a critical procedure in which the shuttle slowly rotated to allow the ISS crew to visually inspect and photograph its heat shield for any potential damage. This inspection, part of post-*Columbia* safety protocols, was completed without issue, and an uneventful docking with the ISS followed.

Shortly after docking, German astronaut Thomas Reiter officially became a member of the ISS Expedition 13 crew. The transfer was marked by the installation of Reiter's personalized Soyuz spacecraft seat liner, an essential cushion for his return to Earth in the event of an emergency.

The Multi-Purpose Logistics Module (MPLM) *Leonardo* was successfully attached to the ISS's Unity module. Concerns arose regarding potential obstructions caused by straps near the docking equipment, but video inspections confirmed that they posed no danger, and the docking proceeded as planned. *Leonardo* delivered over 7,400 pounds of equipment and supplies essential for the continued operation and expansion of the space station.

While docked at the ISS, the *Discovery* crew conducted focused inspections of the shuttle's heat shield. Data collected by the wing's leading-edge sensors revealed six minor impacts, with a maximum force of 1.6 g. Ground tests had shown that impacts of around 10 g were necessary to cause any significant damage, leaving mission control confident in the shuttle's integrity.

Astronauts Michael Fossum and Piers Sellers conducted a 7.5-hour spacewalk, one of three planned EVAs during the mission. They tested the 50-foot Orbital Boom Sensor System (OBSS) as a work platform, assessing its stability for possible in-orbit repairs to the shuttle. Sellers, working first, attached foot restraints to the boom, followed by Fossum. Together, they performed increasingly vigorous operations to simulate potential repair scenarios. Both astronauts reported that the boom's motion dampened rapidly, making it an effective work platform. Sellers commented, "I felt almost no motions at all, just a few inches each way."

However, Fossum encountered a problem with his 85-foot-long safety tether, which was damaged after he mistakenly left it in a locked position, expecting it to automatically retract. Upon realizing the error, Fossum exclaimed, "Oh no! ... That's embarrassing." He quickly replaced the damaged tether with a spare.

Another important task during this EVA was making safe a cable cutter on the ISS's mobile transporter, which was accomplished without issue. Meanwhile, *Discovery*'s flight plan revision requested the crew to keep their email folders clean to minimize the time required to uplink new messages. A press conference from orbit followed, featuring questions from NASA centers and the European Space Agency (ESA).

Robotic operations continued onboard, with Canadarm2 releasing the mobile transporter from one end while remaining attached to the Destiny module, preparing for work on the system during the following day's EVA. The crew also began preparations for the second spacewalk, organizing equipment and cameras.

During the mission's second EVA, Sellers and Fossum worked for 6 hours and 47 minutes, deploying a spare pump module and replacing an umbilical cable reel, which provided power, data, and video to the Mobile Transporter rail car. The spacewalkers also oversaw cargo transfers between *Discovery*, the ISS, and the *Leonardo* module. By the end of the transfer operations, *Leonardo* was packed with more than 4,300 pounds of returned experiments, equipment, and trash.

On July 10, 2006, President George W. Bush personally spoke with the crew, commending them for their contributions to space exploration and service. Pilot Mark Kelly commented during an on-orbit press conference that the crew often did not know the wake-up songs in advance, leading to moments of improvisation when the music began unexpectedly.

The third and final spacewalk, conducted by Sellers and Fossum, focused on testing shuttle repair techniques. Using pre-damaged samples of heat shield materials brought aboard on a special pallet, the astronauts applied NOAX (Non-Oxide Adhesive Experimental), a material designed to repair heat shield damage in space. Coordinated by mission control, the timing of repairs was carefully planned based on the sun's position, as the materials worked best when warm and cooling. During the EVA, Sellers lost one of his spatulas, which floated away over the shuttle's port side. The tool posed no threat to the mission, and mission controllers quickly determined it was not a hazard.

With all three spacewalks successfully completed, the *Discovery* crew enjoyed a well-deserved day off, having transferred thousands of pounds of supplies and equipment earlier in the flight. Texas Governor Rick Perry called Mission Specialist Michael Fossum, a fellow Texas A&M University graduate, congratulating him on becoming the first Aggie in space.

As the mission neared its conclusion, Expedition 13 crewmember Jeffrey Williams oversaw the final closeout procedures for the MPLM, which was transferred back to *Discovery*'s payload bay for return to Earth. The Canadarm2, operated by Stephanie Wilson and Lisa Nowak, completed the transfer.

In preparation for re-entry, the shuttle crew used *Discovery*'s arm and extension boom to inspect the orbiter's wings and nose cap for any signs of damage from micrometeoroid impacts. Discussions on the ground focused on a minor leak in the Auxiliary Power Unit (APU) 1, but tests increased confidence in the system's integrity, and it was cleared for normal use during re-entry.

Discovery undocked from the ISS after a nine-day stay, with Pilot Mark Kelly maneuvering the shuttle to a position above the station before performing the final separation burn. Final inspections of the starboard wing and nosecap were conducted with the shuttle's robotic arm, confirming the orbiter's readiness for re-entry.

Discovery's re-entry proceeded smoothly, with only minor deviations from a nominal landing procedure. The APU 1 was started early, and a manual intervention ensured the heater thermostat on APU 3 functioned properly. The shuttle safely landed at Kennedy Space Center's Shuttle Landing Facility at 09:14:43 EDT on July 17, 2006. During post-landing inspections, Commander Steven Lindsey remarked that this was one of the cleanest inspections he had ever seen.

In a post-mission press briefing, NASA confirmed that the shuttle program was "back in business," with plans to replace *Discovery*'s windows before its next flight, STS-116. The mission's success also bolstered confidence in using the shuttle's arm as a work platform for potential future servicing missions, including to the Hubble Space Telescope.

Throughout the mission, critical support came from the ISS crew, which included Commander Pavel Vinogradov and Flight Engineer Jeffrey Williams. CAPCOM personnel, responsible for radio communications with the shuttle, included Steve Frick and Rick Sturckow during ascent and descent, with Rick Mastracchio, Lee Archambault, and Julie Payette managing orbital operations. Michael D. Leinbach served as launch director, while Steve Stich, Tony Ceccacci, and Norm Knight directed the flight from mission control.

STS-115

STS-115, a pivotal Space Shuttle mission, marked the resumption of ISS assembly following the Columbia disaster. Flown by Space Shuttle *Atlantis*, STS-115 was launched from Launch Complex 39B at Kennedy Space Center on September 9, 2006, at precisely 11:14:55 EDT (15:14:55 UTC). This mission was designated

ISS-12A by the International Space Station program and represented the first ISS assembly mission after two successful "Return to Flight" missions, STS-114 and STS-121.

Originally scheduled for launch in April 2003, the mission faced significant delays due to the Columbia tragedy in February 2003. After several postponements, the revised launch was set for August 27, 2006. However, factors such as Tropical Storm Ernesto and an unprecedented lightning strike at the launchpad pushed the launch date to early September.

The primary objective of STS-115 was to deliver and install the second port-side truss segment (ITS P3/P4) on the ISS. The P3/P4 truss, weighing over 17.5 short tons (approximately 16 metric tons), included two large solar arrays (designated 2A and 4A) and critical batteries to augment the station's power capabilities. Due to the substantial weight of the truss, the crew was reduced from the typical seven to six astronauts.

The crew of STS-115 was led by Commander Brent W. Jett Jr., a veteran astronaut on his fourth and final mission. The pilot, Christopher Ferguson, was embarking on his first spaceflight, while mission specialists included Canadian Space Agency astronaut Steven MacLean, Daniel C. Burbank, Joseph R. Tanner, and Heidemarie Stefanyshyn-Piper. This mission marked the first spaceflights for Ferguson and Stefanyshyn-Piper and the last flights for Jett and Tanner.

A significant achievement of the mission was Steven MacLean's operation of Canadarm2, a Canadian-built robotic arm integral to ISS assembly. MacLean became the first Canadian astronaut to operate this advanced robotic system in space. He also conducted a spacewalk during the mission, becoming only the second Canadian to do so, following Chris Hadfield.

Over the course of the mission, three spacewalks were conducted, primarily to install and connect the truss segments and solar arrays. The crew worked meticulously to remove restraints on the solar arrays and ensure their successful deployment. This task was critical, as the new solar arrays expanded the ISS's power-generating capabilities, enabling future missions and experiments. Additionally, the astronauts performed essential maintenance tasks on the station, ensuring its continued operation and preparing it for the next assembly mission, STS-116.

The mission patch for STS-115, designed by students from York University in Toronto, where MacLean had studied, holds special significance. These students, Graham Huber, Peter Hui, and Gigi Lui not only designed the mission patch but also created MacLean's personal patch for this historic flight.

NASA's decision to advance the STS-115 launch date to August 27, 2006, was driven by the need for optimal lighting conditions to photograph the Space Shuttle's external tank during ascent. This practice had been employed in earlier missions, such as STS-31 and STS-82. Additionally, the launch window had to align with the upcoming Soyuz TMA-9 launch in mid-September, which was delivering a new crew and supplies to the International Space Station (ISS). To avoid operational conflicts, the Soyuz spacecraft would not dock while the Space Shuttle was at the station.

Milestones of STS-115:

- 147th NASA crewed space flight
- 116th space shuttle mission since STS-1
- 27th flight of *Atlantis*
- 91st post-Challenger mission
- 3rd post-Columbia mission
- 1st post-Columbia mission of *Atlantis*

Atlantis safely returned to Earth on September 21, 2006, landing at Kennedy Space Center.

STS-116

STS-116, carried out by Space Shuttle Discovery, was a crucial chapter in the International Space Station (ISS) assembly. Launched on December 9, 2006, from Kennedy Space Center's Launch Pad 39B, this mission was marked by its complexity and the critical tasks involved in reshaping the ISS's power and thermal systems. The mission was commanded by Mark Polansky, with William Oefelein serving as the Pilot. The crew included Mission Specialists Nicholas Patrick, Robert Curbeam, Joan Higginbotham, Christer Fuglesang, and Sunita Williams. Sunita Williams remained on the ISS as part of Expedition 14, replacing European Space Agency (ESA) astronaut Thomas Reiter.

Atlantis began its journey to the launchpad on July 24, 2006, when it was rolled from the Orbiter Processing Facility to the Vehicle Assembly Building (VAB). There, the orbiter was mated to its external tank on July 26 and rolled out to Launch Complex 39B on August 2. Initially scheduled for July 31, the rollout was delayed due to concerns about nearby storms and the risk of a lightning strike. Such an event could potentially cause irreparable damage to the shuttle's systems.

By August 5–6, engineers completed a critical flight readiness check of *Atlantis'* main engines, confirming they were ready for launch. The crew arrived at Kennedy Space Center on August 7 for four days of rehearsals, culminating in a practice countdown on August 10.

NASA's Flight Readiness Review (FRR) took place on August 15–16, where top managers assessed key concerns, including foam loss from the external tank. Foam loss had become a focal issue after the Columbia disaster, which was caused by a piece of foam from the external tank striking the orbiter's wing, leading to a catastrophic breach during re-entry. Engineers also identified potential problems with bolts securing the shuttle's Ku-band antenna, which may not have been threaded correctly. Although the installation had functioned without issue on previous flights, the bolts were replaced by August 20 while *Atlantis* remained on the launch pad. This unprecedented repair was completed without affecting the planned launch date.

On August 25, 2006, a powerful lightning strike, the most intense ever recorded at Kennedy Space Center, hit the lightning rod atop the launch pad. NASA postponed the launch for 24 hours to assess possible damage. The delay extended further on August 27, pushing the earliest possible launch to August 29. Simultaneously, Tropical Storm Ernesto posed a new threat. On August 28, NASA decided to roll *Atlantis* back to the VAB as forecasts suggested Ernesto would regain strength and pass closer to the launch site.

On the morning of August 29, as *Atlantis* began its rollback to the VAB, meteorologists revised their forecast, predicting the storm would not hit Kennedy Space Center as strongly as expected. In an unprecedented move, NASA reversed its decision and returned *Atlantis* to the launch pad, choosing to weather the storm on-site. Winds were predicted to peak below 79 mph (126 km/h), NASA's threshold for keeping the shuttle outdoors. By the early hours of August 31, Tropical Storm Ernesto had passed with minimal impact. Inspection teams found only minor issues that required simple repairs.

NASA set a new target launch date of September 6, after gaining approval from Russian space managers to extend the launch window. However, on the morning of September 6, a problem arose with one of the shuttle's three electricity-producing fuel cells. An apparent internal short was detected, and the launch was postponed with insufficient time to resolve the issue.

NASA initially ruled out September 9 as a possible launch date due to a potential conflict with the scheduled Soyuz TMA-9 mission. As news agencies speculated that this would be the final opportunity for a launch until October, NASA worked to resolve the fuel cell issue, eventually scheduling a new launch attempt for September 8. Ultimately, *Atlantis* lifted off successfully on September 9, 2006, following weeks of weather challenges and technical obstacles. This launch marked a significant step in continuing the assembly of the ISS after the post-Columbia hiatus.

On the morning of September 8, 2006, NASA encountered a setback when one of the four engine cut-off (ECO) sensors in *Atlantis'* external fuel tank malfunctioned. The faulty sensor, known as ECO sensor No. 3, incorrectly indicated that liquid hydrogen remained in the tank, despite the fact that it had been drained. The other three sensors functioned normally, signaling an empty tank. While NASA could technically proceed with three operational ECO sensors, they opted to delay the launch by 24 hours to investigate the issue and drain the remaining fuel. The malfunction was confirmed to be a faulty reading from ECO sensor No. 3, a critical part of the system designed to prevent the shuttle's engines from running dry and potentially damaging the vehicle during ascent.

At 11:40 AM EDT, NASA officially scrubbed the launch for the day. The decision came just half an hour before the scheduled liftoff, during the T-9:00 minute hold. With a weather forecast giving a 70% chance of favorable conditions, all other aspects of the mission were deemed ready for flight, barring the sensor issue.

On September 9, 2006, with all ECO sensors functioning properly, *Atlantis* prepared for launch again. After a smooth countdown, the shuttle lifted off at 15:15 UTC (11:15 EDT) from Kennedy Space Center. As *Atlantis* ascended, the International Space Station (ISS) was orbiting 350 kilometers (220 miles) above the northern Atlantic Ocean, between Greenland and Iceland.

During the shuttle's climb to orbit, Mission Control observed a minor issue with the Flash Evaporator System, a cooling unit that appeared to have ice buildup. The crew was promptly asked to reconfigure the system, and the adjustments cleared the ice, allowing the system to function normally. Ice formation in the Flash Evaporator System is not uncommon and had been noted in previous missions, so the issue was resolved without significant concern.

Approximately 8.5 minutes after launch, *Atlantis'* main engines shut down, successfully placing the shuttle in orbit. Immediately after engine cutoff, astronauts Joseph Tanner and Steven MacLean began documenting the external tank's condition using hand-held video and digital cameras. This procedure was crucial to assess any potential damage that may have occurred during ascent, especially from foam debris shedding off the external tank—a concern ever since the Columbia disaster. The footage and imagery were transmitted to Mission Control for analysis to ensure the shuttle's safe continuation toward the ISS.

With the external tank successfully jettisoned and no major issues identified during launch, *Atlantis* proceeded smoothly to its rendezvous with the ISS, marking a successful first day of the mission.

The primary objectives of STS-116 were delivering and installing the P5 truss segment, reconfiguring the station's electrical and thermal systems, and continuing ISS assembly. The mission was also notable for its complex series of spacewalks, or Extravehicular Activities (EVAs), designed to rewire the ISS to accommodate new modules and future expansions.

After a successful launch and docking with the ISS, the crew set to work on their primary tasks.

On their first full day in space, the crew of *Atlantis* conducted a comprehensive inspection of the shuttle using the Orbiter Boom Sensor System (OBSS), a 15-meter (50-foot) long extension attached to the shuttle's robotic arm, Canadarm. This procedure was critical to assess any potential damage sustained by the shuttle during launch, particularly on the reinforced carbon-carbon panels along the leading edges of both the port and starboard wings, as well as the nose cap. Pilot Chris Ferguson led the inspection, with mission specialists Dan Burbank and Steve MacLean carefully controlling the robotic arm.

As the inspection proceeded, the crew worked ahead of schedule to prepare for *Atlantis'* docking with the International Space Station (ISS) and the three planned extra-vehicular activities (EVAs). Mission specialists Joe Tanner and Heidemarie Stefanyshyn-Piper began checking the spacesuits and tools that would be used during the spacewalks scheduled for Days 4, 5, and 7. These spacewalks would be essential for installing the P3/P4 truss segment, deploying new solar arrays, and configuring the arrays for operation.

Meanwhile, aboard the ISS, Expedition 13 Flight Engineer Jeffrey Williams prepared the station for *Atlantis'* arrival, ensuring that the digital cameras were ready to take high-resolution images of the shuttle's heat shield. Williams also pressurized the Pressurized Mating Adapter 2 (PMA-2), located at the end of the Destiny Laboratory Module, where *Atlantis* would dock. Expedition 13 Commander Pavel Vinogradov assisted by prepacking equipment for return to Earth aboard the shuttle.

On Flight Day 3, *Atlantis* executed a flawless approach to the ISS. Commander Brent Jett guided the shuttle through a carefully choreographed orbital backflip, positioning the shuttle about 180 meters (600 feet) below the station. This maneuver allowed the ISS crew to take high-resolution photographs of the shuttle's heat shield to ensure no significant damage occurred during launch.

At 10:46 UTC, *Atlantis* successfully docked with the ISS, and the crew hatch was opened nearly two hours later at 12:35 UTC. The ISS crew welcomed the shuttle astronauts aboard, marking the official start of joint operations between the crews.

Following the docking, pilot Chris Ferguson and mission specialist Dan Burbank used the shuttle's robotic arm, Canadarm, to lift the 17.5-ton P3/P4 truss segment from *Atlantis'* payload bay. They then handed the truss over to the ISS's Canadarm2, operated by Steven MacLean and ISS Flight Engineer Jeff Williams. MacLean became the first Canadian astronaut to operate Canadarm2 in space, a significant milestone.

Meanwhile, Tanner and Stefanyshyn-Piper began preparations for their first spacewalk by performing a "camp-out" procedure in the Quest Airlock. This new method, designed to prevent decompression sickness, involved sleeping in the airlock at a lower pressure to acclimate their bodies to the conditions they would encounter during the spacewalk.

Flight Day 4 marked the installation of the P3/P4 truss on the ISS. After the Canadarm2 successfully attached the truss, Tanner and Stefanyshyn-Piper began their first spacewalk at 09:17 UTC. Their primary tasks included connecting power and data cables between the P1 and P3/P4 trusses and releasing the launch restraints on the newly installed truss segment.

The spacewalk was highly productive, allowing the astronauts to complete tasks ahead of schedule. As a result, they tackled several objectives planned for later EVAs. The EVA concluded at 15:43 UTC after 6 hours and 26 minutes of work. During this spacewalk, a bolt, spring, and washer assembly from a launch lock floated away into space, but the incident did not impact the overall mission.

Following the spacewalk, the crew began preparing for the next day's EVA, with astronauts Burbank and MacLean camping out in the Quest Airlock to acclimate themselves for their spacewalk.

On Flight Day 5, Burbank and MacLean conducted their first spacewalk, spending 7 hours and 11 minutes outside the station. Their primary objective was to activate the Solar Alpha Rotary Joint (SARJ), a crucial component that allows the station's solar arrays to rotate and track the sun. The spacewalkers released the locks securing the joint and addressed several minor issues, including a malfunctioning helmet camera, a broken socket tool, and a stubborn bolt. The teamwork between the two astronauts resolved these issues, and they completed the SARJ activation successfully.

In addition to their primary tasks, Burbank and MacLean completed several "get-ahead" objectives, advancing work that had been scheduled for future missions. While the activation of the SARJ was successful, engineers on the ground encountered a minor glitch during its checkout. Despite this, they anticipated the timely deployment of the solar arrays on the following day.

On Flight Day 6, the crew continued their work on deploying the new solar arrays attached to the P4 truss. The unfurling process was delayed due to the software issues encountered the previous day, but NASA engineers quickly developed a workaround to address the problem. The panels were deployed incrementally to prevent them from sticking, a precaution learned from previous missions.

Despite minor sticking, the solar arrays were fully deployed without any significant issues. Although the installation was complete, the arrays would not begin generating power until the station's electrical system was rewired during the next shuttle mission, STS-116, scheduled for December 2006.

Other activities included a "double walk-off" maneuver of the station's Canadarm2 from its position on the Mobile Base System to the Destiny Laboratory Module. Interviews were conducted between Commander Jett, MacLean, Canadian Prime Minister Stephen Harper, and students back on Earth.

The third and final spacewalk of the mission took place on Flight Day 7. A minor issue delayed the start of the spacewalk when a circuit breaker tripped, cutting power to the airlock's depressurization pump. After a brief investigation, the breaker was reset, and astronauts Tanner and Stefanyshyn-Piper began their spacewalk at 10:00 UTC.

During the 6-hour and 42-minute EVA, the astronauts removed hardware securing the P3/P4 radiator during launch. The ground team then unfurled the radiator, significantly enhancing the station's ability to dissipate heat. In addition to these tasks, the spacewalkers retrieved a materials exposure experiment, performed maintenance on the P6 truss, installed a wireless TV antenna, and replaced the S-band antenna assembly on the S1 truss.

Several "get-ahead" tasks were also completed, and near the end of the spacewalk, the astronauts conducted a test to evaluate the use of infrared video to detect debris damage on *Atlantis*' wing. After the EVA, the station's mobile transporter was moved to a worksite on the P3 truss for further inspections.

In addition to these major tasks, STS-116 included exchanging ISS crew members. Sunita Williams replaced Thomas Reiter as part of Expedition 14, and Reiter returned to Earth aboard Discovery. The mission also involved delivering supplies, equipment, and scientific experiments to the station.

On Flight Day 8, the final full day of *Atlantis* docked to the ISS, the crew shifted focus to preparing for undocking procedures. After a well-deserved rest following the successful mission milestones, the astronauts began transferring equipment, science experiments, and other cargo from the ISS to *Atlantis* in preparation for the journey back to Earth.

A joint crew press conference between Expedition 13 and STS-115 highlighted the significance of the mission and the challenges ahead for future ISS assembly flights. Mission Commander Brent Jett reflected on the success of STS-115 and emphasized the demanding nature of the construction tasks yet to come. "All of the rest of the assembly missions are going to be challenging. We have similar payloads flying in the future. We are off to a good start on assembly. I think we can pass along a lot of the lessons to the future crews," Jett stated.

Flight Day 9 marked *Atlantis*' departure from the ISS. The day began with farewell ceremonies between the crews of STS-115 and Expedition 13. At 10:27 UTC, the hatch between the shuttle and the station was closed and locked, followed by a series of checks to ensure no leaks. At 12:50 UTC, *Atlantis* undocked from the ISS, beginning a slow, 360-degree flyaround of the station to document its new configuration with the recently installed P3/P4 truss segment.

This visual survey was critical in capturing the expanded ISS structure, including the new solar arrays, for analysis and future mission planning.

On Flight Day 10, the crew inspected *Atlantis*' heat shield to ensure it was ready for re-entry. Using the Orbiter Boom Sensor System, they scanned the shuttle's nose cap and wing leading edges for any signs of damage caused by micrometeoroids or space debris during the mission. *Atlantis* was now orbiting about 80 kilometers (50 miles) behind the ISS as the crew completed the detailed scans.

With no damage detected, the crew spent the rest of the day stowing equipment, preparing for re-entry, and landing.

The crew of *Atlantis* spent Flight Day 11 conducting final preparations for re-entry, including testing the shuttle's reaction control thrusters and practicing landing procedures using onboard computers. During these checks, the crew spotted an unidentified object in a co-orbital path with *Atlantis*, which raised concerns that the object might have come off the shuttle itself.

Using onboard cameras, the astronauts transmitted images of the object to Mission Control, but the resolution was insufficient to identify the debris. NASA engineers speculated it could be something benign, such as ice or a piece of shimstock, but there was also a possibility that it could be a critical part, such as a piece of the shuttle's thermal protection system. As a precaution, NASA delayed the de-orbit burn and landing, scheduling an additional inspection on Flight Day 12 to ensure *Atlantis* was safe for re-entry.

Following the sighting of the unknown object on Flight Day 11, the crew used the shuttle's robotic arm to conduct further inspections of *Atlantis* on Flight Day 12. The Orbiter Boom Sensor System was deployed to recheck the shuttle's heat shield for any potential damage. After thorough scans and analysis by ground flight controllers, no safety concerns were identified. With *Atlantis* cleared for re-entry, NASA confirmed favorable weather conditions for landing the following day.

The crew completed the final preparations for landing, including stowing the Ku-band antenna and packing up gear. During the day, they received confirmation that the Soyuz TMA-9 spacecraft had successfully docked with the ISS, delivering the first half of the Expedition 14 crew.

Flight Day 13 brought *Atlantis'* return to Earth. The landing procedures began early in the morning, starting with the prestart of the Auxiliary Power Units (APUs) at 04:37 EDT. The payload bay doors were closed at 04:45 EDT, and by 04:52 EDT, the crew received the final "go" from Mission Control for the prime re-entry window.

At 05:15 EDT, *Atlantis* initiated the deorbit burn, firing its engines for 2 minutes and 40 seconds to slow the shuttle and begin its descent through Earth's atmosphere. The burn was flawless, with Mission Control confirming the shuttle was on the correct trajectory for landing. As *Atlantis* re-entered the atmosphere, it conducted a series of roll reversals to bleed off speed, dropping from 27,000 kilometers per hour (17,000 mph) to a landing speed of less than 760 kilometers per hour (470 mph).

At 06:21:30 EDT, *Atlantis'* main gear touched down on Runway 33 at Kennedy Space Center, with the nose gear following six seconds later. After traveling over 8,000,000 kilometers (5,000,000 miles), *Atlantis* stopped at 06:22:16 EDT, successfully concluding mission STS-115. The landing took place 48 minutes before sunrise, marking the 21st night landing in the history of the Space Shuttle Program.

After landing, NASA technicians discovered a small hole, approximately 2.7 mm (0.108 inches) in diameter, in one of *Atlantis'* radiator panels. The damage was attributed to a micrometeorite impact during the mission. While the damage posed no risk to the shuttle's safe return, it was a reminder of the hazards spacecraft face in orbit.

During the mission, NASA's Mission Management Team also conducted a detailed analysis of debris events that occurred during launch. A significant debris event was recorded 48 seconds after launch, near the point of maximum aerodynamic pressure (Max Q). The debris originated from the external tank and did not pose a danger to the shuttle, but the event raised concerns for future missions. NASA engineers concluded that no further inspections were required for this flight, and STS-115 was deemed a success in every aspect.

STS-300

The contingency mission designated STS-300 was planned as a rescue mission to provide support if Space Shuttle *Discovery* became disabled during either STS-114 or STS-121, the "Return to Flight" missions following the *Columbia* disaster. This contingency mission, known as the Contingency Shuttle Crew Support mission, was a precautionary plan developed by NASA to ensure that the crew of *Discovery* could be safely returned to Earth if necessary.

STS-300 was a modified version of the STS-115 mission, but with adjustments to the launch timeline and crew size. If *Discovery* encountered issues that left it unable to safely return, STS-300 would have been launched as early as August 17, 2006, to rescue the stranded astronauts. The crew size was reduced to four astronauts to accommodate the needs of the rescue operation, and the mission would have been executed with a streamlined approach.

The four-person crew for STS-300 was a subset of the full STS-115 team and included Brent Jett, commander, Christopher Ferguson, pilot and backup operator for the Remote Manipulator System (RMS), Joseph Tanner, mission specialist 1, prime operator for RMS, and lead spacewalker (Extravehicular 1), and Daniel Burbank, mission specialist 2, and second spacewalker (Extravehicular 2).

This contingency plan was part of NASA's enhanced safety protocols developed after the loss of Columbia, ensuring that no shuttle crew would be stranded in orbit without a reliable rescue option. Fortunately, the need for STS-300 never arose, as both STS-114 and STS-121 missions concluded successfully. However, the planning and readiness for such a mission underscored NASA's commitment to crew safety during the shuttle era.

STS-117

Preparations for STS-117 began on February 7, 2007, when *Atlantis* was transported to the Vehicle Assembly Building (VAB). Here, the orbiter was mated to its external fuel tank and solid rocket boosters in a process completed by February 12. Later that day, the payload canister containing the S3/S4 truss and solar arrays for the ISS was delivered to the launch pad and secured in the Payload Changeout Room.

Despite a minor delay caused by erratic pressure readings from an Operational Pressure Transducer (OPT) on the right-hand Solid Rocket Booster, *Atlantis* rolled out to the launch pad on February 15. The 3.4-mile journey from the VAB to Launch Pad 39A took several hours, with the Shuttle arriving at 15:09 EST. Engineers quickly resolved the transducer issue, replacing all six units.

The STS-117 crew arrived at Kennedy Space Center on February 21, 2007, to participate in the Terminal Countdown Demonstration Test (TCDT). During this crucial phase, the astronauts rehearsed launch procedures, inspected the payload, conducted safety exercises, and completed a simulated main engine cut-off exercise. After the TCDT, the crew returned to Houston to continue preparations for the upcoming launch.

On February 26, a severe hailstorm struck Kennedy Space Center, causing extensive damage to *Atlantis* and its external tank. Hailstones, some as large as golf balls, left thousands of divots in the foam insulation of the external tank, damaged an Ice Frost Ramp, and inflicted minor damage to 26 heat shield tiles on *Atlantis'* left wing. This extensive damage required the Shuttle to be rolled back to the VAB on March 4 for detailed inspections and repairs.

For the next two months, technicians worked meticulously to repair the orbiter and external tank. *Atlantis* returned to the launch pad on May 15, bearing visible marks from the hailstorm. A second Flight Readiness Review on May 30–31 confirmed that all necessary repairs were complete, and the Shuttle was cleared for launch.

The hailstorm not only delayed STS-117 but also disrupted the Shuttle launch manifest. The next mission, STS-118, originally scheduled for June 28, 2007, was pushed back to August 8, 2007. This delay had a ripple effect, moving the launch dates for STS-120 and STS-122 to October 23 and December 6, 2007, respectively.

Prior to launch, NASA identified a potential, though unlikely, risk associated with one of the 24 Composite Overwrapped Pressure Vessels (COPVs) aboard *Atlantis*. These vessels, designed to store gases under high pressure, could pose a threat if one were to burst during launch. Such an event could severely damage the Shuttle and endanger the crew. NASA took precautionary measures, adjusting launch procedures to mitigate the risk and ensure the safety of both the orbiter and the crew.

Pre-launch processing had proceeded smoothly throughout the day. The STS-117 crew began boarding *Atlantis* at 16:17 EDT (20:17 UTC), and by 20:58 UTC, all astronauts were strapped in and ready for liftoff. The hatch was sealed by 21:40 UTC, ensuring all was secure for the launch. Weather conditions were 80% favorable for the launch window, although there were concerns about the weather at Transoceanic Abort Landing (TAL) sites. However, the Istres TAL site in France cleared for emergency landings just in time for the launch. Conditions at Zaragoza, Spain, showed signs of improvement, while the third TAL site, Morón Air Base, was closed for runway maintenance until June 15, 2007.

On June 8, 2007, Space Shuttle *Atlantis* lifted off from Kennedy Space Center's Launch Pad 39A on STS-117, also designated as ISS Assembly Flight 13A. Originally scheduled for March 15, the mission faced delays due to severe damage caused by a hailstorm on February 26. This incident required extensive repairs, pushing the launch back by several months. When *Atlantis* finally ascended at 19:38 EDT, it marked the 250th orbital human spaceflight and set a record as the heaviest Shuttle flight to date.

Commanded by Frederick Sturckow, with Lee Archambault as Pilot, the STS-117 crew included Mission Specialists Patrick Forrester, Steven Swanson, John "Danny" Olivas, James Reilly, and Clayton Anderson. Anderson was assigned to replace Sunita Williams as a member of Expedition 15 aboard the International Space Station (ISS), while Williams returned to Earth on *Atlantis* after spending 195 days in space.

NASA launch commentator George Diller's words as *Atlantis* ascended, "And liftoff of Space Shuttle *Atlantis*, to assemble the framework for the science laboratories of tomorrow!" emphasized the importance of this mission. It was the first Shuttle launch from Pad 39A since the tragic *Columbia* disaster during STS-107 in 2003.

Atlantis's ascent into space lasted eight and a half minutes, largely without incident. The only anomaly noted during the flight was a small piece of debris seen after the separation of the solid rocket boosters, two minutes and five seconds after liftoff. Shuttle Program Manager Wayne Hale later confirmed that the debris posed no threat to the orbiter, as preliminary analysis indicated it did not impact the spacecraft.

As *Atlantis* soared into orbit, the International Space Station (ISS) was positioned 220 miles (350 km) above the southern Indian Ocean, southwest of Australia. Onboard the ISS, cosmonauts Fyodor Yurchikhin and Oleg Kotov, along with NASA astronaut Sunita Williams, watched the shuttle's launch via a live video feed provided by flight controllers in Houston.

After reaching orbit, the *Atlantis* crew quickly began post-launch procedures. They opened the shuttle's payload bay doors and set up essential computers and equipment for the mission. Additionally, the crew powered up the shuttle's robotic arm to ensure it was operational for the upcoming tasks. During the robotic arm checkout, mission controllers observed a small issue: a 4-inch by 6-inch section of insulation blanket on the shuttle's port Orbital Maneuvering System (OMS) pod had become displaced, pulling away from a row of adjacent heat-shield tiles. This observation prompted further inspections to assess the situation before any repair decision could be made.

On the second day of the mission, the crew of Space Shuttle Atlantis focused on inspecting the shuttle's heat shield, a vital component of the vehicle's thermal protection system. Following an extended rest period, which gave the crew an extra half-hour of sleep after a late night spent downloading in-cabin video footage, the astronauts began meticulously ensuring the shuttle's safety in orbit.

Piloted by Lee Archambault, with Mission Specialists Patrick Forrester and Steven Swanson, the crew utilized Atlantis' robotic arm and the Orbiter Boom Sensor System (OBSS) to inspect the shuttle's wing leading edges and nose cap. These inspections had become standard protocol after the Columbia disaster and had been refined over the course of three post-Columbia missions. The procedures used during the STS-117 mission were designed to be more efficient, allowing the crew to scan larger surface areas in less time while still capturing high-resolution data. A camera attached to the end of the OBSS simultaneously recorded close-up photographs while a laser scanner collected precise measurements.

The inspection began with the starboard wing, where the astronauts carefully made multiple passes along the leading edge to capture data from every angle. After completing the survey of the nose cap, they shifted to the port wing and repeated the process. During these operations, the crew downlinked detailed video footage to engineers at Mission Control in Houston, providing them with an up-close view of a section of thermal blanket that had peeled back on the port-side orbital maneuvering system (OMS) pod, which required further analysis.

Meanwhile, Mission Specialists John "Danny" Olivas, James Reilly, and Clayton Anderson conducted a thorough check of the spacesuits that would be used during the planned spacewalks at the International Space Station (ISS). They prepared the suits and associated hardware for transfer to the station, ensuring everything was ready for the upcoming extravehicular activities (EVAs). Additionally, the crew installed a centerline camera, extended the outer ring of the Orbiter Docking System, and tested the rendezvous tools that would be crucial for the upcoming docking with the ISS.

On the third day of the mission, Atlantis continued its approach toward the ISS. The crew was awakened at 13:08 UTC, and soon after, Atlantis performed the Terminal Insertion burn to adjust its trajectory for docking. As the shuttle closed in on the station, Commander Rick Sturckow executed the Rendezvous Pitch Maneuver (RPM) at a distance of 600 feet (180 meters) below the ISS. This intricate backflip maneuver allowed ISS crewmembers Fyodor Yurchikhin and Oleg Kotov to photograph the heat shield tiles on Atlantis' underside using high-resolution cameras, providing valuable imagery to assess the condition of the shuttle's thermal protection system.

At 19:36 UTC, Atlantis successfully docked with the ISS, connecting to the Destiny module's Pressurized Mating Adapter-2 while the two spacecraft orbited 220 miles (350 kilometers) above the northeastern coast of Australia. Once the shuttle was securely docked, hooks and latches activated to firmly pull the two spacecraft together, creating a solid, airtight connection. After completing leak checks to ensure the docking interface was secure, the hatch between Atlantis and the ISS was opened at 21:20 UTC, allowing the crews to meet face-to-face.

One of the key tasks following the docking was the official crew exchange. Flight Engineer Suni Williams, who had been aboard the ISS, was replaced by Clayton Anderson. This transfer was marked by the movement of Anderson's customized Soyuz seat liner into the Russian spacecraft, formally making him a member of the ISS crew while Williams prepared for her return to Earth aboard Atlantis.

After docking, the mission's focus shifted to preparing for the installation of a new truss segment. Pilot Lee Archambault and Mission Specialist Patrick Forrester used Atlantis' Canadarm to grapple the S3/S4 truss segment, carefully lifting it from the shuttle's payload bay and maneuvering it toward the ISS. Once in position, the station's robotic arm, Canadarm2, operated by Suni Williams, took control of the truss from the shuttle's arm at 00:28 UTC. The handover marked a critical step in the mission, though the truss would remain attached to Canadarm2 overnight in preparation for its installation the following day.

To prepare for the next day's spacewalk, James Reilly and Danny Olivas began the "campout" procedure, spending the night in the Quest airlock. This procedure allowed them to pre-breathe oxygen in a lower-pressure environment, helping to purge nitrogen from their bloodstreams and reducing the risk of decompression sickness during the spacewalk.

The fourth day of the mission began with the crew preparing to continue the assembly of the International Space Station (ISS) by installing the S3/S4 truss segment. Pilot Lee Archambault, Mission Specialist Patrick Forrester, and station Flight Engineer Oleg Kotov operated the Space Station Remote Manipulator System (SSRMS) to position and attach the S3/S4 truss to the outboard end of the S1 truss. This installation was a key step, setting the stage for the first spacewalk of the mission.

Mission Specialists James Reilly and John "Danny" Olivas, both in their fully pressurized spacesuits, ventured out of the Quest airlock to begin the spacewalk. Their primary task was to release the launch restraints on the four Solar Array Blanket Boxes, which housed the folded solar arrays. Once the restraints were removed, they secured bolts, cables, and connectors, finalizing the preparations needed for the activation of the newly installed truss segment. The two astronauts also rotated the array canisters into position for the deployment of the solar arrays, scheduled for the following day.

The spacewalk was delayed by an hour due to a temporary loss of attitude control on the station. The station's control moment gyroscopes (CMGs) went offline, causing a brief halt in the operation. However, once the gyroscopes were brought back online by flight controllers, Reilly and Olivas resumed their work, completing the objectives of the spacewalk.

This marked the 84th EVA dedicated to the assembly and maintenance of the ISS and was the fourth spacewalk for Reilly and the first for Olivas. The mission management team at NASA decided to extend the mission by two days and add a fourth spacewalk. The extension was designed to give the crew additional time to complete the ISS assembly tasks and to allow ground engineers to develop a plan to repair the damaged thermal blanket on Atlantis' orbital maneuvering system (OMS) pod. While the torn blanket posed no immediate threat to the shuttle crew during reentry, there was a concern that the damage could affect the internal structure of the OMS pod, necessitating repairs once the shuttle returned to Earth.

On the fifth day of the mission, the crew successfully deployed the solar arrays attached to the S3/S4 truss segment, significantly enhancing the power generation capabilities of the ISS. Prior to the crew's wake-up, ground controllers initiated the process of unfurling the solar arrays. Once the shuttle crew took over, they unfolded the arrays one wing at a time, carefully pausing between stages to allow the sunlight to warm the thin panels. This procedure helped prevent the delicate individual panels from sticking together.

The first wing of the solar array was fully deployed by 11:29 AM CDT, and the second wing followed at 12:58 PM CDT. The successful deployment of the solar arrays was a major milestone in the mission, providing the ISS with increased power for its systems and scientific experiments.

However, during the deployment, the station continued to experience problems with its electrically driven gyroscopes, which are critical for maintaining attitude control. The issue was traced to a malfunction in the Russian navigation computer, which occurred when flight controllers attempted to transition attitude control from the shuttle's computers back to the ISS. The navigation computer failed to accept the command and required a full reboot of the Russian command and control system. This reboot triggered alarms both aboard the station and at Mission Control.

By the end of the day, the issues had been resolved, and the station's gyroscopes resumed attitude control shortly after 8:00 PM CDT. With the situation stabilized, the crew relocated the Mobile Transporter in preparation for the next day's spacewalk. Throughout the troubleshooting efforts, there was no risk to the crew or the shuttle-station combined stack, and the mission proceeded as planned.

On the sixth day of the mission, Space Shuttle Atlantis saw Mission Specialists Patrick

Forrester and Steven Swanson conduct the second of four planned spacewalks, or EVAs. The primary objective of this EVA was to prepare the Solar Alpha Rotary Joint (SARJ), located between the S3 and S4 truss segments, for rotation. This joint is critical to allowing the solar arrays to track the Sun, optimizing power generation for the ISS.

The crew's day began at 13:08 UTC, and Mission Specialists John "Danny" Olivas and Sunita Williams assisted Forrester and Swanson in preparing for the spacewalk. Meanwhile, Commander Rick Sturckow, Pilot Lee Archambault, and Mission Specialist James Reilly worked on retracting the 2B solar array wing on the starboard side of the P6 Truss. They monitored the process and sent commands to begin retracting the array, a crucial step before installing the SARJ.

Forrester and Swanson exited the Quest airlock at 18:03 UTC and made their way to the P6 Truss to oversee and assist with the retraction of the solar array. Positioned in a foot restraint on the station's Canadarm2, Forrester was able to use specially prepared tools to aid in the folding of the array's photovoltaic panels. Despite the challenges presented by the sticky panels, flight controllers managed to fold seven and a half of the 31.5 solar array bays, while Forrester and Swanson manually helped an additional five and a half bays (about 45 feet) fold properly before they turned their attention to the SARJ.

The second EVA yielded mixed results. The spacewalkers successfully removed all of the launch locks holding the SARJ in place, but encountered an issue when attempting to install a drive-lock assembly. It was discovered that the motor control circuits for the SARJ had been wired in reverse, sending commands to the drive-lock assembly installed during EVA 1 rather than the one currently being worked on. As a precaution, one launch lock was left in place to prevent any unintended rotation of the joint. Despite this setback, the majority of the EVA objectives were accomplished.

This spacewalk marked the 85th dedicated to the assembly and maintenance of the ISS. As EVA 2 unfolded, mission managers approved plans for a repair task to fix the damaged thermal blanket on Atlantis' orbital maneuvering system (OMS) pod during the next spacewalk, EVA 3. Meanwhile, Russian flight controllers worked overnight to resolve the issues with the Russian segment's computers, which had experienced malfunctions earlier in the mission. Throughout the day, the shuttle provided propulsion backup while the station's control moment gyroscopes maintained attitude control.

Flight Day 7 began with a significant challenge for the crew when, at 06:30 UTC, a computer malfunction on the Russian segment of the ISS left the station without orientation control. This malfunction triggered a series of events, including a false fire alarm that woke the crew at 11:43 UTC. Engineers hypothesized that the issue might have been caused by the new S4 solar array, or perhaps by changes in the electrical circuitry delivering power from the new arrays to the Russian segment. The malfunction required a reboot of the Russian command computers, and although the situation was quickly brought under control, it highlighted the complexity of integrating new hardware into the ISS systems.

Throughout the day, both the Atlantis and ISS crews focused on retracting the solar array blanket atop the P6 Truss. This task, essential for clearing the way for future station assembly, was conducted with precision. Commander Rick Sturckow, Pilot Lee Archambault, Mission Specialist Sunita Williams, and Flight Engineer Clayton Anderson carefully retracted the solar array by three more bays. By the end of the day, 15.5 of the 31.5 bays remained to be folded into the protective box. The retraction would need to be completed before the mission's end.

In preparation for EVA 3, Mission Specialists James Reilly and John Olivas conducted a thorough review of procedures, ensuring that all tasks were fully understood. Later, the crew took part in interviews with radio and television stations, sharing details about their mission and experiences in space. Before turning in for the night, Reilly and Olivas entered the Quest Airlock for the pre-breathe campout protocol, spending the night in a low-pressure environment to prepare for the next day's spacewalk.

The eighth day of the mission focused on the

third spacewalk (EVA 3) to repair the thermal blanket on Space Shuttle Atlantis and assist in folding a solar array on the International Space Station (ISS). The crew began their day with a wake-up call at 7:41 AM CDT, preparing for a day of critical tasks.

Mission Specialists James Reilly and John "Danny" Olivas commenced EVA 3 after switching their spacesuits to internal battery power and exiting the Quest airlock. This marked the 86th EVA dedicated to the assembly and maintenance of the ISS, and the ninth of 2007. During the spacewalk, Olivas, anchored to the end of Atlantis' robotic arm, worked on repairing the thermal blanket on the shuttle's port-side orbital maneuvering system (OMS) pod. Using his helmet camera, Olivas transmitted close-up views of the damaged insulation and surrounding areas to flight controllers in Houston, allowing them to assess the damage. Over the course of two hours, he carefully stapled and pinned down the torn thermal blanket, ensuring it was secure.

Meanwhile, Reilly worked inside the Destiny laboratory, installing an external hydrogen vent for the oxygen generation system. Midway through the spacewalk, flight controllers instructed Reilly to disconnect the P-12 connector, which had been installed during EVA 1. Russian controllers were planning to restart the troubled computers aboard the ISS, and although the connector was not in use, engineers wanted to eliminate any possibility that it was causing electrical interference.

Once these tasks were complete, the two spacewalkers, along with their colleagues on the shuttle and station, completed the final retraction of the starboard-side P6 solar array. The retraction required 28 commands and was finalized seven hours and 15 minutes into the spacewalk, at 7:40 PM CDT. The successful folding of the array cleared the way for the P6 truss to be relocated to its permanent position during the upcoming STS-120 mission. EVA 3 was completed successfully, lasting seven hours and 58 minutes, with all objectives met.

Earlier in the day, U.S. and Russian engineers had isolated the Russian computers from U.S. electrical power, but this did not resolve the issue. Russian flight controllers managed to bring the primary computers back online by bypassing a faulty circuit, though secondary systems remained offline for further troubleshooting.

On Flight Day 9, the crew of Atlantis had a relatively light day after the intense activities of the previous spacewalks. The highlight of the day was Mission Specialist Sunita Williams setting a new space endurance record for female astronauts. At 5:47 UTC, Williams reached 188 days and 4 hours in space, matching the record set by astronaut Shannon Lucid during her 1996 flight to the Mir space station. By coincidence, Williams set her new record on the 44th anniversary of the launch of Valentina Tereshkova, the first woman to fly in space, in 1963.

Throughout the day, the crew continued transferring supplies between Atlantis and the ISS. Mission Specialists James Reilly, John Olivas, Steven Swanson, and Patrick Forrester spent time reviewing procedures for the upcoming EVA 4 and preparing their spacesuits and tools. Reilly and Olivas focused on post-spacewalk reconfiguration tasks, while Forrester and Swanson worked on configuring their suits for the next spacewalk.

The crew also participated in a space-to-ground news conference, sharing updates on the mission with reporters. Later in the day, flight controllers in Moscow successfully restarted several systems on the Russian segment of the ISS. Following the bypass of a faulty power switch on Flight Day 8, cosmonauts Fyodor Yurchikhin and Oleg Kotov applied the same fix to the remaining channels, bringing all six processors online—four active and two in standby.

The crew concluded the day by reviewing plans for the final spacewalk, ensuring that all tasks were ready for the following day's activities.

On June 17, 2007, Flight Day 10 of the STS-117 mission aboard the Space Shuttle Atlantis began with the crew receiving their wake-up call at 6:38 a.m. Central Daylight Time. The day was marked by the successful completion of the mission's fourth and final spacewalk (EVA 4), a critical milestone in the ongoing assembly and maintenance of the International Space Station

(ISS).

Mission Specialists Patrick Forrester and Steven Swanson prepared meticulously for their extravehicular activity. At 11:25 a.m. UTC, they exited the Quest airlock, embarking on the 87th spacewalk dedicated to ISS construction and upkeep. Inside Atlantis, Astronaut James Reilly choreographed the spacewalk from the flight deck, with ISS Flight Engineer Oleg Kotov serving as the intravehicular crew member, providing essential support and coordination.

The primary objectives for EVA 4 included several vital tasks to enhance the ISS's operational capabilities. Forrester and Swanson first retrieved a television camera and its support structure from a stowage platform attached to the Quest airlock. They then expertly installed the equipment onto the newly added S3 truss segment, significantly improving the station's external monitoring and communication functions.

Next, the astronauts focused on the Solar Alpha Rotary Joint (SARJ), a critical component that allows the station's solar arrays to rotate and track the sun, maximizing energy absorption. They verified the configuration of Drive Lock Assembly (DLA) 2 and removed the last six SARJ launch restraints. This action enabled the joint to rotate freely for the first time, a significant step in the ISS's power generation enhancements.

With their primary tasks completed ahead of schedule, Forrester and Swanson proceeded to tackle several "get-ahead" tasks—additional objectives that mission managers hoped could be addressed if time permitted. They installed a computer network cable on the Unity Node, a crucial upgrade that would allow astronauts in the U.S. segment of the station to command systems within the Russian segment, thereby improving operational integration between the two sections.

The astronauts then turned their attention to the Destiny Laboratory Module, where they opened a hydrogen vent valve that had been installed by James Reilly during EVA 3. This valve is part of the station's oxygen generation system, essential for supporting long-duration missions by providing a sustainable oxygen supply for the crew.

Forrester and Swanson also attempted to reattach two debris shield panels on the Zvezda Service Module that had been previously removed during an earlier spacewalk. Despite their efforts, they were unable to secure the panels due to alignment issues. To ensure safety, they tethered the panels securely to the station's exterior, preventing them from becoming hazards.

After six hours and 29 minutes, EVA 4 concluded successfully. The completion of these tasks marked the fulfillment of STS-117's major objectives, significantly advancing the ISS's assembly and functionality. That evening, ground controllers initiated a test of the newly operational SARJ, commanding a five-degree rotation to verify its performance. The test was successful, confirming that the joint was functioning as intended.

June 18, Flight Day 11, provided the crew with some well-deserved off-duty time during the first half of the day. They spent the remainder of the day transferring the last of the cargo between Atlantis and the ISS, ensuring that all equipment and supplies were correctly stowed for the continuation of the station's operations and the shuttle's return journey.

Meanwhile, flight controllers on Earth conducted rigorous tests on the ISS's Russian central and terminal computers, which had experienced issues earlier in the mission. The computers remained stable throughout the day, and successful tests confirmed their reliability. The shuttle performed a maneuver to adjust the complex's orientation, allowing for a scheduled water and waste dump. The attitude control was successfully switched to and from Russian command, further validating the restored functionality of the station's systems.

Preparations for Atlantis's departure began in earnest. The crews of both Atlantis and the ISS gathered for a heartfelt farewell ceremony, expressing gratitude and camaraderie forged through their collaborative efforts. At the end of the day, the hatches between the shuttle and the station were sealed, and Atlantis was cleared for undocking on Flight Day 12.

On June 19, Flight Day 12, the crew initiated

the undocking procedures. Pilot Lee Archambault and Mission Specialist Patrick Forrester powered on the shuttle's systems that had been conserved during the docked phase. At 9:42 a.m. UTC, the docking mechanisms were disengaged, and Atlantis gently separated from the ISS as they orbited above New Guinea. The shuttle had been attached to the station for eight days and 19 hours, during which substantial progress had been made on the ISS's assembly.

Following undocking, Archambault maneuvered Atlantis to a distance of 450 feet directly in front of the ISS before commencing a full fly-around at 3:07 p.m. UTC. This maneuver allowed the crew to photograph the station from multiple angles, providing valuable imagery for engineers to assess the condition of the ISS's exterior structures.

During the separation, the crew observed several small objects floating away from the shuttle, which appeared to be harmless pieces of ice. However, they also noted a larger, more distinct piece of debris. Commander Rick Sturckow reported the sightings to Mission Control, describing one object as resembling "little phenolic-looking, kind of tan-looking washers with four slots," which are used to secure Multi-Layer Insulation blankets in the payload bay. The origin of these objects—whether from Atlantis or the ISS—was not immediately clear, prompting careful monitoring.

Later in the day, Archambault, Forrester, and Swanson utilized the shuttle's robotic arm to conduct a comprehensive inspection of Atlantis's thermal protection system. Using the Orbiter Boom Sensor System, they scanned the leading edges of the wings and the nose cap to detect any potential damage from micrometeoroids or orbital debris—a critical safety measure before re-entry into Earth's atmosphere.

Meanwhile, Astronaut Sunita Williams, who was returning to Earth after a record-breaking mission aboard the ISS, engaged in additional exercise routines to prepare her body for the transition back to Earth's gravity. Williams had spent 195 days in space, the longest duration for any female astronaut at that time.

Flight Day 13, June 20, was dedicated to preparing Atlantis for landing. The crew performed various hardware tests, verified the shuttle's flight control systems, and engaged in interviews with major television networks, sharing insights about their mission experiences.

On June 21, Flight Day 14, the crew faced challenges due to unfavorable weather conditions at Kennedy Space Center. Despite closing the payload bay doors and configuring the shuttle for landing, both landing opportunities were scrubbed. Thunderstorms, showers, and low clouds posed unacceptable risks, necessitating an additional day in orbit.

The following day, June 22, Flight Day 15, presented improved prospects for landing, albeit not at the originally planned location. With weather conditions still poor in Florida, Mission Control redirected Atlantis to Edwards Air Force Base in California, where conditions were ideal. The crew prepared for re-entry, performing the deorbit burn at 11:43:47 a.m. UTC. Commander Sturckow and Pilot Archambault fired the shuttle's Orbital Maneuvering System engines for two minutes and 33 seconds, reducing the shuttle's velocity by approximately 200 miles per hour—a critical adjustment to begin their descent.

As Atlantis re-entered the Earth's atmosphere, it endured intense heat and friction, with temperatures on the thermal protection system reaching up to 3,000 degrees Fahrenheit. At an altitude of 50,000 feet, Sturckow took manual control of the orbiter, skillfully guiding it through a series of banking turns to dissipate speed and align with the runway approach path.

At 12:49:38 p.m. UTC, Atlantis touched down smoothly on Runway 22 at Edwards Air Force Base, concluding its successful STS-117 mission. The shuttle had completed 219 orbits around the Earth, traveling more than 5.8 million miles. Flight surgeons were on hand to assist Sunita Williams, who was carefully extracted from the shuttle to begin her readjustment to Earth's gravity after her extended stay in space.

Post-landing procedures commenced promptly. Atlantis was secured and prepared for its cross-country journey back to Kennedy Space Center, where it would undergo processing for future missions. The shuttle was mounted atop a modified Boeing 747 Shuttle Carrier Aircraft for the ferry flight, an operation that included refueling stops—such as at Offutt Air Force Base in Omaha, Nebraska—before reaching Florida.

On June 23, the STS-117 crew returned to Houston, where they were greeted by NASA officials, colleagues, family, and friends during a welcoming ceremony at NASA Hangar 276 at Ellington Field. The mission had not only achieved its primary objectives but also demonstrated the effectiveness of the shuttle program's ability to adapt to challenges, such as the unexpected computer failures aboard the ISS and weather-related delays.

Throughout the mission, the crew conducted four spacewalks totaling over 27 hours, during which they installed the S3/S4 truss segments, retracted a problematic solar array, and performed essential maintenance tasks. The successful deployment of the new truss segments and the activation of the SARJ significantly enhanced the ISS's power generation capabilities, paving the way for future expansion and scientific research.

This mission was crucial for the ongoing construction of the International Space Station (ISS). Its primary payload was the second starboard truss segment, designated S3/S4, along with associated energy systems, including a pair of solar arrays. Once delivered to the ISS, the S3/S4 truss was attached to the starboard side of the station. The crew retracted an old solar array and deployed the new one, enhancing the station's energy capacity. This expansion supported ISS operations and set the stage for future additions to the station's framework.

Another key objective of the mission was crew rotation. STS-117 transported Expedition 15 crew member Clayton Anderson to the ISS, while returning astronaut Sunita Williams to Earth after her tenure aboard the station.

Midway through the mission, NASA extended the timeline by two days, allowing for a fourth spacewalk, or extra-vehicular activity (EVA), to repair a gap in the thermal blanket on *Atlantis*' Orbital Maneuvering System (OMS). This extension was confirmed on June 11, 2007, and the repair was successfully completed during EVA 3.

After completing all mission objectives, Atlantis undocked from the ISS on June 19, 2007, and began preparations for reentry. The Shuttle safely returned to Earth, landing at Edwards Air Force Base in California on June 22, 2007, as the weather at Kennedy Space Center prevented a Florida landing.

STS-118

The orbiter was moved from the Orbiter Processing Facility to the Vehicle Assembly Building on July 2, 2007, before making its final journey to Launch Pad 39A on July 10. The crew completed their terminal countdown demonstration test on July 19, and NASA's Flight Readiness Review on July 25-26 declared the mission "GO" for launch. A minor valve repair delayed the launch by one day, but by August 8, all systems were ready.

On the day of launch, fueling of the external tank began at 8:11 a.m. EDT, and the crew proceeded to the Operations & Checkout (O&C) building, departing for the launch pad by mid-afternoon. Despite minor technical glitches, including a small crack in the external tank's foam and issues with the crew hatch switches, Endeavour lifted off as scheduled at 6:36 p.m. EDT. The launch marked another successful milestone in NASA's continued efforts to maintain the shuttle program's reliability and safety.

However, STS-118 was not without challenges. During liftoff, a piece of insulation foam detached from the external fuel tank and punctured Endeavour's heat shield. This incident brought back painful memories of the foam impact that had caused the loss of Columbia. In this case, the damage was smaller and located in a non-critical area, but NASA remained cautious, conducting detailed inspections during the mission. Ultimately, KSC Launch Director Michael Leinbach reported that Endeavour was the "cleanest" orbiter in terms of post-flight condition since the Return to Flight missions. By August 31, 2007, engineers confirmed that there was no heat-related damage to the orbiter.

STS-118 marked Space Shuttle Endeavour's return to flight after nearly five years, launching on August 8, 2007, from Kennedy Space Center (KSC), Florida. This mission was particularly significant as it was Endeavour's first flight since STS-113 in November 2002 and the first successful shuttle mission following the tragic loss of Columbia during reentry on STS-107 in February 2003. Originally, Columbia had been assigned to this mission, which would have been its 29th and final flight, but Endeavour stepped in after the Columbia disaster, carrying with it the weight of the shuttle program's recent history.

STS-118 was initially planned as Space Shuttle Columbia's 29th mission and its first visit to the International Space Station (ISS). However, following the tragic loss of Columbia during reentry in 2003, NASA reassigned the mission to Endeavour. This marked Endeavour's return to flight after a five-year hiatus during which the orbiter underwent a comprehensive refit, including over 200 modifications aimed at enhancing its safety and performance. One of the most significant upgrades introduced on this flight was the Station-Shuttle Power Transfer System (SSPTS). This system allowed Endeavour to draw up to eight kilowatts of power directly from the ISS, converting the station's 120-volt direct-current (120VDC) to the 28VDC required by the orbiter. The SSPTS extended the mission duration by conserving the cryogenic hydrogen and oxygen used in the orbiter's fuel cells, permitting an additional three to four days of docked operations.

The mission also showcased several other critical system upgrades. A new three-string GPS system replaced the older Tactical Air Navigation units, improving the shuttle's ability to calculate its position during reentry and landing. The Advanced Health Management System, another first for STS-118, provided enhanced monitoring of the orbiter's three main engines (SSMEs), with the capability to shut them down if signs of catastrophic failure emerged. These systems reflected NASA's unwavering commitment to safety and innovation, especially in the wake of the Columbia disaster.

A Crew of Experience and Symbolism

STS-118's crew, commanded by veteran astronaut Scott Kelly, included pilot Charles O. Hobaugh, mission specialists Tracy Caldwell, Richard Mastracchio, Dafydd Williams of the Canadian Space Agency, Barbara Morgan, and Alvin Drew. While each crewmember played a vital role in the mission's success, Barbara Morgan's presence carried deep significance. A former teacher and the backup to Christa McAuliffe, the teacher who perished in the Space Shuttle Challenger disaster in 1986, Morgan represented the continuation of NASA's vision of space exploration through education. Although Morgan flew as a standard mission specialist, her role symbolized NASA's dedication to educational outreach and to honoring the legacy of teachers in space.

Morgan's journey to space captured widespread media attention due to her connection to the Challenger tragedy and the Teacher in Space Project. After Challenger's loss, NASA redirected its focus toward the Educator Astronaut Project, selecting Morgan in 1998 as the first mission specialist educator. Morgan, having served as a key figure in NASA's educational outreach since the late 1980s, completed two years of astronaut training and officially began her duties in 2000. Unlike McAuliffe, who was classified as a spaceflight participant, Morgan was a fully trained mission specialist, capable of performing all the responsibilities of a regular astronaut. Her participation in STS-118 embodied NASA's enduring commitment to inspiring future generations through education.

The symbolism of Morgan's mission was powerfully captured in the official STS-118 patch, which featured a flame representing the importance of knowledge and education. The flame's tip touched Morgan's name—a poignant tribute to the legacy of teachers like Christa McAuliffe and a reminder of the role educators play in space exploration.

As Morgan reflected on her dual roles as a teacher and astronaut, she highlighted the shared dedication between both professions:

"Whether you're teaching school, or whether you're training as an astronaut, you put all you can into it, and get the most out of it."

Following the launch, the crew completed standard procedures for post-ascent operations, which included opening the payload bay doors, activating the Spacehab module, and powering up the Remote Manipulator System. These tasks were essential for enabling the mission's primary objectives and ensuring that all systems were functioning correctly. After completing these activities, the crew entered their scheduled sleep period, marking the end of a successful first day in orbit.

On Flight Day 2 – Thursday, August 9, the Space Shuttle crew spent the second day of their mission meticulously inspecting the orbiter's outer hull and heat shield. These routine checks were essential for ensuring the shuttle's safety during re-entry, particularly in light of past shuttle missions, where damage to the heat shield had caused catastrophic results. During a mission status briefing, Deputy Shuttle Program Manager and Mission Management Team Chairman, John Shannon, reported that approximately nine pieces of foam were observed breaking off the external fuel tanks during launch. Of these, three struck the shuttle. Fortunately, all three strikes were classified as minor, and initial assessments indicated no immediate threat to the shuttle's integrity.

On the third day, Space Shuttle *Endeavour* successfully docked with the International Space Station (ISS) at 18:02 UTC (14:02 EDT), marking a key milestone in the mission. Before docking, *Endeavour* executed a critical maneuver known as the Rendezvous Pitch Maneuver (RPM), during which the shuttle performed a one-degree-per-second backflip. This maneuver allowed the ISS crew to capture high-resolution digital images of the shuttle's heat shield, which were promptly transmitted to NASA's Image Analysis Team and the Mission Management Team for evaluation.

Once the docking was completed, the crews conducted leak checks before opening the hatches at 20:04 UTC (16:04 EDT), at which point the Expedition 15 crew warmly welcomed the STS-118 astronauts aboard the station.

As part of the analysis of the RPM photos, an area of concern was identified on the underside of *Endeavour*. A section of black silica tiles near the right landing gear door showed visible damage—a gouge measuring approximately 3.5 inches by 2 inches (8.9 cm by 5.1 cm). While the tile was penetrated, the felt backing beneath it remained intact. A foam strike during launch was identified as the likely cause of the damage. Although this particular area was less vulnerable than the reinforced carbon-carbon leading-edge tiles implicated in the *Columbia* disaster, the damage prompted concern among the mission managers. A focused inspection was scheduled for August 12 to determine whether repairs would be required during an extravehicular activity (EVA).

NASA officials acknowledged that foam shedding from the external tank was a known issue, with up to 300 pieces of foam sometimes striking the orbiter during ascent. While this was not unusual, each instance required careful evaluation to ensure the safety of the orbiter during re-entry.

Additionally, the Station-to-Shuttle Power Transfer System (SSPTS) was activated after docking. This system allowed the shuttle to draw electrical power from the ISS, reducing its reliance on its own power-generating resources. The successful operation of the SSPTS hinged on the mission's extension from 11 to 14 days.

On the fourth day of the mission, astronauts Rick Mastracchio and Dave Williams conducted the first spacewalk of the mission. Beginning at 21:45 UTC, the spacewalk lasted 6 hours and 17 minutes, during which the astronauts successfully installed the S5 truss segment onto the ISS. The addition of this truss increased the station's total mass to 232,693 kilograms (513,000 pounds), continuing the steady expansion of the ISS's structure.

Following the successful EVA, Lead ISS Flight Director Joel Montalbano reported that the SSPTS was functioning well, and the recommendation would be made to extend the mission to the full 14 days.

John Shannon provided additional information regarding the foam strike observed earlier. Detailed analysis indicated that a piece of foam had detached from the external tank's feed line area and ricocheted off a nearby strut before hitting the shuttle's underside. Interestingly, a similar piece of foam had been lost during mission STS-115, but it did not impact the orbiter on that occasion.

Shannon emphasized that while initial reviews suggested the damage did not fully penetrate the thermal tile, a focused inspection remained scheduled for flight day 5 to gather more information. The inspection would concentrate on five specific areas of the shuttle, and the resulting data would be used for thermal testing to determine whether repairs or additional actions were necessary.

Shannon also addressed concerns about a potential gap filler protruding from the shuttle, which had been noted during the earlier inspection. After further examination, it was determined that the protrusion was "shim stock," a material that would harmlessly burn away upon re-entry, posing no risk to the shuttle or its crew.

The crew's progress in addressing the foam strike and performing key mission objectives was closely monitored as they prepared for the extended mission duration.

On Flight Day 5 – Sunday, August 12, the crew conducted a focused inspection of the shuttle's thermal protective tiles, using high-resolution imaging to assess potential damage areas. The inspection proceeded smoothly, and by 20:56 UTC, both the *Endeavour* and Expedition 15 crews were informed that the Station-to-Shuttle Power Transfer System (SSPTS) was operating as expected, confirming the mission's extension to the full 14 days. Commander Scott Kelly's response captured the crew's relief and optimism: "That's great news, thanks."

During the daily mission status briefing, Lead Flight Director Matt Abbott officially announced the extension of the mission. He noted that the SSPTS was functioning as planned, and the data gathered from the focused inspection was encouraging. The Mission Management Team, led by Chairman John Shannon, unanimously supported the extension of the mission and the addition of a fourth extravehicular activity (EVA). Shannon emphasized the strategic importance of the SSPTS, which allowed for longer shuttle missions and enhanced the ability to supply the ISS with vital resources such as oxygen and water.

Shannon provided additional details on the focused inspection. The team had gathered excellent laser data and imagery, and several areas of concern were resolved. However, two adjacent thermal plates on the shuttle's underside revealed a gouge deeper than initially anticipated. NASA planned to use this data for thermal analysis and testing at the Johnson Space Center, simulating the damage to determine the best course of action. While decisions were pending further review, the inspections confirmed that no immediate threat to the shuttle existed.

Shannon also mentioned that NASA would thoroughly review foam loss from the external tank dating back to mission STS-114. This review would help guide future actions and ensure that no speculative decisions were made until all relevant data had been fully analyzed.

On the sixth day of the mission, astronauts Rick Mastracchio and Dave Williams completed the second EVA of the mission. They successfully removed a new Control Moment Gyroscope (CMG) from *Endeavour*'s payload bay and transported it to the Z1 truss, where they replaced the failed CMG. The failed unit was secured on an external stowage platform for future return to Earth aboard mission STS-122. Initial tests confirmed that the new CMG was functioning normally, ensuring continued control of the ISS's orientation.

During the EVA, Mastracchio experienced an alarm in his suit indicating elevated levels of carbon dioxide. After reviewing sensor data, NASA determined the issue was due to faulty instrumentation, and the suit itself was in good condition. The suit was cleared for future use without concerns.

At the mission status briefing, John Shannon provided an update on *Endeavour*'s heat shield. Initial modeling suggested that the majority of heat upon re-entry would concentrate on the backside of the gouge, away from the more vulnerable filler bar, which was a favorable situation. However, Shannon stressed the complexity of the damage's aerodynamic shape and emphasized the need for continued analysis. Engineers would run additional flow modeling tests and proceed to arc jet testing that night to simulate the high-speed airflow that would affect the damaged tiles during re-entry.

Shannon also revealed that a specialized "team four" had been assembled to work alongside operations and engineering teams to analyze the data and explore repair options. The team was tasked with evaluating potential solutions and making recommendations to the Mission Management Team. Shannon expressed confidence that, should a repair be deemed necessary, the crew would be capable of executing it without significantly disrupting the mission timeline. He reiterated that the damage was not catastrophic but rather an issue they preferred to address, comparing it to the Orbital Maneuvering System (OMS) blanket damage encountered on mission STS-117 in June 2007.

The team had three on-orbit repair techniques at their disposal, all of which had been thoroughly tested in post-*Columbia* Return to Flight missions. The STS-118 crew had been trained in these procedures, ensuring they were well-prepared to handle any necessary repairs.

On Flight Day 7 – Tuesday, August 14, the crew of *Endeavour* began their day with a special birthday greeting for astronaut Tracy Caldwell, as her family sang to her over the radio. The day's activities centered around successfully removing and installing the External Stowage Platform (ESP-3) from *Endeavour*'s payload bay onto the P3 truss of the International Space Station (ISS). Alongside these technical operations, the crew also engaged in transfer activities and participated in two Public Affairs events, including a Q&A session with children at the Discovery Center in Boise, Idaho.

During an interview with CBS, Commander Scott Kelly addressed concerns about the tile damage on *Endeavour*, expressing his confidence in NASA's decision-making process. "My understanding is this tile damage is not an issue of the safety of the crew... I'm not concerned with our safety," Kelly stated, emphasizing that any repair would primarily be for ease of processing once the orbiter returned to the Kennedy Space Center.

In the Mission Management Team briefing, Kirk Shireman, Deputy International Space Station Program Manager, confirmed that the installation of the External Stowage Platform had gone smoothly. The new Control Moment Gyroscope (CMG) was functioning properly, and the handover of attitude control from the shuttle to the station was completed without issues. Shireman also reported on the successful installation of a new Russian computer aboard the ISS, with system testing scheduled for late August. He also highlighted the milestone achievement of the Zarya module completing its 50,000th orbit around Earth.

John Shannon provided an update on the orbiter's condition, reporting that all systems were in excellent shape. He confirmed that thermal analysis of the damaged tile had been completed, and computational fluid dynamics testing at Ames Research Center had shown promising results. Arc jet testing would continue into Tuesday night, and Shannon announced plans to develop an EVA scenario in the Neutral Buoyancy Lab, should repairs become necessary. The team remained "cautiously optimistic" that repairs would not be required.

On Flight Day 8 – Wednesday, August 15, Rick Mastracchio and Clayton Anderson embarked on the third spacewalk of the mission, which commenced at 14:37 UTC. Over the course of the EVA, the two astronauts successfully relocated a CETA Cart, retrieved the P6 Transponder, moved the S-band antenna from P6 to P1, and installed a new S-Band Baseband Signal Processor and Transponder on the P1 truss. However, during a routine glove inspection, Mastracchio noticed a potential tear on the thumb of his left glove. Out of an abundance of caution, NASA managers decided to end the spacewalk early at 20:05 UTC, although the crew had accomplished nearly all of their objectives, save for retrieving the MISSE experiment.

At the Mission Management Team briefing, Joel Montalbano reassured the public that Mastracchio's suit had never been in any real danger, and the early termination of the EVA was merely a precautionary measure. Extended photography of the glove was conducted during suit removal, and further analysis was scheduled to be completed before the fourth EVA, now postponed until at least August 18, 2007.

Lead spacewalk officer Paul Boehm concurred that the spacewalk had gone well, and emphasized that at no point was Mastracchio's safety compromised. EVA Office Manager Steve Doering explained that the glove inspection procedure had been implemented following the discovery of a cut in astronaut Robert Curbeam's glove during mission STS-116. Mastracchio's glove had sustained damage to the second layer of the suit, called vectran, which is a high-strength material designed to resist tearing. A video analysis of Mastracchio's path during the EVA was planned to determine where the tear may have occurred, and an additional review of his gloves from the previous two EVAs would be conducted.

John Shannon reported that no definitive decision had been made regarding repairs to the tile damage on *Endeavour*. Preliminary arc jet testing revealed some erosion on the backside of the adjacent tile, but it had not penetrated through the entire layer. These initial results were encouraging, but further tests were scheduled for Wednesday night. Shannon anticipated that a final decision on whether to perform repairs would likely be made on Thursday, remarking that he remained "cautiously optimistic that repairs will not be needed."

On Flight Day 9 of Space Shuttle Endeavour's mission, the shuttle and station crews continued their key transfer activities and preparations for potential extravehicular activity (EVA). This included a review of the Thermal Protection System (TPS) repair procedures, should NASA decide that a repair to the orbiter was necessary. Such measures were standard in the event that damage to the spacecraft's heat shield was detected, ensuring the crew's safety during re-entry.

Mission Specialists Barbara Morgan and Alvin Drew also participated in an educational event that morning. In collaboration with students at the Challenger Center for Space Science Education in Alexandria, Virginia, this event highlighted NASA's ongoing commitment to educational outreach, particularly after the Challenger tragedy. The event was hosted by Dr. June Scobee Rodgers, widow of Challenger's commander, Dick Scobee, and the Founding Chairman of the Challenger Center. Morgan, as NASA's first educator astronaut, and Drew also conducted interviews with major media outlets, including Associated Press, Reuters, and Idaho Public Television.

During the interview, Drew addressed concerns about the divot found on Endeavour's underside, noting that the engineers on the ground had analyzed the issue extensively. He expressed confidence that the damage would not impede a safe re-entry, stating, "The biggest danger is more to just being able to reuse Endeavour once it gets back on the ground." Morgan echoed these sentiments, affirming the crew's trust in NASA's engineers and their decision-making.

At 01:00 UTC on August 17, CAPCOM Shane Kimbrough officially notified Commander Scott Kelly that NASA's Mission Management Team (MMT) had decided that no repair was needed for the damaged tile on the orbiter's belly. This decision was based on extensive analysis, including arc jet tests that simulated the most extreme possible damage during re-entry. The tests showed that even under the worst conditions, the orbiter's structural integrity remained intact, thus confirming that a repair was unnecessary.

John Shannon, the chair of the MMT, emphasized the team's unanimous conclusion that conducting an EVA to repair the damage would introduce more risks than benefits. Despite recommendations from the JSC Engineering Independent Group that an in-flight repair might expedite the turnaround time for Endeavour's next mission, the potential hazards of conducting such a repair outweighed the time-saving benefits. Shannon also highlighted that normal post-flight processing would involve replacing around 60 tiles, which meant that repairing the single damaged tile was not critical for the shuttle's return.

When pressed by reporters about the likelihood of shuttle or crew loss, Shannon expressed complete confidence in the analysis, noting that over 200 engineers and experts from 30 different organizations—including NASA Ames, Langley Research Center, Jet Propulsion Laboratory, and others—contributed to the decision. The MMT also reported a minor strike by micro-meteoroid debris on the orbiter's commander's window, which would be analyzed further, though initial assessments suggested it was consistent with previous missions' damage.

The only other concern was Hurricane Dean, which had intensified in the Caribbean and was projected to move toward the Gulf of Mexico. Shannon assured that contingency plans were in place, although no immediate impact on the mission's timeline was expected.

On August 17, 2007, Flight Day 10 marked a relatively calm day aboard the space shuttle and the International Space Station (ISS). The crew focused on transfer operations and troubleshooting a communication issue between the shuttle and the station. The day's main event was a joint press conference held with U.S. and Canadian news agencies, followed by the traditional combined crew portrait. During the press conference, Commander Scott Kelly reaffirmed the crew's agreement with NASA's decision not to repair Endeavour's damaged tile, stating, "We agree absolutely 100 percent with the decision to not repair the damage."

In the mission status briefing, Lead Flight Director Matt Abbott provided an update on Hurricane Dean. NASA was closely monitoring the storm, and while no changes to the mission timeline had been made, contingency plans were in place should the situation worsen.

Meanwhile, Deputy ISS Program Manager Kirk Shireman reported on the success of the S-band communication system that had been relocated during the third EVA, noting that the Station-to-Shuttle Power Transfer System (SSPTS) had successfully replenished the station's oxygen reserves. He added that transfer operations between the shuttle and station were about 75% complete.

The EVA Office Manager Steve Doering provided a comprehensive report on preparations for the final EVA scheduled for the following day. After analyzing video footage from the third spacewalk, NASA determined that there were no sharp edges or excessive wear that might pose a risk to the astronauts' gloves, a concern raised during the previous EVA. A review of glove manufacturing processes confirmed there were no issues, and the astronauts' routes during the spacewalk were examined to identify areas requiring additional inspection. The final EVA was expected to be less hand-intensive than previous spacewalks, reducing the risk of damage to the gloves.

With all analyses complete, the team concluded that the planned EVA could proceed as scheduled.

On Flight Day 11, Mission Specialists Barbara Morgan and Tracy Caldwell posed for a commemorative photograph, holding an image of Expedition 15 crewmembers. Behind them were tributes to their classmate Patty Hilliard Robertson, who tragically lost her life during training, and the fallen crew of STS-107. This moment reflected the deep sense of camaraderie and respect that permeated the space program.

NASA managers, monitoring Hurricane Dean's progress, decided overnight to shorten the fourth EVA (extravehicular activity) by two hours as a precaution. The hurricane's trajectory was heading into the Gulf of Mexico, and while it posed no immediate threat to the mission, NASA chose to err on the side of caution. As a result, the EVA lasted five hours and two minutes, bringing the total EVA time for the mission to 23 hours and 15 minutes. During this spacewalk, three primary objectives and one additional task were accomplished. However, the planned task of securing debris shields on the Destiny laboratory and relocating a toolbox was deferred.

Astronauts Dave Williams and Clayton Anderson marveled at the sight of Hurricane Dean from space during their EVA. "Holy smoke," exclaimed Anderson, capturing the awe of witnessing such a powerful force of nature from orbit. Williams, equally impressed, replied, "Man, that's impressive," to which Anderson added, "They're only impressive when they're not coming towards you."

Following the spacewalk, transfer activities between the shuttle and station were completed ahead of schedule. Both crews worked efficiently to ensure everything was transferred back to Endeavour after the EVA.

Canadian astronaut Dave Williams set two national records during this mission. He became the Canadian astronaut with the most spacewalks—completing his third EVA—and surpassed Chris Hadfield in total EVA time, with 17 hours and 47 minutes logged.

In the mission status briefing, LeRoy Cain, Launch Integration Manager, reported that the Mission Management Team was taking extra precautions due to Hurricane Dean. It was decided that STS-118 would be shortened by a day to mitigate potential risks, and the EVA was truncated accordingly. As a further precaution, the crew would close the hatch Saturday evening rather than Sunday as originally planned, and undocking was scheduled for 11:57 UTC on Sunday. The first landing opportunity at Kennedy Space Center (KSC) was projected for August 21, 2007.

At 19:46 UTC, a brief farewell ceremony was held between the shuttle and station crews, followed by the closure of the hatch at 20:10 UTC, marking the beginning of Endeavour's departure from the ISS.

On Flight Day 12, Endeavour successfully undocked from the ISS at 11:56 UTC. NASA managers had decided the previous day to forgo the station fly-around typically performed after undocking, allowing the shuttle crew some much-needed rest following several intense workdays.

After undocking, two four-second firings of Endeavour's Reaction Control System (RCS) created a safe distance between the shuttle and station, positioning Endeavour above the ISS. The crew then deployed the shuttle's robotic arm to conduct the Late Inspection of the reinforced carbon-carbon (RCC) tiles on the orbiter's nose cap and wing leading edges, a critical step in ensuring the vehicle's readiness for re-entry.

During the mission status briefing, Matt Abbott, the lead flight director, reported that the undocking was flawless, and the late inspection was completed without issue. Preparations for re-entry would proceed as scheduled, with the Entry Flight Director Steve Stich and his team taking over mission duties in preparation for Tuesday's landing.

John Shannon, chair of the Mission Management Team, confirmed that all teams were "go" for landing. Despite earlier concerns about Hurricane Dean, the storm's track now appeared more favorable, shifting farther south than anticipated. As a result, NASA's contingency plans for a hurricane evacuation in Houston would likely not be necessary. By Monday, White Sands Space Harbor would be removed as a backup landing site, leaving Edwards Air Force Base and Kennedy Space Center as the primary landing options.

On August 20, 2007 (Flight Day 13), with Endeavour trailing the ISS by approximately 68 miles (109 kilometers), the crew spent Flight Day 13 conducting a series of check-out tests in preparation for entry. The Spacehab module was configured for re-entry, and the crew completed last-minute stowage tasks. Commander Scott Kelly and Pilot Charles Hobaugh practiced landing procedures using the shuttle's onboard landing simulator. The crew also engaged in an educational event, speaking with students at La Ronge school in Saskatchewan, Canada, reinforcing NASA's commitment to inspiring future generations.

During the mission status briefing, Entry Flight Director Steve Stich reported that Hurricane Dean's latest trajectory posed no threat to the mission, and contingency plans would not be activated. The weather forecast for Tuesday's landing at Kennedy Space Center remained favorable, with two Florida landing opportunities available. The first required a deorbit burn at 15:25 UTC, leading to a landing at 16:32 UTC, and the second called for a burn at 17:00 UTC with landing at 18:16 UTC. If neither opportunity was viable due to weather, NASA would consider delaying the landing by a day, or opting for Edwards Air Force Base as a backup site.

On Flight Day 14, the crew of Endeavour began their deorbit preparations in earnest. After closing the payload bay doors, they transitioned the shuttle's computer systems to landing mode and donned their reentry suits. At 14:30 UTC, the crew received the "go" for fluid loading, a procedure designed to help astronauts readjust to gravity and prevent post-landing hypotension.

At 15:08 UTC, the crew was given final clearance for the deorbit burn, and the auxiliary power units were activated. The 4-minute engine burn at 15:28 UTC slowed the shuttle by 252 mph (406 km/h), setting its trajectory for re-entry. Endeavour passed the point of peak heating at 16:20 UTC, and the main landing gear touched down at Kennedy Space Center at 16:32:16 UTC, followed by nose gear touchdown at 16:32:29 UTC. The orbiter came to a complete stop at 16:33:20 UTC.

Upon wheel stop, CAPCOM Christopher Ferguson congratulated the crew, humorously stating, "Congratulations, you've given a new meaning to the term 'higher education.'"

Initial post-landing inspections revealed minimal damage to the thermal tiles, despite concerns raised after the foam strike during launch. Although Barbara Morgan did not join the traditional post-landing "walkaround" with the crew, as she remained in the crew transport vehicle for additional medical evaluations, the other six astronauts took time to inspect Endeavour before returning to crew quarters.

During the post-landing press conference, NASA Administrator Michael D. Griffin highlighted that the International Space Station was now approximately 60% complete following STS-118, stressing the mission's vital contributions to space exploration. Associate Administrator William H. Gerstenmaier confirmed that the damage to the orbiter was less severe than what had been observed during arc jet testing, validating NASA's decision to forgo in-flight repairs.

At the conference's close, Commander Scott Kelly addressed the tile damage, commenting that while he had anticipated questions about it, he was "underwhelmed" by the extent of the damage. Barbara Morgan, reflecting on her time in space, remarked on the profound experience of participating in a mission that pushed humanity forward, stating, "When you look down and see our Earth... you realize what we are trying to do as a human race, it's pretty profound."

Contingency Mission: STS-322

STS-322 was the designated Contingency Shuttle Crew Support (CSCS) mission, prepared to launch in the event Endeavour became disabled during STS-118. This mission would have been a modified version of STS-120, with its launch date moved forward. If required, STS-322 would have launched no earlier than September 22, 2007. The crew for the contingency mission would have consisted of a four-person subset of the full STS-120 crew, tasked with rescuing the STS-118 crew if Endeavour had been unable to return to Earth. Fortunately, the mission was not needed as Endeavour completed its flight successfully.

STS-120: Building the Future of Space Exploration

Following the completion of STS-118 in August 2007, NASA managers made the critical decision to modify the external tank intended for use on the upcoming STS-120 mission. These modifications focused on the liquid oxygen feed-line brackets and involved the application of new low-density foam and the use of thinner gaskets. This decision was made in response to the foam shedding issues observed since the "Return to Flight" mission of STS-114 in 2005. Notably, during STS-118, a piece of foam had impacted the orbiter's belly, causing a small gouge, which raised concerns about the need for further refinements.

As preparations for STS-120 progressed, a hydraulic fluid leak in Discovery's right main landing gear strut caused a delay in the orbiter's rollover to the Vehicle Assembly Building (VAB). The leak was repaired by September 19, 2007, and Discovery was then successfully transferred to the VAB. There, it was mated to its external tank and solid rocket boosters. On

September 30, 2007, Discovery, mounted on the mobile launcher platform, was transported to Launch Complex 39A, a significant step in preparing for the mission.

The STS-120 crew arrived at Kennedy Space Center on October 7, 2007, to begin their final pre-flight preparations. This included the terminal countdown demonstration test, a critical rehearsal of the launch sequence, which ran from October 7 to October 10.

During this time, NASA's Engineering and Safety Center, established in the aftermath of the 2003 Columbia disaster, raised concerns about the integrity of three of Discovery's reinforced carbon-carbon (RCC) panels on the orbiter's wing leading edge. The panels had shown minor deterioration in the silicon-carbide protective coating. NASA had been closely monitoring this issue, conducting extensive tests using thermography to evaluate the condition of the panels. The results of these tests, reviewed in August, indicated that the panels had not worsened during Discovery's previous flights, leading engineers to conclude that the orbiter was safe to fly without panel replacement. Nonetheless, had NASA opted for repairs, Discovery would have needed to be rolled back to the VAB, causing further delays.

On October 16, 2007, NASA managers held the Flight Readiness Review, a crucial assessment before launch, and confirmed that the mission could proceed as planned. The concerns raised by the NESC were deemed non-critical, and NASA officially announced that Discovery's RCC panels were fit for the mission.

The launch of STS-120, also known as ISS-10A, took place as scheduled on October 23, 2007, at 11:38 EDT. Commander Pamela Melroy led the crew of STS-120, making her the second woman to command a Space Shuttle mission. This mission also marked the first time in history that two female commanders were in space simultaneously, as Expedition 16 aboard the ISS was commanded by Peggy Whitson, the first female commander of the space station.

The STS-120 crew, comprising seven astronauts, each with crucial roles, included Commander Pamela Melroy, making it her third and final spaceflight, Pilot George D. Zamka, who was on his first spaceflight, Mission Specialist 1 Douglas H. Wheelock, also on his first spaceflight, Mission Specialist 2/Flight Engineer Stephanie Wilson, who had completed her second spaceflight, Mission Specialist 3 Scott E. Parazynski, who was on his fifth and final spaceflight, Mission Specialist 4 Paolo A. Nespoli, who was on his first spaceflight and represented the European Space Agency, Mission Specialist 5 Daniel M. Tani, who was on his second and final spaceflight and was part of Expedition 16, and Clayton Anderson, an ISS Flight Engineer from Expedition 16, who was also aboard the station and played a vital role in the mission's success.

Shortly before liftoff, a small piece of ice, approximately 4 inches by 0.5 inches, was observed on the hydrogen umbilical between the orbiter and the external tank. The inspection team quickly assessed the situation and determined that the ice posed no significant threat to the thermal protection system, allowing the launch to proceed. Despite earlier concerns about weather conditions, the skies remained clear, and the launch unfolded without significant issues. The mission was further highlighted by the attendance of filmmaker George Lucas, who was present to witness the launch of a *Star Wars* lightsaber into space as part of the payload.

On their first full day in space, the crew of STS-120 conducted a detailed inspection of the Shuttle's reinforced carbon-carbon (RCC) panels using the Shuttle Remote Manipulator System (SRMS) and the Orbiter Boom Sensor System (OBSS). This RCC survey was a routine safety measure, ensuring that the Shuttle's critical heat shield elements were intact following the rigors of launch. In addition to the RCC survey, the astronauts inspected the Extravehicular Mobility Units (EMUs), the spacesuits that would be used during their upcoming extravehicular activities (EVAs).

Throughout the day, the crew also performed preliminary tasks in preparation for docking with the International Space Station (ISS), scheduled for the following Thursday. During a status briefing, Deputy Shuttle Program Manager and Mission Management Team (MMT) Chairman John Shannon reported that preliminary data from the ascent imagery showed no major issues of concern. The imagery taken before the separation of the Solid Rocket Boosters (SRBs) indicated no significant foam losses. Shannon noted that there were still a few areas awaiting higher-resolution imagery, including a possible protruding gap filler and some losses from the external tank's foam insulation. However, none of these observations indicated any immediate problems.

Shannon also addressed a vapor trail seen coming off the external tank during launch. This phenomenon, which had been observed on previous missions such as STS-114 and STS-121, was identified as condensate forming on the backside of the tank due to the specific combination of temperature, humidity, and the angle of the Sun. Although this plume appeared more dramatic in the imagery, it was deemed an expected condition and posed no concern.

On the morning of Flight Day 3, the crew of STS-120 worked through a series of complex rendezvous procedures, leading up to the Shuttle's approach to the ISS. At 11:34 UTC, Discovery began the rendezvous pitch maneuver (RPM), a critical part of docking preparations. This maneuver allowed the ISS crew to photograph Discovery's underside for a detailed inspection of its heat shield. After completing the RPM, the Shuttle was cleared for docking, which was successfully achieved at 12:40 UTC.

Once docked, the joint crew of the ISS and STS-120 conducted a safety review, ensuring that all systems were functioning as expected. As part of the crew exchange between the station and the

Shuttle, Daniel Tani's custom seat liner was transferred from Discovery to the Soyuz TMA-11 spacecraft, marking the beginning of his stay as part of Expedition 16. At the same time, Clayton Anderson's seat liner was moved from the Soyuz to Discovery, officially ending his role on Expedition 16 and preparing him for his return to Earth aboard the Shuttle.

Just before the crew signed off for their rest period, they received confirmation that no focused inspection of Discovery's heat shield would be required. Preliminary review of the RPM photography revealed that the Shuttle's heat shield was in excellent condition, with no areas requiring further examination. During the MMT briefing, John Shannon confirmed that Discovery was a "pretty clean vehicle" with no major issues identified. He also reported that the piece of ice observed before launch had dislodged during main engine ignition and grazed the orbiter's underside during ascent. However, the impacted area was in good condition, and the ice did not cause any damage. Shannon assured the team that they would continue to review data and imagery but that Discovery was in a safe and stable configuration for the remainder of the mission.

On Flight Day 4, the crews of STS-120 and the ISS focused on preparations for the mission's first extravehicular activity (EVA). The spacewalk, performed by astronauts Scott Parazynski and Douglas Wheelock, began at 10:02 UTC and concluded at 16:16 UTC. This EVA marked a significant milestone as the crew successfully completed all the planned tasks, including the critical preparation of the Harmony module for its removal from Discovery's payload bay.

While Parazynski and Wheelock carried out the EVA, astronauts Stephanie Wilson, Daniel Tani, and Clayton Anderson operated the station's robotic arm to maneuver Harmony from the Shuttle's bay and attach it to the port side of the Unity node. At 15:38 UTC, Harmony was officially connected to the ISS, adding 2,666 cubic feet (75.5 cubic meters) of living space to the station. This expansion increased the station's overall volume by almost 20%, from 15,000 cubic feet (420 cubic meters) to 17,666 cubic feet

(500.2 cubic meters), significantly enhancing the crew's living and working conditions.

Later in the day, during the Mission Management Team (MMT) meeting, managers confirmed that Discovery's heat shield was cleared for reentry, with no focused inspections required. However, a late inspection was still scheduled for Flight Day 13 after undocking. John Shannon, the MMT chairman, also mentioned that a new task had been added to the second EVA: a visual inspection of the Solar Alpha Rotary Joint (SARJ) on the starboard side. For the past month and a half, the SARJ had been registering some intermittent vibrations, though they were not considered severe. Since Tani would be working near the area during the EVA, he was tasked with removing the insulation covers and photographing the swing bolts for further analysis.

The crews began Flight Day 5 by continuing work on Harmony, this time preparing the module for crew entry. Power and data cables were connected, and at 12:24 UTC, the hatch to Harmony was opened for the first time. Expedition 16 commander Peggy Whitson, flight engineer Yuri Malenchenko, and mission specialist Paolo Nespoli were the first to enter the new module. Wearing protective masks, they installed a temporary air duct to ensure that the air inside Harmony circulated through the station's filters.

Until the air in Harmony was fully exchanged with the station's atmosphere, access to the module was limited to a few crew members at a time, and they were not permitted to remain inside for extended periods. The crew had several tasks to complete within the node, including removing approximately 700 screws and bolts that secured equipment and panels during launch but were no longer needed in microgravity.

After the ventilation system had run for several hours, both crews participated in a joint press conference from inside the newly installed Harmony module. They fielded questions from CBS, Fox News, and WHAM-TV, marking a celebratory moment for the mission as Harmony became fully integrated into the ISS.

On October 28, 2007, the crews aboard the Space Shuttle Discovery and the International Space Station awoke at 05:08 UTC, preparing for the second extravehicular activity (EVA) of the mission. Astronauts Scott Parazynski and Daniel Tani began their spacewalk at 09:32 UTC, ahead of schedule by thirty minutes. Their primary task was to disconnect the P6 truss from the Z1 truss, enabling the robotic arm, operated by Wilson and Wheelock, to move the truss into a parked position for the night. In a light-hearted moment, Parazynski quipped to Wilson, "Don't drop it!" as the large structure was maneuvered.

Afterward, Parazynski installed handrails on the Harmony module, while Tani carried out a series of inspections requested by mission managers. One of these was a close examination of handrails on the Crew and Equipment Translation Aid (CETA) cart, which was suspected of causing damage to Rick Mastracchio's glove during the previous STS-118 mission. Tani confirmed that the handrails appeared intact, with no sharp edges that could have caused the tear.

Tani then proceeded to inspect the Solar Alpha Rotary Joint (SARJ), which had exhibited unusual vibrations for over a month. Upon removing the joint's cover, he discovered metal shavings, wear patterns, and discoloration on one of the internal rings. His observation, "It's like the result that you get with metal, iron filings, and you put a magnet under it, and they stand straight up," indicated the severity of the wear. Tani carefully collected samples of the shavings, which would later be returned to Earth for analysis. Based on these findings, mission managers adjusted plans for the next EVA, assigning Parazynski to inspect the port SARJ for comparison.

The following day, October 29, began with a joint news conference and a crew photo. The teams then continued their work, moving the P6 truss from its overnight position using the station's robotic arm to the shuttle's robotic arm. The station's arm was repositioned to an outboard worksite, preparing to attach the P6 truss to the P5 truss on the next EVA.

In parallel, Peggy Whitson conducted a magnet test on the metal shavings Tani had collected. By placing a magnet under a slip of paper with the shavings, Whitson confirmed they were ferrous in nature, ruling out the possibility that they originated from thermal covers made of aluminized Mylar. These findings led mission planners to alter the mission timeline, adding an extra day to inspect the starboard SARJ more thoroughly. The new plan called for Discovery to undock from the station on November 5, 2007, with a landing scheduled for November 7.

On October 30, the crew successfully completed the installation of the P6 solar arrays onto the P5 truss during their third EVA, which commenced at 08:45 UTC and concluded at 15:53 UTC. The deployment of the first set of solar arrays, 2B, proceeded without incident. However, while deploying the second set, 4B, the crew noticed an anomaly. Upon closer inspection, they discovered a tear in a small section of the array. Despite the damage, the array was 90% unfurled and still generated 97% of its expected power. ISS Program Manager Mike Suffredini assured the public that the situation was under review, with high-resolution images taken by the crew being analyzed overnight to determine the best course of action.

On October 31, the crew continued working on various tasks, including preparing for the fourth EVA and installing a Zero Gravity Stowage Rack in the Harmony module. They also participated in public affairs events, fielding questions from news agencies and receiving calls from dignitaries, including Italian President Giorgio Napolitano. In the afternoon, mission control informed the crew that the EVA would be delayed to allow more time to develop a plan to address the solar array issue.

By November 1, NASA managers decided to delay the EVA further to conduct additional analysis and give the crew more time to prepare tools and review procedures. Former President George H. W. Bush and his wife, Barbara, visited Mission Control at Johnson Space Center and spoke with the crew via video link. The day was spent configuring tools for the solar array repair, including assembling a device referred to as a "cufflink." The cufflink, made of two wires with tabs at each end, was designed to offload structural stress from the damaged hinge of the solar array by transferring the load through reinforced holes in the array panels. The repair

plan involved using the station's robotic arm and the shuttle's Orbital Boom Sensor System (OBSS) to extend Parazynski's reach as he worked from a foot restraint at the end of the system, with Wheelock assisting from the robotic arm controls.

On November 2, both shuttle and station crews reviewed the EVA procedures with ground teams and completed final preparations. The upcoming EVA would mark the first operational use of the OBSS as a work platform—a technique tested during STS-121 but never before used in an actual repair operation. The crews were ready to attempt the complex and delicate task of repairing the solar array, an essential component of the station's power system.

On October 28, 2007, the crew aboard Space Shuttle Discovery and the International Space Station (ISS) awoke at 05:08 UTC, ready to embark on their second extravehicular activity (EVA) of the mission. At 09:32 UTC, astronauts Scott Parazynski and Daniel Tani began their spacewalk, thirty minutes ahead of schedule. Their primary task was to disconnect the P6 truss from the Z1 truss, allowing the ISS robotic arm, operated by Stephanie Wilson and Doug Wheelock, to move the truss to a parking position for the night. Parazynski lightened the mood by jokingly telling Wilson, "Don't drop it!" as the massive structure was delicately maneuvered.

Following this, Parazynski moved on to install handrails on the newly delivered Harmony module, while Tani was assigned to inspect components as directed by mission managers. One of his inspections focused on the Crew and Equipment Translation Aid (CETA) cart, suspected of causing damage to Rick Mastracchio's glove during the previous mission, STS-118. After close examination, Tani confirmed that the handrails were intact and had no sharp edges that could explain the glove tear.

Tani's next task was to investigate the Solar Alpha Rotary Joint (SARJ), a critical mechanism responsible for rotating the station's solar arrays to track the Sun. The joint had been experiencing unusual vibrations for over a month. After removing the protective cover, Tani observed metal shavings, wear patterns, and discoloration on one of the rings within the joint. He collected samples of the metal shavings, noting that they resembled iron filings reacting to a magnet. These samples were returned to Earth for further analysis. Based on Tani's findings, mission managers revised plans for the following EVA, assigning Parazynski to inspect the port SARJ for comparison.

The following day, October 29, the crew participated in a joint news conference and posed for a group photo before continuing their work. Using the station's robotic arm, they transferred the P6 truss from its overnight location to Discovery's robotic arm. The station's arm was then repositioned in preparation for attaching the P6 truss to the P5 truss during the next EVA.

Meanwhile, ISS Commander Peggy Whitson conducted an experiment on the metal shavings collected by Tani. Using a magnet under a slip of paper, she confirmed that the particles were ferrous, ruling out the possibility that they came from aluminized Mylar thermal covers. As a result, mission planners extended the mission timeline by one day to allow for a more detailed inspection of the SARJ. The new plan called for Discovery to undock from the ISS on November 5, with a landing on November 7.

On October 30, the crew successfully completed the installation of the P6 solar arrays onto the P5 truss during their third EVA. The spacewalk, which began at 08:45 UTC and ended at 15:53 UTC, saw the deployment of the 2B solar array without any issues. However, while deploying the 4B array, the crew noticed a tear in a small section of the array. Despite this, the array was 90% unfurled and still generated 97% of its expected power. ISS Program Manager Mike Suffredini announced that high-resolution images of the array were being analyzed overnight to determine the best course of action.

On October 31, the crew continued their work, including preparing for the fourth EVA and installing a Zero Gravity Stowage Rack in the Harmony module. They also participated in two public affairs events, fielding questions from news agencies and receiving calls from dignitaries, including Italian President Giorgio Napolitano. Later in the day, mission control informed the crew that the fourth EVA would be delayed to allow more time to develop a plan for repairing the damaged solar array.

By November 1, NASA managers decided to postpone the EVA further, giving the crew additional time to prepare the necessary tools and review procedures. That day, former President George H. W. Bush and his wife, Barbara, visited Mission Control and spoke with the crew via video link. The rest of the day was spent assembling and configuring tools for the solar array repair. The key tool, a "cufflink," was designed to relieve structural stress on the damaged hinge of the array. Made of two wires with tabs on each end, the cufflink would be threaded through reinforced holes in the array panels, transferring the load from the hinge to the cufflink.

On November 2, both the shuttle and ISS crews reviewed the EVA procedures and completed final preparations. This EVA would mark the first operational use of the Orbiter Boom Sensor System (OBSS) as a work platform, a technique previously tested during STS-121. The crew was ready to undertake the complex and delicate repair of the solar array, which was critical to the station's power supply.

November 3 was a significant day for the mission. Awakened by the theme music from *Star Wars*, the crew began preparations for the fourth EVA, which started at 10:03 UTC. Parazynski, secured to the OBSS, was carefully maneuvered to the damaged solar array by Wheelock. Working slowly and methodically, Parazynski installed five cufflinks to relieve the strain on the array's hinges. By 15:23 UTC, the crew inside the ISS successfully deployed the array, and Parazynski completed additional inspections of both the 2B and 4B arrays before concluding the EVA at 17:22 UTC.

This marked the end of the mission's spacewalks, with Parazynski completing his seventh career EVA and achieving a total spacewalking time of 47 hours and 5 minutes, placing him fifth overall in EVA duration. Wheelock, who had completed three EVAs during the mission, logged a total of 20 hours and 41 minutes. The total EVA time for STS-120 was 27 hours and 14 minutes.

With the final spacewalk completed, the crew

spent the next day, November 4, transferring cargo and preparing for Discovery's undocking from the ISS. After a farewell ceremony, the hatches were closed at 20:03 UTC. Clayton Anderson, who had spent 137 days aboard the ISS as part of Expedition 16, expressed his gratitude to the ground teams for their support.

During the STS-120 mission, astronaut Paolo Nespoli played a crucial role in the assembly of the International Space Station (ISS) while also conducting valuable scientific research as part of the European Space Agency's (ESA) Esperia mission. This mission, named after the ancient Italian name for the western land, reflected the collaboration between the European and international space communities. Nespoli's contributions included a series of experiments aimed at advancing scientific knowledge in various fields, many of which were sponsored by ESA and the Italian Space Agency (ASI).

Among the experiments conducted, three focused on human physiology: Chromosome-2, Neocytolysis, and HPA (Hypobaric Hypoxia Adaptation). These studies were designed to deepen our understanding of the effects of space travel on the human body, particularly in the context of long-term space habitation. Chromosome-2 aimed to explore the effects of space radiation on chromosomes, potentially uncovering crucial information about genetic stability during extended missions. Neocytolysis studied the destruction of newly formed red blood cells in microgravity, providing insights into anemia experienced by astronauts. HPA examined how the human body adapts to low oxygen environments, mimicking conditions astronauts may experience in space.

In addition to these physiology experiments, Nespoli also conducted two biological studies: FRTL-5 and SPORE. Sponsored by the Italian Space Agency, these experiments were essential for understanding cellular behavior in space. FRTL-5, focused on thyroid cells, aimed to observe how microgravity influences cellular processes critical to human health. SPORE investigated the resistance of spores to the extreme conditions of space, shedding light on potential risks to astronauts and spacecraft equipment from microbial contamination.

Simultaneously, NASA prepared for contingency operations in the event of an emergency during the STS-120 mission. The Contingency Shuttle Crew Support mission, designated STS-320, was developed as a rescue plan in case Space Shuttle *Discovery* became disabled. STS-320 would have been a modified version of the STS-122 mission, with a launch planned no later than January 5, 2008, if needed. A subset of four astronauts from the STS-122 crew was designated to conduct the rescue. This contingency plan reflected NASA's commitment to ensuring crew safety during critical space missions, demonstrating the inherent risks and meticulous planning required for human spaceflight.

On November 5, Discovery undocked from the ISS at 10:32 UTC. The shuttle performed a fly-around to photograph the station's new configuration before completing a final separation burn. The following day, the crew prepared for landing, checking the orbiter's systems and refining the trajectory for the landing opportunities at Kennedy Space Center.

On November 7, Discovery made a successful landing at Kennedy Space Center, touching down at 13:01 EST. The mission, which had lasted 15 days, 2 hours, and 23 minutes, covered 6.25 million miles in 238 orbits. The successful delivery and installation of the Harmony module marked a significant step in the continued expansion of the ISS, setting the stage for future assembly missions

STS-122: Expanding the International Space Station's Capabilities

STS-122, flown by Space Shuttle *Atlantis*, marked the 24th mission to the International Space Station (ISS) and the 121st overall flight of the Space Shuttle program. Launched on February 7, 2008, after multiple delays, the primary objective of this mission, designated ISS-1E by the ISS program, was to deliver the European Space Agency's (ESA) *Columbus* laboratory to the ISS. This mission represented a significant milestone in European space exploration, as the *Columbus* module became a vital platform for scientific research aboard the station.

From the left (front row) are astronauts Stephen N. Frick, commander; European Space Agency's (ESA) Leopold Eyharts; and Alan G. Poindexter, pilot. From the left (back row) are astronauts Leland D. Melvin, Rex J. Walheim, Stanley G. Love and European Space Agency's (ESA) Hans Schlegel, all mission specialists. Eyharts will join Expedition 16 in progress to serve as a flight engineer aboard the International Space Station.

The mission also facilitated crew rotations for Expedition 16. French astronaut Léopold Eyharts, representing ESA, replaced NASA astronaut Daniel M. Tani, who returned to Earth aboard *Atlantis* after his extended stay on the ISS.

Preparations for the mission began with the arrival of the external tank (ET-125) at Kennedy Space Center on September 14, 2007, after traveling by barge from the Michoud Assembly Facility in Louisiana. The tank underwent inspection and modifications to its liquid oxygen feedline bracket before being attached to the solid rocket boosters on October 18, 2007. *Atlantis* was then moved to the Vehicle Assembly Building (VAB) and rolled out to launch pad 39A on November 10, 2007. Two days later, the *Columbus* module was loaded into the orbiter's payload bay.

A terminal countdown demonstration test was successfully completed on November 20, 2007, but complications soon followed. During the final Flight Readiness Review on November 30, NASA confirmed the shuttle's launch date for December 6, 2007. However, during fueling on that day, two of the four liquid hydrogen engine cutoff (ECO) sensors failed to respond correctly,

leading Launch Director Doug Lyons to postpone the launch. This system was critical, as it prevented the engines from running dry during ascent.

Further troubleshooting revealed wiring issues within the external tank, prompting NASA to consider flying with only two out of the four ECO sensors operational. However, after continued sensor malfunctions during testing, NASA managers decided to delay the launch further to avoid any risks.

The second launch attempt on December 9, 2007, also failed when a third ECO sensor malfunctioned during fueling. This led to a complete scrub of the December launch window, and NASA rescheduled the mission for January 2008. In an effort to diagnose the sensor issue, engineers conducted a tanking test on December 18, utilizing time-domain reflectometry (TDR) equipment to gather data. This testing pinpointed the problem to the liquid hydrogen feed-through connector.

Meanwhile, an unrelated issue arose with the shuttle's radiator retract hose, which was found to be bent in an unusual shape. Although not deemed a mission-critical issue, NASA engineers closely monitored the situation and designed a tool to guide the hose back into place during pre-launch procedures.

After resolving the technical difficulties, *Atlantis* successfully launched on February 7, 2008, at 14:45 EST (19:45 UTC). The mission unfolded smoothly, with main engine cutoff (MECO) occurring at 19:54 UTC, followed by the separation of the external tank and an OMS-2 engine burn to circularize the shuttle's orbit and align it with the ISS. The crew then opened the payload bay doors, deployed the Ku-band antenna, activated the robotic arm, and downlinked external tank separation footage to NASA for analysis.

The STS-122 crew, commanded by Stephen Frick, included several astronauts on their first spaceflights, such as pilot Alan G. Poindexter and mission specialists Leland D. Melvin and Stanley G. Love. ESA astronaut Hans Schlegel, also on his second and final spaceflight, was instrumental in the installation and activation of the *Columbus* laboratory.

The primary payload, the *Columbus* laboratory, weighed over 12,000 kilograms (26,625 pounds) and housed several advanced science modules. These included the Biolab, the Fluid Science Laboratory (FSL), the European Drawer Rack (EDR), and the European Physiology Modules (EPM), all essential for conducting biological and fluid physics experiments. STS-122 also carried the Solar Monitoring Observatory (SOLAR), the European Technology Exposure Facility (EuTEF), and a new Nitrogen Tank Assembly (NTA). These payloads were stored in the ICC-Lite payload rack and mounted in the shuttle's cargo bay. Additionally, the mission carried a spare Drive Lock Assembly (DLA) to support potential repairs to the malfunctioning Solar Alpha Rotary Joint (SARJ) on the ISS.

On board *Atlantis* during the STS-122 mission, the Official Flight Kit (OFK) carried three symbolic green starting flags provided by NASCAR. These flags commemorated two significant milestones: the 50th running of the Daytona 500 on February 17, 2008, and NASA's 50th anniversary on October 1, 2008. Once returned to Earth, one of these flags would be publicly displayed at the Daytona International Speedway in Florida, another would be presented to Ryan Newman, the winner of the 2008 Daytona 500, and the third would be used by NASA in its anniversary celebrations.

The crew's seat assignments during launch and landing were as follows:

Flight Deck:
Seat 1: Stephen Frick (Commander)
Seat 2: Alan G. Poindexter (Pilot)
Seat 3: Leland D. Melvin (Mission Specialist)
Seat 4: Rex J. Walheim (Mission Specialist)
Mid-Deck:
Seat 5: Hans Schlegel (Mission Specialist)
Seat 6: Stanley G. Love (Mission Specialist)
Seat 7: Léopold Eyharts (Flight Engineer) during launch and Daniel M. Tani for landing.

The second day of the mission was dedicated to preparing *Atlantis* for docking with the ISS, scheduled for the following day. The crew installed the centerline camera and extended the orbiter docking system ring, essential for a smooth connection with the space station. A significant portion of the day was spent inspecting the shuttle's thermal protection system using the Orbiter Boom Sensor System (OBSS), which played a vital role in ensuring that the shuttle's heat shield was intact after launch.

To adjust their orbit and prepare for docking, the crew executed a burn using the Orbital Maneuvering System (OMS) engines early in the day. During morning interviews with CBS and NBC, Expedition 16 Commander Peggy Whitson shared her excitement about the upcoming arrival of the *Columbus* module, coincidentally coinciding with her birthday. She remarked, "My present is a new module that we're going to install on the station, I'm really looking forward to it."

In the afternoon mission status briefing held at Johnson Space Center, Lead Shuttle Flight Director Mike Sarafin confirmed that the mission was proceeding without any technical issues. The shuttle was on track for a scheduled docking at 17:25 UTC on February 9. Sarafin also noted that *Atlantis* had enough consumables to extend the mission if needed, though a final decision would be made after reviewing the inspection data collected by flight day five. If approved, the extra day would be added after the third spacewalk, scheduled for flight day nine.

John Shannon, Chairman of the Mission Management Team (MMT), gave the official "go" for docking after the first on-orbit meeting. Initial imagery from the launch ascent showed no significant concerns, with only one small piece of foam potentially coming into contact with the shuttle approximately 440 seconds after launch. Shannon assured that this was not a major issue, as the foam fragment would not have had enough energy to cause any substantial damage to the orbiter.

Looking ahead, Shannon explained that the team would continue to analyze data from the upcoming Rendezvous Pitch Maneuver (RPM), which would be conducted before docking, along with imagery from the OBSS survey. A decision on whether further inspections of the shuttle's thermal protection system would be necessary was expected to be made during the Mission Management Team meeting on flight day three. Shannon expressed optimism, stating that improvements to the external tank's design had been "phenomenal," and the initial data was promising.

On February 9, 2008, the crew of Space Shuttle Atlantis worked through the morning rendezvous timeline as they approached the International Space Station (ISS) on the third day of their mission. The day involved several precise adjustment burns using the orbiter's engines to refine the shuttle's path toward the ISS. Between 16:24 and 16:31 UTC, Atlantis executed the Rendezvous Pitch Maneuver, a critical step that allowed the Expedition 16 crew aboard the station to use high-resolution cameras to inspect Atlantis' thermal protection system. During this inspection, additional images were taken of the starboard Orbital Maneuvering System (OMS) pod due to the appearance of a raised thermal blanket in that area, deemed an "area of interest."

At 17:17 UTC (12:17 EST), Atlantis docked with the ISS at the new forward position of the Pressurized Mating Adapter on the Harmony module, marking the shuttle's first docking at this location. Following the docking, the crew conducted leak checks before opening the hatches between the shuttle and station at 18:40 UTC. A warm exchange of greetings took place between the crews, followed by a mandatory safety briefing. Afterward, the teams began their tasks, including using the station's robotic arm to grapple the Orbiter Boom Sensor System (OBSS) and hand it off to Atlantis' robotic arm for future activities. The crew swap between Daniel Tani and Léopold Eyharts was officially completed that evening as Tani became a member of the STS-122 crew, and Eyharts took on his new role as a flight engineer for Expedition 16.

During the day's mission status briefing, Flight Director Mike Sarafin reported a minor issue with one of Atlantis' three General Purpose Computers (GPC), which failed to start up correctly before the rendezvous. However, this malfunction did not affect the operation. Additionally, NASA was investigating a slight tear in the thermal protection blanket on the starboard OMS pod.

At 20:14 UTC, ground controllers informed the crew of a 24-hour delay to the first extravehicular activity (EVA-1), initially scheduled for Flight Day 4. Stanley Love would replace Hans Schlegel for this spacewalk, slated for Monday. Mission Management Team (MMT) Chairman John Shannon cited a "crew medical issue" for the change, though no details were disclosed for privacy reasons. Schlegel was reported to have lost his voice, and since communication is vital during an EVA, Love was selected to replace him. The shuttle had enough consumables to extend the mission by up to two days, and NASA was exploring power conservation measures to accommodate a potential extension.

The crew's first full day aboard the ISS was dedicated to performing a focused inspection of the thermal blanket on the starboard OMS pod, preparing for the upcoming spacewalk, and transferring supplies between the shuttle and station. ESA confirmed that Hans Schlegel was the crewmember with the medical issue but assured the public that it was not serious and did not affect the health of any other crewmembers.

During the day, Tani and Eyharts spent hours going through station familiarization procedures, helping Eyharts become acquainted with the location of key items on the station. Meanwhile, Love, Walheim, and Schlegel reviewed EVA procedures. In preparation for the upcoming spacewalk, Love and Walheim spent the night in the Quest airlock.

In the mission status briefing, MMT Chairman John Shannon reported that the thermal protection system inspections were progressing rapidly, with no concerns regarding the orbiter's underside or reinforced carbon-carbon surfaces on the wings and nose. The starboard OMS pod blanket was set to be further evaluated, with a decision expected at Monday's MMT meeting.

February 11 marked the mission's first spacewalk. Love and Walheim suited up and began the EVA slightly ahead of schedule at 14:13 UTC (09:13 EST). Inside the shuttle and station, pilot Alan Poindexter and Mission Specialist Hans Schlegel provided support.

By 19:53 UTC (14:53 EST), the spacewalkers had completed preparations for the unberthing of the Columbus module from the shuttle's payload bay. Using the station's robotic arm, operated by Leland Melvin, the European laboratory was carefully extracted from the payload bay and installed on the ISS. At 21:44 UTC, Eyharts and Melvin declared Columbus officially part of the ISS. The EVA concluded at 22:11 UTC, lasting 7 hours and 58 minutes.

The crew spent February 12 focused on activating and outfitting the newly installed Columbus module. After completing leak checks overnight, the module was powered up in "Berth Survival Mode," a minimal configuration that maintains basic functionality. Schlegel and Eyharts were the first to enter the module, marking a proud moment for Europe. Eyharts acknowledged the contributions of all those involved in the development of Columbus, from space agencies to industry and the public.

Throughout the day, crewmembers continued working on the module's water, thermal control, and command systems. Ground controllers later informed Commander Frick that the MMT had cleared the OMS pod blanket for reentry as-is, with no safety concerns.

On February 13, both crews prepared for the second EVA of the mission. Assisted by Commanders Whitson and Frick, Walheim and Schlegel suited up for their spacewalk, which officially began at 14:27 UTC (09:27 EST). During the EVA, they replaced a depleted Nitrogen Tank Assembly on the P1 truss with a new tank brought up by Atlantis. The spacewalk ended at 21:12 UTC, after 6 hours and 45 minutes of work.

Lead ISS Flight Director Sally Davis announced an additional docked day extension, allowing the crew to continue their work on the station. The entire thermal protection system was cleared for reentry, with no outstanding issues.

February 14 was a relatively light day, providing the crew with some rest after a busy week of activities. In addition to performing maintenance tasks, both crews participated in media interviews, including a special call from German Chancellor Angela Merkel, who congratulated Eyharts and Schlegel on the successful delivery and installation of the Columbus module.

That evening, the crews reviewed procedures for the final EVA, and Walheim and Love spent the night in the airlock to prepare for the next day's spacewalk.

The final EVA of the mission took place on February 15. Walheim and Love exited the airlock at 13:07 UTC (08:07 EST) and worked on tasks that included outfitting Columbus and preparing the station for future operations. The EVA concluded at 20:32 UTC (15:32 EST).

While the crew continued outfitting the Columbus module, Atlantis performed a 36-minute burn of its propulsion system, raising the station's altitude by 2.2 kilometers in preparation for the next mission, STS-123.

On February 17, the hatches between Atlantis and the ISS were closed at 18:03 GMT, marking the end of the joint mission. The crew prepared for their departure the following day.

Atlantis undocked from the ISS at 09:24 UTC (04:24 EST) on February 18, following a smooth departure from the station.

On February 19, the crew made final preparations for landing. The payload bay doors were closed at 10:14 UTC on February 20, and NASA cleared Atlantis for its first landing opportunity at Kennedy Space Center (KSC). The deorbit burn began at 13:00 UTC, with entry interface occurring at 13:35 UTC.

Atlantis touched down on Runway 15 at KSC at 9:07 EST (14:07 UTC) on February 20, concluding the mission after 12 days, 18 hours, and 21 minutes in space.

STS-123

STS-123, flown by Space Shuttle *Endeavour*, was a pivotal mission to the International Space Station (ISS), marking the 1J/A ISS assembly mission. Initially scheduled for launch on February 14, 2008, the mission was delayed due to the postponement of STS-122, and *Endeavour* ultimately launched on March 11, 2008. This was the 25th shuttle mission to the ISS, with a duration of 15 days, 18 hours, making it the longest stay by a shuttle at the ISS at the time. The mission delivered critical components to the station, including the first module of Japan's Kibo laboratory, the Japanese Experiment

Logistics Module - Pressurized Section (ELM-PS), and the Canadian-built Special Purpose Dexterous Manipulator (SPDM), commonly known as Dextre.

The mission was commanded by Dominic L. Pudwill Gorie, on his fourth and final spaceflight. Pilot Gregory H. Johnson, and Mission Specialists Robert L. Behnken and Michael Foreman, all made their first spaceflights, with Behnken and Foreman later performing several crucial spacewalks during the mission. Veteran astronauts Richard M. Linnehan and Takao Doi of Japan's JAXA rounded out the crew, with Linnehan conducting his fourth and final spaceflight. Additionally, Garrett Reisman joined the crew as part of ISS Expedition 16, marking his first spaceflight. French astronaut Léopold Eyharts, also a member of Expedition 16, flew on his second and final mission, assisting with the assembly and operations aboard the ISS.

Mission Objectives and Payload

STS-123 had the significant task of delivering and installing essential ISS components. Among the key payloads was the pressurized section of Kibo, the first Japanese module attached to the ISS, weighing approximately 8,484 kilograms (18,704 lbs). Additionally, the mission delivered the Special Purpose Dexterous Manipulator (SPDM) or Dextre, a robotic system designed to perform delicate tasks on the station's exterior. This mission also marked the first full use of the Station-to-Shuttle Power Transfer System (SSPTS), allowing the shuttle to draw power from the ISS, which extended its stay in orbit.

On March 11, 2008, at 2:28 AM EDT, Space Shuttle *Endeavour* launched from Kennedy Space Center's Launch Complex 39A. During the ascent, minor anomalies occurred, including a switch to the backup controller for the Flash Evaporator System and a control card failure affecting thruster instrumentation. Despite these issues, the mission proceeded without further complications.

Upon reaching orbit, the crew conducted a thorough inspection of *Endeavour's* thermal protection system using the shuttle's Orbiter Boom Sensor System (OBSS), a post-*Columbia* disaster protocol. Images sent back to mission control revealed no damage. As the shuttle closed in on the ISS, Commander Gorie executed a precise 360-degree backflip, known as the Rendezvous Pitch Maneuver, to allow for photographic surveillance of the shuttle's underside.

Docking and Key Installations

On March 13, *Endeavour* docked with the ISS at 3:49 UTC, and the hatches between the shuttle and station were opened approximately two hours later. The Canadian-built Dextre, temporarily docked on a pallet, was transferred to the Mobile Base Station for assembly over the next few days. Astronauts Linnehan and Reisman embarked on the first of five spacewalks to begin the complex assembly process of Dextre.

On March 14, the Japanese Logistics Module (JLP) was successfully removed from *Endeavour's* cargo bay and attached to the Harmony module of the ISS. The JLP provided a critical new capacity for storage and experimentation on the station.

STS-123 featured five spacewalks (EVAs), totaling nearly 30 hours of extravehicular activity. These spacewalks were crucial for installing and activating Dextre, which encountered initial power-up issues that ground engineers resolved. Despite some difficulties, including the need for a prybar to remove transport bolts, the astronauts successfully assembled the robot.

Throughout the mission, the shuttle and station crew worked collaboratively on various tasks, including outfitting the new Japanese module, calibrating Dextre, and installing spare parts on the station's truss structure. The astronauts also tested a new repair method for shuttle thermal tiles, using a tool designed to dispense a specialized ablative material into test tiles, which were later returned to Earth for analysis.

The mission saw challenges, including issues with Dextre's temporary power cable and minor difficulties during spacewalks, such as stubborn bolts and misfitting equipment. Nevertheless, the astronauts adapted and completed their tasks, including transferring critical supplies to the ISS. One notable event was the replacement of a failed Remote Power Control Module on the station's truss during the fourth spacewalk, showcasing the intricate work required to maintain the station's functionality.

After a successful mission, *Endeavour* undocked from the ISS on March 25, 2008, following a brief delay due to solar array issues on the station. The next day, the shuttle prepared for its return to Earth, with the crew testing the orbiter's thrusters and control surfaces. Léopold Eyharts, having completed his stay aboard the ISS, joined the shuttle crew for the return journey, occupying a special recumbent seat designed to ease the physical stress of reentry after extended periods in microgravity.

On March 27, 2008, *Endeavour* made its 16th night landing at Kennedy Space Center, marking the end of a successful mission. The landing was accompanied by the characteristic glow of the shuttle's auxiliary power units (APUs), a familiar sight for those monitoring the shuttle's descent. STS-123 had achieved its objectives, delivering vital components to the ISS, conducting significant spacewalks, and setting the stage for future missions to continue the assembly and expansion of the International Space Station.

STS-124: Completing Japan's Kibo Laboratory

On April 26, 2008, the Space Shuttle *Discovery* was transferred from its processing bay in the Orbiter Processing Facility (OPF) to the Vehicle Assembly Building (VAB) at Kennedy Space Center. Inside the VAB,

Discovery was lifted vertically and, on April 28, 2008, mated with its external fuel tank and solid rocket boosters, completing a critical step in the launch preparation process. Following a week of further preparations, on May 2, 2008, at 23:47 EDT, the entire shuttle stack was carefully transported to Launch Complex 39A (LC-39A) atop the Mobile Launch Platform. The shuttle arrived and was secured at the pad by 06:06 EDT on May 3, 2008, transported by the massive Crawler Transporter, a vehicle designed specifically for moving the shuttle and its platform to the launch pad.

Concurrent with these activities, the payload for the mission—the Japanese Experiment Module (JEM), named *Kibō*—was transported to the Payload Changeout Room at the launch pad on April 29, 2008. On May 5, *Kibō* was installed into *Discovery*'s payload bay, marking the mission's payload integration phase. The STS-124 crew arrived at Kennedy Space Center on May 6, 2008, for a three-day Terminal Countdown Demonstration Test (TCDT). After completing this critical launch rehearsal, the crew returned to Johnson Space Center on May 9, 2008. Following successful reviews and testing, *Discovery* was cleared for launch, scheduled for May 31, 2008.

On May 31, 2008, at precisely 21:02 UTC (17:02 EDT local time), *Discovery* lifted off from LC-39A. This marked the beginning of the STS-124 mission, which was also referred to as ISS-1J, under the International Space Station (ISS) program. The launch was executed with minimal debris shed from the external fuel tank, a key concern following the *Columbia* disaster in 2003.

Speaking before liftoff, Commander Mark Kelly emphasized the mission's significance, stating, "While we've all prepared for this event today, the discoveries from *Kibō* will definitely offer hope for tomorrow." The launch proceeded smoothly, but an inspection of the launch pad revealed significant damage to one of the trenches designed to channel flames away from the shuttle during lift-off. The damage was attributed to the carbonation of epoxy and corrosion of steel anchors holding the trench's refractory bricks in place, exacerbated by hydrochloric acid—a byproduct of the shuttle's solid rocket boosters.

The STS-124 mission was vital to the continued expansion of the ISS. Its primary objective was to deliver *Kibō*, the largest laboratory module for the station, built by the Japan Aerospace Exploration Agency (JAXA). The mission also included the rotation of an ISS crew member, with astronaut Garrett Reisman returning to Earth, replaced by Gregory Chamitoff as part of Expedition 17.

On June 1, 2008 (Flight Day 2): the crew performed a limited inspection of *Discovery*'s thermal protection system using the shuttle's robotic arm. This was a standard post-launch procedure to ensure no critical damage had occurred during ascent. The crew also prepared the orbiter for docking with the ISS.

On June 2, 2008 (Flight Day 3), at 18:03 UTC, *Discovery* successfully docked with the ISS. The hatch between the shuttle and the station was opened at 19:36 UTC, allowing the crew to transfer between the two vehicles. Gregory Chamitoff officially replaced Garrett Reisman on the ISS crew, marking the start of his role in Expedition 17.

On June 3, 2008 (Flight Day 4), the mission's first spacewalk, led by Mike Fossum and Ron Garan, lasted six hours and forty-eight minutes. During this extravehicular activity (EVA), the astronauts retrieved the Orbiter Boom Sensor System (OBSS), inspected components of the Solar Alpha Rotary Joint, and prepared the *Kibō* module for installation on the station. Inside the ISS, astronauts Karen Nyberg and Akihiko Hoshide used the station's robotic arm to move *Kibō* from *Discovery*'s payload bay and attach it to the Harmony node of the ISS.

On June 4, 2008 (Flight Day 5), the hatch to the newly installed *Kibō* laboratory was opened at 21:05 UTC, marking a major milestone in the mission. The crew also undertook a crucial repair of the ISS's malfunctioning toilet, ensuring the continued functionality of the station's life-support systems.

On June 5, 2008 (Flight Day 6), Fossum and Garan conducted their second EVA, which lasted seven hours and eleven minutes. During this spacewalk, they replaced a nitrogen tank used for maintaining the station's ammonia cooling system and collected debris from the solar arrays for analysis. The EVA was a key step in maintaining the ISS's operational health.

On June 6, 2008 (Flight Day 7), the crew completed a delicate maneuver to relocate the *Kibō* Logistics Module from the Harmony node to its final position attached to the Pressurized Module, further advancing the construction of the ISS.

On June 7, 2008 (Flight Day 8), astronauts Hoshide and Nyberg activated *Kibō*'s robotic arm, moving its joints for the first time. This system was essential for conducting future experiments and handling cargo outside the station. With the mission at its midpoint, the crew reflected on the fast pace of the mission, with Nyberg commenting, "The week has gone way too fast."

On June 8, 2008 (Flight Day 9), Fossum and Garan completed the mission's third and final spacewalk. The pair replaced an empty nitrogen tank and collected samples of debris from the solar array, further ensuring the station's ongoing capability to support long-duration missions.

On June 9, 2008 (Flight Day 10), *Kibō*'s robotic arm was extended to its full length of 33 feet, and the crew performed tests on all six joints. Additionally, the storage unit of *Kibō* was opened, completing the module's full activation.

On June 11, 2008 (Flight Day 12), *Discovery* undocked from the ISS at 11:42 UTC, concluding its mission to the station. The shuttle then performed a flyby of the ISS to allow for photo documentation of the completed work, including the installation of *Kibō*. Commander

Kelly expressed gratitude to the station crew and left with the hope that they had left "a better, more capable space station than when we arrived."

On June 14, 2008 (Flight Day 15, Landing), after completing deorbit preparations, *Discovery* performed a flawless re-entry and landed at Kennedy Space Center at 11:15 EDT on June 14, 2008. This concluded STS-124, a mission that significantly advanced the capabilities of the ISS and marked a major milestone in international space collaboration.

The STS-124 mission was a resounding success, delivering *Kibō*, the largest laboratory module of the ISS, and performing essential maintenance to ensure the station's continued functionality.

STS-126

STS-126 marked the 124th NASA Space Shuttle mission and the 22nd flight of *Endeavour* (OV-105) to the International Space Station (ISS). The mission, launched in November 2008, played a critical role in resupplying the ISS and conducting crucial repairs to its systems. Commanded by Christopher J. Ferguson, the crew also included Pilot Eric A. Boe and Mission Specialists Donald R. Pettit, Stephen G. Bowen, Heidemarie M. Stefanyshyn-Piper, Robert S. Kimbrough, and Sandra H. Magnus. Magnus was part of the ISS Expedition 18 crew, joining Gregory E. Chamitoff, who had been aboard the ISS since the previous mission.

Originally, Joan E. Higginbotham had been scheduled to fly on STS-126 but left NASA in 2007 for a private-sector position. Stephen G. Bowen replaced her and was reassigned from STS-124 to allow for crew rotations, including the swap of ISS Flight Engineer Gregory E. Chamitoff with Garrett E. Reisman.

Endeavour began its journey to launch pad 39B on September 11, 2008, when it was moved from the Orbiter Processing Facility to the Vehicle Assembly Building at Kennedy Space Center. It rolled out to pad 39B on September 18 and was positioned as the standby rescue mission, or Launch On Need (LON) flight, for STS-125, which was tasked with servicing the Hubble Space Telescope. This marked the 18th instance of two flight-ready shuttles positioned at launch pads simultaneously, with *Atlantis* at 39A.

However, an issue with the Hubble Space Telescope led NASA to postpone STS-125. This delay bumped STS-126 into the next available flight slot. On October 23, 2008, *Endeavour* was moved from pad 39B to 39A, where it was prepared for its new mission. The payload, including the Multi-Purpose Logistics Module (MPLM) *Leonardo*, was delivered to pad 39A on October 22.

The official countdown for STS-126 began on November 11, 2008. The crew arrived at Kennedy Space Center on November 12 to finalize preparations. On November 13, the Mission Management Team, chaired by LeRoy Cain, gave the official "go" for launch. Weather conditions were favorable, with a 70% chance of acceptable conditions. The next day, the external tank was filled with liquid hydrogen and oxygen, and the crew suited up for launch. After a slight delay due to a minor issue with the white room closeout door, NASA Launch Director Michael Leinbach gave the green light, wishing the crew "Good luck, Godspeed and have a happy Thanksgiving in orbit."

At precisely 00:55 UTC on November 14, *Endeavour* lifted off, marking the start of the STS-126 mission. External tank separation occurred smoothly eight minutes into the flight.

STS-126 was scheduled as a 16-day mission with four planned spacewalks, primarily focused on servicing the Solar Alpha Rotary Joints

(SARJ), which had exhibited anomalies since August 2007. The SARJs are critical for the ISS's ability to rotate its solar arrays to track the Sun. Both the port and starboard SARJs were lubricated, and the remaining 11 trundle bearings on the starboard SARJ were replaced. This was vital to restoring the station's ability to generate power efficiently.

In addition to the repairs, the *Leonardo* MPLM carried over 6,400 kg (14,100 lb) of supplies, including two new crew quarters racks, a second galley, a new Waste and Hygiene Compartment, and the advanced Resistive Exercise Device (aRED) for crew physical health. These upgrades improved the living conditions for long-term space habitation. The module also contained scientific equipment, including the GLACIER cryogenic freezer for preserving experiments, as well as various hardware and spare parts needed to maintain ISS systems.

STS-126 carried the signatures of over 500,000 students from the 2008 Student Signatures in Space program, a collaboration between NASA and Lockheed Martin. This initiative, active since 1997, celebrates Space Day each year by allowing students from more than 500 schools to sign giant posters. These signatures are then scanned onto a disk, which is flown aboard a space shuttle mission. The STS-126 mission marked the eighth time student signatures were sent into space, with the program's inaugural flight beginning on STS-86. This unique outreach effort inspired a generation of students to connect with space exploration and contributed to broader educational goals.

The Agricultural Camera (AgCam) was also aboard STS-126, installed in the Destiny module of the International Space Station. Developed by students and faculty at the University of North Dakota, AgCam serves both scientific and educational purposes. The camera, which captures visible and infrared light images of Earth, focuses on monitoring crops, grasslands, forests, and wetlands, primarily in the Great Plains and Rocky Mountain regions. From their university campus, students operate the AgCam, collaborating with NASA engineers and astronauts to capture data that supports agricultural producers. The information provided assists farmers in managing their land, monitoring the environment, and improving disaster response, including flood management and wildfire mapping. AgCam is an excellent example of how space technology can be applied to practical, Earth-based issues while simultaneously providing valuable learning opportunities for students.

STS-126 also carried an important biological experiment: the first bovine embryos and porcine embryonic stem cells ever flown on an American spacecraft. This experiment, a joint effort by ZeroGravity Inc., the University of Florida, and the USDA Agricultural Research Service, aimed to study the effects of microgravity on embryonic development. The data collected from this experiment would offer valuable insights into how space environments impact biological processes, furthering scientific understanding of space biology and its potential applications for future long-duration space missions.

The second day of the mission, November 15, 2008, was dedicated to inspecting *Endeavour's* heat shield. Using the shuttle's robotic arm (Canadarm) and the Orbiter Boom Sensor System (OBSS), the crew conducted a thorough survey of the shuttle's exterior. The collected images were sent to the image analysis team on the ground for detailed evaluation. Additionally, the crew extended the docking ring, installed the centerline camera, and organized the necessary tools in preparation for docking with the ISS the following day. The crew also performed checks on the spacesuits that would be used in the mission's spacewalks.

During a briefing, LeRoy E. Cain, chair of the Mission Management Team, reported that a small piece of thermal blanket had come loose under the left Orbital Maneuvering System (OMS) pod during ascent. However, this area was not exposed to extreme heat during reentry, and no damage was observed. The Ku-band antenna, used for communications, also exhibited two issues, one related to automatic handover from Ku to S-band and another involving antenna "lock" on satellite targets. Both issues were minor, and ground teams were able to manually control the handover and address the antenna drift.

On November 16, 2008, the crew awoke and began final preparations for rendezvous and docking with the ISS. At 19:27 UTC, *Endeavour* conducted its final engine burn to fine-tune its approach to the station. By 21:00 UTC, the shuttle was in position below the station, allowing the ISS crew to photograph the shuttle's underside using 400 mm and 800 mm cameras as part of the standard thermal protection inspection. These high-resolution images were transmitted to NASA's analysis team for further assessment.

Commander Christopher Ferguson manually guided the shuttle through the Rendezvous Pitch Maneuver (RPM), a 360-degree backflip that exposed *Endeavour*'s heat shield to the station's cameras. Once the maneuver was complete, Ferguson smoothly docked *Endeavour* with the ISS at 22:01 UTC. The hatches between the two spacecraft were opened at 00:16 UTC, marking the official start of joint operations. Upon entering the ISS, Ferguson jokingly commented, "Hey, we figured we'd go for a 10-year anniversary party for the space station, so that's what we showed up for," referring to the 10th anniversary of the launch of *Zarya*, the ISS's first module, which was launched on November 20, 1998.

Following greetings and a safety briefing, the crews began cargo transfers and robotic operations. At 02:50 UTC, a symbolic exchange took place as Gregory Chamitoff and Sandra Magnus officially swapped positions. Chamitoff became a mission specialist aboard *Endeavour*, while Magnus transitioned to ISS Flight Engineer for Expedition 18, following the exchange of their Soyuz seatliners.

During a mission status briefing, Cain noted that *Endeavour*'s ascent had been remarkably clean, with no significant damage observed. The thermal blankets and tiles were intact, confirming the earlier suspicion that a piece of ice had detached during launch without posing any risk to the shuttle. The Ku-band antenna issue, while present, had not affected the mission and docking operations proceeded without complications.

After docking *Endeavour* with the International Space Station (ISS), the Expedition 18 members warmly welcomed the crew of STS-126. The two crews immediately began their first major task: transferring the *Leonardo* Multi-Purpose Logistics Module (MPLM) from the shuttle's payload bay to the Earth-facing port of the Harmony module. Using Canadarm2, astronauts Donald Pettit and Robert Kimbrough maneuvered the 12,247 kg (27,000 lb) *Leonardo* module into position. The container was securely attached to Harmony at 18:04 UTC, and after leak checks, the hatch was opened at 23:43 UTC. The crew then took air samples and inspected *Leonardo* for condensation or cargo shifts, finding no issues, and began transferring its contents to the ISS.

Among the cargo were supplies, scientific equipment, and experiments, including a unique study involving spiders and butterflies. This experiment, designed by schoolchildren from Florida, Texas, and Colorado, aimed to compare the webs spun by spiders in microgravity to those created on Earth, alongside observing the behavior of butterflies in space. The insects would later return to Earth for further study.

Later in the day, the crew reviewed procedures for the first spacewalk (EVA), planned for the following day. Mission Specialists Heidemarie Stefanyshyn-Piper and Stephen Bowen prepared for their EVA by performing the "campout" protocol. This involved spending the night in the Quest airlock at a lower air pressure (10.2 psi), allowing their bodies to acclimate to the reduced pressure required for their spacesuits.

During a Mission Management Team briefing, LeRoy E. Cain confirmed that the shuttle's wing leading edge was in good condition, eliminating the need for a focused inspection that had been scheduled for flight day six. This freed up valuable time to work on the station's new water reclamation unit, a critical component for long-duration space missions. Cain also reported that *Endeavour*'s external tank had shown minimal foam loss during ascent, with only three small areas of concern, none of which posed a risk to the orbiter.

As the ISS approached its 10-year anniversary on 20 November 2008, ISS Deputy Program Manager Kirk Shireman took time to reflect on the progress and milestones achieved. By that date, the station had hosted 30 Progress resupply vehicles, 17 Soyuz spacecraft, 27 space shuttle missions, and one Automated Transfer Vehicle (ATV). The station had orbited Earth 57,509 times, covering over 2.1 billion kilometers (1.3 billion miles), and 167 people from 14 nations had visited the station, including the STS-126 crew.

The crew awoke on November 18, 2008, ready to undertake the first spacewalk of the mission. Stefanyshyn-Piper and Bowen suited up early and entered the Quest airlock ahead of schedule. The EVA began at 18:09 UTC, with Stefanyshyn-Piper becoming the first female Lead Spacewalker.

Their primary task was to work on the starboard Solar Alpha Rotary Joint (SARJ), which had been malfunctioning for some time. However, shortly after beginning the work, Stefanyshyn-Piper discovered an issue: a significant amount of grease had leaked inside her tool bag. "I think we had a grease gun explode in the large bag, because there's grease in the bag," she reported. While attempting to clean up the grease, one of her crew lock bags floated away from the station. Despite this, Mission Control reassured her that the floating bag posed no danger to the station or shuttle.

After inventorying the contents of the lost bag, the ground team determined that Bowen's tool bag contained duplicates of the necessary items, allowing the two astronauts to share equipment and continue the EVA as planned. Although this extended the duration of the spacewalk, all major objectives were completed. The lost tool bag, valued at approximately $100,000, later became visible to amateur astronomers as it orbited Earth before eventually burning up upon re-entry.

During the post-EVA briefing, lead ISS Flight Director Ginger Kerrick commented that the exact cause of the bag's loss was unknown, stating, "We don't know if perhaps the hook just came loose inside the bag." She reminded everyone that space operations, despite rigorous precautions, are subject to human error. Kerrick emphasized that future spacewalks would include extra measures to prevent grease guns from inadvertently activating by securing them to the outside of tool bags.

Flight day six was primarily focused on cargo transfer operations. The crews of STS-126 and Expedition 18 worked efficiently throughout the morning to complete the installation of two new crew quarters racks into the Harmony node. These racks, essential for the long-term habitation of the ISS, bore the signatures of the ground team members who helped design and construct them. Simultaneously, another rack containing equipment slated for return to Earth was loaded into the *Leonardo* Multi-Purpose Logistics Module (MPLM).

During the day's Mission Status briefing, Lead ISS Flight Director Ginger Kerrick reported that all racks were successfully installed, and about 25% of the overall cargo transfers had been completed—slightly ahead of schedule. The crews also began the process of activating the Water Recovery System, a vital technology that converts urine and condensate into potable water. After initial system checkouts, samples of processed water would be collected and sent back to Earth with *Endeavour* for analysis.

Later in the day, the crew reviewed procedures for the second spacewalk (EVA), scheduled for the following day. Astronauts Heidemarie Stefanyshyn-Piper and Robert Kimbrough slept in the airlock as part of the "campout" protocol, acclimating their bodies to lower pressure in preparation for the EVA.

On the tenth anniversary of the International Space Station, Stefanyshyn-Piper and Kimbrough conducted the second spacewalk of the mission, which lasted 6 hours and 45 minutes. During this EVA, they completed all planned tasks without complications. These tasks included relocating two crew equipment carts to make room for the station's final set of solar arrays, lubricating the station's robotic arm, and continuing repairs on the starboard Solar Alpha Rotary Joint (SARJ).

Inside the station, the rest of the crew continued transferring cargo between *Leonardo* and the ISS, as well as continuing to activate the Water Recovery System. This system is essential for long-duration missions, as it provides a sustainable water supply by recycling urine and condensate. The day concluded with a joint crew news conference, during which the shuttle and station crews answered questions from reporters around the world and posed for a group photo to commemorate the station's decade of operation.

Following the morning wake-up call, the crews set to work on various planned activities. ISS Commander Mike Fincke and Flight Engineer Sandra Magnus focused on testing the latches on the Exposed Facility Berthing Mechanism for the Japanese *Kibo* laboratory, while Magnus also worked on installing the Total Organic Carbon Analyzer (TOCA). This device would play a critical role in analyzing water purity aboard the station.

Meanwhile, engineers on the ground continued troubleshooting the Urine Processor Assembly (UPA), a key component of the Water Recovery System. The system had initially run on the evening of 20 November but shut down after two hours of operation. Engineers suspected that a sensor issue or a problem with the centrifuge motor could be to blame.

Later in the day, the shuttle's engines were used to reboost the station's altitude by 1.9 km (1.2 mi) in preparation for the arrival of the next Progress resupply vehicle. Cargo transfers between the shuttle and station continued, with approximately 75% of the transfers completed. That evening, Stefanyshyn-Piper and Stephen Bowen spent the night in the airlock as part of the campout procedure in preparation for the next EVA.

The third EVA of the mission began at 18:01 UTC, with Stefanyshyn-Piper and Bowen focusing on completing repairs to the starboard SARJ. They cleaned, lubricated, and replaced the trundle bearings in the joint. Although installing the final trundle bearing assembly was deferred to the fourth EVA, all other tasks were successfully completed.

With her fifth career spacewalk, Stefanyshyn-Piper moved into 25th place in cumulative EVA time, logging 33 hours and 42 minutes.

Inside the station, the crew continued transfer operations and made progress on the water reclamation system. A sample was collected from the Water Processor Assembly, containing 10% urine and 90% condensate, which would be returned to Earth for analysis. Engineers on the

ground also continued troubleshooting the UPA, investigating whether a sensor was touching part of the centrifuge, causing it to slow down.

Crewmembers Fincke and Pettit spent the day reconfiguring the Urine Processor Assembly in an attempt to reduce vibrations that might be contributing to its shutdowns. Despite these efforts, the unit continued to operate intermittently, shutting down after two to three hours. Ground engineers continued evaluating potential causes and solutions.

Meanwhile, the crew had several hours of off-duty time, during which they participated in media interviews. Managers on the ground were considering extending *Endeavour*'s mission by one additional day to give the crew more time to troubleshoot the Water Recovery System.

The day began with preparations for the mission's final spacewalk. Bowen and Kimbrough officially began the EVA at 18:24 UTC, shortly after mission managers confirmed that an extra docked day had been approved, extending the mission to a total of 16 days. This extension provided more time to address the issues with the Water Recovery System. The EVA lasted 6 hours and 7 minutes, bringing the total EVA time for the mission to 26 hours and 41 minutes.

Transfer operations dominated the day as the crew worked to move supplies from the ISS to *Endeavour* and *Leonardo*. The starboard SARJ was tested and tracked the Sun for the first time in over a year during a three-hour, two-orbit trial, showing promising results.

ISS Program Manager Mike Suffredini noted during the Mission Status briefing that the water recycling system was functioning normally after modifications by the crew. While the ISS crew would not be drinking the recycled water until it was thoroughly tested on Earth, approximately six liters of sample water would be brought back for analysis. Suffredini also commented that the SARJ maintenance performed during the EVAs appeared successful, though further testing was needed to confirm the joint's long-term performance.

Using Canadarm2, astronauts Pettit and Kimbrough removed *Leonardo* from the Harmony module and securely stowed it in

Endeavour's cargo bay at 21:52 UTC. Meanwhile, Stefanyshyn-Piper packed up the equipment and supplies used during the four spacewalks for return to Earth. Magnus continued work on the ISS's new regenerative life support system, draining a condensate collection tank and collecting additional water samples for testing.

On their final full day of joint operations, the shuttle and station crews enjoyed some off-duty time, participated in interviews with reporters, and shared a special Thanksgiving meal. Afterward, they completed last-minute cargo transfers and gathered in the Harmony node to say their farewells. At 23:31 UTC, the hatches between the two spacecraft were closed, and the *Endeavour* crew began preparing for undocking the next day.

At 14:47 UTC, *Endeavour* undocked from the ISS, concluding a docked mission time of 11 days, 16 hours, and 46 minutes—making it the second-longest docked shuttle mission after STS-123. Pilot Eric Boe guided the shuttle through a flyaround inspection of the ISS. However, the final separation burn was delayed to avoid debris from a Russian Cosmos satellite that had broken apart earlier in the year. The burn was safely completed at 23:23 UTC.

The crew then conducted a late inspection of *Endeavour*'s heat shield using Canadarm. During the Mission Status briefing, Flight Director Mike Sarafin explained that delaying the burn was a precautionary measure to ensure the shuttle remained a safe distance from the debris field.

The Mission Management Team officially cleared *Endeavour* for re-entry after the heat shield inspection revealed no issues. The crew spent the day preparing for landing by inspecting the shuttle's flight control surfaces and reaction control system thrusters. Near the end of the day, the crew deployed a Department of Defense satellite, Picosat, which tested two new types of photovoltaic solar cells for space applications.

Due to poor weather forecasts at Kennedy Space Center, mission managers called up Edwards Air Force Base as a backup landing site. The forecast for Florida included thunderstorms and high crosswinds, conditions that violated landing weather constraints. Entry Flight Director Bryan Lunney indicated that if the weather did not improve, the shuttle would land at Edwards the next day.

After evaluating the weather conditions at Kennedy Space Center, the decision was made to land *Endeavour* at Edwards Air Force Base. The deorbit burn was initiated at 20:19 UTC, and the shuttle touched down at 21:25 UTC. This marked the final landing of *Endeavour* at Edwards, and the only shuttle mission to land on the temporary runway 04 due to refurbishment of the main runway.

The landing required new braking and rollout techniques due to the shorter runway length, which was 910 meters (2,990 feet) shorter than the main runway. The shuttle was successfully recovered and later transported back to Kennedy Space Center aboard the Shuttle Carrier Aircraft, concluding the successful STS-126 mission.

Chapter 7 - The Final Years and End of an Era (2009–2011)

STS-119

On the morning of March 11, 2009, *Space Shuttle Discovery* was poised for its STS-119 mission, a key flight in the ongoing construction of the International Space Station (ISS). The shuttle had undergone extensive preparation leading up to this launch, having been moved from its Orbiter Processing Facility to the Vehicle Assembly Building on January 7, 2009. Shortly thereafter, on January 11, the payload consisting of the S6 truss segment, solar arrays, and batteries was delivered to Launch Pad 39A. *Discovery* itself was rolled out to the launch pad on January 14, 2009, beginning its move at 05:17 EST and completing it by 12:16 EST.

From January 19–22, 2009, the STS-119 crew gathered at Kennedy Space Center for the Terminal Countdown Demonstration Test, a final practice run before launch. During this time, mission managers conducted a Flight Readiness Review (FRR), where they evaluated the shuttle's systems and set a target launch date for February 19, 2009. However, concerns emerged regarding the shuttle's hydrogen flow control valves. A breakage in one of these valves during the previous flight, STS-126, prompted further testing to ensure *Discovery's* valves were safe for flight. These valves are crucial for managing the flow of gaseous hydrogen between the external fuel tank and the main engines, ensuring a smooth and even fuel flow. After testing, engineers replaced the suspect valves with ones that had fewer flight hours, delaying the launch.

Once the valves were replaced and deemed safe, the Mission Management Team approved a new launch date of March 11, 2009. The STS-119 crew arrived at Kennedy Space Center on March 8 to make final preparations for the mission. However, just hours before the scheduled launch, a leak in the liquid hydrogen vent line between the shuttle and the external tank forced mission managers to scrub the March 11 launch attempt. This leak mirrored a similar issue that occurred during the later STS-127 mission and was traced to a misalignment in the Ground Umbilical Carrier Plate (GUCP). After addressing the problem, the shuttle was cleared for another attempt on March 15.

On March 15, 2009, at 19:43 EDT, *Discovery* successfully lifted off from Launch Pad 39A. The launch was visually striking, with mission managers reporting no concerns regarding debris during ascent. Bill Gerstenmaier, Associate Administrator for Space Operations, noted that the initial review of launch imagery showed no anomalies. Launch Director Michael D. Leinbach echoed the sentiment, calling it "the most visibly beautiful launch I've ever seen."

Once in orbit, the STS-119 crew immediately began their tasks, opening the payload bay doors, deploying the Ku-band antenna, and activating the shuttle's robotic arm. The crew also downlinked critical imagery taken during the separation of the external fuel tank. These operations were vital to ensure the smooth delivery of the mission's primary payload—the S6 truss segment.

In a curious and widely reported occurrence, a bat was seen resting on the external fuel tank of *Discovery* during the countdown. Initially thought to be a fruit bat, it was later identified as a free-tailed bat. NASA observers believed the bat would fly off as the shuttle began its ascent, but the bat remained attached to the tank as

Discovery launched. It is presumed that the bat was either shaken off during ascent or incinerated by the rocket's exhaust. Experts who analyzed photos suggested the bat may have had a broken wing, preventing it from flying away.

Despite this unusual stowaway, the launch proceeded without further incident, and STS-119 continued on its mission to the ISS. The 13-day mission, including installing the S6 truss segment, concluded successfully when *Discovery* landed safely at Kennedy Space Center on March 28, 2009. The mission not only advanced the ISS's power generation capabilities but also contributed valuable data through scientific experiments conducted during the flight.

The STS-119 mission was commanded by Lee Archambault, embarking on his second and final spaceflight. Joining him as pilot was Dominic A. Antonelli, who made his first trip to space. The mission specialists included Joseph M. Acaba, Steven Swanson, Richard R. Arnold, and John L. Phillips. Acaba and Arnold were making their first flights, while Swanson and Phillips were experienced astronauts, with Phillips completing his third and final spaceflight. Koichi Wakata, a Japanese astronaut from the Japan Aerospace Exploration Agency (JAXA), also participated in the mission as a mission specialist and flight engineer for Expedition 18 aboard the ISS. Wakata's involvement marked his third spaceflight.

Additionally, Sandra Magnus, an ISS flight engineer from Expedition 18, returned to Earth aboard *Discovery*, having completed her second spaceflight. The crew exhibited a blend of seasoned veterans and new astronauts, contributing to the smooth execution of both assembly tasks and scientific experiments.

A primary objective of STS-119 was the delivery and installation of the S6 Truss and its associated solar arrays. The S6 Truss, weighing over 14,000 kilograms, was a massive structure essential for providing additional power to support the ISS's expanding needs. The deployment of these solar arrays marked the completion of the station's Integrated Truss Structure, a key milestone in ISS construction.

In addition to the truss assembly, STS-119 facilitated several scientific experiments. One such experiment was the Shuttle Ionospheric Modification with Pulsed Local EXhaust (SIMPLEX), which studied the effects of shuttle exhaust on the ionosphere. Another, the Shuttle Exhaust Ion Turbulence Experiments (SEITE), focused on analyzing turbulence caused by the exhaust in space, while the Maui Analysis of Upper Atmospheric Injections (MAUI) experiment provided data on atmospheric conditions. These experiments expanded the understanding of the space environment, benefiting future missions and space exploration.

Another critical experiment, the "Boundary Layer Transition Detailed Test Objective," was conducted during reentry. One of the thermal protection system tiles on *Discovery* was deliberately raised by 0.25 inches (6.4 mm), with the aim of initiating a boundary layer transition at Mach 15. The goal of this test was to observe the heat and aerodynamic effects caused by this transition, which could provide insights into improving future spacecraft design. The experiment was deemed a success and later repeated during STS-128 with a larger tile displacement.

The STS-119 payload was extensive, with the majority of the cargo focused on the S6 Truss and associated components. The truss segment alone weighed over 14,000 kilograms, making it one of the largest payloads delivered to the ISS. Additionally, the mission carried the Orbiter Docking System, which was essential for securely connecting *Discovery* to the ISS during the mission. The Shuttle Power Distribution Unit (SPDU) and Orbiter Boom Sensor System (OBSS) were also part of the payload, along with the Canadarm, a robotic arm used to assist with cargo handling and external maintenance on the ISS.

In total, STS-119 transported approximately 16,957 kilograms of cargo, showcasing the space shuttle's capacity and versatility in supporting space station assembly and scientific research.

On the second day of the STS-119 mission, the crew of *Discovery* began their day with a wake-up call and immediately set to work inspecting the shuttle's thermal protection system. Using *Discovery's* robotic arm and the Orbiter Boom Sensor System (OBSS), the crew conducted a detailed five-hour inspection of the orbiter's heat shields, scanning for any potential damage that might have occurred during launch. The image assessment team transmitted the images and video from the survey to Earth for analysis. Ensuring the thermal protection system was intact was critical for the shuttle's safe return to Earth.

In preparation for docking with the International Space Station (ISS) on flight day three, the crew performed a series of checks on the spacesuits that would be used for upcoming spacewalks, ensuring they were fully operational. They also extended the ring of the orbital docking system and installed the centerline camera, both essential for the docking procedure.

During the day's Mission Management Team briefing, chairman LeRoy Cain reported that initial reviews of the launch imagery showed no significant issues, with no foam loss or debris strikes. The hydrogen flow control valves, which had been a point of concern leading up to the mission, had performed flawlessly. Cain described the launch as "picture perfect" and noted that *Discovery* was in excellent condition.

On the third day, the crew focused on preparations for the shuttle's rendezvous and docking with the ISS. As *Discovery* approached the space station, Commander Archambault executed the Rendezvous Pitch Maneuver (RPM), during which the shuttle performed a backflip to allow the ISS crew to photograph its underside. This maneuver provided crucial imagery to ensure no damage had occurred to the shuttle's heat shields during ascent.

At 21:20 UTC, *Discovery* successfully docked with the ISS, marking another smooth and precise operation. After completing the standard hatch leak checks, the shuttle and the ISS hatches were opened at 23:09 UTC. The two crews greeted each other warmly and then conducted a mandatory safety briefing to review emergency procedures aboard the station.

Following the briefing, the crew began the initial task of transferring supplies and equipment. One of the key transfers was the exchange of Sandra Magnus' Soyuz seat liner for Koichi Wakata's. This swap officially marked Wakata's transition to the Expedition 18 crew as a flight engineer, while Magnus assumed her role as a mission specialist for STS-119.

During the Mission Status briefing later that day, Lead Flight Director Paul Dye praised the crew for executing a flawless docking. No significant issues were reported, though the image analysis team continued reviewing both ascent and RPM imagery to determine if any additional inspections would be necessary.

The fourth day of the mission was dedicated to preparing to install the S6 truss segment, a critical component that would expand the ISS's solar power capabilities. After the crew completed their post-sleep routines, they began moving the large S6 truss from *Discovery's* payload bay. Due to the size and location of the truss, a coordinated effort between the station's and the shuttle's robotic arms was required.

Mission specialists Phillips and Magnus controlled the ISS robotic arm, using it to grapple the S6 truss and move it out of the payload bay. However, because the station's arm could not reach the installation site directly, the truss was handed off to the shuttle's robotic arm, operated by Antonelli. This intricate series of handoffs ensured that the truss could be positioned correctly for installation.

Once the truss was secured, the station's robotic arm was moved along its mobile base to a worksite closer to the final installation point on the far right side of the ISS. The shuttle's robotic arm then returned the truss to the station's arm, which held it in place overnight in preparation for installation during the next day's spacewalk.

In addition to the truss work, the crews participated in a media event with Channel One News, answering questions and discussing their mission. They also reviewed procedures for the first extravehicular activity (EVA). Mission specialists Steven Swanson and Richard Arnold, who would be conducting the EVA, spent the night "camping out" in the Quest airlock. This procedure, which involved sleeping in a reduced-nitrogen atmosphere, helped prepare their bodies for the spacewalk by reducing the risk of decompression sickness.

During the day's Mission Management Team briefing, Lead ISS Flight Director Kwatsi Alibaruho announced that the Damage

Assessment Team had completed their review of the launch and flight day two imagery. The team determined that no focused inspection of *Discovery* would be required, confirming the shuttle was in excellent condition for the remainder of the mission.

Following their wake-up call, the combined crews of *Discovery* and the International Space Station set to work preparing for the mission's first spacewalk. At 16:22 UTC, mission specialists Steven Swanson and Richard Arnold exited the Quest airlock to begin the installation of the S6 truss segment, a crucial component for the station's power generation system. Once Swanson and Arnold were in position, crew members John Phillips and Koichi Wakata operated the station's robotic arm from inside the ISS, carefully maneuvering the S6 truss into its final position.

Swanson and Arnold worked together to bolt the truss into place, connecting power and data cables that allowed the ground team to begin activating the segment remotely. In addition to the installation, the spacewalkers performed several tasks to prepare the truss for operation. They removed launch locks, stowed a keel pin, jettisoned four thermal covers, and deployed the blanket boxes that housed the solar arrays during launch. After six hours and seven minutes of work, the spacewalk concluded at 21:11 UTC.

Initially, the deployment of the solar arrays was scheduled for flight day 8, but mission managers on the ground decided to move the task up to flight day 6. This decision was made after determining that a focused inspection of *Discovery* would not be necessary, allowing the crew more time to resolve any potential issues with the arrays before the second spacewalk.

The next day, both crews focused on deploying the solar wings of the newly installed S6 truss. To ensure the smooth extension of the arrays, the ISS was repositioned to allow constant sunlight to warm them up, preventing what is known as "stiction"—a sticky friction caused by long-term storage in the blanket boxes.

At 15:06 UTC, the deployment of the 1B solar array began. John Phillips, commanding the operation from inside the ISS, paused the unfurling process at the halfway point to allow the array to warm in the sunlight for about 45 minutes. After this break, the crew completed the extension without issue. Following this success, the 3B array deployment commenced at 16:35 UTC. This array was expected to be more difficult to deploy due to having been packed for eight years. As with the first array, the astronauts paused halfway through the extension to allow the array to warm up. Despite minor stiction, the 3B array fully unfurled by 17:17 UTC, and all the slats flattened out as expected.

The successful deployment of these solar arrays brought the ISS's total power output to 120 kilowatts, doubling the station's scientific power generation capacity to 30 kilowatts. With the new arrays in place, the surface area of the ISS's solar panels now covered nearly one acre, or approximately 38,400 square feet.

Later in the day, crew members Michael Fincke, Yuri Lonchakov, Koichi Wakata, and Sandra Magnus participated in a media event with *Reuters*, *Voice of America*, and the *Pittsburgh Post-Gazette*.

During the Mission Status briefing, ISS Flight Director Kwatsi Alibaruho expressed satisfaction with the successful deployment of the arrays, praising the crew for their efforts. Mission Management Team Chairman LeRoy Cain noted that the ground team was working on a revised flight schedule to ensure that critical experiment samples could be returned to Earth under the proper conditions. The samples required cold storage, and in the event of weather-related delays, the team evaluated the best way to preserve them. As a result, the flight plan was adjusted so that the hatch closure and undocking could be delayed slightly, allowing the samples to remain inside the station's freezer for as long as possible without compromising the scheduled March 28 landing.

Reflecting on the day's achievements, Dan Hartman, chairman of the Space Station Mission Management Team, remarked:

"It was a truly fantastic day in space. The International Space Station team and its partnerships are on cloud nine with the completion of the integrated truss assembly, as well as the finalization of our electrical power grid on the space station. It took years to get here. We had some struggles along the way, but it's a major accomplishment for NASA and the partnership team."

On the seventh day of the mission, Swanson and Joseph Acaba performed the second spacewalk of STS-119, exiting the Quest airlock at 16:51 UTC. Their primary tasks involved preparing for future upgrades and maintenance. They worked on loosening bolts and installing foot restraints and tools to make it easier for the spacewalkers on the upcoming STS-127 mission to replace the Port 6 truss batteries.

Additionally, Swanson and Acaba installed a second Global Positioning Satellite (GPS) antenna on the Japanese Kibo laboratory module, enhancing the station's navigation capabilities. They also photographed radiator panels extended from the Port 1 and Starboard 1 trusses and reconfigured connectors on the Zenith 1 truss to power the Control Moment Gyroscopes, which help stabilize the station's orientation.

The spacewalkers faced a challenge with the Unpressurized Cargo Carrier Attachment System (UCCAS), a device used for securing cargo outside the station. A pin prevented the system from fully deploying, but Swanson and Acaba resolved the issue by safely tying it in place. The spacewalk concluded at 23:21 UTC, after six hours and thirty minutes of work.

On flight day 8, at 20:31 UTC, *Discovery* performed a critical maneuver to avoid a piece of orbital debris. The shuttle-station complex was rotated 180 degrees to ensure its safety. Later, at 23:23 UTC, the shuttle began rotating the station back to its normal attitude, with *Discovery* in a "back" orientation relative to the station.

While the maneuver was being conducted, ISS Commander Michael Fincke worked on the Urine Processor Assembly, a vital part of the station's life support system. At the same time, mission specialists Joseph Acaba and Richard Arnold entered the Quest airlock to begin preparations for the mission's third spacewalk, scheduled for the following day.

On March 23 (Flight Day 9: Spacewalk 3), Acaba and Arnold embarked on the mission's third spacewalk at 15:37 UTC, which lasted six hours and 27 minutes, concluding at 22:04 UTC. The spacewalk's primary objectives involved relocating the Crew Equipment Translation Aid (CETA) cart from the Port 1 to the Starboard 1 truss segment and installing a new coupler on the CETA cart. They also worked on lubricating the snares of the space station's robotic arm, ensuring its continued operational efficiency.

One of the spacewalk's tasks was to deploy the Port 3 unpressurized cargo carrier attachment system (UCCAS), but the team encountered difficulties with the mechanism. Despite multiple attempts, they were unable to deploy the system and instead secured it in place for future evaluation by engineers. As a result, Mission Control canceled the installation of a similar system on the starboard side. The issue with the Port 3 UCCAS was later resolved during STS-127, when a custom-made tool was used to release the stuck pin.

On March 24 (Flight Day 10), at 17:05 UTC, all members of the *Discovery* and ISS crews gathered in the Harmony module for a special event. The astronauts participated in a live video conference with President Barack Obama, members of Congress, and schoolchildren. This joint news conference highlighted the mission's accomplishments and promoted STEM education by directly engaging with students.

The crews of *Discovery* and the ISS prepared for undocking on flight day 11. At 17:59 UTC, they closed the hatches between the space shuttle and the ISS. Shortly thereafter, at 19:53 UTC, *Discovery* successfully undocked from the ISS, marking the conclusion of its time attached to the station.

On March 26 (Flight Day 12), as part of the routine post-undocking procedure, pilot Dominic "Tony" Antonelli used the shuttle's robotic arm to grapple the Orbiter Boom Sensor System (OBSS). The OBSS enabled the crew to scan *Discovery*'s thermal protection system for any potential damage caused by orbital debris during its stay in space. The sensor's cameras and lasers carefully checked the shuttle's exterior to ensure it was safe for reentry.

On March 27 (Flight Day 13), the penultimate day of the mission, the crew focused on stowing items within the shuttle's crew cabin and conducting a final checkout of the orbiter's flight control surfaces. These systems ensured

Discovery could safely navigate during reentry into Earth's atmosphere.

March 28, the crew of *Discovery* began preparing for reentry early on flight day 14. However, the first landing opportunity was waved off due to concerns about high winds at Kennedy Space Center. After assessing the weather conditions, mission managers gave the crew the green light to proceed with the second landing opportunity. Following a successful deorbit burn, *Discovery* re-entered Earth's atmosphere and landed smoothly at 15:13 EDT, concluding the STS-119 mission.

The crew, led by Commander Lee Archambault, emerged from *Discovery* after a successful mission that enhanced the International Space Station's capabilities and paved the way for future scientific research and exploration.

STS-125: The Final Servicing Mission to the Hubble Space Telescope

STS-125, also known as Hubble Space Telescope Servicing Mission 4 (HST-SM4), marked the fifth and final Space Shuttle mission dedicated to the Hubble Space Telescope (HST).

The crew of STS-125 was led by Commander Scott Altman, a veteran astronaut on his fourth and final spaceflight. Pilot Gregory C. Johnson, making his first and only trip to space, accompanied Altman. The mission specialists were a mix of seasoned astronauts and first-time flyers. Michael T. Good, on his maiden voyage, served as Mission Specialist 1, while Megan McArthur, also a first-time astronaut, filled the role of Flight Engineer. John M. Grunsfeld, a prominent figure in the Hubble program, embarked on his fifth and last spaceflight as Mission Specialist 3. Mike Massimino, a devoted New Yorker and second-time astronaut, joined as Mission Specialist 4, while Andrew J. Feustel, another first-time astronaut, rounded out the crew as Mission Specialist 5. Massimino famously brought the home plate from Shea Stadium, later returning it to the New York Mets and throwing the first pitch at Citi Field upon his return to Earth.

Originally, HST-SM4 was scheduled for launch between late 2005 and early 2006. However, following the Columbia disaster in 2003, NASA Administrator Sean O'Keefe implemented new safety protocols requiring that all future shuttle missions have the ability to reach the International Space Station (ISS) as a safe haven in case of emergencies. As Hubble's orbit made such a diversion impossible, O'Keefe canceled the mission in January 2004. The decision, made independently of other departments, was met with widespread criticism from the scientific community, the media, and even within NASA. Senator Barbara Mikulski, an advocate for space science, was among the most vocal opponents, promising to work toward reversing the decision.

Public and political pressure grew as Representative Mark Udall introduced a bill calling for an independent review of O'Keefe's decision. Around the same time, the Space Telescope Science Institute (STScI) released data from the Hubble Ultra-Deep Field survey, showcasing the telescope's unparalleled capability by revealing thousands of distant galaxies. This data reinforced the importance of Hubble, prompting renewed efforts to save the mission.

NASA Chief Scientist John Grunsfeld, a veteran of two previous Hubble servicing missions, became a key figure in the fight to restore the mission. Although he initially considered retiring after the cancellation, Grunsfeld decided to stay and explore alternatives, including a possible robotic servicing mission. When O'Keefe resigned in December 2004, hope for the mission's reinstatement was rekindled.

In 2005, Michael D. Griffin succeeded O'Keefe as NASA Administrator. An engineer who had worked on Hubble's construction, Griffin swiftly reversed O'Keefe's decision. He concluded that a robotic servicing mission was not feasible and, following the successful "Return to Flight" shuttle missions STS-114 and STS-121, announced that a shuttle mission to repair Hubble would go ahead. On October 31, 2006, Griffin confirmed that the mission was reinstated, with a tentative launch date in 2008. The announcement thrilled the scientific community and public, especially Senator Mikulski, who called Hubble the greatest telescope since Galileo's invention.

STS-125, originally slated for launch aboard Space Shuttle *Discovery*, faced a series of scheduling and logistical changes that ultimately led to its reassignment to *Atlantis*. Initially planned for a launch no earlier than May 2008, the mission was meant to precede STS-119, a flight dedicated to the International Space Station (ISS) assembly. However, a series of delays affecting multiple shuttle missions necessitated a reshuffling of the launch order. On January 8, 2007, NASA officially switched the orbiter from *Discovery* to *Atlantis*.

In early July 2008, the crew of STS-125 traveled to Kennedy Space Center for the Crew Equipment Interface Test, allowing them to familiarize themselves with *Atlantis* and the specialized hardware they would use during the mission. This hands-on experience was crucial, as STS-125 was designated as the final servicing mission to the Hubble Space Telescope (HST), requiring precision and careful preparation.

On August 22, 2008, after a brief delay due to Tropical Storm Fay, *Atlantis* was transferred from the Orbiter Processing Facility to the Vehicle Assembly Building (VAB). There, it was mated with its external fuel tank and solid rocket boosters. However, issues during the mating process and inclement weather caused by Hurricane Hanna delayed the subsequent rollout to Launch Pad 39A, which typically occurs seven days after the initial rollover.

Compounding the delays, Lockheed Martin encountered manufacturing setbacks while producing new external tanks for future missions. These tanks required modifications in line with the recommendations of the Columbia Accident Investigation Board. The upgrades resulted in an inability to produce two tanks—one for *Atlantis* and one for *Endeavour*, which was standing by for a potential rescue mission—in time for the original August 2008 launch date. As a result, NASA postponed the STS-125 launch to October 2008.

Atlantis made its first rollout to Launch Pad 39A on September 4, 2008. However, just weeks later, on September 27, the Science Instrument Command and Data Handling (SIC&DH) Unit aboard the Hubble Space Telescope failed. Given the importance of this unit to the telescope's continued operation, NASA opted to delay the mission once more, pushing the launch to 2009 to allow for the replacement of the failed unit. Consequently, *Atlantis* was rolled back to the VAB on October 20, 2008.

On October 30, NASA decided to remove *Atlantis* from its rocket stack and return the orbiter to the Orbiter Processing Facility, targeting a new launch date of May 12, 2009. The rocket stack originally assembled for STS-125 was reassigned to the STS-119 mission. On March 23, 2009, *Atlantis* was mated to a new stack and rolled out to Launch Pad 39A on March 31. As the new launch date approached, NASA managers requested a slight adjustment, moving the launch to May 11, 2009, at 2:01 p.m. EDT to extend the launch window.

On May 11, 2009, Space Shuttle *Atlantis* successfully lifted off from Kennedy Space Center at 2:01 p.m. EDT, marking the beginning of STS-125. Despite a smooth countdown, flight systems immediately registered issues during ascent. Alarms indicated a malfunction with a hydrogen tank transducer and a circuit breaker. However, mission control quickly advised the crew to disregard these alerts and continue their climb into orbit.

Post-launch inspections revealed no significant debris-related damage to the orbiter during ascent, but a thorough analysis was conducted to ensure *Atlantis* remained in safe condition. Upon reaching orbit, the crew opened the payload bay doors, deployed the Ku-band antenna, and began robotic operations, which included a survey of both the payload bay and crew cabin using the orbiter's robotic arm.

At Launch Pad 39A, a 25-foot section of the flame deflector showed damage where the heat-resistant coating had come off during launch. Although this was not as severe as the damage observed during the previous STS-124 mission, NASA engineers assured that it would not affect the upcoming STS-127 launch, scheduled for June 2009.

After the crew's morning wake-up call, they immediately set to work on the day's primary tasks, focusing on the inspection of the Space Shuttle's heat shield. Using the shuttle's robotic arm and the Orbiter Boom Sensor System (OBSS), the crew carried out a detailed survey of the orbiter's thermal protection system (TPS) tiles and the Reinforced Carbon-Carbon (RCC) surfaces, which protect critical areas such as the wing leading edges and nose cap. Engineers on the ground closely monitored the inspection and identified a small area of damage on the forward section of the right wing's TPS tiles, likely sustained during ascent.

Mission managers promptly alerted Commander Scott "Scooter" Altman to the find. They relayed that one of the orbiter's wing leading-edge sensors had recorded a debris impact event approximately 104 to 106 seconds after liftoff, possibly correlating with the observed damage. CAPCOM Dan Burbank informed the crew that the damage did not appear serious initially, but the image analysis team would conduct a more thorough review. Ground engineers would analyze the data to determine if a focused inspection was necessary to ensure the integrity of the orbiter's heat shield.

Meanwhile, as part of the Flight Day 2 Execute Package, ground engineers provided additional details on a circuit breaker failure that occurred during launch. The breaker, responsible for Channel 1 of the shuttle's Aerosurfaces (ASA 1), was a component of the Flight Control Systems (FCS), a vital subsystem of the shuttle's Guidance, Navigation, and Control (GNC)

systems. Fortunately, the failure posed no threat to the mission due to the built-in redundancy of the shuttle's systems.

The crew also performed checks of the equipment for the upcoming spacewalks, including the Extravehicular Activity (EVA) tools and spacesuits. They ensured that the Flight Support System (FSS) was ready for the crucial task of berthing the Hubble Space Telescope, scheduled for the following day.

On May 13, Flight Day 3, the crew shifted focus to the rendezvous and capture of the Hubble Space Telescope. After completing their post-sleep activities, they initiated the rendezvous operations, using the shuttle's engines to fine-tune their approach to the telescope. Despite minor communication delays, Commander Altman and Pilot Gregory "Ray-J" Johnson expertly guided the shuttle to within fifty feet of Hubble. At 17:14 UTC, Mission Specialist Megan McArthur successfully grappled the telescope using the shuttle's robotic arm, and by 18:12, Hubble was securely berthed in the payload bay of Atlantis.

With Hubble safely stowed, astronauts John Grunsfeld, Drew Feustel, Mike Massimino, and Michael Good began preparing for the next day's spacewalk. They gathered and inspected the EVA tools, conducted checks on the spacesuits, and ensured that all necessary equipment was in place for the mission's critical repairs and upgrades to the telescope.

During the Mission Management Team (MMT) briefing, Chairman LeRoy Cain provided a reassuring update on the damage assessment. The team had cleared all of the orbiter's TPS tiles and thermal blankets and expected to clear the RCC surfaces by the following day. Cain confirmed that no additional focused inspections would be required. Although the wing's leading-edge sensors had detected a debris event, the impact was well below the threshold that would suggest significant damage, and the mission was not at risk. Cain noted that a routine late-stage inspection before re-entry would provide further confirmation, but he expressed confidence that the orbiter had not sustained any critical damage.

Later in the day, during the Mission Status briefing, Lead Flight Director Tony Ceccacci reported that a camera survey of the payload bay had revealed fine particulate matter near the box containing the Wide Field Camera 3, a key instrument for Hubble. The crew was instructed to take additional high-resolution images of the area for further analysis. Engineers on the ground suspected the particles were insulation debris dislodged during the shuttle's launch. While the team advised caution during the upcoming spacewalks, the debris posed no significant risk to the mission's objectives.

On the morning of May 14, the crew of Atlantis prepared for their first spacewalk, with Mission Special. Mission Specialists John Grunsfeld and Andrew Feustel, assisted by Mike Massimino and Michael Good, donned their spacesuitsacewalk officially began as Grunsfeld and Feustel switched their suits to battery power, marking the start of a carefully planned Extravehicular Activity (EVA).

One of Feustel's first tasks was to visually inspect the particulate matter that had been observed earlier around the Wide Field Camera 3 (WFC3) storage box. "I don't really see any of those particles... It's almost imperceptible. I can see a few particles on the front of the W-SIPE, little, whitish, grey-looking, real small. It's low density, too," he reported to ground control. With the inspection complete, the two astronauts proceeded with their primary objective: replacing the old Wide Field and Planetary Camera 2 (WFPC2), which had been installed during the first Hubble servicing mission in 1993, with the more advanced WFC3.

The camera replacement proved challenging when Feustel encountered difficulty removing bolts that had become tight after years in the vacuum of space. Despite repeated attempts, the bolts refused to budge. Mission managers on the ground instructed Grunsfeld to retrieve a contingency torque limiter from the airlock, a tool designed to apply more force without risking damage. Even with this tool, the bolts remained stuck. After assessing the situation, ground controllers authorized Feustel to apply as much force as he felt necessary, and with one final effort, the bolts finally gave way.

The installation of WFC3 significantly upgraded Hubble's capabilities, enabling it to capture larger, clearer, and more detailed images across a wider spectrum of colors than its predecessor. Once the camera was in place, the Space Telescope Operations Control Center at the Goddard Space Flight Center performed an aliveness test, confirming that the new camera was installed correctly and functioning as expected.

The next major task was to replace Hubble's Science Instrument Command and Data Handling Unit (SIC&DH). This vital computer controls the operation of the telescope's science instruments and formats data for transmission back to Earth. The SIC&DH had failed in September 2008, delaying the STS-125 mission while a replacement was prepared and the crew trained for the task. Although the failure had not disabled Hubble, the replacement restored the system's redundancy, ensuring the telescope's continued reliability.

The final objective of the spacewalk was to install the Soft-Capture Mechanism (SCM), which included the Low Impact Docking System (LIDS). This 72-inch-wide docking port would allow future spacecraft to dock with Hubble and safely deorbit it at the end of its operational life. Feustel also installed two of four Latch Over Center Kits (LOCKs) designed to make opening and closing Hubble's large access doors easier for future spacewalks.

The spacewalk ended at 20:12 UTC, lasting seven hours and twenty minutes. It was the nineteenth spacewalk dedicated to servicing Hubble, bringing the total time spent on Hubble EVAs to 136 hours and thirty minutes. Although the EVA had run over an hour longer than scheduled due to the bolt removal issue, the crew proceeded through their post-EVA tasks and evening activities without further delays. During the mission status briefing, Hubble Project Senior Scientist David Leckrone expressed his relief at the successful replacement of the camera, noting, "I don't normally reveal my age, and I'm not going to here, but I can tell you I'm five years older now than I was when I came to work this morning. We can sleep pretty well tonight, knowing that's been accomplished."

The following day, the Atlantis crew woke up and prepared for the mission's second spacewalk.

This time, it was Mike Massimino and Michael Good's turn to suit up, with Grunsfeld and Feustel assisting. As they readied for their EVA, the ground team relayed encouraging news: the Wide Field Camera 3 had passed all functional tests overnight, confirming it was in excellent working order.

Meanwhile, Commander Scott Altman and Flight Engineer Megan McArthur conducted a robotic survey of a small section of the shuttle's heat shield tiles, which had not been fully imaged during the initial inspection. Once the analysis was complete, mission managers cleared all of the shuttle's Thermal Protection System (TPS) components, deferring any additional inspections until the pre-landing check.

The second spacewalk began at 12:49 UTC, and Massimino and Good focused on replacing Hubble's three gyroscope rate-sensing units (RSUs), critical components that enable the telescope to accurately orient itself. Each RSU contains two gyroscopes. The first replacement, RSU 2, went smoothly, but problems arose when they attempted to install RSU 3. The unit would not align correctly on the guide pins, preventing it from seating properly in the equipment bay.

After several failed attempts, ground controllers decided to place the RSU intended for Bay 1 into the RSU 3 slot, which succeeded. However, when Massimino and Good tried to install the final RSU, they encountered the same seating issue. Ultimately, the team opted to use a spare unit originally removed from Hubble during the STS-103 mission. This spare had been refurbished on Earth and met most of the newer units' specifications. The final installation was successful, and ground controllers at Goddard confirmed that all six gyroscopes, as well as the telescope's new battery, passed preliminary tests.

The issues with seating the RSUs set the team back by two hours, but after confirming they felt physically capable, Massimino and Good continued their work. Flight controllers evaluated their spacesuit consumables and determined that with a quick oxygen recharge in the airlock, the pair could safely proceed. Moving to the battery installation site, they removed an original battery module from Bay 2 and replaced it with a new unit. These batteries power Hubble when its solar arrays are not exposed to sunlight during its orbital passes through Earth's shadow.

The spacewalk concluded at 20:45 UTC, lasting seven hours and fifty-six minutes. It marked the twentieth EVA dedicated to Hubble, bringing the total servicing time to 144 hours and twenty-six minutes.

During the post-EVA briefing, Tomas Gonzalez-Torres, the Lead Spacewalk Officer, and Hubble Program Manager Preston Burch reassured the press that the use of the spare RSU would not impact the telescope's operational life. Burch remarked, "I would say the difference in the projected longevity of the observatory in the out years is very small. We don't see this as a significant detriment at all to the observatory. This was a tremendous accomplishment for us." Lead Flight Director Tony Ceccacci added that while the EVA's length had shifted the crew's schedule, they would adjust their sleep cycles accordingly to ensure they received adequate rest before the next day's activities.

On May 16, Flight Day 6 of the mission, astronauts John Grunsfeld and Andrew Feustel prepared for the third spacewalk aboard *Atlantis*. This spacewalk was anticipated to be the most challenging and uncertain of the mission, with high-priority tasks to accomplish. The crew was tasked with removing the now-obsolete Corrective Optics Space Telescope Axial Replacement (COSTAR) instrument, which had been installed during the STS-61 mission to correct a spherical aberration in the Hubble Space Telescope's mirror. In its place, they would install the Cosmic Origins Spectrograph (COS) and attempt to repair the Advanced Camera for Surveys (ACS), which had failed twice due to electrical issues between 2006 and 2007.

The ACS repair presented significant challenges, as the camera had not been designed to be serviced in space. Following earlier spacewalks that had encountered unexpected issues, ground managers braced for potential complications. However, the spacewalk began smoothly at 13:35 UTC, with Grunsfeld and Feustel working so efficiently that they were more than an hour ahead of schedule. After successfully removing COSTAR and stowing it in *Atlantis'* payload bay, the astronauts installed the COS instrument. They then proceeded with the ACS repair, using specially designed tools to remove an access panel, replace four circuit boards, and install a new power supply.

The spacewalk, lasting six hours and thirty-six minutes, was a success. The ACS passed initial "aliveness" tests, marking the twenty-first Hubble servicing spacewalk and Grunsfeld's seventh extravehicular activity (EVA), placing him fourth in the record books for spacewalking time. Dave Leckrone, Senior Project Scientist for the Hubble Space Telescope, had predicted the smoothness of this spacewalk despite its complexity, humorously attributing it to Murphy's Law.

After the spacewalk, functional tests on the ACS revealed that while the wide-field channel of the camera was operational, issues persisted with the high-resolution channel, suggesting that the electrical short was located in a part of the circuit that the repair had not addressed. Although the high-resolution channel remained in question, the wide-field channel's functionality was considered a major victory, as it was responsible for the bulk of ACS's scientific output. The third channel, the solar-blind channel, passed overnight tests without issue.

On May 17, Flight Day 7, astronauts Michael Massimino and Mike Good embarked on the fourth spacewalk of the mission at 13:45 UTC. Their objective was to repair the Space Telescope Imaging Spectrograph (STIS), which had failed in 2004 due to a blown power supply. Like the ACS, the STIS was not designed for space-based repairs, presenting a difficult challenge. The astronauts needed to remove a cover plate fastened with over 100 screws using a fastener-capture plate, a specially designed tool to trap screws and washers to prevent them from floating away in microgravity.

Massimino encountered a major obstacle when he discovered that a handrail, essential for securing the fastener-capture plate, had a stripped bolt. Ground managers authorized him to use brute force to remove the handrail after tests conducted at Goddard Space Flight Center confirmed that it could be done safely. Massimino, guided by Feustel from inside the orbiter, successfully removed the handrail and

proceeded with the repair, though the battery in one of his power tools failed, causing a delay. Massimino returned to the airlock to retrieve a spare tool and recharge his oxygen supply, ensuring they could complete the repair.

Despite these challenges, the STIS repair was completed successfully, though the astronauts were nearly two hours behind schedule. As a result, ground controllers postponed the installation of New Outer Blanket Layers (NOBLs) onto the telescope's exterior. The spacewalk, originally planned for six hours and thirty minutes, lasted eight hours and two minutes, becoming the sixth longest spacewalk in history at the time. It was also the twenty-second spacewalk devoted to Hubble servicing, bringing Massimino's total EVA time to thirty hours and forty-four minutes.

Jennifer Wiseman, Chief of Exoplanet and Stellar Astrophysics at Goddard Space Flight Center, emphasized the significance of the STIS repair during the mission status briefing, calling it a major victory for both the mission and the scientific community. STIS performed unique and critical functions, including analyzing the materials that make up planets and observing the motion of stars around black holes. After initial tests showed no issues, STIS was placed into functional testing. However, the telescope unexpectedly entered safe mode due to a low thermal limit sensor. Ground controllers at Goddard planned to restart the tests once the sensor returned to normal range, with expectations that the instrument was in good condition despite the sensor warning.

On May 18, Flight Day 8, astronauts John Grunsfeld and Andrew Feustel embarked on the fifth and final spacewalk of the mission. Beginning at 12:20 UTC, the pair worked swiftly and efficiently, completing several high-priority tasks and even managing to tackle additional objectives ahead of schedule. Their first task was to remove an aging battery module from the Hubble Space Telescope and replace it with a new pack. This, combined with the battery replacement during the second spacewalk, gave Hubble a full set of newly installed nickel-hydrogen batteries, ensuring a more reliable power supply for the telescope.

Next, Grunsfeld and Feustel replaced the Fine Guidance Sensor (FGS) number three, an essential instrument for maintaining the telescope's stability and focus during imaging. NASA engineers likened the precision of the new FGS to keeping a laser beam focused on a U.S. dime from 320 kilometers away. The installation of the new batteries and the FGS were crucial upgrades for Hubble, and both passed initial "aliveness" and functional tests, confirming their operational status.

Working so efficiently that they were over an hour ahead of their timeline, the astronauts used the extra time to remove degraded insulation panels from three of Hubble's bays and install three New Outer Blanket Layers (NOBLs), which would provide additional thermal protection for the telescope's sensitive instruments.

The spacewalk concluded at 19:22 UTC, lasting seven hours and two minutes. This final EVA brought the total time spent on spacewalks during the mission to thirty-six hours and fifty-six minutes, while the overall time spent on Hubble servicing over the years reached a total of 166 hours and six minutes. This was the twenty-third and final spacewalk devoted to Hubble, marking the end of an era. As Lead Flight Director Tony Ceccacci noted, this EVA was also the last planned spacewalk to be conducted from a Space Shuttle airlock. For Grunsfeld, it was likely his last spacewalk, bringing his career total to fifty-eight hours and thirty minutes—just two minutes shy of Jerry L. Ross, who ranked third for the most spacewalking time.

With the successful completion of all major objectives, as well as several additional tasks, Hubble was upgraded to its most advanced state since its launch nineteen years prior. The enhancements made the telescope more powerful than ever, enabling it to peer deeper into the universe and farther back in time, closer to the Big Bang. Hubble's contributions to science were immense, not only in terms of the breathtaking images it captured but also in the wealth of scientific data it generated. On average, 14 scientific papers were published each week based on Hubble's data. The upgrades completed during this mission were expected to extend the telescope's operational life through at least 2014, though Hubble Senior Scientist David Leckrone expressed optimism that, if all went well, the telescope could continue functioning far beyond that date. With the James Webb Space Telescope set to launch in December 2021, Hubble's continued operation, providing ultraviolet, visible, and near-infrared observations, would remain a valuable asset to the scientific community.

On May 19, Flight Day 9, the crew began preparations to release Hubble back into space. After waking at 08:31 UTC, the astronauts set to work using the Shuttle's robotic arm to lift Hubble out of *Atlantis'* payload bay. Astronaut Megan McArthur grappled Hubble with the robotic arm at 10:45 UTC and carefully maneuvered it into position for deployment. Standing by in case a spacewalk became necessary, Mike Good and Michael Massimino were ready to respond to any issues during the telescope's release.

Once the ground team completed the final checks, the go-ahead was given to Commander Scott Altman. At 12:57 UTC, McArthur successfully released the Hubble Space Telescope as the Shuttle orbited over Africa. After confirming the successful deployment of Hubble, Altman addressed the ground team, saying, "And Houston, Hubble has been released, it's safely back on its journey of exploration as we begin steps to conclude ours. Not everything went as we planned, but we planned a way to work around everything and with the whole team pulling together... we've been able to do some incredible things. And now Hubble can continue on its own, exploring the cosmos, and bringing it home to us as we head for home in a few days. Thank you."

Following Hubble's release, the crew performed a small separation burn, backing the Shuttle away from the telescope. With the mission's primary objective accomplished, the crew then began the standard late inspection of *Atlantis'* thermal protection system. Using the Shuttle's robotic arm, the astronauts meticulously examined the wing leading-edge panels, the reinforced carbon-carbon nose cap, and the heat shield tiles to ensure everything was in proper condition for reentry.

As the crew worked through their final procedures, ground controllers analyzed weather conditions at the landing site. To avoid potential showers that could interfere with the scheduled landing, the decision was made to bring the Shuttle home one orbit earlier than planned. This adjustment meant that the crew would land at 10:01 am EDT on Friday, concluding their remarkable mission.

On May 20, during Flight Day 10, Space Shuttle Atlantis gently released the Hubble Space Telescope into free orbit after a week of intense servicing. With the primary mission objectives complete, the crew was granted a much-needed rest day, allowing them to recuperate and prepare for their return to Earth. During the day, they captured their traditional in-flight portrait and participated in a live news conference, engaging with reporters worldwide. They also had the opportunity to speak with the Expedition 19 crew aboard the International Space Station (ISS), who congratulated them on their successful mission. The Atlantis crew expressed their gratitude for the ongoing work of the ISS team during their long-duration missions.

Despite the mission's success, concerns about the weather at Kennedy Space Center (KSC) raised the possibility of delaying Atlantis's landing. The Spaceflight Meteorology Group at Johnson Space Center forecasted unfavorable conditions for the scheduled landing on Friday. As a precaution, mission managers asked the crew to power down non-critical systems to conserve power in case the landing was delayed until Saturday. Meanwhile, the ground team carefully reviewed imagery from the late inspection and officially cleared Atlantis' thermal protection system for reentry. NASA managers, however, delayed releasing the backup rescue mission, STS-400, flown by Endeavour, until after Atlantis performed its deorbit burn.

As the crew settled in for the night, they received a special call from President Barack Obama. The president congratulated them on their successful mission and lightheartedly asked if they could spot his house in Chicago from space. Obama praised the crew's dedication and highlighted their contribution to America's spirit of exploration and innovation.

On May 21, Flight Day 11, preparations for landing began in earnest. Pilot Gregory Johnson practiced landing maneuvers using the orbiter's Portable In-Flight Landing Operations Trainer (PILOT) program, while Commander Scott Altman and other key crew members conducted flight control surface checks and reaction control system tests. The rest of the crew busily stowed away equipment used during their mission. Ground teams and the astronauts participated in a deorbit briefing, and Altman and Johnson ran through landing simulations.

In an unprecedented moment in space history, the crew became the first shuttle astronauts to testify live from orbit before the United States Senate. Senator Barbara Mikulski and former astronaut Senator Bill Nelson led a discussion on the importance of spaceflight and the Hubble repair mission. The historic testimony added another layer of significance to an already remarkable mission.

The day ended with growing concerns about weather conditions at KSC, as forecasts indicated that the chance of a Florida landing on Friday was slim. Ground managers decided to proceed with deorbit preparations, but advised the crew that backup landing sites, such as Edwards Air Force Base in California, could be activated if necessary. Despite these uncertainties, the crew remained focused and ready for any outcome.

On May 22, Flight Day 12, as the Atlantis crew went through their entry procedures and checklists, ground teams closely monitored weather patterns over Florida. Unfortunately, conditions remained unfavorable, with low clouds and thunderstorms breaching the landing criteria. As a result, the first and second landing opportunities were waived. Entry Flight Director Norm Knight called up Edwards Air Force Base, initiating preparations for a potential landing in California on Saturday if conditions in Florida remained unfavorable.

By May 23, Flight Day 13, the Atlantis crew had six potential landing opportunities, three at each of the designated landing sites. The first attempt at KSC was waived due to continuing poor weather conditions, and the second opportunity was also passed over after further evaluation. Given the favorable weather forecasts at Edwards Air Force Base, the decision was made to attempt a landing there if weather in Florida did not improve by Sunday.

Finally, on May 24, Flight Day 14, after assessing weather conditions at both KSC and Edwards Air Force Base, mission managers opted for a landing in California. The deorbit burn was initiated at 14:24 UTC, bringing Atlantis out of orbit for reentry. At 8:39 a.m. PDT, Atlantis touched down smoothly on Runway 22 at Edwards Air Force Base, concluding the STS-125 mission after completing 197 orbits and traveling approximately 5.2 million miles. This would be Atlantis's final landing at Edwards.

Following the landing, the crew completed post-landing checklists and performed the traditional walk-around of the orbiter. Commander Scott Altman humorously remarked, "I didn't realize it was going to be so hard to get back to the Earth!" during a brief meeting with NASA personnel.

In the post-landing press conference, Associate Administrator for Space Sciences Ed Weiler declared the mission a resounding success, noting the long journey from its initial cancellation in 2004 to the triumphant conclusion in 2009. He dubbed the mission "Hubble's Great American Comeback Story, chapter two," reflecting on the telescope's renewed lease on life.

After post-landing processing at Edwards, Atlantis was prepared for its ferry flight back to Florida. On June 1, the orbiter was lifted onto the Shuttle Carrier Aircraft (SCA), a modified Boeing 747, using the Mate-Demate device. Over the next two days, the SCA made several stops for refueling and crew changes, including an overnight stop at Biggs Army Airfield in El Paso, Texas, before arriving back at Kennedy Space Center on June 2. The shuttle's payload bay still carried the mission's equipment, making the vehicle one of the heaviest to be ferried, weighing in at approximately 600,000 pounds when combined with the 747. Upon arrival, Atlantis was towed to the Orbiter Processing Facility to be readied for its next mission, STS-129.

The success of STS-125 marked the completion of the last planned shuttle mission to service the Hubble Space Telescope. Unlike previous missions, Atlantis' trajectory did not allow for a safe haven at the ISS in the event of a critical failure, which necessitated the standby status of the STS-400 rescue mission. Following a successful inspection and reentry clearance, Endeavour was officially released from its standby state on May 21.

STS-127

STS-127, also known as ISS Assembly Flight 2J/A, was a critical mission in the assembly and expansion of the International Space Station (ISS). Launched on July 15, 2009, from Kennedy Space Center in Florida, this mission was the 23rd flight of Space Shuttle Endeavour and marked the continuation of international collaboration in space. The primary objective of STS-127 was to deliver and install the final components of the Japanese Experiment Module (JEM), also known as Kibo, further enhancing the ISS's scientific capabilities.

Mark Polansky commanded the mission, with Douglas Hurley serving as the pilot. The crew included Mission Specialists Christopher Cassidy, Tom Marshburn, David Wolf, Julie Payette from the Canadian Space Agency, and Tim Kopra, who served as the mission's flight engineer and replaced Koichi Wakata on the ISS. Wakata had been aboard the ISS since March 2009 and returned to Earth with the STS-127 crew.

STS-127's key tasks revolved around installing the Exposed Facility (EF) and the Experiment Logistics Module Exposed Section (ELM-ES) in the Kibo laboratory. The EF, often called "Kibo's porch," provided an external platform for experiments exposed to the harsh space environment, allowing for studies that require direct exposure to the vacuum, radiation, and other space conditions. This addition made Kibo the largest laboratory module on the ISS, significantly expanding the station's research capabilities.

Cassidy, Marshburn, and Wolf conducted five spacewalks, successfully installing these critical components. The spacewalks also included replacing batteries on the ISS's Port 6 truss, essential for maintaining the station's power supply. These extravehicular activities (EVAs) were complex and physically demanding, showcasing the crew's expertise and the intricate choreography required for successful ISS assembly missions.

In addition to the hardware installation, STS-127 carried several scientific experiments, including the Shuttle Exhaust Ion Turbulence Experiments (SEITE) and the Japanese-made Investigating the Structure of Paramagnetic Aggregates from Colloidal Emulsions (InSPACE-2). These experiments contributed to a wide range of scientific research, from studying Earth's atmosphere to advancing materials science.

The mission experienced several delays before launch due to weather conditions and technical issues, including a hydrogen leak that had to be repaired. Despite these challenges, Endeavour successfully lifted off on July 15, 2009, and the mission proceeded smoothly.

STS-127 concluded with a successful landing at Kennedy Space Center on July 31, 2009. The mission was a testament to international cooperation and the ongoing efforts to expand the ISS's capabilities. By completing the Japanese Experiment Module, STS-127 fulfilled a significant milestone in the ISS's construction. It strengthened the foundation for future scientific research in space, contributing to humanity's understanding of the universe and developing new technologies.

STS-128

After the foam on the external tank caused damage to the orbiter during the STS-127 mission, engineers implemented stringent testing to ensure the safety of future flights. Discovery's external tank underwent rigorous pull tests to confirm the integrity of the foam insulation before being cleared for use. The results indicated that the foam issues from STS-127 were likely due to surface contamination prior to foam application, and were deemed an isolated incident. With no further concerns, Discovery was moved from the Orbiter Processing Facility to the Vehicle Assembly Building (VAB), where it was mated with the external tank in preparation for the upcoming STS-128 mission.

A notable change implemented for STS-128 involved the Ground Umbilical Carrier Plate (GUCP) vent housing, which had caused issues during previous missions, STS-119 and STS-127. The vent system had exhibited leaks due to a misalignment, prompting the replacement of the one-part rigid seal in the external tank with a two-part flexible seal to address the problem.

On August 4, 2009, Discovery began its slow rollout from the VAB to Launch Complex 39A. The shuttle traveled atop the Crawler-transporter, covering the 3.4-mile (5.5-kilometer) journey. The move, which started at 02:07 EDT, took longer than expected due to adverse weather, including lightning warnings. The crawler had to make several stops to remove mud from its treads and bearings, finally arriving at the launch pad at 13:50 EDT. Once in place, technicians quickly prepared Discovery for the Terminal Countdown Demonstration Test (TCDT), a critical countdown dress rehearsal.

Commanded by veteran astronaut Frederick W. "Rick" Sturckow, the STS-128 crew included Pilot Kevin A. Ford and Mission Specialists José M. Hernández, Patrick G. Forrester, John "Danny" Olivas, European Space Agency (ESA) astronaut Christer Fuglesang, and Nicole Stott. Stott was part of an ISS crew rotation, replacing astronaut Timothy Kopra, who had been aboard the station since July 2009. Kopra returned to Earth with the STS-128 crew after completing his mission as a flight engineer on the ISS.

Nicole Stott was originally scheduled to return to Earth aboard *Soyuz TMA-15*. However, changes in the flight plan were made due to concerns over potential delays in future Space Shuttle missions. Such delays could have extended Canadian astronaut Robert Thirsk's mission aboard the International Space Station (ISS) beyond the preferred six-month duration for station crew members. To avoid this, Stott's return was rescheduled, and she would instead return on the next shuttle mission.

STS-128 became significant as the final Space Shuttle flight used for ISS crew rotation. During this mission, Nicole Stott replaced astronaut Tim Kopra as a member of the station's crew. Stott's stay on the ISS lasted until her return on *STS-129*, although that mission did not carry her replacement to the station, highlighting the shift away from the shuttle as a crew transport vehicle for the ISS.

Another notable aspect of the STS-128 mission was the role of Christer Fuglesang, a Swedish astronaut with the European Space Agency (ESA). The ESA named his mission *Alissé*, a title proposed by Jürgen Modlich from Baierbrunn, Germany. The name draws inspiration from 15th-century explorers who used the trade winds to follow Christopher Columbus across the Atlantic to the New World, symbolizing the spirit of exploration embodied by Fuglesang's journey.

STS-128 also marked a historic moment for Hispanic representation in space exploration. For the first time, two Hispanic Americans flew on the same mission. John "Danny" Olivas, originally from El Paso, Texas, made his second spaceflight, while José M. Hernández, from Stockton, California, made his first. Both astronauts, of Mexican heritage, contributed to the growing diversity within NASA's astronaut corps, serving as an inspiration to the Hispanic community and symbolizing NASA's commitment to inclusion.

Discovery's crew of seven astronauts arrived at Kennedy Space Center on August 5, 2009, to participate in the TCDT. This training exercise, concluding later in the week, involved a full practice countdown, short of the actual liftoff, to familiarize both the crew and the ground team with the launch process. Meanwhile, in an unusual move, engineers modified the left Solid Rocket Booster (SRB) while it was still on the pad. A check valve filter assembly in the booster had broken and was replaced.

Concerns arose when X-ray testing revealed voids in the foam insulation on the external tank. These voids, thought to have formed during the foam's injection molding process, were flagged as a potential risk for foam shedding, similar to what occurred during STS-127. Engineers suspected that trapped air in the voids might expand during ascent due to high temperatures, causing the foam to break away. While a rollback was considered to address the defect, the decision was made to proceed without additional inspections, and the tank was cleared for launch.

The first launch attempt for STS-128, scheduled for August 25, 2009, was scrubbed due to multiple weather violations, including lightning and precipitation near the launch and landing areas. This initial delay pushed the launch back by 24 hours.

The second attempt was aborted on August 26, 2009 due to a technical issue. During fueling, a sensor failed to detect the proper closure of a liquid hydrogen (LH2) fill-and-drain valve in Discovery's aft compartment. Engineers initially suspected the issue lay with the sensor rather than the valve itself. After draining the orbiter's fuel tank, tests showed that the valves functioned normally. Nevertheless, the launch was delayed to give the team more time to evaluate the system and allow the engineers involved in the tests to rest.

After thorough analysis, NASA opted to delay the launch to August 28, 2009, at 23:59 EDT. A flight rule waiver was granted for cycling the LH2 valve, and discussions took place regarding a test failure of an Ares I booster that shared similarities with the SRBs used in Discovery's mission. As the new launch window approached, weather once again threatened to interfere, with storms forming near Kennedy Space Center. However, the weather cleared just in time, and Discovery successfully launched at 23:59 EDT on August 28, 2009.

The primary payload for STS-128 was the Multi-Purpose Logistics Module (MPLM) Leonardo, a crucial component for enhancing the International Space Station's capacity to support a six-member crew. This mission delivered vital supplies, scientific equipment, and life-support systems to the station. Among the key items brought aboard the ISS were three life-support racks, a new crew quarter for installation in the Kibo module, and a treadmill named COLBERT (Combined Operational Load-Bearing External Resistance Treadmill). The treadmill was temporarily placed in Node 2, with plans to move it to Node 3 later. Additionally, an Air Revitalization System (ARS) was temporarily installed in Kibo, to be relocated to Node 3 at a future date.

The Leonardo module itself was a significant piece of equipment. Measuring 21 feet (6.4 meters) in length and 15 feet (4.6 meters) in diameter, it had a launch payload mass of 27,510 pounds (12,480 kilograms) and a return payload mass of 16,268 pounds (7,379 kilograms). The module's empty weight was 9,810 pounds (4,450 kilograms), highlighting its capability to carry large amounts of cargo necessary for the station's operations.

Payload Distribution:

Orbiter Docking System: 1,800 kilograms (4,000 lbs)

Shuttle Power Distribution Unit (SPDU): ~17 kilograms (37 lbs)

APC/MISSE Carriers: 57 kilograms (126 lbs) each

ROEU umbilical: ~79 kilograms (174 lbs)

Leonardo (MPLM FM-1): 12,131 kilograms (26,744 lbs)

Lightweight Multi-Purpose Experiment Support Structure Carrier (LMC): 1,780 kilograms (3,920 lbs)

Orbiter Boom Sensor System: ~382 kilograms (842 lbs)

Canadarm: 410 kilograms (900 lbs)

In total, the shuttle carried 16,973 kilograms (37,419 lbs) of cargo, crucial to ISS operations.

Lightweight Multi-Purpose Carrier (LMC): The Discovery shuttle also transported a Lightweight Multi-Purpose Experiment Support Structure Carrier (LMC), which housed the Ammonia Tank Assembly (ATA). This assembly was necessary to replace an empty ammonia tank on the ISS, a task completed during an extravehicular activity (EVA). Ammonia is vital for the station's cooling systems, making this replacement essential for the long-term operation of the station.

TriDAR: STS-128 marked the first test flight of the TriDAR, a 3D dual-sensing laser camera designed to assist with autonomous rendezvous and docking operations. The TriDAR tracked the ISS's position and orientation during docking, demonstrating its potential for future use in docking procedures for spacecraft like Orion.

Scientific Payloads: The mission also delivered three science racks to the ISS. These included the Fluids Integrated Rack (FIR) and the first Materials Science Research Rack (MSRR-1) for installation in the Destiny laboratory, as well as MELFI-2 (Minus Eighty Laboratory Freezer for ISS), which was installed in Kibo. The FIR enabled experiments studying the behavior of liquids in microgravity, a critical aspect of understanding chemical reactions in space. One of the key experiments involved studying colloids, mixtures where particles remain suspended in liquid without the interference of gravity-driven sedimentation. The Light Microscopy Module (LMM) enabled the observation of heat pipes, important for fluid transfer without the distortions of gravity.

Mission Experiments: STS-128 also participated in a series of crew seat vibration tests, similar to those conducted on STS-125 and

STS-127. These tests aimed to help engineers understand how astronauts experience the physical vibrations of launch and reentry, data that would be critical for designing crew seats in future NASA spacecraft, including Orion.

Another critical experiment repeated during this mission was the Boundary Layer Transition (BLT) Detailed Test Objective (DTO). This experiment, previously conducted on STS-119, involved raising one of the shuttle's thermal protection system (TPS) tiles to trigger a boundary layer transition, where airflow changes from smooth (laminar) to turbulent at high speeds. The tile was raised by 0.35 inches (8.9 mm), compared to 0.25 inches (6.4 mm) during STS-119, to create turbulence at Mach 18, resulting in more heat during reentry.

Additionally, the shuttle tested a catalytic coating for future use in the Orion spacecraft. Two TPS tiles coated with this material were placed downstream from the BLT tile to collect data on entry heating performance. The tiles were equipped with sensors to gather extensive data on how this coating behaved during the intense heat of reentry.

On August 29, 2009, during Flight Day 2, Space Shuttle Discovery embarked on its first full day in orbit, with a crucial task at hand: inspecting the Thermal Protection System (TPS). The crew utilized the Shuttle Remote Manipulator System (SRMS) to grapple the Orbiter Boom Sensor System (OBSS) and carefully survey the wing leading edges, nose, and the Orbital Maneuvering System (OMS) pods. This inspection ensured that no damage had occurred during launch. Meanwhile, other crew members were preparing the space suits for the upcoming three Extra-Vehicular Activities (EVAs) and setting up vital tools for the shuttle's docking with the International Space Station (ISS), including the installation of the Centerline Camera and the extension of the Orbiter Docking System Ring Extension.

Flight Day 3, on August 30, marked a significant milestone as Discovery successfully docked with the Pressurized Mating Adapter (PMA)-2 on the forward end of the Harmony module. Prior to docking, Commander Rick Sturckow executed the Rendezvous Pitch Maneuver, a maneuver that allowed Expedition 20 Commander Gennady Padalka and Flight Engineer Michael Barratt to photograph the shuttle's heat shield. These images were relayed to Mission Control for analysis. After docking, an important crew exchange occurred: Nicole Stott and Tim Kopra switched Soyuz seat liners, officially making Stott a Flight Engineer for Expedition 20 while Kopra resumed his role as STS-128 Mission Specialist. The crews then began transferring supplies and equipment, checking the pressure inside the Multi-Purpose Logistics Module (MPLM) Leonardo, and preparing for the next phase of the mission.

On Flight Day 4, August 31, the MPLM Leonardo was berthed to the Earth-facing port of the Harmony module using the Space Station Remote Manipulator System (SSRMS). After securely attaching the MPLM, the crews opened the hatch and began transferring essential items from Discovery's mid-deck. These included the MDS experiment and the space suits needed for the first spacewalk. Danny Olivas and Nicole Stott, with help from Tim Kopra, prepared the necessary tools for the upcoming EVA. Meanwhile, the ground controllers vented the nitrogen lines of the Port 1 (P1) Ammonia Tank Assembly (ATA) in preparation for its removal during the spacewalk.

Flight Day 5, on September 1, saw the completion of the mission's first spacewalk (EVA 1), carried out by Danny Olivas and Nicole Stott. During the spacewalk, they removed the empty Ammonia Tank Assembly and stowed the EuTef and MISSE 6 experiments. Simultaneously, the crew inside the ISS transferred several critical items, including new crew quarters, the C.O.L.B.E.R.T. treadmill, and the Node 3 Air Revitalization System (ARS). While the treadmill and ARS were stowed temporarily, the new crew quarters were installed in the Kibo module and activation of the system began.

The activities continued on Flight Day 6, September 2, with joint crews finalizing the activation of the new crew quarters. They also transferred the remaining major items from MPLM Leonardo, including the Fluids Integrated Rack (FIR), Materials Science Research Rack, and the Minus Eighty Degree Laboratory Freezer ISS 2 (MELFI-2). Astronauts Danny Olivas and José Hernández engaged the public by answering questions submitted via YouTube and Twitter. As the day progressed, Olivas and Christer Fuglesang prepared for the second EVA by "camping out" in the airlock at a reduced pressure, allowing their bodies to acclimate for the upcoming spacewalk.

Flight Day 7, on September 3, featured the mission's second spacewalk (EVA 2), conducted by Olivas and Fuglesang. They successfully installed a new Ammonia Tank Assembly and accomplished two additional tasks, including installing protective lens covers on the SSRMS End B cameras. As the new tank was integrated into the station's cooling loop, other crew members continued the ongoing transfer of items between Discovery's mid-deck and the ISS.

On Flight Day 8, September 4, the crews took some well-deserved off-duty time to rest and engage in a Public Affairs Office (PAO) event, where they shared a meal and took a group photo. Despite the lighter schedule, progress continued, with the space station crew calibrating the Oxygen Generation System's (OGS) hydrogen sensor and Timothy Kopra and Nicole Stott continuing their handover process. Danny Olivas and Christer Fuglesang spent the night in the Quest airlock, acclimatizing to the lower pressure in preparation for their final spacewalk.

Flight Day 9, on September 5, saw the third and final spacewalk (EVA 3) of the mission, where Olivas and Fuglesang completed all their scheduled tasks. They installed two GPS antennas, deployed the Starboard 3 (S3) Payload Attach System (PAS), and routed cables for Node 3 avionics. The joint crew also focused on transferring return cargo to Discovery, including a bolt replacement on the Common Berthing Mechanism (CBM) to ensure smooth detachment of the MPLM and preparation for future operations with the Japanese H-II Transfer Vehicle (HTV).

On Flight Day 10, September 6, the crew enjoyed more off-duty time but also transferred scientific samples from the ISS to Discovery's Glacier freezer. These samples were destined for analysis on Earth, with the aim of advancing research on bone and muscle loss in space as well as discovering treatments for terrestrial illnesses. The crew also completed the final closeout of MPLM Leonardo, preparing it for its return to Earth.

Flight Day 11, on September 7, marked the closure of hatches between the ISS and Discovery. After completing all transfers, the crew deactivated, demated, and berthed the MPLM Leonardo back into Discovery's payload bay. Before the day's end, the two crews bid farewell during a formal ceremony. The hatches between the two spacecraft were then sealed, and the Pressurized Mating Adapter (PMA)-2 was depressurized in preparation for the shuttle's undocking the following day.

On Flight Day 12, September 8, Space Shuttle Discovery undocked from the ISS at 19:26 UTC. After backing away from the station, the shuttle performed a complete flyaround, providing the crew with an opportunity to photograph the ISS. After two separation burns, astronauts Kevin Ford, José Hernández, and Christer Fuglesang conducted a final inspection of the shuttle's TPS using the OBSS. With the inspection completed, the OBSS was secured, and the SRMS was powered down.

During Flight Day 13, on September 9, the crew of Discovery began preparations for re-entry. The day was spent stowing items and performing system checks, including testing the Flight Control Systems (FCS) and Reaction Control System (RCS) jets. Commander Frederick Sturckow and pilot Kevin Ford also deactivated the Wing Leading Edge System (WLES) and stowed the Ku-band antenna. The crew concluded the day by reviewing landing procedures, anticipating their return to Earth.

However, Flight Day 14, on September 10, saw a delay in Discovery's landing at Kennedy Space Center due to adverse weather conditions. Both the primary and secondary landing windows were waved off, extending the mission by another day.

Finally, on Flight Day 15, September 11, Discovery landed safely at Edwards Air Force Base in California at 20:53 EDT. The mission's conclusion at Edwards marked the final time a shuttle would land at the base and necessitate a ferry flight back to Kennedy Space Center. After refueling stops in Amarillo, Fort Worth, and Barksdale, Discovery returned to its home at Kennedy on September 21, 2009. This landing represented the last in a series of historic milestones for the Space Shuttle program as it entered its final phase of operations.

STS-129

STS-129, designated as ISS Assembly Flight ULF3, was a vital mission in supporting and maintaining the International Space Station (ISS). Launched on November 16, 2009, from Kennedy Space Center in Florida, this mission marked the 31st flight of Space Shuttle Atlantis. The mission's primary objectives were to deliver critical spare parts and equipment to the ISS, ensuring the station's ability to function and sustain operations long into the future.

Commanded by veteran astronaut Charles O. Hobaugh, the STS-129 crew included Pilot Barry E. Wilmore and Mission Specialists Leland D. Melvin, Randolph J. Bresnik, Michael J. Foreman, and Robert L. Satcher Jr. The mission also included ISS crew rotation, with Bresnik and Wilmore serving as flight engineers aboard the ISS during the mission.

Robert Thirsk from CSA was originally slated to return from his stay aboard the station with STS-129, but due to flight delays, it was announced that Stott and Thirsk would swap return seats, with Stott returning aboard STS-129, and Thirsk returning on Soyuz TMA-15.

STS-129 was the 2nd flight to carry two African-American astronauts, Leland Melvin and Robert Satcher. The first was STS-116, which included Robert Curbeam and Joan Higginbotham. This was a particularly significant mission for NASA as STS-129 was the final Space Shuttle crew rotation flight to or from the ISS, the last few flights of the Space Shuttle program, scheduled to end in 2011.

In November 2009, the Space Shuttle Atlantis was scheduled to launch on mission STS-129, but its launch window, set between November 16 and 20, faced multiple challenges. A critical factor was the launch of the Russian Mini Research Module-2 (MRM-2) aboard a Soyuz-U rocket from the Baikonur Cosmodrome in Kazakhstan. Additionally, the Eastern Range at Cape Canaveral had scheduling conflicts with two other satellite launches—Intelsat 14 on November 14 and 15 aboard an Atlas V rocket and a Wideband Global SATCOM satellite on November 19 aboard a Delta IV rocket. These overlapping events complicated Atlantis's launch prospects.

On November 10, 2009, MRM-2 launched successfully and docked with the International Space Station (ISS) two days later. This cleared the way for Atlantis, as the Delta IV launch was delayed, giving Atlantis extra opportunities toward the end of its launch window. Atlas V's scheduled launch with Intelsat 14 was scrubbed on November 14 due to a technical issue requiring a rollback, further reducing the risk of a postponement for Atlantis, which ultimately set its launch for November 16.

Atlantis began its final preparations on October 6, 2009, when it was towed from its hangar in Orbiter Processing Facility-1 to the Vehicle Assembly Building (VAB) at Kennedy Space Center. Known as the "rollover," the process started at 07:00 EDT and was completed by 08:25 EDT. Originally, Atlantis was scheduled to roll out to Launch Pad 39A on October 13, but an issue with a crane during the shuttle's attachment to its external fuel tank and solid rocket boosters caused a delay, pushing the rollout to October 14.

At 06:38 EDT on October 14, Atlantis slowly moved the 3.4 miles to Launch Complex 39A atop a Crawler-Transporter. By 13:31 EDT, the shuttle was securely positioned on the launch platform. Final preparations continued with the installation of the payload and a flight readiness review (FRR) in late October. During the FRR, NASA approved the installation of a special minicam to film Atlantis's external fuel tank during ascent, allowing engineers to study any potential issues with the tank's Ice Frost Ramps.

The FRR also addressed two technical concerns: the potential impact of vibrations from main engine ignition on a bolt structure at the aft of the shuttle, and an aluminum bracket installed to anchor the shuttle's toilet to the crew module. NASA planned to replace the bracket with a titanium version for future missions. On October 29, Atlantis's payload was moved to the launch pad, and the payload was installed into the shuttle's bay by November 4.

By the time of the final post-launch interview on November 16, 2009, Shuttle Launch Director Mike Leinbach noted that Atlantis had achieved a record low number of Interim Problem Reports (IPRs), with only 54 issues recorded—a testament to the efficiency and professionalism of the processing team.

Final launch preparations began in earnest on November 13, 2009, when Atlantis's payload bay doors were closed. At 1:00 PM that day, the official launch countdown clock started, and the crew arrived at Kennedy Space Center's Shuttle Landing Facility. The crew spent the following days conducting practice landings and preparing for launch. On the evening of November 15, the Rotating Service Structure, which protected Atlantis from the elements, was retracted.

Early on November 16, the Mission Management Team gave the green light for tanking operations to begin. By 08:00 EST, the external tank was fully loaded with cryogenic fuel. All teams involved in the mission provided their final "go" for launch during a countdown hold at T-minus nine minutes.

The initial weather forecast called for a 90% chance of favorable launch conditions. However, as launch day progressed, lower cloud ceilings reduced the probability of a successful launch to 70%, but conditions improved to 80% by liftoff.

Atlantis successfully launched on November 16, 2009, at 14:28 EST. Launch commentator George Diller announced the liftoff, stating, "Liftoff of Space Shuttle Atlantis, on a mission to build, re-supply, and to do research on the International Space Station." The shuttle's powered ascent followed the standard mission timeline, with main engine cutoff (MECO) occurring eight minutes and 24 seconds into the flight, and the external tank separating 14 seconds later.

A further boost from Atlantis's Orbital Maneuvering System (OMS) engines was unnecessary, as the shuttle had already reached its planned preliminary orbit. An NC-1 engine firing later adjusted the shuttle's trajectory to match the ISS's, settling it into a 147 by 118 statute mile orbit.

NASA reported three instances of foam loss from the external tank during the shuttle's ascent, but the debris was determined to have occurred after the critical aerodynamic phase of flight and was not a cause for concern. Upon reaching orbit, the crew opened Atlantis's payload bay doors, deployed the radiators, and activated communications with the ground using the Ku-band antenna. The protective doors covering the star trackers were also opened, and the crew prepared the onboard computer network for operations.

To ensure the shuttle was ready for docking with the ISS, the crew conducted a thorough checkout of Atlantis's robotic arm in preparation for a detailed survey of the shuttle's wing leading edge panels and nose cap, scheduled for the next day. After completing these initial tasks, the crew ended their first day in space with a well-earned rest.

The primary mission payload for STS-129 aboard *Atlantis* included the ExPRESS (Expedite the Processing of Experiments to the Space Station) Logistics Carriers (ELC-1 and ELC-2), along with several critical systems, scientific experiments, and unique memorabilia.

ary payload consisted of two ExPRESS Logistics Carriers, ELC-1 and ELC-2. These carriers were designed to transport spare hardware to the International Space Station (ISS) and were essential for ensuring the station's long-term operational capacity. Each carrier was a steel framework with a mass capacity of approximately 4,400 kilograms (9,800 pounds) and a total volume of 30 cubic meters. ELC-1 weighed 6,280 kilograms (13,850 pounds), while ELC-2 weighed 6,100 kilograms (13,400 pounds).

ELC-1 carried vital spare parts, including an Ammonia Tank Assembly, Battery Charger Discharge Unit, a station robotic arm Latching End Effector, a Control Moment Gyroscope, Nitrogen Tank Assembly, Pump Module, Plasma Contactor Unit, and two Passive Flight Releasable Attachment Mechanisms (PFRAMs). It was installed on the Unpressurized Cargo Carrier Attachment System #2 (UCCAS 2) on the port (P3) side of the ISS's main truss.

ELC-2 transported an Oxygen-filled High-Pressure Gas Tank (HPGT), a Cargo Transport Container (CTC-1), a Mobile Transporter Trailing Umbilical System Reel Assembly (MT TUS-RA), a Control Moment Gyroscope, a Nitrogen Tank Assembly, and a Pump Module. It also carried hardware for the Materials International Space Station Experiment (MISSE) 7. ELC-2 was installed on the Starboard 3 (S3) truss segment.

MISSE-7 was a critical experiment designed to expose various materials and coatings to harsh space conditions. The experiment evaluated how materials respond to atomic oxygen, ultraviolet radiation, extreme temperatures, and other factors in the space environment. This data would help scientists develop and test materials for future spacecraft capable of withstanding the rigors of space.

MISSE-7 consisted of two suitcase-sized Passive Experiment Containers (PECs), designated MISSE 7A and MISSE 7B. Once installed on the exterior of the ISS during a spacewalk, these containers opened to expose the materials. MISSE 7A was oriented to face both space and Earth, while MISSE 7B was oriented forward and backward relative to the ISS orbit. The experiment contained both active and passive elements, with passive components evaluated pre- and post-flight in ground laboratories.

Another significant item delivered to the ISS was a repaired S-band Antenna Sub-Assembly (SASA), which had been returned to Earth for repairs during STS-120 in 2007. The SASA was essential for the ISS's communication capabilities, specifically for transmitting and receiving signals between the station and ground control via Tracking and Data Relay Satellites.

The SASA package consisted of three main components: the Assembly Contingency Radio Frequency Group (ACRFG), the SASA boom, and the Avionics Wire Harness. The ACRFG handled the transmission and reception of radio signals, while the boom assembly provided the structural interface for mounting the SASA to the Zenith 1 truss of the ISS. The Avionics Wire Harness supplied power to the ACRFG and facilitated communication signals. During the mission, spacewalkers installed the repaired SASA on the ISS's Zenith 1 truss.

In anticipation of future SpaceX cargo missions to the ISS, *Atlantis* carried a Commercial Orbital Transportation Services (COTS) Ultra High Frequency (UHF) Communication Unit (CUCU) developed by SpaceX and NASA. This unit would allow communication between the ISS and SpaceX's Dragon spacecraft, enabling telemetry transmission and command relays. The CUCU would eventually be integrated into the ISS for future resupply missions.

Along with the CUCU, the Crew Command Panel (CCP) was also flown to the ISS. The CCP would provide astronauts with feedback about the Dragon spacecraft's status during approach to the station and allow them to send basic commands to the spacecraft as needed.

STS-129 also carried items of historical and cultural significance. Astronaut Randolph Bresnik brought a scarf worn by the pioneering aviator Amelia Earhart. The scarf, originally displayed at the Ninety-Nines Museum of Women Pilots in Oklahoma City, had special significance to Bresnik's family. His grandfather, Albert Louis Bresnik, had been Earhart's personal photographer from 1932 until her disappearance in 1937. After returning from the mission, the scarf would be re-displayed at the museum alongside photographs taken by Bresnik's grandfather.

In addition to Earhart's scarf, *Atlantis* carried

memorabilia related to the National Football League (NFL). Items included the official coin for the coin toss of Super Bowl XLIV, a football signed by Pro Football Hall of Fame inductees, and various other NFL-related memorabilia, all of which were flown as part of the mission's cargo.

On November 17, 2009, the crew of *Space Shuttle Atlantis* began their first full day in space at 09:28 UTC, with the primary objective of inspecting the shuttle's thermal protection system (TPS). This inspection was crucial for detecting any signs of damage sustained during launch. Using the shuttle's robotic arm and the Orbiter Boom Sensor System (OBSS), the crew meticulously examined the reinforced carbon nose cap and the wing leading edge panels in a six-hour operation.

The data collected, including images and videos from the TPS survey, were downlinked to mission control for review by an image analysis team. Preliminary assessments revealed no significant damage to the heat shield, as confirmed by Mission Management Team (MMT) chairman LeRoy Cain during the day's briefing.

In addition to the TPS inspection, preparations were made for the upcoming docking with the International Space Station (ISS) scheduled for Flight Day 3. The crew extended the orbital docking system's ring and tested the spacesuits for the planned extravehicular activities (EVAs). Commander Charles Hobaugh installed the docking system's centerline camera, a critical tool for the shuttle's precise rendezvous with the ISS. Two scheduled burns, NC-2 and NC-3, adjusted Atlantis' trajectory for the upcoming docking. The NC-3 burn, lasting 12 seconds, was completed later in the day.

Flight Day 3 saw Atlantis closing in on the ISS as rendezvous operations began. After waking up at 09:28 UTC, the crew initiated a series of orbital maneuvers, including the NH and NC-4 burns, to refine the shuttle's approach trajectory. At 15:52 UTC, Commander Hobaugh performed the Rendezvous Pitch Maneuver (RPM), a backflip that allowed ISS astronauts Nicole Stott and Jeffrey Williams to photograph Atlantis' heat shield from the station. These photos, taken with high-resolution 400mm and 800mm lenses, were sent to mission control for detailed analysis.

At 16:51 UTC, *Atlantis* docked with the ISS at the Harmony module's Pressurized Mating Adapter-2, while orbiting 220 miles above Earth over Australia. Following leak checks, the hatches between the shuttle and the station were opened at 18:28 UTC. The crews greeted one another in a traditional welcome ceremony, followed by a safety briefing, as joint operations began. Notably, Nicole Stott's role as a station flight engineer officially ended as she transitioned to Atlantis' crew.

Later that day, at 19:52 UTC, *Atlantis'* robotic arm, operated by Mission Specialists Melvin and Bresnik, handed over ExPRESS Logistics Carrier 1 (ELC-1) to the station's Canadarm2. Piloted by shuttle Pilot Wilmore and station Flight Engineer Williams, the Canadarm2 attached the carrier to the Port 3 truss at 21:27 UTC. Meanwhile, spacewalkers Michael Foreman and Robert Satcher spent the night in the Quest airlock as part of the "campout" procedure, helping them acclimate to the lower pressure environment in preparation for the next day's spacewalk.

Flight Day 4 commenced with the crew's wakeup call at 09:28 UTC. Mission Control confirmed that no further inspections of Atlantis' heat shield would be necessary, allowing the crew to focus on cargo transfers between the shuttle and the station. NASA also cleared the shuttle's heat shield for re-entry.

The day's main event was the first extravehicular activity (EVA 1), conducted by Foreman and Satcher. Foreman, donning a suit with red stripes, and Satcher in an all-white suit, completed the scheduled tasks almost two hours ahead of time. This allowed them to perform additional tasks, such as lubricating the Kibo robotic arm's snares and routing cables on Zarya. A challenging task involved deploying the Payload Attach System (PAS) on the Starboard 3 truss. After some difficulty loosening a steel bolt, they secured the PAS. The EVA marked the 228th U.S. spacewalk, 134th in support of space station assembly, and was the fourth EVA for

Foreman and the first for Satcher.

Nicole Stott celebrated her 47th birthday in space. However, a false depressurization alarm interrupted the crew's sleep, causing them to extend their sleep period by 30 minutes to compensate for the lost rest.

The crew awoke at 09:28 UTC and resumed their busy schedule of cargo transfers. Over half of the mission's transfer activities were completed by the end of the day. Inside the Unity node, crew members routed cables and hoses in preparation for the arrival of the Tranquility node on a future shuttle mission.

In addition to the transfer tasks, Commander Hobaugh and Mission Specialist Melvin operated the shuttle's robotic arm to remove the Express Logistics Carrier 2 (ELC-2) from the payload bay. The Canadarm2, operated by Melvin and Stott, later attached the carrier to the Outboard Payload Attachment System on the station's truss.

Throughout the day, various crew members participated in media interviews, sharing their experiences in space. The crew also began preparations for the second spacewalk, scheduled for the following day.

After being awakened 30 minutes later than scheduled, the crew's primary focus was on the second spacewalk (EVA 2), conducted by Foreman and Bresnik. The spacewalk began later than planned due to false alarms the previous night, but the pair completed all assigned tasks, as well as several additional tasks originally scheduled for EVA 3.

The spacewalkers worked with precision, completing the installation of components on the ISS truss without any major issues. This marked the 229th U.S. spacewalk and the 107th spacewalk in support of the space station. Foreman, having completed five spacewalks by this mission, accumulated a total of 32 hours and 19 minutes of EVA time.

Earlier in the day, Melvin and Satcher operated the shuttle's robotic arm to hand off the ExPRESS Logistics Carrier 2 to Canadarm2. The carrier was successfully attached to the station's truss at 14:14 UTC. Meanwhile, ground engineers confirmed that a minor misalignment in Atlantis' Orbiter Docking System was not a concern, clearing it for undocking and redocking if necessary.

The crew of *Space Shuttle Atlantis* began Flight Day 7 with an early wakeup call at 07:58 UTC. The day brought exciting news for Mission Specialist Randolph Bresnik, who learned of the birth of his daughter, Abigail Mae Bresnik, at 17:04 UTC the previous day. Bresnik received the news through a private phone patch from Mission Control in Houston, making him the second astronaut to become a father while in space—preceded by astronaut Michael Fincke, who welcomed a baby girl during his time aboard the ISS in 2004.

This milestone added a special touch to a relatively relaxed day for the crew, who were granted a half-day off. They took part in interviews with media outlets such as WTTG-TV in Washington, D.C., Bay News 9 in Tampa, Florida, and WBBM Radio in Chicago. Astronauts Wilmore, Melvin, Satcher, and Stott also participated in an educational event with Tennessee Technological University, where Tennessee students, the University President, and Representative Barton Gordon (D-Tennessee) interacted with the crew. Wilmore, a Tennessee Technological University alumnus, enjoyed a personal connection to the event, with his parents in attendance.

Despite the downtime, some astronauts continued working part-time, transferring equipment between the shuttle and the ISS and investigating false alarms that had occurred earlier in the mission. Preparations were also underway for the third and final spacewalk of the mission. Bresnik and Satcher, assisted by Foreman, prepared their tools and began the "campout" procedure in the Quest airlock, acclimating themselves to the lower pressure for the upcoming EVA.

The crew woke at 07:28 UTC on Flight Day 8, with the primary focus being the mission's third and final spacewalk (EVA 3). Mission Specialists Robert Satcher and Randolph Bresnik were tasked with completing a series of essential tasks outside the ISS. Satcher wore an all-white spacesuit, while Bresnik's suit featured broken red stripes for easy identification. Foreman coordinated the spacewalk activities from inside the shuttle, while Melvin and Wilmore operated the station's robotic Canadarm2.

The spacewalk began slightly later than scheduled due to Satcher needing to reinsert a detached valve in his spacesuit's drink bag, which provides water during the EVA. Despite the delay, the spacewalkers worked ahead of schedule, completing both their primary objectives and several "get-ahead" tasks. EVA 3 marked the 230th spacewalk by U.S. astronauts and the 136th in support of space station assembly, totaling 849 hours and 18 minutes. It was also the 108th spacewalk conducted from the ISS itself, adding another 662 hours and 3 minutes to the space station's EVA record. Notably, it was the second EVA for both Satcher and Bresnik.

Meanwhile, other crew members focused on supporting the spacewalk and completing cargo transfers. Commander Hobaugh and ISS Commander De Winne worked together to transfer nitrogen gas from *Atlantis* to the ISS and also dealt with the broken Urine Processor Assembly/Distillation Assembly (UPA DA), which was packed and transferred to the shuttle for return to Earth.

On November 24, the crew awoke at 06:58 UTC, preparing for the end of their time aboard the ISS. Early in the day, *Atlantis* used its thrusters to boost the station to a higher orbit, increasing its altitude by 1.5 kilometers (0.93 miles) over the course of a 27-minute burn.

Final internal transfers between the shuttle and the station continued throughout the day. In total, around 1,400 pounds of water were transferred from *Atlantis* to the ISS, and 2,100 pounds of experiments and equipment were packed for return to Earth.

A brief interruption occurred at 12:00 UTC when a false fire alarm triggered in the Japanese Kibo laboratory. The alarm was attributed to dust particles stirred up during transfer operations, marking the third false alarm of the mission. Despite these minor issues, both crews continued with their planned activities.

At 13:00 UTC, the shuttle and station crews participated in a joint news conference broadcasted live to NASA centers and international media outlets. During the event, Expedition 21 astronaut Robert Thirsk remarked

that the ISS was "nearly complete," with construction at 86% completion.

Later, at 15:00 UTC, a historic change of command ceremony took place aboard the Destiny laboratory, where European Space Agency astronaut Frank De Winne handed over ISS command to Jeffrey Williams. This marked the first command handover during a shuttle mission. After the ceremony, *Atlantis* crew members took a brief off-duty period before preparing to undock.

At 17:43 UTC, the astronauts bid farewell to the ISS crew in the Harmony module, and the hatches between *Atlantis* and the ISS were closed at 18:12 UTC. This concluded nearly six days of joint operations, with a total of 5 days, 23 hours, and 44 minutes spent working together. In the final hour before undocking, *Atlantis'* crew prepared the shuttle for departure, conducting leak checks and setting up the centerline camera for undocking.

On November 25 (Flight Day 10 – Undocking), the crew awoke at 06:29 UTC to prepare for the shuttle's undocking from the ISS. At 09:53 UTC, *Atlantis* separated from the station, concluding a docked period of 6 days, 17 hours, and 2 minutes. Pilot Wilmore then maneuvered the shuttle to a point 450 feet in front of the station and performed a flyaround to capture imagery of the ISS.

Following the flyaround, *Atlantis* executed separation burns, increasing its distance from the station. The crew encountered a minor issue during the routine waste water tank flush, with the flow reduced due to a potential blockage. Despite this, the crew continued their late inspection of the shuttle's thermal protection system (TPS) using the Orbital Boom Sensor System (OBSS) to survey the right wing leading edge, nose cap, and left wing leading edge. The inspection lasted more than five hours.

Flight Day 11 marked the crew's final full day in space, with a 06:28 UTC wakeup call. The day was significant for being the eighth shuttle mission to coincide with the U.S. Thanksgiving holiday. "Atlantis is in great shape," remarked Entry Flight Director Bryan Lunney, as the crew prepared for their return to Earth.

The crew spent the day stowing items in the

shuttle's cabin and testing re-entry systems. Commander Hobaugh and Pilot Wilmore, assisted by Bresnik, checked out the Flight Control System (FCS) and fired the Reaction Control System (RCS) thrusters. The crew also carried out two scientific experiments involving exhaust plumes observed by ground and satellite stations.

After a midday meal, the astronauts participated in a media interview before stowing the Ku-band antenna and setting up the recumbent seat for Nicole Stott, who had spent 91 days aboard the ISS.

The crew's final wakeup call came at 05:28 UTC on November 27, as they prepared for re-entry and landing. With favorable weather conditions at Kennedy Space Center, Flight Director Bryan Lunney gave the "go" to close the shuttle's payload bay doors at 10:52 UTC. The crew completed their pre-landing checks and, at 13:37 UTC, initiated the deorbit burn over the Indian Ocean, decelerating the shuttle enough to re-enter Earth's atmosphere.

Atlantis made its final descent, entering Earth's atmosphere at 14:12 UTC. At 14:44 UTC, the shuttle's main landing gear touched down on Runway 33 at Kennedy Space Center, followed by the nose gear seconds later. The mission officially ended with wheels stop at 14:45:05 UTC, marking the 72nd shuttle landing at Kennedy Space Center.

After powering down the orbiter, the crew disembarked and performed the traditional walk-around of *Atlantis*, meeting with NASA employees. Speaking briefly to the press, Commander Hobaugh expressed gratitude for the mission's success, stating, "We really had an amazing mission."

The crew returned to Johnson Space Center in Houston on November 30, where they were welcomed with a traditional homecoming celebration at Ellington Field.

STS-130

STS-130, designated as ISS Assembly Flight 20A, was a significant mission in constructing and enhancing the International Space Station (ISS). Launched on February 8, 2010, from Kennedy Space Center in Florida, this mission marked the 24th flight of Space Shuttle Endeavour. The primary objective of STS-130 was to deliver and install the Tranquility module and the Cupola, two critical components that would greatly enhance the ISS's habitability and functionality.

On January 6, 2010, Space Shuttle *Endeavour* was carefully transported to Launch Pad 39A at Kennedy Space Center, marking a critical milestone in the preparation for the STS-130 mission. This mission, scheduled to deliver the Tranquility node and Cupola module to the International Space Station, involved months of meticulous preparation and handling to ensure the shuttle's readiness.

Endeavour's journey toward the launch pad began earlier on December 11, 2009, when the orbiter was moved from the Orbiter Processing Facility 2 (OPF-2) to the Vehicle Assembly Building (VAB) High Bay 1. This transfer, known as "rollover," started at 13:00 EST and took just over an hour, concluding at 14:05 EST. Once inside the VAB, *Endeavour* was attached to its external fuel tank and solid rocket boosters, which had been previously stacked and assembled.

After undergoing further inspection and assembly inside the VAB, *Endeavour* was cleared for its move to Launch Pad 39A. The "rollout" process began early on January 6, 2010, at 04:13 EST. Engineers at Kennedy Space Center took extra precautions due to unusually cold weather, which had extended the time required to prepare the shuttle for transport. The 3.4-mile (5.5 km) journey from the VAB to the launch pad took 6 hours and 24 minutes, with *Endeavour* reaching the pad at 10:37 EST. This careful process ensured that all systems were protected from the elements and that the shuttle was in prime condition for its upcoming mission.

The first launch attempt for *Endeavour*'s STS-130 mission was scheduled for February 7, 2010, at 04:39 EST. Weather forecasts initially predicted a 70% chance of favorable conditions, but as the launch window approached, low clouds over the Kennedy Space Center forced a postponement. The weather, especially concerning the Return-to-Launch-Site (RTLS) abort landing site, degraded significantly, with the chance of favorable conditions dropping to just 30%. After careful monitoring, the launch was scrubbed during the final hold at T-minus 9 minutes.

Despite the setback, preparations continued, and the next launch attempt was set for the following day. On February 8, 2010, at precisely 04:14:08 EST (09:14:08 UTC), *Endeavour* successfully lifted off from Launch Pad 39A. The countdown and launch proceeded smoothly, with a 60% chance of favorable weather, and the crew was safely en route to the International Space Station. This successful launch followed a 23-hour, 34-minute turnaround from the initial scrub, demonstrating the efficiency and expertise of the shuttle launch teams at NASA.

Commanded by George D. Zamka, the STS-130 crew included Pilot Terry W. Virts and Mission Specialists Nicholas J. M. Patrick, Robert L. Behnken, Stephen K. Robinson, and Kathryn P. Hire.

On February 9, 2010, the STS-130 crew dedicated most of their day to the thorough inspection of *Endeavour's* thermal protection system (TPS), a critical safety measure to ensure the shuttle's heat shield was intact for re-entry. All six astronauts participated at different points in this inspection, meticulously scanning the TPS, with particular attention given to the port wing. Meanwhile, astronauts Bob Behnken and Nick Patrick began preparing the spacesuits that would be used for the three spacewalks scheduled during the mission.

As Behnken and Patrick worked on the spacesuits, fellow crew members Stephen Robinson and Kay Hire began checking and preparing the tools essential for the upcoming rendezvous with the International Space Station (ISS). These tools included a hand-held LIDAR gun used to measure the shuttle's closing rate and distance from the ISS, as well as the Orbiter Docking System (ODS), which would physically connect *Endeavour* to the ISS. A centerline camera in the ODS would aid Commander George Zamka during docking procedures. Once Behnken completed his suit preparations, he joined Robinson and Hire in finalizing the docking tools.

The crew's workday on February 10 began with a series of orbital burns designed to close the gap between *Endeavour* and the ISS. At 600 feet (180 meters) below the station, Commander George Zamka initiated the Rendezvous Pitch Maneuver (RPM), slowly rotating the shuttle to allow ISS Commander Jeff Williams and flight engineer Oleg Kotov to photograph *Endeavour's* thermal protection system, adding an additional layer of inspection. At 05:26 UTC (00:06 EST), Space Shuttle *Endeavour* successfully docked with the ISS, marking the start of the mission's most critical operations.

Following routine leak checks, the hatches between *Endeavour* and the ISS were opened at 06:26 UTC (01:26 EST). The two crews, comprising members of Expedition 22 and STS-130, held a standard welcome ceremony followed by a safety briefing. Once formalities were completed, Commander Zamka, Behnken, and Robinson began transferring the spacesuits that Behnken and Patrick would use for the upcoming spacewalks. During this time, Nick Patrick and ISS flight engineer T.J. Creamer used the station's robotic arm, the Canadarm2, to transfer the Orbiter Boom Sensor System (OBSS) back to the shuttle's robotic arm, operated by Kay Hire and pilot Terry Virts.

Flight Day 4 focused on final preparations for the first spacewalk scheduled for the following day. Astronauts Nick Patrick and Bob Behnken organized their tools, while Commander Zamka and ISS flight engineer Soichi Noguchi replaced the Hard Upper Torso (HUT) on Behnken's spacesuit. A problem with a wire harness had prevented the suit's Wireless Video System (WVS) and glove and boot heaters from functioning correctly. After the HUT swap and subsequent testing, the suit was deemed operational.

The crews also engaged in transfer-related activities in the morning, followed by a joint meal. A Public Affairs Office (PAO) event connected the STS-130 and Expedition 22 crews with television stations in Sacramento, California, and Mobile, Alabama, as well as a radio station in St. Louis, Missouri. The day concluded with a spacewalk procedures review before Patrick and Behnken entered the Quest Airlock to sleep in reduced pressure conditions (10.2 psi) to prevent decompression sickness, part of their preparation for the spacewalk.

On February 12, Behnken and Patrick conducted the mission's first spacewalk, beginning at 02:17 UTC. Their tasks included moving through *Endeavour's* payload bay to release the launch locks on the newly delivered Tranquility module and its Cupola. Once cleared, the Tranquility module was carefully relocated to the port side of the Unity node using the station's robotic arm. Before Tranquility's installation, the spacewalkers also removed and secured Dextre's ORU platform on a truss segment, which would serve as a backup platform for a future mission.

After Tranquility was installed, Behnken and Patrick connected temporary heater and data cables between Unity and Tranquility, completing their tasks ahead of schedule. The spacewalk, lasting six and a half hours, concluded at 08:49 UTC, followed by other crew members completing various transfer activities, pushing the transfer process past the halfway mark.

On Flight Day 6, the crews opened the hatches to the new Tranquility module for the first time. *Endeavour* crew members George Zamka, Terry Virts, Stephen Robinson, and Kay Hire assisted in the initial outfitting of the node, preparing it for its operational role on the ISS. During this time, Virts and Hire worked on preparing the Cupola for its move from the end of Node 3 to its permanent location.

Meanwhile, Behnken and Patrick

reconditioned Behnken's original spacesuit for use by Patrick, after a small issue with the suit's fan was identified. The pair then prepared tools for the second spacewalk. PAO events continued, with Bob Behnken and Nick Patrick answering questions collected by Capcom Mike Massimino from Twitter, followed by another session where Kay Hire and Terry Virts answered questions from the Associated Press, CBS News, and Reuters. The day concluded with another spacewalk procedures review.

The second spacewalk of the mission, completed by Behnken and Patrick, commenced on February 14 and lasted 5 hours and 54 minutes. The astronauts' tasks included installing ammonia coolant loops and thermal blankets to protect the hoses, outfitting Tranquility's Earth-facing port for the Cupola, and installing handrails and a non-propulsive vent valve. During the installation of an ammonia hose, a small leak occurred, but procedures were in place to mitigate contamination, requiring a "bake-out" at the end of the spacewalk to neutralize any lingering ammonia.

Simultaneously, Virts, Hire, and other ISS crew members worked on further outfitting the Tranquility module, including setting up its ventilation system, connecting electrical and computer cables, and configuring racks. They successfully activated the ammonia cooling system, ensuring the node's systems were fully operational. In light of the crew's progress, NASA extended the mission by one additional day.

On Flight Day 8, the crew successfully relocated the Cupola from its temporary launch location to its permanent spot on Tranquility's Earth-facing port. The move, executed using the space station's robotic arm, was overseen by Kay Hire and Terry Virts, with ISS Commander Jeff Williams assisting in the release and securement of the module. Although delayed slightly due to tighter-than-expected bolts, the relocation was completed without issue.

Once in place, the crew continued outfitting the Cupola, though the window covers remained closed until after the third and final spacewalk. Meanwhile, Behnken and Patrick began preparations for their third spacewalk, scheduled for Flight Day 10, including resizing another spacesuit due to communication issues in Behnken's original suit.

On the ninth day of the STS-130 mission, the crew carried out the critical task of relocating the Pressurized Mating Adapter 3 (PMA-3). Originally attached to the Harmony node as a temporary measure, PMA-3 was moved to its permanent location at the end of the Tranquility module, where it would remain until 2017. The PMA-3 provides essential shielding against micro-meteoroids and orbital debris, a vital function in maintaining the safety of the International Space Station (ISS). Astronauts Bob Behnken and Nick Patrick, with assistance from Jeff Williams and Soichi Noguchi, executed the release and relocation of PMA-3. Meanwhile, inside the ISS, Kay Hire and Terry Virts continued their work outfitting the Cupola module, which was gradually being prepared for its role as the station's premier observation and robotics platform.

The combined Expedition 22 and STS-130 crews shared a meal and enjoyed some off-duty time later in the day. Before concluding the day's activities, they conducted a review of the procedures for the upcoming third and final spacewalk of the mission, ensuring they were prepared for the tasks ahead.

Flight day 10 marked the mission's final spacewalk, during which astronauts Bob Behnken and Nick Patrick completed a series of vital tasks. These included connecting heater and data cables to PMA-3, removing thermal covers and launch locks from the Cupola, and installing handrails on Tranquility. They also laid a video cable in preparation for the installation of a new base on the Russian segment of the ISS. One of the significant milestones of this spacewalk was Behnken connecting the second ammonia cooling loop to Tranquility and disconnecting a temporary power cable that had been providing interim support.

Once the locks were removed, pilot Terry Virts opened the Cupola's windows for the very first time, unveiling stunning views of Earth below. Kay Hire and Terry Virts also worked on setting up the Cupola's robotics station, positioning it for future use. The day was filled with a mix of tasks, all contributing to the ongoing transformation of the ISS into a more advanced and capable station.

On flight day 11, the crew received a special call from U.S. President Barack Obama, who spoke to them alongside a group of schoolchildren. Following this uplifting conference, the crew focused on transferring Environmental Control and Life Support System (ECLSS) racks into the Tranquility module, an essential step in making the module fully operational. This labor-intensive task, led by ISS Commander Jeff Williams, flight engineer T.J. Creamer, Shuttle Commander George Zamka, and mission specialist Stephen Robinson, took most of the day. Simultaneously, Terry Virts continued configuring the Cupola for future robotics operations, although he encountered minor issues with the installation of some corner panels needed to hold the workstation in place.

Bob Behnken and Nick Patrick shifted their focus to reconfiguring the airlock for future use by the station crew, and by the end of the day, they had transferred their spacesuits and tools back to the shuttle in preparation for their return to Earth. With over 75% of the mission's planned transfers complete, the day concluded with the Shuttle performing a reboost of the ISS using its vernier thrusters.

On flight day 12, the joint STS-130 and Expedition 22 crews completed their remaining transfer tasks and conducted a press conference with journalists from NASA centers and Japan. A significant ceremonial moment occurred when ISS Commander Jeff Williams and Shuttle Commander George Zamka officially opened the Cupola module for use, marking a new era for the ISS's observation capabilities. After the ribbon-cutting ceremony, the crews gathered for a final meal together in the Unity module, celebrating the successful collaboration.

Later, the two crews conducted a farewell ceremony, closing the hatches between the Space Shuttle Endeavour and the ISS as the STS-130 crew prepared for their journey home. The remainder of the day was spent stowing equipment and checking rendezvous tools in anticipation of undocking.

Space Shuttle Endeavour undocked from the

ISS at 00:54 UTC on flight day 13. Pilot Terry Virts carefully backed Endeavour away from the station to a distance of 400 feet (120 meters), allowing for a full fly-around to capture images of the newly upgraded ISS. After completing the fly-around, Virts used Endeavour's thrusters to move the Shuttle behind the station. With the separation burns completed, the crew shifted their focus to a late inspection of the Shuttle's thermal protection system, a critical safety measure before reentry. Commander George Zamka and Terry Virts led this procedure, assisted by rotating crew members. Zamka also performed wastewater and condensate dumps, essential for preparing Endeavour for reentry.

As flight day 14 began, the crew turned their attention to preparing Space Shuttle Endeavour for its final descent. Commander George Zamka and pilot Terry Virts, with assistance from Stephen Robinson, checked out the Flight Control System (FCS) and conducted a hot-fire test of the Reaction Control System (RCS). Communication checks with mission control centers at Merritt Island, White Sands Space Harbor, and Edwards Air Force Base were also successfully completed. The crew, meanwhile, spent the remainder of the day stowing equipment and deactivating the Space Shuttle's robotic arm, all while participating in interviews with CNN, CNN Español, and Univision.

On the final day of the mission, the STS-130 crew awoke early to begin their landing preparations. After closing the payload bay doors and activating the Auxiliary Power Units, they donned their launch and entry suits. At 21:14 UTC, Commander George Zamka and pilot Terry Virts fired Endeavour's orbital maneuvering system (OMS) engines for a deorbit burn, slowing the Shuttle by 200 feet per second. The Shuttle re-entered Earth's atmosphere and touched down safely at Kennedy Space Center's Shuttle Landing Facility on runway 15 at 22:22 EST.

Following the successful landing, the crew exited the orbiter to inspect it before addressing the press on the runway, concluding another successful Space Shuttle mission and marking the end of STS-130.

STS-131

STS-131, designated as ISS Assembly Flight 19A, was a crucial mission in the ongoing resupply and maintenance of the International Space Station (ISS). Launched on April 5, 2010, from Kennedy Space Center in Florida, this mission marked the 33rd flight of Space Shuttle Discovery and played a vital role in sustaining the ISS's operations and scientific capabilities. The mission's primary objectives included delivering essential supplies, scientific experiments, and equipment to the ISS and conducting maintenance tasks to ensure the station's continued functionality.

Commanded by Alan G. Poindexter, the STS-131 crew included Pilot James P. Dutton Jr. and Mission Specialists Richard M. Mastracchio, Dorothy M. Metcalf-Lindenburger, Stephanie D. Wilson, Naoko Yamazaki of the Japan Aerospace Exploration Agency (JAXA), and Clayton C. Anderson. This mission was particularly notable for including three female astronauts, the most women ever flown on a single shuttle mission.

One of the central tasks of STS-131 was the delivery and installation of the Multi-Purpose Logistics Module (MPLM) Leonardo, loaded with over 17,000 pounds of supplies, including scientific experiments, crew provisions, and spare parts. The MPLM, essentially a large pressurized cargo container, was carefully berthed to the ISS using the station's robotic arm,

Canadarm2, and the crew gradually unloaded its contents during the mission.

Most of the mission focused on performing maintenance tasks essential for the ISS's long-term operations. Throughout three spacewalks conducted by Mastracchio and Anderson, the crew successfully replaced an old ammonia tank assembly on the station's truss structure. This task was critical for maintaining the ISS's thermal control system, which relies on ammonia to dissipate heat generated by the station's electronic systems and solar arrays.

The spacewalks also involved retrieving and replacing a Rate Gyro Assembly (RGA), part of the station's guidance, navigation, and control system. This component is vital for maintaining the station's orientation in space. Additionally, the astronauts installed a new tool platform on the station's Dextre robotic arm, enhancing its capability to perform complex robotic tasks in the future.

In addition to its logistical and maintenance objectives, STS-131 carried out multiple scientific experiments to advance microgravity research. One of the key experiments was the Muscle Atrophy Research and Exercise System (MARES), designed to study the effects of microgravity on muscle function and help develop countermeasures for astronauts during long-duration spaceflights. Another experiment, the Window Observational Research Facility (WORF), provided a new, dedicated facility for Earth observation from the station, allowing for more detailed and systematic monitoring of the Earth's surface.

The mission also facilitated a crew rotation on the ISS. Astronaut Timothy Creamer, flight engineer on Expeditions 22 and 23, returned to Earth after nearly six months in space. His place was taken by Tracy Caldwell Dyson, who arrived on a subsequent Soyuz flight.

STS-131 was one of the final flights of the Space Shuttle program, as NASA was winding down the shuttle fleet in preparation for its retirement. The mission highlighted the shuttle's indispensable role in supporting the ISS by delivering critical supplies and equipment and enabling complex maintenance tasks essential for the station's long-term sustainability.

After completing their mission, the STS-131 crew returned to Earth, with Discovery landing safely at Kennedy Space Center on April 20, 2010. The mission was deemed a success, with all objectives met, further solidifying the ISS's status as a premier platform for international collaboration and scientific space research.

STS-132

STS-132, designated as ISS Assembly Flight ULF4, was a landmark mission in the history of the Space Shuttle program, marking the penultimate flight of Space Shuttle Atlantis. Launched on May 14, 2010, from Kennedy Space Center in Florida, this mission was critical in the final stages of assembling and outfitting the International Space Station (ISS). The mission's primary objectives were to deliver and install the Russian Mini-Research Module-1 (MRM-1), spare parts and scientific equipment crucial for the ISS's continued operation.

Commanded by Kenneth T. Ham, the STS-132 crew included Pilot Dominic A. "Tony" Antonelli and Mission Specialists Garrett E. Reisman, Michael T. Good, Stephen G. Bowen, and Piers J. Sellers. This mission was particularly significant as it was the last scheduled flight for Atlantis, a shuttle that had served NASA's human spaceflight program for more than 25 years.

On February 24, 2010, the external tank for Space Shuttle Atlantis, designated ET-136, began its six-day, 900-mile journey across the Gulf of Mexico. It departed from NASA's Michoud

Assembly Facility in New Orleans, Louisiana, en route to the Kennedy Space Center (KSC) in Florida. The massive tank, measuring 154 feet in length and 28 feet in diameter, was transported by the solid rocket booster retrieval ship *Liberty Star* while enclosed in the barge *Pegasus*. After arriving at the KSC turn basin on March 1, 2010, the tank was offloaded and carefully driven to the Vehicle Assembly Building (VAB), where it would undergo final preparations for Atlantis's upcoming mission.

By March 29, 2010, workers in the VAB had lifted ET-136 into high bay No. 1, securing it to Atlantis's twin solid rocket boosters. This day-long operation marked a significant milestone, with the tank firmly bolted in place by 6:00 p.m. EDT, ready to support the shuttle's next flight.

On April 13, 2010, Space Shuttle Atlantis rolled out of its processing bay, Orbiter Processing Facility-1 (OPF-1), at 7:00 a.m. EDT, marking what was then believed to be its final mission. The orbiter paused during its journey to the VAB, allowing engineers and technicians to take photographs with Atlantis as it commemorated its 25th anniversary at KSC—Atlantis had first arrived at the space center in 1985 following its cross-country journey from the shuttle factory in Palmdale, California. The orbiter was then attached to ET-136 and the solid rocket boosters, completing its final major assembly stage.

Two days later, on April 15, 2010, the STS-132 mission payload arrived at Launch Pad 39A. The payload, housed inside a canister shaped like the shuttle's 60-foot-long payload bay, included the Mini-Research Module-1 (MRM-1) and the ICC-VLD cargo-carrying pallet. On the evening of April 21, Atlantis began its rollout to the launch pad, a process that took over six hours to complete. Despite delays due to wet weather, the shuttle stack and mobile launch platform were securely positioned by 6:03 a.m. on April 22, 2010.

As engineers prepared Atlantis for flight, they noticed paint peeling from the MRM-1 module. While the issue was declared not to affect the module's performance, it posed a potential debris hazard in space. This was reminiscent of alarms that occurred during Atlantis's prior STS-129 mission in 2009, where false alerts from the MRM-2 (Poisk) module disrupted the shuttle crew.

On May 5, 2010, NASA held a Flight Readiness Review (FRR) at KSC to finalize Atlantis's launch preparations. At the end of the review, top managers gave the green light for a May 14 launch at 2:20 p.m. EDT. During a news conference that followed, NASA's Associate Administrator for Space Operations, William Gerstenmaier, and Shuttle Program Manager John Shannon discussed technical concerns addressed during the review. One key issue involved ceramic inserts around Atlantis's windows, tested after a similar insert loosened during Discovery's re-entry on STS-131. The inserts were reinstalled on Atlantis using a thicker braided cord to prevent a recurrence.

Atlantis's solid rocket boosters also carried a piece of history, as one segment had previously flown on Atlantis's maiden flight in 1985 (STS-51-J). This segment, now supporting STS-132, reflected the durability and reuse of key shuttle components across multiple missions.

On May 10, 2010, Atlantis's crew arrived at KSC from Johnson Space Center, Houston, to begin final launch preparations. The astronauts, arriving in four T-38 Talon jets, were greeted with anticipation as the countdown to launch neared. The official countdown began on May 11 at 4:00 p.m. EDT, with clocks ticking back from T-43 hours.

By May 12, 2010, program managers held a Mission Management Team (MMT) meeting to assess readiness, officially clearing Atlantis for launch. The favorable weather forecast, with a 70 percent chance of favorable conditions, further boosted expectations for the upcoming liftoff.

Early on May 14, 2010, Atlantis's external tank was loaded with liquid oxygen and liquid hydrogen, marking the start of launch day operations. The shuttle's six-member crew began their preparations at 5:00 a.m. EDT with medical checkups and suiting up. By 11:00 a.m., all crew members were seated aboard the orbiter, performing final communication checks with ground controllers.

Atlantis launched on schedule at 2:20 p.m. EDT on May 14, 2010, with commentator

George Diller noting the shuttle's "historic achievements in space." The shuttle followed its standard ascent timeline, reaching main engine cutoff (MECO) at 8 minutes and 32 seconds into flight. ET-136 separated from Atlantis 15 seconds later, with the orbiter settling into its planned orbit without the need for additional engine boosts. The shuttle's path was adjusted by a brief NC-1 engine firing to align its orbit with that of the International Space Station (ISS).

More than 39,000 spectators witnessed the launch, including notable guests such as television host David Letterman, Apollo astronaut Buzz Aldrin, and former NASA administrator Michael Griffin. Russian dignitaries, including Sergei Ivanov and Anatoly Perminov, were also in attendance.

Once in orbit, the crew deployed the shuttle's payload bay doors and activated critical systems, including the Ku-band antenna and radiators. The crew also conducted inspections, sending imagery back to Earth for analysis. Preliminary reports indicated that ET-136 had performed well, with minimal foam shedding.

Atlantis's STS-132 mission was originally slated to be its final flight. The mission patch depicted the shuttle flying into the sunset, symbolizing its retirement. However, following the mission, Atlantis was prepared as a contingency rescue vehicle for future flights. Ultimately, Atlantis was selected for one final mission—STS-135—launching in July 2011 and marking the true end of the Space Shuttle era.

The payload aboard STS-132 was a vital part of Atlantis's mission to the International Space Station (ISS), carrying scientific equipment, replacement components, and unique memorabilia. The central cargo of the mission was the Russian Rassvet Mini-Research Module 1 (MRM-1), which means "dawn" in Russian. The module was constructed by Energia, a Russian aerospace company, and it arrived at Kennedy Space Center on December 17, 2009, aboard an Antonov 124 cargo plane. Once at Cape Canaveral, the module was transferred to an Astrotech processing bay, where it underwent final preparations for launch.

The MRM-1 was an important addition to the ISS, not only for its research capabilities but also for its role in future expansion. It was outfitted with an airlock and a radiation heat exchanger intended for the Russian Nauka module, which was scheduled for launch in 2021. Additionally, it carried a spare elbow joint for the European Robotic Arm (ERA) and a portable work platform that allowed astronauts to conduct experiments in the vacuum of space. Externally, MRM-1 was equipped with standard grapple fixtures, enabling it to be unloaded from Atlantis's payload bay using the ISS's robotic arm.

Alongside the Rassvet module, Atlantis carried the Integrated Cargo Carrier-Vertical Light Deployable (ICC-VLD2) pallet. This lightweight structure was loaded with six new battery Orbital Replacement Units (ORUs), a Ku-band Space-to-Ground Antenna (SGANT), the SGANT boom assembly, an Enhanced ORU Temporary Platform (EOTP) for the Canadian robotic arm Dextre, and Power & Data Grapple Fixtures (PDGF). The new batteries were installed on the P6 truss of the ISS, replacing older units. The used batteries were loaded back onto the ICC-VLD for return to Earth. The ICC-VLD2, constructed of aluminum, measured 8 feet in length, 13 feet in width, and 10 inches thick, with a launch weight of approximately 8,330 pounds. Upon its return, the pallet weighed 2,933 kilograms (6,466 pounds).

Additional items aboard Atlantis included the Orbiter Boom Sensor System (OBSS), located on the starboard sill of the payload bay, and the Canadarm robotic arm on the port sill. These essential tools were vital for conducting inspections and repairs during the mission.

A more symbolic payload was also carried on STS-132. The mission included memorabilia such as a compact disc containing digital copies of all entries submitted to NASA's Space Shuttle Program Commemorative Patch Contest. The winning patch, designed by Blake Dumesnil of Johnson Space Center, commemorated the conclusion of the Shuttle Program. Seventeen handcrafted beads from artists across North America were also on board as part of the Beads of Courage initiative, aimed at inspiring children with serious illnesses.

In a nod to history, Atlantis also carried a 4-

inch long piece of wood from the famous apple tree that inspired Sir Isaac Newton's theory of gravity. Astronaut Piers Sellers took this artifact, along with a portrait of Newton, into space. The wood was part of the Royal Society's collection in London and was returned after the mission.

Other personal items included a flag from Clarkson University in honor of STS-132 lead flight director Michael Sarafin, an alumnus of the university. A variety of memorabilia was flown inside the shuttle's Official Flight Kit (OFK), ensuring that the legacy of Atlantis's missions would be remembered in many personal and symbolic ways.

During Atlantis' mission, the crew engaged in several scientific experiments, contributing to both short-term and long-term research initiatives. The shuttle transported new experimental equipment to the International Space Station (ISS) while returning completed experiments to Earth at the mission's end, demonstrating the shuttle's vital role in space research.

One of the key short-term experiments conducted was Micro-2, designed by researchers from Rensselaer Polytechnic Institute. This experiment involved sending microorganisms into space to study new ways of preventing the formation and spread of bacterial clusters, known as biofilms. Biofilms pose a potential health risk to astronauts during long-term space missions. After the shuttle returned to Earth, scientists analyzed the biofilms to assess how microgravity affected their growth and development. This research holds promise for improving astronaut health on future missions.

Another short-term experiment was Hypersole, a Canadian study that explored sudden changes in skin sensitivity experienced by astronauts in space. Specifically, the research focused on how the sensitivity of the soles of astronauts' feet affects their balance. Three members of the Atlantis crew participated in identical trials both before the launch and immediately after landing. This experiment was also conducted on five astronauts from subsequent missions STS-133 and STS-134. The findings from Hypersole are expected to contribute to ongoing studies related to aging and

provide insights that could benefit elderly individuals and those suffering from balance disorders.

(SIMPLEX) was another significant study conducted during the mission. On Flight Day 12, the crew performed the SIMPLEX burn, which investigated plasma turbulence caused by the shuttle's exhaust in the ionosphere. Ground-based radars were used to measure the turbulence, helping scientists understand the processes by which chemical releases from the shuttle can induce plasma turbulence. This data is particularly relevant for military operations that rely on radio-based navigation and communication, as plasma turbulence can interfere with these systems.

The crew's first full day in space, starting at 08:20 UTC, was dedicated primarily to inspecting Atlantis' thermal protection system. Using the shuttle's robotic arm and the Orbiter Boom Sensor System (OBSS), the astronauts surveyed the shuttle's exterior for any damage caused during launch. A challenge arose when the crew discovered a snagged cable affecting the Laser Dynamic Range Imager (LDRI) and the Intensified TV Camera (ITVC), tools critical for the inspection. Mission Control switched to the backup sensor package, which consisted of a laser camera and digital camera mounted on the OBSS. Though less advanced, this system provided valuable images for analysis, scanning the shuttle's right wing, nose cap, and left wing, which were sent to the ground for detailed examination.

Commander Kenneth Ham also installed a center-line camera in the Orbiter Docking System (ODS) to assist with Atlantis' approach to the ISS. Meanwhile, astronauts Stephen Bowen and Michael Good spent time preparing spacesuits for transfer to the ISS, while Garrett Reisman assisted with both the thermal protection survey and equipment checkouts.

Two course correction burns were executed on Flight Day 2 to refine Atlantis' trajectory toward the ISS. The first burn, called NC-2, lasted 10 seconds and altered the shuttle's speed by 8 feet per second (2.4 m/s), raising both the apogee and perigee of its orbit by approximately 1 mile (1.6 km). The second burn, NC-3, lasted 8 seconds, adjusting the shuttle's velocity by about 2 feet per second (0.61 m/s).

While these maneuvers were being carried out, Mission Control assessed a potential collision risk with a piece of orbital debris. However, updated tracking information revealed that no avoidance maneuver would be necessary, as the debris was not expected to pass dangerously close to the ISS.

The crew of STS-132 began their third day in space at 07:20 UTC, preparing for one of the mission's most critical operations: docking with the International Space Station (ISS). Under the command of Ken Ham, a series of precisely timed rendezvous burns—NH, NC4, and TI—were executed to gradually align Atlantis' orbit with that of the ISS. The NH burn, the longest of these maneuvers, lasted 1 minute and 24 seconds and increased the shuttle's velocity by 132 feet per second. This burn elevated Atlantis into a 212-by-145-mile orbit. The subsequent NC4 circularization burn, lasting 63 seconds, placed the shuttle into a near-circular orbit at 214-by-210 miles. The final terminal initiation (TI) burn, lasting 12 seconds, brought the shuttle into close proximity with the ISS, leaving just 9 miles (14 km) between them.

At 13:26 UTC, Ken Ham, now flying from the shuttle's aft flight deck, positioned Atlantis beneath the ISS for the Rendezvous Pitch Maneuver (RPM), a 360-degree flip designed to allow the ISS crew to photograph the shuttle's thermal protection system. ISS crew members Oleg Kotov, Timothy Creamer, and Soichi Noguchi used high-powered cameras with 400mm and 800mm lenses to capture 398 images of Atlantis' underside, which would later be analyzed for potential damage from launch.

By 14:28 UTC, Atlantis docked with the ISS Pressurized Mating Adapter-2 (PMA-2) as the two spacecraft orbited 220 miles (350 km) above the South Pacific Ocean. After docking, Atlantis' vernier thrusters were fired to reorient the ISS and minimize the risk of Micro-Meteoroid Orbital Debris (MMOD) impacts. Crews from both Atlantis and the ISS then conducted a series of leak checks before opening the hatches at 16:18 UTC.

Following a brief welcome ceremony and safety briefing from the ISS crew, the astronauts quickly transitioned into work mode, beginning equipment and supply transfers. One of the first items moved to the station was the crew's spacesuits. In addition, high-priority Japanese Aerospace Exploration Agency (JAXA) experiments were transferred to the Kibo module by station crew member Soichi Noguchi.

Meanwhile, astronauts Piers Sellers and Tracy Caldwell Dyson began work to relocate the Integrated Cargo Carrier-Vertical Light Deployable (ICC-VLD) platform. They used the station's robotic arm to transfer the pallet from Atlantis to the ISS's mobile base system in preparation for the upcoming spacewalks. To wrap up the day, the entire Atlantis crew participated in a thorough review of the next day's spacewalk procedures. Mission Specialists Garrett Reisman and Stephen Bowen camped out overnight in the Quest airlock, where the air pressure was reduced to help them acclimate to the space environment and prevent decompression sickness, known as "the bends."

The crew started Flight Day 4 with the news from Mission Control that no detailed flight inspections would be required that day, allowing them to focus on the first of three planned extravehicular activities (EVAs). Mission Specialists Garrett Reisman and Stephen Bowen prepared to exit the airlock to install new components on the ISS and perform essential maintenance.

Their first task was the installation of a spare Space to Ground Antenna (SGANT). Additionally, they worked on adding an enhanced tool platform for the Special Purpose Dexterous Manipulator (SPDM), commonly known as Dextre. They also released the launch locks on six new batteries that were scheduled to be installed on the Port 6 (P6) truss segment. These batteries were critical for maintaining the station's power supply.

As Reisman and Bowen prepared for their spacewalk, they were assisted by Expedition 23 Flight Engineer Timothy Creamer, who helped with suit-up preparations. Meanwhile, Michael Good and Pilot Dominic Antonelli, acting as the intravehicular officers, coordinated spacewalk operations from inside Atlantis. Station Flight

Engineer Tracy Caldwell Dyson, alongside Piers Sellers, operated the station's robotic arm during the EVA. Commander Ken Ham closely supervised the entire extravehicular activity from inside Atlantis.

However, the spacewalk was not without its challenges. While installing the SGANT, the astronauts noticed a small gap between the antenna dish and its mounting pole. After adjusting the bolts and applying a higher torque setting, they reduced the gap, though a final determination about its acceptability would be left to engineers on the ground. Another issue occurred when Bowen removed a cover from a connector, which triggered a shutdown of one of the station's Command and Control (CNC) computers. This temporary glitch caused a two-minute loss of communication and halted the operation of Canadarm2, the station's robotic arm, as both the ground and onboard teams reconfigured the system.

Despite these challenges, the spacewalk concluded successfully at 19:19 UTC. Reisman and Bowen conducted a full inventory of their tools before reentering the Quest airlock. This marked the second spacewalk for Reisman and the fourth for Bowen. It was also the 237th spacewalk conducted by U.S. astronauts and the 144th in support of ISS assembly and maintenance.

Concurrently, the shuttle's robotic arm successfully grappled the Mini-Research Module-1 (MRM-1) in preparation for its installation on the Zarya module the following day. This milestone set the stage for continued ISS expansion and scientific progress.

On Flight Day 5, the focus of the Atlantis crew was the installation of the Mini-Research Module-1 (MRM-1) onto the International Space Station (ISS). As the shuttle orbited over Tampa, Florida, Commander Ken Ham and Pilot Tony Antonelli operated Atlantis' robotic arm to unberth MRM-1 from the shuttle's payload bay at 09:49 UTC. By 10:14 UTC, the module was handed off to the station's Canadarm2, operated by Mission Specialists Garrett Reisman and Piers Sellers from the station's Cupola. Canadarm2 then delivered MRM-1 to its final position on the Earth-facing port of the Zarya service module,

where it docked at 12:20 UTC while the ISS orbited over Argentina.

During the docking procedure, Sellers reported to Mission Control that the expected "capture 1" signal didn't appear, to which CAPCOM Steve Swanson humorously responded, "Garrett did too good of a job flying. He went right down the middle and got a hole in one." The docking marked a milestone as it was the first time the Russian automated docking system had been used in conjunction with the station's robotic arm.

Later in the day, at 17:20 UTC, shuttle crew members Ham, Reisman, and Sellers, along with station crew members Oleg Kotov, Alexander Skvortsov, and Tracy Caldwell Dyson, participated in a press conference with reporters from MSNBC, Fox News, and CNN. Topics discussed included the crews' experiences in space, ongoing medical experiments aboard the ISS, and the Gulf of Mexico oil spill.

As the day progressed, Reisman and Sellers used Canadarm2 to unberth the Orbiter Boom Sensor System (OBSS) from Atlantis' cargo bay and hand it off to the shuttle's robotic arm, operated by Ham and Antonelli. Meanwhile, Mission Specialists Stephen Bowen and Michael Good prepared for the next day's second spacewalk, configuring tools and preparing their spacesuits. At 21:45 UTC, Bowen and Good began their "campout" in the Quest airlock, where the pressure was reduced to 10.2 psi to help prevent decompression sickness.

The second spacewalk of the mission, **EVA 2**, commenced on Flight Day 6, with the primary objective being to remove and replace batteries on the Port 6 (P6) truss of the ISS. The spacewalk began at 10:38 UTC, ahead of schedule. Stephen Bowen's first task was to fix a cable snag in the OBSS's pan and tilt mechanism, which he completed in under 30 minutes. Meanwhile, Michael Good began work on replacing the batteries, originally installed in November 2000.

The plan was to replace three batteries, but the efficient work of Bowen and Good allowed them to install a fourth battery as well. After battery replacement was complete, the astronauts moved on to secure the backup Ku-band antenna on the Z1 truss, tightening bolts that had caused issues during the first EVA. A "wiggle test" confirmed that the antenna was now firmly in place, and the launch locks were removed, leaving the antenna operational.

During the spacewalk, Commander Ken Ham provided photo and television support, while Pilot Tony Antonelli served as the spacewalk choreographer. ISS crew member Tracy Caldwell Dyson assisted with preparations inside the station. EVA 2 was the 238th spacewalk conducted by U.S. astronauts and marked Bowen's fifth and Good's third spacewalk. It was also the 145th spacewalk in support of ISS assembly and maintenance.

On Flight Day 7, the crew took some time for rest but remained focused on preparing for the final spacewalk. Earlier in the day, at 10:52 UTC, ISS Commander Oleg Kotov and Flight Engineer Alexander Skvortsov opened the hatch to the newly installed MRM-1 module. Wearing eye and breathing protection as a precaution, they discovered some metal filings drifting inside the module, but otherwise reported that it appeared clean. Ground controllers in Houston and Moscow worked with the crew to develop a plan for safely removing the debris.

At 12:25 UTC, Commander Ham, Antonelli, Sellers, and Caldwell Dyson participated in interviews with the Associated Press, Fox News Radio, and CBS News. Ham also joined Mission Control to celebrate Lonnie J. Schmitt, the first flight controller to complete 100 shuttle missions.

In addition to the media engagements, the crew spent time transferring equipment, supplies, and experiments between Atlantis and the ISS. Mission Specialists Good and Reisman continued preparations for EVA 3, and the crew conducted a final review of the spacewalk procedures. As part of their preparations, Good and Reisman spent the night in the Quest airlock, with its air pressure again reduced to 10.2 psi.

On Flight Day 8, the crew executed EVA 3, the final spacewalk of the STS-132 mission. Michael Good and Garrett Reisman successfully connected two ammonia jumpers on the P4/P5 truss segment, then proceeded to the P6 truss to complete the battery replacement. They installed the last two new batteries and retrieved an old battery that had been temporarily stowed on the truss during the previous EVA.

After finishing the battery work, Good and Reisman moved to Atlantis' payload bay to remove a grapple fixture and return it to the Quest airlock. They then fixed insulation on the Dextre robot and stored tools in an external toolbox on the Z1 truss. Tony Antonelli choreographed the spacewalk from inside the shuttle, while Ken Ham and Steve Bowen focused on continuing the transfer work between Atlantis and the ISS. EVA 3 was the 239th U.S. spacewalk, marking Good's fourth and Reisman's third, and the 146th in support of ISS assembly.

Flight Day 9 provided the crew with some well-deserved off-duty time in the afternoon, but the morning was spent reinstalling the ICC-VLD platform into Atlantis' payload bay. The platform had served its purpose for the mission and was returned using Canadarm2, operated by Piers Sellers, Garrett Reisman, and Tracy Caldwell Dyson. The reinstallation began just after 04:30 EDT and was completed by 05:50 EDT.

In addition to their work, the combined shuttle-station crew answered questions from students at 12 NASA Explorer Schools across the U.S. During this educational outreach event, the astronauts shared their experiences aboard the ISS and discussed the importance of space exploration. Following this, the crew enjoyed a joint meal before the Atlantis astronauts took a break with two and a half hours of off-duty time

On flight day 10 of the STS-132 mission, the joint crews of Space Shuttle *Atlantis* and Expedition 23 aboard the International Space Station (ISS) awoke to begin the final hours of their shared mission. This day marked the culmination of their collaborative work, including the transfer of critical time-sensitive scientific research samples requiring careful temperature control. With the transfers successfully completed, both crews gathered for a joint press conference, capturing the historic moment with a group photo. Before parting ways, they held a farewell ceremony filled with camaraderie and mutual appreciation.

After the ceremony, the hatches between *Atlantis* and the ISS were closed, followed by a thorough leak check to ensure all systems were secure. At 15:22 UTC, the space shuttle undocked from the ISS as the two spacecraft orbited 220 miles (350 kilometers) above the Southern Ocean, southwest of Perth, Australia. Guided by Pilot Tony Antonelli, *Atlantis* slowly backed away to a distance of approximately 400 feet (120 meters), allowing Antonelli to conduct a picturesque fly-around of the space station. This maneuver enabled both the ISS and shuttle crews to photograph the spacecraft in orbit, capturing breathtaking views of the two engineering marvels against the backdrop of space. Following the fly-around, the shuttle executed two separation burns, gently propelling *Atlantis* away from the ISS as it began its return journey to Earth.

The crew of *Atlantis* started flight day 11 with a brief period of personal time before embarking on the crucial task of conducting a late inspection of the shuttle's Thermal Protection System (TPS), specifically the wing leading edges and nose cap. The inspection, carried out using the shuttle's robotic arm and its Orbiter Boom Sensor System (OBSS) extension, ensured the shuttle was free from damage that might have occurred during its mission in space. By 09:50 UTC, the right wing inspection was complete, followed by the nose cap at 10:52, and finally the left wing at 11:17.

While the inspection proceeded smoothly and ahead of schedule, other crew members busied themselves with post-undocking activities, such as stowing items no longer required for the mission. Spacewalkers Mike Good and Steve Bowen carefully cleaned and stored their spacesuits in preparation for the upcoming landing. The rest of the day was spent with the crew enjoying some well-deserved off-duty time, a brief respite as they approached the final phases of their mission.

As the mission neared its conclusion, the astronauts dedicated flight day 12 to readying the shuttle for re-entry and landing. Commander Ken Ham, Pilot Antonelli, and Mission Specialist Mike Good conducted a series of critical tests on *Atlantis'* flight control system (FCS), including hot-fire checks of the rudder and flaps, which would steer the shuttle through the atmosphere toward the runway. Afterward, Ham and Antonelli tested the shuttle's 44 attitude control thrusters, ensuring precise orientation for the upcoming descent from space.

Throughout the day, all crew members worked to secure the cabin, methodically stowing equipment in preparation for the shuttle's return to Earth. The crew also gathered for a deorbit briefing at 5:40 am EDT, where they reviewed procedures for their final hours in orbit. Late in the day, Mission Specialists Garrett Reisman and Piers Sellers stored the Ku-band antenna in *Atlantis'* cargo bay, completing the final preparations for landing.

The final day of the STS-132 mission began early for the crew, with wake-up calls at 12:20 EDT (4:20 UTC). At 7:40 UTC, they initiated deorbit preparations, securing the payload bay doors by 9:01 UTC. At 11:42 UTC, *Atlantis* fired its engines for the deorbit burn, a precise three-minute maneuver that initiated the shuttle's descent from orbit as it soared 220 miles (350 kilometers) above Indonesia. As the spacecraft re-entered Earth's atmosphere at an altitude of 400,000 feet (120,000 meters) and traveling at Mach 25, the crew executed a series of s-turns to reduce speed, reaching Mach 22 by the time they were 2,000 miles (3,200 kilometers) from the Kennedy Space Center (KSC).

At 12:34 UTC, *Atlantis* was approximately 180,000 feet (55,000 meters) above the Earth, hurtling towards the KSC runway at 9,200 miles per hour. This phase of re-entry saw the shuttle endure intense heating, peaking at 2,900 degrees Fahrenheit for a short duration. As the shuttle descended further, long-range cameras at KSC captured its approach, gliding towards runway 33. At 12:44 UTC, Commander Ham manually took control of the orbiter, guiding it safely to the ground.

Touchdown occurred at 08:48:11 EDT (12:48:11 UTC) on *Atlantis'* main wheels, with the nose gear touching down 10 seconds later. The shuttle came to a full stop at 08:49:18 EDT, concluding the mission after 11 days, 18 hours, 29 minutes, and 9 seconds. During the mission, *Atlantis* traveled a total of 7,724,851 kilometers (4,800,000 miles) in orbit, completing its journey with a flawless landing at the Shuttle Landing Facility at Kennedy Space Center.

The crew of STS-132 departed Florida for Houston on May 27, where they were welcomed home with a ceremony held at 5 pm EDT at NASA's Ellington Field Hangar 276.

STS-133

STS-133 was the 39th and final mission of the Space Shuttle Discovery, marking a significant moment in NASA's storied Shuttle program. Launched on February 24, 2011, from Kennedy Space Center in Florida, this mission was initially scheduled for November 2010 but faced multiple delays due to technical issues.

STS-133's crew consisted of six veteran astronauts: Commander Steven Lindsey, Pilot Eric Boe, and Mission Specialists Alvin Drew, Steve Bowen, Michael Barratt, and Nicole Stott. Notably, Steve Bowen replaced astronaut Timothy Kopra, injured in a bicycle accident just weeks before the mission. Bowen, already a seasoned spacewalker from previous missions, performed admirably on short notice, exemplifying the flexibility and readiness of NASA's astronaut corps.

STS-133, originally scheduled for launch on September 16, 2010, faced multiple delays that pushed its launch to February 24, 2011. In June 2010, NASA rescheduled the launch to late October, placing it ahead of the next planned mission, STS-134. This delay marked the longest vertical processing period for a shuttle—170 days—since STS-35's 185-day span.

On September 9, 2010, Space Shuttle *Discovery* was transferred from its hangar in Orbiter Processing Facility (OPF)-3 to the Vehicle Assembly Building (VAB). Emerging at 6:54 AM EDT, the shuttle began its quarter-mile rollover to the VAB at 10:46 AM EDT, marking its 41st such journey. This rollover was originally scheduled for the previous day but was delayed due to a broken water main that disrupted fire suppression systems. The transfer completed smoothly, setting the stage for the shuttle's integration with its external fuel tank and solid rocket boosters (SRBs).

Inside the VAB, the assembly process continued. The SRBs, designated as flight set 122 by Alliant Techsystems, were a combination of one new segment and segments reused from 54 previous shuttle missions dating back to STS-1. After *Discovery* was rotated vertically and lifted into the high bay, engineers encountered a minor complication: a pre-positioned nut inside the aft compartment slipped out of place. This initially raised concerns that the shuttle would need to be returned to a horizontal position for repairs, but engineers were able to resolve the issue on-site, completing the attachment of the shuttle to its external tank, known as the "hard mate," on September 11.

On September 20, 2010, *Discovery* began its 3.4-mile rollout to Launch Pad 39A at 7:23 PM EDT, arriving at the pad approximately six hours later. More than 700 shuttle workers and their families watched the shuttle's journey as it was secured on the pad by early morning the following day. This marked *Discovery*'s 44th rollout to a launch pad in its long career.

As launch preparations continued, engineers discovered several technical issues that caused further delays. On October 14, 2010, a vapor leak was detected in *Discovery*'s Orbital Maneuvering System (OMS) during routine inspections. The leak, caused by a faulty propellant line in the shuttle's aft compartment, emitted monomethyl hydrazine (MMH), a highly toxic fuel used in the OMS engines. Despite replacing an Air Half Coupling (AHC) flight cap, the leak persisted. Engineers ultimately drained the OMS tanks and conducted in-situ repairs at the launch pad to avoid rolling *Discovery* back to the VAB. After replacing faulty seals, further testing confirmed that the issue was resolved.

However, another issue emerged on November 2, when engineers identified an electrical fault in the backup Main Engine Controller (MEC) for engine No. 3. Initial troubleshooting suggested the problem was caused by transient contamination in a circuit breaker, but further tests revealed a more significant issue. As a result, NASA decided to delay the launch for at least 24 hours to allow for a comprehensive review.

On November 5, during the fueling process for another launch attempt, a hydrogen leak was detected at the Ground Umbilical Carrier Plate (GUCP), an attachment point for a pipe that vented gaseous hydrogen from the external tank. Despite efforts to stop the leak by adjusting the vent valve, the problem persisted, and the leak spiked to dangerous levels. After purging the tank and disconnecting the vent arm, engineers inspected the GUCP and found that an internal seal had been asymmetrically compressed during installation. The seal and hardware were replaced, and new measurements ensured proper alignment before continuing with launch preparations.

Further complications arose when cracks were discovered in the foam insulation of *Discovery*'s external tank during inspections after the November 5 launch attempt. The cracks appeared in the flange between the liquid oxygen tank and the intertank, a section of the external tank. To investigate, NASA conducted a full tanking test in December, loading the external tank with cryogenic propellants and using Optical Strain systems to measure displacements and strains in the tank structure. These tests confirmed that the cracks were caused by thermal stress during fueling, exacerbated by the use of lightweight aluminum-lithium alloy in the tank's construction.

Technicians repaired the damaged sections by installing "doublers," thicker metal reinforcements designed to prevent further cracking. These repairs were completed while *Discovery* remained on the launch pad. By November 23, scanning of the stringers and flanges revealed no additional damage, allowing NASA to move forward with launch preparations.

Despite efforts to launch in December 2010, NASA ultimately postponed the mission to February 2011 due to ongoing repairs and conflicts with the International Space Station's schedule. The launch was officially set for February 24, 2011, after the Flight Readiness Review confirmed the shuttle's safety and the resolution of earlier technical issues.

On February 24, 2011, Space Shuttle *Discovery* lifted off from Kennedy Space Center's Launch Pad 39A at precisely 16:53:24 EST, marking its final journey into space. The liftoff was initially scheduled for 16:50:24 EST but was delayed by three minutes due to a minor glitch in the Range Safety Officer's computer system. Despite this delay, *Discovery* ascended smoothly, taking 8 minutes and 34 seconds to reach orbit. During the ascent, a piece of foam was observed breaking away from the external tank at approximately four minutes into the flight. This foam shedding was not considered a threat, as it occurred after the shuttle had passed through the most sensitive phases of the atmosphere. NASA engineers attributed the foam loss to a phenomenon known as "cryo-pumping," where trapped air in the foam liquefies during fueling and re-expands during ascent, causing small sections of foam to break away.

Once *Discovery* reached orbit, the crew opened the shuttle's payload bay doors and activated the Ku-band antenna for high-speed communication with Mission Control. At the same time, Pilot Eric Boe and Mission Specialist Alvin Drew deployed the Shuttle Remote Manipulator System (SRMS), or Canadarm, to begin operations. Later that day, imagery of the external tank captured during launch was downlinked to NASA engineers for analysis.

On Flight Day 2, the crew began preparations for docking with the International Space Station (ISS). Early in the day, the shuttle's Orbital Maneuvering System (OMS) engines performed the NC2 burn to adjust *Discovery*'s trajectory and bring it closer to the ISS. Commander Steve Lindsey, alongside Boe and Drew, conducted a detailed inspection of the shuttle's Re-enforced Carbon-Carbon (RCC) panels using the Orbital Boom Sensor System (OBSS). This inspection focused on the starboard wing, nose cap, and port wing, taking about six hours to complete.

Meanwhile, Drew, along with Mission Specialists Michael Barratt and Steve Bowen, prepared the Extravehicular Mobility Units (EMUs) that would be used for the two planned spacewalks. In the evening, another OMS burn, the NC3 burn, further aligned the shuttle for rendezvous with the space station.

Discovery docked with the ISS at 19:14 UTC on February 26, marking the 13th time it had visited the station. Although the docking occurred on schedule, a delay in achieving a "hard mate" between the shuttle and ISS due to relative motion between the two spacecraft pushed the crew's timeline back by 40 minutes. Once the hatches were opened at 21:16 UTC, the Expedition 26 crew warmly greeted the STS-133 crew.

The day's primary task was the transfer of the ExPRESS Logistics Carrier-4 (ELC-4) from *Discovery*'s payload bay to its permanent position on the ISS. This operation required the coordinated use of both the Space Station Remote Manipulator System (SSRMS), operated by Nicole Stott and Michael Barratt, and the Shuttle Remote Manipulator System, operated by Boe and Drew. The transfer was completed successfully when ELC-4 was installed on the S3 truss at 03:22 UTC on February 27.

On Flight Day 4, Stott and Barratt used the Canadarm2 to grapple the OBSS from *Discovery*'s payload bay and pass it to the Shuttle Remote Manipulator System. This handoff allowed the OBSS to be moved out of the way, making room for the Permanent Multipurpose Module (PMM) to be removed from the payload bay. The crew also participated in media interviews with various news outlets and conducted additional cargo transfers between the shuttle and ISS.

Later in the day, Bowen and Drew prepared tools for their upcoming spacewalk on Flight Day 5. To prevent decompression sickness, they spent the night in the Quest airlock, where the air pressure was lowered to 10.2 psi, a standard procedure before extravehicular activities.

The first of two spacewalks, or Extra-Vehicular Activities (EVAs), took place on Flight Day 5. Bowen and Drew exited the Quest airlock at 10:46 EST, switching their suits to internal battery power to begin their six-hour and 34-minute spacewalk. Their first task was to install a power cable between the Unity and Tranquility modules, providing a contingency power source if needed. They then relocated a failed ammonia pump to External Stowage Platform 2.

Other tasks included installing a wedge under a camera on the S3 truss and performing a Japanese experiment called "Message in a Bottle" to collect a sample of vacuum. Despite minor delays due to issues with the Cupola module's robotic control station, the spacewalk was a success, and the crew returned to the airlock at 17:20 EST.

Flight Day 6 saw the installation of the Leonardo Permanent Multipurpose Module (PMM) to the nadir port of the Unity module. This marked the transition of the Leonardo module from a cargo module into a permanent part of the ISS. Bowen and Drew also reviewed procedures for their second spacewalk and began their pre-EVA campout in the Quest airlock.

Bowen and Drew conducted their second spacewalk on Flight Day 7. Drew removed thermal insulation from a platform, while Bowen installed a camera assembly on the Dextre robotic arm and swapped out an attachment bracket on the Columbus module. Additionally, Drew installed a light on a cargo cart and repaired insulation on a valve on the station's truss. Meanwhile, the shuttle and ISS crew began the internal outfitting of the Leonardo PMM.

During these days, the crew continued transferring cargo from the Leonardo PMM into the ISS, while also completing additional mission objectives. On Flight Day 9, the crew enjoyed some off-duty time before reconfiguring equipment used during the spacewalks. The outfitting of the Leonardo module continued, and the crew conducted a final inspection of *Discovery*'s Thermal Protection System using the OBSS on Flight Day 11.

On March 7, *Discovery* undocked from the ISS for the final time, performing a fly-around of the station before final separation. The shuttle crew then conducted a late inspection of the shuttle's heat shield using the OBSS, confirming that it was safe for re-entry.

The crew spent Flight Day 13 stowing equipment and preparing for re-entry. A checkout of the shuttle's flight control system was conducted, along with a hot-fire test of the Reaction Control System. The Ku-band antenna was stowed, and the final deorbit preparation briefing was carried out.

On March 9, *Discovery* finally returned to Earth, closing the shuttle's payload bay doors and performing a successful deorbit burn. The shuttle re-entered the atmosphere and touched down at Kennedy Space Center's Shuttle Landing Facility at 11:58:14 EST. This marked *Discovery*'s final landing, ending its illustrious 27-year career. It was the last shuttle landing to occur in daylight, as the two remaining missions of the program would land at night.

STS-134

STS-134 was the penultimate mission of NASA's Space Shuttle program, flown by the Shuttle Endeavour. Launched on May 16, 2011, from Kennedy Space Center in Florida.

The STS-134 crew was led by Commander Mark Kelly, an experienced astronaut who had flown on several previous Shuttle missions. The crew included Pilot Gregory H. Johnson and Mission Specialists Michael Fincke, Greg Chamitoff, Andrew Feustel, and European Space Agency (ESA) astronaut Roberto Vittori. Kelly's participation in the mission drew considerable public attention, as he was the husband of U.S. Congresswoman Gabrielle Giffords, who was recovering from an assassination attempt earlier in the year.

The Space Shuttle program was initially scheduled to retire after the completion of mission STS-133, but controversy arose surrounding the cancellation of several crucial International Space Station (ISS) components, particularly the Alpha Magnetic Spectrometer (AMS), to meet the Shuttle's retirement deadlines. This prompted the United States Government to reconsider, leading to discussions of an additional mission to ensure the AMS and other science experiments reached the ISS.

On June 19, 2008, the U.S. House of Representatives passed the NASA Authorization Act of 2008, providing NASA with the necessary funding for an additional mission. This act explicitly directed NASA to deliver scientific experiments to the ISS, ensuring that critical research components, including the AMS, would be part of the station's scientific agenda. The U.S. Senate included a similar mandate, with unanimous approval from the Senate Committee on Commerce, Science, and Transportation on June 25, 2008. The full Senate amended and passed the act on September 25, 2008, and the House subsequently approved it on September 27. President George W. Bush, despite his previous opposition to additional shuttle missions due to potential delays in transitioning to Project Constellation, signed the bill into law on October 15, 2008.

By the spring of 2009, the Obama administration incorporated funds for an additional mission, STS-134, into its proposed NASA budget for 2010. Originally planned as the final scheduled mission of the Shuttle program, the mission underwent several rescheduling efforts due to delays in the Shuttle's launch schedule. It was initially set to coincide with Expedition 26, which would have marked a historic moment in space exploration as brothers Mark Kelly, commander of STS-134, and Scott Kelly, Expedition 26 commander, would have become the first siblings to be in space simultaneously. However, delays pushed the mission beyond Expedition 26.

The launch of STS-134 carried additional emotional significance due to the presence of U.S. Representative Gabrielle Giffords, the wife of Commander Mark Kelly. Giffords had survived an assassination attempt in January 2011 and traveled to Kennedy Space Center (KSC) in Florida to witness the first launch attempt in April 2011. She returned to KSC on May 16, 2011, to attend the rescheduled launch, which was one of the most anticipated launches in years. Her recovery and presence added a poignant layer to the mission's already intense public interest.

U.S. President Barack Obama also planned to attend the April 29, 2011 launch, but despite the last-minute scrub, he toured the Orbiter Processing Facility at Launch Complex 39, where he met with the Shuttle crew and Representative Giffords. The Shuttle Endeavour began its rollout to the launch pad on March 10, 2011, at 19:56 EST, completing the process by 03:49 EST the following morning.

Tragically, on March 14, 2011, during the STS-134 pre-launch preparations, a fatal incident occurred at Launch Pad 39A. James Vanover, an engineer with United Space Alliance, committed suicide by jumping from the launchpad. Endeavour was already in position at the time, and work was suspended for the day to allow for grief counseling for NASA employees. This marked the first fatality at a shuttle launchpad since 1981.

STS-134's mission timeline commenced on May 16, 2011, when Space Shuttle Endeavour lifted off from Kennedy Space Center at 08:56 EDT. The successful launch followed an on-time tanking process, during which the Shuttle's external tank was filled with over 1.9 million liters (500,000 gallons) of liquid oxygen and liquid hydrogen. Once in orbit, the crew, led by Commander Mark Kelly, set to work on preparing the shuttle for its critical mission. Their immediate tasks included opening the payload bay doors, activating the Ku-band antenna for communications, and deploying the Shuttle Remote Manipulator System (Canadarm). Commander Kelly and Pilot Greg Johnson completed several engine burns, including the OMS-2 burn to circularize Endeavour's orbit and the NC-1 burn to align the Shuttle's trajectory with the ISS.

The crew also activated the Alpha Magnetic Spectrometer, a particle physics experiment designed to search for dark matter and antimatter. This marked a significant scientific milestone, as the AMS was one of the key payloads for this mission, and its data would be crucial for ongoing space-based scientific research. Additionally, mission specialist Mike Fincke downlinked video footage of the external tank, further contributing to NASA's analysis and review processes for Shuttle launches.

On May 17, 2011 (Flight Day 2 of STS-134), the crew of *Endeavour* embarked on a series of preparatory tasks for their upcoming docking with the International Space Station (ISS). The most critical of these was the inspection of the Shuttle's Thermal Protection System (TPS), a routine but vital procedure conducted to ensure the orbiter's heat shield integrity for its eventual return to Earth. Using the Orbiter Boom Sensor System (OBSS), the crew meticulously surveyed the leading edges of *Endeavour's* wings and nose cone, while the Shuttle Remote Manipulator System (SRMS, or Canadarm) was employed to inspect the thermal tiles around the Orbital Maneuvering System (OMS) pods. Meanwhile, Greg Johnson, Roberto Vittori, and Greg Chamitoff operated the OBSS, while the remainder of the crew focused on preparing for docking by installing critical tools such as the center-line camera in the Orbiter Docking System. They also readied the Extravehicular Mobility Units (EMUs), or spacesuits, for the mission's four planned spacewalks.

On May 18, 2011 (Flight Day 3), *Endeavour* successfully docked with the ISS's Pressurized Mating Adapter (PMA-2) at 10:14 UTC. This docking marked a key moment as the six astronauts of STS-134 joined the Expedition 27 crew aboard the ISS. Following a series of safety checks, the hatches between the spacecraft were opened at 11:38 UTC. A welcome ceremony followed, during which the two crews participated in a safety briefing. One of the mission's primary tasks, the transfer and installation of the Express Logistics Carrier (ELC) 3, began soon after docking. Using

Endeavour's SRMS, the crew unberthed ELC 3 from the payload bay and handed it off to the ISS's Canadarm2, which secured it to the Port 3 (P3) truss segment at 16:18 UTC. Oxygen was transferred from *Endeavour* to the ISS, and preparations continued for the mission's spacewalks, with the two EMUs moved to the Quest Airlock in anticipation of extravehicular activities.

On May 19, 2011 (Flight Day 4), the crew installed the Alpha Magnetic Spectrometer-02 (AMS-02), a significant scientific instrument designed to detect cosmic particles and help scientists investigate dark matter and antimatter. Operated by Drew Feustel and Roberto Vittori using the Canadarm, AMS-02 was lifted from *Endeavour's* payload bay and transferred to Canadarm2. Greg Chamitoff and Greg Johnson then used Canadarm2 to install the AMS-02 on the S3 truss segment at 09:46 UTC. The installation of AMS-02 completed the U.S. Orbital Segment of the ISS. Later in the day, the crew conducted a review of the tools and procedures needed for the mission's first spacewalk, which would take place the following day. Feustel and Chamitoff, who were scheduled to perform the spacewalk, camped out in the Quest Airlock overnight to purge nitrogen from their bloodstreams, thus preventing decompression sickness.

On May 20, 2011 (Flight Day 5), Drew Feustel and Greg Chamitoff conducted the first spacewalk of STS-134. The spacewalk, which lasted 6 hours and 19 minutes, saw the installation of the Materials International Space Station Experiment (MISSE) and the initial setup of a new wireless video system. A malfunction in Chamitoff's suit—specifically a CO2 sensor failure—led to an adjustment in the planned activities, including the installation of an ammonia jumper between the P3 and P6 truss segments. Despite the issues, the spacewalkers successfully installed a new light on the Crew Equipment Translation Aid (CETA) cart and placed a cover on the Starboard Solar Alpha Rotary Joint (SARJ). Meanwhile, the rest of the crew continued transferring equipment between *Endeavour* and the ISS, as Expedition 27 prepared for the departure of three of its crew

members.

On May 21, 2011 (Flight Day 6), the crew focused on inspecting a damaged section of *Endeavour's* thermal protection tiles, which had been struck during launch. Using Canadarm2, the crew grappled the OBSS and passed it to *Endeavour's* Canadarm, operated by Greg Johnson, Mike Fincke, and Roberto Vittori. After a detailed two-hour inspection, NASA cleared the Shuttle for re-entry. Feustel and Fincke began preparations for the second spacewalk, conducting another campout in the Quest Airlock overnight. In a significant moment of international and personal connection, the joint crew participated in a live video call with Pope Benedict XVI. The Pope spoke to the astronauts, offering prayers for Gabby Giffords, who had recently undergone surgery, and condolences to Paolo Nespoli, who had just lost his mother.

On May 22, 2011 (Flight Day 7), Feustel and Fincke conducted the second EVA of STS-134. The spacewalk, lasting 8 hours and 7 minutes, became the second-longest in spaceflight history. During the excursion, the astronauts transferred 5 pounds (2.3 kg) of ammonia to the Port 6 Photovoltaic Thermal Control System (PVTCS), lubricated the Solar Alpha Rotary Joint (SARJ) and Dextre's "hand," and installed a stowage beam on the Starboard 1 (S1) truss. A minor issue arose when one of the thermal blanket bolts came loose, but the task was completed successfully. Back inside the ISS, the rest of the crew continued cargo transfers while Russian cosmonaut Dmitri Kondratyev ceremonially handed over command of Expedition 27 to Andrei Borisenko, signaling the start of Expedition 28.

On May 23, 2011 (Flight Day 8), the crew took some off-duty time. Commander Mark Kelly and Mike Fincke participated in a live video interview with students from Mesa Verde Elementary School in Tucson, Arizona. Roberto Vittori and Paolo Nespoli later spoke with Italian President Giorgio Napolitano. Later that day, the Expedition 27 crew members Dmitri Kondratyev, Paolo Nespoli, and Catherine Coleman undocked from the ISS aboard the Soyuz TMA-20 spacecraft. Before re-entry, Soyuz TMA-20 performed a fly-around of the ISS, capturing photographs of both the station and *Endeavour*. The Soyuz safely landed in Kazakhstan at 02:27 UTC on May 24, 2011, marking the successful conclusion of Expedition 27 and the beginning of Expedition 28.

On May 24, 2011, during the ninth day of STS-134, the crew of *Endeavour* remained busy with media engagements, technical work, and mission preparations. Mission specialists Greg Chamitoff and pilot Greg Johnson began the day with a series of interviews conducted with television stations across the United States, including KPIX-TV, KGO-TV, and KFBK. Later, Commander Mark Kelly, along with mission specialists Mike Fincke and Chamitoff, participated in interviews with media outlets such as *The Daily*, *KDKA*, *Pittsburgh Tribune-Review*, and KTRK-TV.

While media interactions continued, the crew also focused on crucial tasks aboard the International Space Station (ISS). Johnson and mission specialist Roberto Vittori conducted equipment transfers between the ISS and the shuttle, cleaning and organizing the Permanent Multipurpose Module (PMM) *Leonardo*, which had been brought to the station earlier in the mission. Additionally, the crew worked on the ISS Oxygen Generator System (OGS) and the Carbon Dioxide Removal Assembly (CDRA), essential life support systems for the station.

Preparations for the next day's extravehicular activity (EVA) were also underway. Drew Feustel, Fincke, and Chamitoff dedicated time to organizing and testing tools that would be used during the upcoming spacewalk. At the end of the day, the shuttle crew and Expedition 28 flight engineer Ron Garan gathered for a review of the EVA procedures for the third spacewalk, scheduled for the following day.

Flight day 10 marked the third spacewalk of the STS-134 mission, a critical day for advancing the station's infrastructure. Astronauts Drew Feustel and Mike Fincke performed the spacewalk using a new protocol known as In-Suit Light Exercise (ISLE). Unlike the traditional "campout" pre-breathe protocol, ISLE involved the astronauts breathing pure oxygen for 60 minutes at reduced airlock pressure (10.2 psi), followed by light exercise and an additional 50 minutes of rest, all while continuing to breathe pure oxygen. This method helped reduce the risk of decompression sickness.

Feustel and Fincke exited the Quest Airlock and began their tasks, which included installing the Power Data Grapple Fixture (PDGF), a critical component for the station's robotic arm, Canadarm2. Although the PDGF was installed successfully, the associated data cable would be installed at a later time. The pair then moved on to route power cables between the Unity and Zarya modules, providing redundant power to the Russian segment of the ISS. Their next task involved completing the installation of a wireless video system, which Feustel and Chamitoff had begun during the first EVA.

Commander Mark Kelly documented the spacewalk, while inside the ISS, pilot Greg Johnson and Roberto Vittori worked alongside Ron Garan to store new supplies and equipment. By the end of the spacewalk, Feustel and Fincke had also captured infrared video of an experiment aboard the Express Logistics Carrier (ELC) 3, contributing to ongoing research efforts.

The crew began flight day 11 by conducting a late inspection of *Endeavour's* Thermal Protection System (TPS), a critical task typically performed post-undocking. However, with plans to leave the Orbiter Boom Sensor System (OBSS) aboard the ISS, the crew completed the inspection while still docked. Following this, the combined STS-134 and Expedition 28 crew held a press conference, speaking to media representatives from NASA centers and ISS partner agencies.

Later in the day, Commander Mark Kelly conducted interviews with local television stations from Tucson, Arizona, while Mike Fincke and Greg Chamitoff prepared for their upcoming spacewalk by spending the night in the Quest Airlock, where the air pressure was reduced to 10.2 psi. This "campout" procedure helped the astronauts acclimate to lower pressures and reduce the risk of decompression sickness during the next EVA.

Flight day 12 marked the final spacewalk of STS-134, as well as the final spacewalk of the entire Space Shuttle program. Mike Fincke and Greg Chamitoff led this historic EVA, beginning by installing the OBSS on the Starboard 1 (S1) truss of the ISS. After completing this task, the pair replaced an End Effector Grapple Fixture (EFGF) with a Power Data Grapple Fixture (PDGF), enabling Canadarm2 to perform future grappling operations more effectively.

With the installation complete, Fincke and Chamitoff moved to the Express Logistics Carrier 3 (ELC-3), where they performed torque-release operations on bolts securing a spare arm for Dextre, the station's robotic arm system. By the end of the EVA, the cumulative time spent on ISS-related spacewalks had surpassed 1,000 hours, a significant milestone in the station's construction and maintenance.

While the EVA was in progress, the rest of the crew continued with equipment transfers between *Endeavour* and the ISS. Flight engineer Ron Garan and mission specialists Greg Johnson and Roberto Vittori assisted in stowing the supplies, ensuring the station remained well-stocked for future operations. Commander Mark Kelly, as on previous EVAs, documented the spacewalk with photos and video.

On flight day 13, the STS-134 crew completed another crucial task aboard the ISS by replacing an absorbent bed in the Carbon Dioxide Removal Assembly (CDRA), ensuring the system could continue filtering carbon dioxide from the station's air. While Fincke and Chamitoff worked on the CDRA, Kelly and Feustel resized two spacesuits for upcoming spacewalks by Expedition 28 crew members Ron Garan and Mike Fossum.

Commander Kelly, pilot Greg Johnson, and Ron Garan later engaged with students and faculty at the University of Arizona via a live broadcast, while other crew members continued the final transfers between the shuttle and the station.

The final full day of joint operations, flight day 14, focused on completing the transfer of supplies and equipment. This included transferring four bags of water from *Endeavour* to the ISS, a vital resource for the station's crew. Fincke completed the CDRA maintenance that had begun the previous day, while Feustel and Chamitoff stowed tools used in the mission's spacewalks.

Toward the end of the day, the Expedition 28 crew held a farewell ceremony for the departing STS-134 crew. After emotional farewells, the crews initiated procedures to close the hatches between the two spacecraft, marking the end of nearly 11 days of joint operations.

On May 30, after 11 days, 17 hours, and 41 minutes, *Endeavour* undocked from the ISS for the final time. Pilot Greg Johnson carefully maneuvered the shuttle away from the station, flying it around the ISS to capture a full lap of photographic documentation. Once the maneuver was complete, Commander Mark Kelly performed a series of separation burns to move the shuttle safely away from the station.

During this period, the crew tested the Sensor Test for Orion Relative Navigation Risk Mitigation (STORRM) system, a key technology for future spacecraft. With all operations complete, the crew began preparations for their return to Earth.

The penultimate day of STS-134 was dedicated to final preparations for reentry. The crew tested *Endeavour's* Flight Control Systems (FCS) and Reaction Control System (RCS), ensuring that all systems were functioning properly for landing. They stowed equipment, including the Ku-band antenna, and performed eye exams and experiments, while also taking part in interviews with major news outlets.

The day ended with a deorbit briefing, as the crew finalized the procedures for their descent and landing.

On June 1, 2011, Space Shuttle *Endeavour* completed its final mission. The crew initiated the deorbit burn at 1:29 a.m. EDT, and the shuttle re-entered Earth's atmosphere shortly after 2:03 a.m. At 2:25 a.m., *Endeavour* crossed the Florida coastline and touched down at Kennedy Space Center at 2:35 a.m. EDT, marking the end of STS-134 and the shuttle's 25th mission.

STS-135

STS-135 was the final mission of NASA's Space Shuttle program, marking the end of an era in human spaceflight that had spanned three decades. The funding and approval process for

STS-135, the final Space Shuttle mission, was a complex and politically charged endeavor. Initially, the mission was designated as a launch-on-need rescue flight for the final planned shuttle mission, STS-134. However, efforts to convert STS-135 into an operational flight began gaining momentum in mid-2010, with bipartisan support from both the U.S. House of Representatives and the Senate.

On July 15, 2010, Senator Bill Nelson of Florida introduced and secured the passage of the 2010 NASA reauthorization bill in a Senate committee. This bill directed NASA to conduct an additional Space Shuttle mission—STS-135—contingent upon a review of safety concerns. This milestone signaled strong support in the Senate, but the full chamber's approval was still required. Meanwhile, a draft reauthorization bill in the House of Representatives, under consideration by the Science & Technology Committee, did not initially include provisions for an additional mission.

Just days later, on July 22, 2010, U.S. Representative Suzanne Kosmas, also from Florida, successfully introduced an amendment to the House version of the bill, adding the extra shuttle flight to the program's manifest. With momentum building, the Senate acted swiftly, passing its version of the NASA reauthorization bill on August 5, 2010, just before Congress adjourned for the August recess.

By August 20, 2010, NASA managers began formal mission planning for STS-135, targeting a June 28, 2011, launch date. On September 29, 2010, the House of Representatives passed the Senate-approved version of the bill by a wide margin, voting 304–118 in favor. This legislation was subsequently sent to President Barack Obama for his approval.

On October 11, 2010, President Obama signed the reauthorization bill into law, granting NASA the authority to proceed with STS-135. However, despite the legislative approval, the mission still lacked specific funding. Historically, a Space Shuttle mission cost approximately $450 million, raising concerns about the financial feasibility of STS-135.

Nevertheless, on January 20, 2011, NASA formally redesignated the mission from STS-335 (its contingency designation) to STS-135, confirming its operational status. In a decisive move, NASA managers announced on February 14, 2011, that STS-135 would proceed as planned, regardless of the unresolved funding situation in Congress. This decision set the stage for the final flight of the Space Shuttle program, ensuring that STS-135 would become a historic moment in NASA's legacy.

The STS-135 crew was smaller than usual, consisting of only four astronauts: Commander Christopher Ferguson, Pilot Douglas Hurley, and Mission Specialists Sandra Magnus and Rex Walheim. This reduced crew size was part of the contingency planning, as the mission was originally conceived as a rescue mission. All four astronauts were veterans of previous Shuttle flights, bringing a wealth of experience to this historic mission.

The preparation for STS-135, the final mission of the Space Shuttle Atlantis, began in earnest with the arrival of External Tank 138 (ET-138) at the Kennedy Space Center. ET-138 was produced at the Michoud Assembly Facility in New Orleans and was delivered to Kennedy on the Pegasus barge. After offloading on July 14, 2010, the tank was transported to a checkout cell inside the Vehicle Assembly Building (VAB) for detailed inspections.

Originally, ET-138 was slated for use with STS-134 (Endeavour), while the refurbished ET-122, damaged during Hurricane Katrina but certified flightworthy after repairs, was reserved for the Launch-On-Need (LON) STS-335 mission. This mission, which would only fly if a rescue of Endeavour's crew was necessary, was considered unlikely. However, when NASA decided to fly Atlantis on the full STS-135 mission, the tank assignments were reversed. ET-138, posing a lower risk, was assigned to Atlantis, while ET-122 was allocated to Endeavour. This ensured the safest possible configuration for the final Space Shuttle mission.

Atlantis underwent significant preparatory work as well. In early December 2010, ground technicians began installing the shuttle's main engines. The center engine was installed on December 7, followed by the right and left engines on December 8 and 9, respectively, inside Orbiter Processing Facility-1 (OPF-1). These were the last main engines ever installed on a Space Shuttle, marking the beginning of the end of NASA's storied Shuttle Program.

By March 29, 2011, technicians inside the VAB had commenced stacking operations for Atlantis's Solid Rocket Boosters (SRBs). The first segment was carefully maneuvered into High Bay 1 and placed onto the mobile launch platform. Completed in mid-April, the boosters featured a mix of refurbished and unflown components, adding a unique historical element. For example, the forward dome on the right-hand booster was new, while the left booster's upper cylinder had flown aboard STS-1, the maiden flight of Space Shuttle Columbia in 1981.

The assembly of the shuttle continued into late April 2011, when ET-138 was mated to the SRBs on April 25, marking a significant milestone in the preparation process. Soon after, on April 29, 2011, President Barack Obama, First Lady Michelle Obama, and their daughters Malia and Sasha visited Kennedy Space Center. During their tour of OPF-1, they stood beneath the wings of Atlantis and received an informal tutorial from United Space Alliance tile technician Terry White and astronaut Janet Kavandi.

On May 17, 2011, Atlantis made its historic final trip from OPF-1 to the VAB, where it was to be mated with ET-138. The transfer took longer than usual, allowing shuttle workers to pose for photographs with the spacecraft, and providing an opportunity for the four STS-135 astronauts to greet the workers. The next day, Atlantis was lifted vertically inside the VAB and carefully lowered into position with the external tank and SRBs, completing the mating process by May 19.

NASA formally announced July 8, 2011, as the intended launch date for STS-135. On May 31, Atlantis began its rollout to Launch Pad 39A. The first motion occurred at 20:42 EDT, but a minor hydraulic leak delayed the operation by 40 minutes. After a 3.4-mile journey, Atlantis arrived at the pad at 03:29 EDT on June 1. Large crowds, including families of NASA employees, gathered to witness this significant event. Simultaneously, Endeavour was landing at Kennedy Space Center's Shuttle Landing Facility, completing its final mission, STS-134.

Preparations for the final launch continued with an external tank fueling test on June 15, 2011. Technicians detected a hydrogen leak in the No. 3 main engine's fuel valve, which was replaced by June 21. Engineers also conducted X-ray inspections of ET-138's stringers to ensure there were no issues, following concerns raised during the STS-133 mission. All tests concluded successfully.

Atlantis's payload was loaded into the shuttle on June 20, following the arrival of the STS-135 crew at Kennedy Space Center. Commander Chris Ferguson, Pilot Doug Hurley, and Mission Specialists Sandra Magnus and Rex Walheim participated in countdown dress rehearsals and emergency training drills. During their training, the crew practiced pad evacuation procedures and simulated a full launch countdown.

On July 8, 2011, the final flight of the Space Shuttle era began. Despite concerns about weather, which initially left only a 30% chance of a successful launch, conditions improved to 60% just an hour before liftoff. At T-minus 31 seconds, the countdown clock briefly halted due to an issue with the Gaseous Oxygen Vent Arm, which had not properly retracted. After verification through a camera feed, the countdown resumed.

The launch of STS-135, the final flight of Space Shuttle Atlantis, was initially threatened by unfavorable weather conditions. Early forecasts gave only a 30% chance of a successful launch, with concerns over clouds, thunderstorms, and lightning near the Kennedy

Space Center. However, as the countdown progressed, the weather improved, raising the chances of launch to 60% just an hour before liftoff.

Launch Director Mike Leinbach conducted the final series of GO/NO GO polls, ensuring that all systems were ready for launch. Shuttle Launch Integration Manager Michael P. Moses issued a waiver regarding Return to Launch Site (RTLS) weather conditions, explaining later that although showers were present within a 20-nautical-mile radius of the Shuttle Landing Facility, they would clear in time for a potential RTLS landing 35 minutes after launch, if necessary.

With all systems cleared, Leinbach addressed the crew with an emotional farewell: "Good luck to you and your crew on the final flight of this true American icon. Good luck, godspeed, and have a little fun up there." Commander Chris Ferguson responded on behalf of the crew: "Thanks to you and your team, Mike. Until the very end, you all made it look easy. The Shuttle will always be a reflection of what a great nation can do when it dares to be bold and commits to follow through. We're completing a chapter of a journey that will never end. The crew of Atlantis is ready to launch."

At T-minus 31 seconds, just before Atlantis's onboard computers were scheduled to take control of the flight, the countdown unexpectedly stopped. The hold was caused by a failure to confirm that the Gaseous Oxygen Vent Arm, known as the "beanie cap," had fully retracted and latched. This issue had never occurred during the history of the Shuttle program. The launch team quickly verified the Vent Arm's position using a closed-circuit camera, allowing the countdown to resume after a 2-minute and 18-second delay.

At 11:29:03.9 EDT on July 8, 2011, Space Shuttle Atlantis lifted off from Kennedy Space Center for the final time. Launch commentator George Diller announced the moment with the words: "All three engines up and burning, two, one, zero, and liftoff! The final liftoff of Atlantis—on the shoulders of the space shuttle, America will continue the dream." Nearly one million spectators gathered at the Kennedy Space Center and surrounding areas, cheering as the Shuttle soared into the sky, a symbolic conclusion to NASA's Shuttle Program.

Atlantis's ascent proceeded according to plan, with the two Solid Rocket Boosters separating from the External Tank (ET) after 2 minutes and 5 seconds of flight. Main Engine Cutoff (MECO) occurred at 8 minutes and 24 seconds into the flight, at 15:37:28 GMT. The ET-138 separated from the Shuttle at 15:37:49 GMT, and a modification to the tank's onboard camera allowed it to transmit live video of its disintegration as it re-entered Earth's atmosphere. A further boost from the Orbital Maneuvering System (OMS-1) was unnecessary due to the nominal MECO, and Atlantis settled into an initial orbit of 225 by 58 kilometers with an inclination of 51.6 degrees.

Throughout Flight Day 1, the crew of Atlantis conducted several orbital adjustments. The OMS-2 burn, lasting 64 seconds, placed Atlantis into a 230 by 158 kilometer orbit, while the NC-1 engine burn, lasting 94 seconds, fine-tuned the Shuttle's trajectory to match that of the International Space Station (ISS). The NC-1 firing altered the Shuttle's velocity by 144.7 feet per second (44.1 meters per second), setting the stage for a successful docking with the ISS.

NASA held a post-launch press conference at 12:10 CDT with senior officials, including Bill Gerstenmaier, Robert Cabana (Director of NASA's John F. Kennedy Space Center), Mike Moses, and Mike Leinbach, discussing the success of the launch and the significance of the final Shuttle mission.

Later that day, at 17:03:20 GMT, the crew opened Atlantis's payload bay doors and began configuring the Shuttle for its on-orbit operations. The deployment of the Ku-band antenna followed, and a self-test confirmed its functionality. CAPCOM astronaut Barry Wilmore communicated with the crew from mission control in Houston, reporting that a preliminary analysis of the launch showed no significant debris or impact damage to Atlantis during ascent. Commander Ferguson and Pilot Doug Hurley also powered up the Shuttle's Robotic Arm, verifying its operational readiness for the planned thermal protection system inspection on the following day.

STS-135, the final mission of the Space Shuttle program, delivered critical supplies and equipment to sustain the International Space Station (ISS) through 2012, marking the end of NASA's Shuttle resupply missions. Following the extension of the ISS program to 2024 (and later to 2030), resupply operations were handed over to private companies through NASA's Commercial Orbital Transportation Services (COTS) program. While some advocated for extending the Shuttle program beyond STS-135, NASA officially ended the Shuttle era with the completion of this mission.

The centerpiece of the STS-135 payload was the Multi-Purpose Logistics Module (MPLM) Raffaello, making its fourth journey to the ISS. Raffaello carried essential supplies to provision the ISS, packed with 16 resupply racks—the maximum capacity for the MPLM. These racks included eight Resupply Stowage Platforms (RSPs), two Integrated Stowage Platforms (ISPs), six Resupply Stowage Racks (RSRs), and one Zero-G Stowage Rack (ZSR), carefully organized to maximize space for the delivery of critical supplies.

On Flight Day 4, the ISS's robotic arm, Canadarm2, lifted Raffaello from Atlantis's payload bay and berthed it to the nadir port of the Harmony node. Over the next few days, the crew unloaded the supplies and reloaded Raffaello with nearly 5,700 pounds (2,600 kg) of unneeded equipment and station waste for the return trip to Earth. On Flight Day 11, the MPLM was detached from Harmony and secured back in the shuttle's payload bay for the journey home.

Also aboard STS-135 was the Lightweight Multi-Purpose Carrier (LMC), which served two key purposes. First, it carried the External Thermal Cooling System (ETCS) Pump Module, a piece of hardware that had been replaced on the ISS in August 2010 due to a failure. The pump module was returned to Earth for failure analysis. Second, the LMC transported the Robotic

The Robotic Refueling Mission (RRM), developed by NASA's Satellite Servicing Capabilities project at the Goddard Space Flight Center, was designed to demonstrate technologies for refueling satellites in orbit using robotic tools. Once installed on the ISS's ExPRESS Logistics Carrier-4 (ELC-4), the RRM aimed to prove the feasibility of extending satellite lifespans through remote refueling—a technology that could eventually be transferred to the commercial sector.

The RRM was equipped with four specialized tools, each outfitted with electronics, cameras, lights, and components like pumps, valves, and sensors. These tools would allow the ISS's robotic arm, Dextre, to perform refueling operations. Initial demonstrations took place in 2012 and 2013, laying the groundwork for future advancements in satellite servicing technology.

Atlantis also carried a miniaturized satellite known as PSSC-2 (Picosatellite Solar Cell Testbed 2). Deployed on Flight Day 13, PSSC-2 became the 180th and final payload to be launched into orbit by a Space Shuttle. This small but significant satellite was designed to test solar cell technologies in space, marking the end of an era for Shuttle-launched payloads.

The mission also included the third flight of the TriDAR sensor package, designated as Detailed Test Objective (DTO) 701A. Developed by Neptec Design Group in partnership with NASA and the Canadian Space Agency, TriDAR was a dual-sensing laser camera system intended for autonomous rendezvous and docking operations. Unlike traditional systems that required reference markers on target spacecraft, TriDAR used a laser-based 3D sensor and a thermal imager to calculate position and orientation in real-time. This advanced technology could one day assist with planetary landings, robotic rover navigation, and spacecraft inspections. Installed in Atlantis's payload bay in April 2011, TriDAR was used during STS-135 to demonstrate its capabilities for future space missions.

In addition to delivering supplies, STS-135 also returned several critical items to Earth. The failed ammonia Pump Module, replaced on the ISS in 2010, was stowed in Atlantis's payload bay for analysis. Another piece of hardware, the Common Cabin Air Assembly (CCAA) Heat Exchanger, which had experienced issues aboard the ISS, was also brought back inside the MPLM. Additional experiments and materials were secured in the middeck lockers of the shuttle, taking advantage of the extra storage space made available due to the smaller four-member crew of STS-135.

STS-135 also carried technology aimed at advancing scientific and operational capabilities aboard the ISS. For example, astronauts utilized an iPhone to log experiments, which remained on the ISS for future use. Additionally, two Nexus S smartphones were integrated into the SPHERES (Synchronized Position Hold, Engage, Reorient, Experimental Satellites) system, allowing astronauts to pilot these small, free-flying satellites for various in-orbit experiments.

On July 9, Flight Day 2 of the STS-135 mission, the crew aboard *Atlantis* focused on conducting a detailed inspection of the shuttle's thermal protection system (TPS). This crucial task was performed to ensure that no damage had occurred to the heat shield during the launch. Commander Chris Ferguson, Pilot Doug Hurley, and Mission Specialist Sandra Magnus used the shuttle's robotic arm and the Orbiter Boom Sensor System (OBSS) to carefully examine the reinforced carbon-carbon panels on the shuttle's nose cap and wing leading edges.

At 6:58 a.m. EDT, the robotic arm grappled the OBSS, and the crew activated the onboard camera and laser sensor package to begin scanning the starboard wing. The survey then moved to the nose cap, followed by the port wing. The gathered imagery and electronic data were downlinked to the ground via Ku band communication for analysis by a team of experts. The survey, which began at 11:00 UTC, was completed five hours later, during which the crew worked efficiently, often ahead of schedule. In his NASA TV commentary, Public Affairs Officer Rob Navias commended the crew's efficiency, and Mission Control Houston, led by CAPCOM astronaut Stephen Robinson, praised their efforts.

Meanwhile, Mission Specialist Rex Walheim worked on the shuttle's middeck, preparing equipment for transfer to the International Space Station (ISS). Later in the day, he joined Hurley to check the rendezvous tools in preparation for *Atlantis's* docking with the ISS on Flight Day 3. Ferguson and Magnus also installed the center-line camera on the shuttle's hatch window, which would aid in aligning *Atlantis* with the ISS during docking.

Two course correction burns (NC2 and NC3) were executed to adjust *Atlantis's* flight path. The NC3 burn, lasting seven seconds, altered the shuttle's velocity by 1.5 feet per second (0.46 m/s), further aligning its trajectory with the ISS. Aboard the station, the Expedition 28 crew prepared the Pressurized Mating Adapter (PMA-2) for the upcoming docking. Crew members Sergei Volkov, Mike Fossum, and Satoshi Furukawa also met with ground experts to plan the photography session for *Atlantis's* rendezvous pitch maneuver (RPM), a critical component of the docking process.

At the Johnson Space Center, Flight Director Kwatsi Alibaruho reported that the mission was progressing smoothly, stating that *Atlantis* was off to one of the best starts in the 30-year history of the Space Shuttle program.

The next day, July 10, marked Flight Day 3 and the long-awaited docking of *Atlantis* with the ISS. The crew began their day at 7:29 UTC, but encountered a minor setback when one of the shuttle's General Purpose Computers (GPC3) failed. This issue did not affect the rendezvous or docking procedures, as the two remaining GPCs were sufficient for the task.

Commander Ferguson and Pilot Hurley executed a series of burns, including the NH, NC4, NCC, MC1-4, and TI burns, to align *Atlantis's* orbit with that of the ISS. The final terminal initiation burn, lasting 12 seconds, placed the shuttle about 1,000 feet (300 meters) below the ISS at 12:29 UTC. By 13:26 UTC, Ferguson, operating from the aft flight deck, positioned the shuttle 600 feet (180 meters) beneath the station to begin the 360-degree RPM. As the shuttle's underbelly came into view, Volkov, Fossum, and Furukawa, positioned aboard the ISS, photographed the thermal protection system using cameras equipped with 1000 mm, 800 mm, and 400 mm lenses. These images were sent to mission control for analysis to ensure that *Atlantis's* heat shield was intact.

At 15:07 UTC, *Atlantis* docked with the ISS's Pressurized Mating Adapter-2 while orbiting 220 miles (350 kilometers) above the

South Pacific Ocean, east of New Zealand. Hurley radioed, "Houston, station, *Atlantis*, capture confirmed and we see free drift," signaling the successful docking. In response, Ron Garan aboard the ISS ceremonially rang the station's bell and welcomed *Atlantis* for its final visit: "Welcome to the International Space Station for the last time."

Following the docking, both crews conducted leak checks before opening the hatches at 16:47 UTC. The shuttle crew floated into the station's Harmony module at 16:55 UTC, where they were greeted by the Expedition 28 crew for a welcoming ceremony and safety briefing. The crew quickly got to work, with Ferguson and Hurley using the shuttle's robotic arm to retrieve the OBSS from the station's Canadarm2, operated by Garan and Furukawa. This handoff prepared the OBSS for any future inspections of the shuttle's heat shield, if necessary. Magnus focused on setting up TV equipment, while Walheim transferred spacewalk gear.

As Flight Day 3 progressed, flight controllers monitored reports from the U.S. Strategic Command regarding a piece of orbital debris from the Russian satellite COSMOS 375. Although it was projected to pass near the shuttle and station complex around noon the next day, updated tracking data determined that no course correction was necessary.

On July 11, Flight Day 4 of STS-135, the crew of *Atlantis* focused on the installation of the Raffaello Multi-Purpose Logistics Module (MPLM) onto the nadir port of the International Space Station's (ISS) Harmony module. The day began at 7:02 a.m., with the astronauts being awakened by the song "Tubthumping" by Chumbawamba, setting the tone for a busy day ahead.

Mission Specialist Sandra Magnus and Pilot Doug Hurley initiated the MPLM installation process at 9:09 UTC, using the station's robotic arm, Canadarm2, to carefully extract Raffaello from the shuttle's payload bay. By 10:46 UTC, they had successfully maneuvered and attached the module to the Harmony node, following precise steps to ensure a secure connection. After conducting leak checks to confirm the integrity of the seal, the hatches between Raffaello and the

ISS were opened before noon, allowing crew members to begin transferring cargo.

With *Atlantis* having launched on time and benefiting from power-saving operations in the early days of the mission, NASA managers approved a one-day extension to the mission. This extra time allowed the crew to focus on cargo transfers between the MPLM and the station. CAPCOM Megan McArthur informed Commander Chris Ferguson that the Mission Management Team had also decided against performing a Focused Inspection of *Atlantis*'s heat shield. According to Damage Assessment Team Chairman Leroy Cain, only one tile ding and four minor areas of damage to insulating blankets were detected, and none posed a threat to the mission.

In preparation for the next day's spacewalk, the STS-135 crew and Expedition 28 crew members Ron Garan, Mike Fossum, and Satoshi Furukawa met toward the end of the workday to review spacewalk procedures. This coordination was critical, as it ensured a smooth execution of the tasks planned for the following day.

On July 12, Flight Day 5, Expedition 28 Flight Engineers Mike Fossum and Ron Garan embarked on a six-hour, 31-minute spacewalk to perform critical maintenance on the ISS. Due to the short mission duration and the reduced crew size aboard *Atlantis*, NASA opted not to utilize STS-135 crew members for the spacewalk, instead relying on the Expedition 28 astronauts.

The primary objectives of the spacewalk included retrieving a failed ammonia pump module from the station's external stowage platform, installing two new experiments, and performing repairs to a base for the station's robotic arm, Canadarm2. The spacewalk began at 13:22 UTC, with Fossum and Garan donning their Extravehicular Mobility Units (EMUs). Fossum's suit was marked with red stripes for identification, while Garan's suit had no markings.

STS-135 Pilot Doug Hurley and Mission Specialist Sandra Magnus, stationed in the cupola, operated Canadarm2 to maneuver Garan to the stowage platform where the failed pump module, which had malfunctioned in 2010, was located. After successfully removing the module,

Garan, still on the arm, carried it to *Atlantis's* payload bay, where Fossum bolted it into place on the Lightweight Multi-Purpose Experiment Support Structure Carrier (LMC).

The next task involved the installation of the Robotics Refueling Mission (RRM) experiment. Fossum, now positioned on Canadarm2, transported the experiment from *Atlantis's* payload bay to a temporary platform on Dextre, the station's robotic hand. As Fossum worked with the RRM, Garan cleared tools and equipment from the shuttle's payload bay, marking the final extravehicular activity (EVA) inside a shuttle payload bay in NASA's history. Recognizing the historical moment, Mission Specialist Rex Walheim, serving as the intra-vehicular officer from *Atlantis's* flight deck, radioed to Garan, "Take a look around, Ronny. You're the last EVA person in the payload bay of a shuttle."

Fossum then moved to the Zarya module, where he freed a wire that had become stuck in a latch door of a data grapple fixture. This fixture, installed during STS-134, serves as a base for Canadarm2 and extends its range to the Russian segment of the station. Garan also deployed the Materials International Space Station Experiment-8 (MISSE-8) on the station's starboard truss. This experiment, focusing on the durability of optical reflector materials, was installed on the Express Logistics Carrier-2 during STS-134.

The final task of the spacewalk involved installing an insulating cover over the Pressurized Mating Adapter-3 (PMA-3) on the Tranquility node. This cover protects the adapter from solar exposure and ensures it remains in optimal condition for future use.

The spacewalk concluded at 19:53 UTC, marking the 160th EVA in support of ISS assembly and maintenance, as well as the 249th spacewalk conducted by U.S. astronauts. Inside the shuttle and station, the crew continued to transfer cargo from the Raffaello MPLM.

Meanwhile, a technical issue arose with the U.S. toilet in the Tranquility module. On Flight Day 4, the crew had reported a strong odor emanating from the urine processor, prompting mission control to deactivate the equipment for investigation. Despite this minor setback, the crew pressed on with their duties, successfully advancing the mission objectives as *Atlantis* approached the final days of its historic flight.

On July 16, 2011, Flight Day 9 of the STS-135 mission began with a sense of accomplishment as the Mission Management Team had added this bonus day to the schedule. This extra time was made possible by Atlantis conserving enough cryogenic oxygen and hydrogen to power its fuel cells for an additional day. The STS-135 crew, consisting of Commander Chris Ferguson, Pilot Doug Hurley, and Mission Specialists Sandra Magnus and Rex Walheim, focused their efforts on transferring supplies and equipment between the International Space Station (ISS) and the Raffaello Multi-Purpose Logistics Module (MPLM).

The day's activities commenced with Ferguson and Hurley successfully repairing the door that allowed access to the lithium hydroxide (LiOH) canisters, essential for air purification. Meanwhile, Sandra Magnus spent about an hour and a half collecting microbial air samples from various locations aboard the ISS. These samples would later be returned to Earth for analysis, contributing to ongoing studies of microbial life in space environments.

Magnus also operated the Japanese Experiment Module Remote Manipulator System (JEMRMS), demonstrating her proficiency with robotic systems. Concurrently, Walheim and ISS crew member Mike Fossum worked on preparing spacewalking equipment in the Quest airlock. Some of this equipment would remain on the ISS for future use, including an upcoming Russian spacewalk scheduled for August 3, 2011. Hurley and ISS crew member Ron Garan, on the other hand, stored cargo in Atlantis's middeck for its return to Earth. The middeck was fully packed with 1,564 pounds of cargo, making efficient use of the space since no astronauts would be riding in the middeck during re-entry. Among the cargo transferred to the ISS, 2,281 pounds of equipment were also stowed in the middeck.

As a tribute to the Space Shuttle Program, the crew recorded a message in honor of Atlantis and the team behind the program. Commander Ferguson highlighted the significance of a U.S.

flag that had flown on the very first shuttle mission, STS-1. This flag was presented to the ISS crew and was left onboard to be retrieved by the next U.S. crew launched to the ISS. This symbolic flag was eventually brought back to Earth by the Crew Dragon Demo-2 crew on June 1, 2020.

Later in the day, Ferguson and Hurley paid tribute to U.S. naval aviators in a video celebrating the centennial of naval aviation. Hurley mentioned notable figures who had contributed to human space exploration, including Alan Shepard, the first American in space; John Glenn, the first American to orbit the Earth; Neil Armstrong and Eugene Cernan, the first and last humans to walk on the Moon; and John Young and Robert Crippen, the first pilots of the Space Shuttle.

Before retiring for the night, CAPCOM Megan McArthur informed the crew that the failure of General Purpose Computer-4 (GPC-4) earlier in the mission was likely caused by a single event upset, possibly due to a Coronal Mass Ejection. Ground teams confirmed that GPC-4 was still functioning well, and plans were made to reassign systems management to GPC-4 the following day, with the intention of using it for undocking.

Flight Day 10 marked the completion of cargo transfer operations between Atlantis and the ISS. The Raffaello MPLM was officially packed with all the return cargo destined for Earth. Space Station Lead Flight Director Chris Edelen noted that the crew had achieved a significant milestone, completing the transfer of 9,403 pounds (4,265 kg) of cargo to the ISS and packing 5,666 pounds (2,570 kg) of cargo for return to Earth inside the MPLM. Additionally, the crew installed the control and power assemblies in the hatch leading to the MPLM, preparing it for its release from Node 2 the following day.

At 10:10 UTC, Doug Hurley and Rex Walheim engaged in an interactive session with students from NASA Explorer Schools across the United States, marking the final educational event conducted by a Space Shuttle crew. Following this, Sandra Magnus and Chris Ferguson continued transferring experiments and equipment to and from Atlantis's middeck. Among the notable items transferred was a new science refrigerator, GLACIER, and the mass spectrometer for the station's Mass Constituent Analyzer. These transfers marked 84% completion of middeck operations.

As the afternoon drew on, the crew had some downtime. NASA TV aired a recorded video featuring Magnus and ISS Flight Engineer Satoshi Furukawa cheering for their respective teams in the 2011 FIFA Women's World Cup final. The United States and Japan would later face off in a match that saw Japan win in a penalty shootout after a 2–2 draw in extra time.

On Flight Day 11, July 18, the STS-135 crew prepared for the undocking of Atlantis from the ISS by returning the Raffaello MPLM to the shuttle's payload bay. The hatches between the ISS and the MPLM were sealed at 5:03 UTC, and shortly afterward, Sandra Magnus and Doug Hurley used the Canadarm2 to unberth Raffaello from Node 2. By 11:48 UTC, the MPLM was securely stowed in Atlantis's payload bay, marking the final MPLM transfer in Space Shuttle history.

At 14:28 UTC, the Atlantis and ISS crews said their farewells, concluding nearly eight days of joint operations. Commander Ferguson presented the ISS crew with a small U.S. flag that had flown on the maiden shuttle flight, STS-1, along with a model of the shuttle signed by program officials. In his speech, Ferguson paid tribute to the thousands of people who had contributed to the success of the Space Shuttle Program over its 30-year history.

After returning to Atlantis, the shuttle crew carried out the final tasks in preparation for undocking, including installing the centerline camera and performing hatch leak checks. With everything in place, the stage was set for Atlantis's departure from the ISS, marking the end of an era in human spaceflight.

On July 19, 2011, Atlantis undocked from the International Space Station (ISS) for the last time, marking the end of an era in Space Shuttle visits to the orbital outpost. With Pilot Doug Hurley at the controls, the undocking occurred at 6:28 UTC as the shuttle and station flew through orbital night over the Pacific Ocean near

Christchurch, New Zealand. Shortly afterward, following a long-standing naval tradition, ISS Flight Engineer Ron Garan rang the station's bell in the Harmony module, announcing, "Atlantis, departing the International Space Station for the last time."

Once separated from the station, Atlantis moved to a station-keeping position approximately 600 feet (180 meters) ahead of the ISS. Before completing a final half-lap fly-around of the station, Hurley paused the shuttle by firing its thrusters, giving the crew a unique opportunity to capture images of the space station as it rotated 90 degrees. This maneuver allowed the Atlantis crew to photograph parts of the station that had rarely been documented in previous shuttle fly-arounds, providing valuable data to experts on the ground. The half-lap fly-around commenced at 7:30 UTC and lasted 25 minutes, with the images serving as a final visual assessment of the ISS's condition.

Inside NASA's flight control rooms in Houston, teams were overseeing their last Space Shuttle mission. Commander Chris Ferguson expressed his gratitude to the Orbit 1 team, led by Flight Director Kwatsi Alibaruho, encouraging them to "make a memory" of this momentous occasion. CAPCOM Daniel Tani, speaking from the ISS flight control room, shared his pride in supporting the mission, noting that "the ISS wouldn't be here without the shuttle." Ferguson responded, "It's been an incredible ride. We're really appreciative we had the opportunity to work with you on this pivotal mission."

After completing the fly-around, Atlantis executed two separation burns, the second at 8:18 UTC, to move away from the ISS and begin its journey home. Later in the day, the crew conducted a thorough inspection of Atlantis's heat shield using the shuttle's robotic arm and the 50-foot Orbiter Boom Sensor System (OBSS), focusing on the reinforced carbon-carbon of the wing leading edges and the nose cap. The inspection data, reviewed by Sandra Magnus and ground engineers, confirmed that the shuttle's thermal protection system (TPS) had sustained no damage during the mission. The crew completed their inspections by 2:30 UTC and

ended the day by going to bed at 4:59 UTC.

July 20, 2011, marked the final full day in space for the STS-135 crew. With landing preparations underway, the crew checked Atlantis's flight control surfaces and hot-fired the shuttle's reaction control system (RCS) jets to ensure everything was in proper working order for re-entry. Following a detailed review of the data collected during the previous day's heat shield inspection, mission managers cleared Atlantis for re-entry.

In a historic moment, the crew deployed the last payload ever released from a Space Shuttle— a technology demonstration satellite known as Pico-Satellite Solar Cell (PSSC-2). The tiny 8.2-pound (3.7 kg) picosatellite, ejected from a spring-loaded canister in the shuttle's payload bay at 7:54 UTC, entered low Earth orbit to test new solar cell technology for future satellites. This deployment, the 180th and final payload release of the shuttle program, was commemorated with a poem recited by Mission Specialist Rex Walheim, who reflected on the shuttle's contributions to space exploration.

Later in the day, the crew participated in their final round of interviews with reporters from ABC, CBS, CNN, Fox, and NBC, answering questions about their mission and the legacy of the Space Shuttle Program. Commander Ferguson and Pilot Hurley also practiced landing procedures using the Portable Inflight Landing Operations Trainer (PILOT), a video game-like simulator that helped them prepare for Atlantis's descent through Earth's atmosphere. The crew then verified the functionality of Atlantis's flight control surfaces and stowed the Ku-band antenna in preparation for re-entry. That night, the Empire State Building in New York City lit up in red, white, and blue to honor 30 years of Space Shuttle flights.

The final day of the Space Shuttle era began with the crew waking to "God Bless America" at 12:29 UTC. The song, played by CAPCOM Shannon Lucid, was dedicated to the entire crew and to the thousands of men and women who had contributed to the success of the Space Shuttle Program over the previous three decades.

With weather conditions near perfect—10-mile visibility and minimal crosswinds—flight controllers opted not to delay Atlantis's landing. The crew was given a "go" to start fluid loading, a process involving the consumption of liquids and salt tablets to help the astronauts readjust to Earth's gravity after their time in space.

Atlantis's deorbit burn took place at 4:49:04 a.m. EDT, lasting three minutes and 17 seconds. This burn decelerated the shuttle over the Indian Ocean, and the spacecraft reoriented itself for re-entry. Atlantis entered Earth's atmosphere at 5:25 a.m. EDT, and the crew experienced the glowing plasma generated by the intense heat of re-entry. The shuttle continued its descent, slowing to 223 miles per hour (359 km/h) as it approached the Kennedy Space Center.

Atlantis touched down on runway 15 at the Kennedy Space Center at 5:57 a.m. EDT, marking the final landing of the Space Shuttle Program. Mission Commentator Rob Navias's voice rang out as the shuttle came to a stop: "Having fired the imagination of a generation, a ship like no other, its place in history secured, the space shuttle pulls into port for the last time. Its voyage, at an end." Commander Ferguson, recognizing the significance of the moment, declared, "Mission complete, Houston. After serving the world for over 30 years, the shuttle has earned its place in history, and it has come to a final stop." CAPCOM Barry Wilmore, speaking for the ground team, added, "We congratulate you, Atlantis, as well as the thousands of passionate individuals across this great spacefaring nation who truly empowered this incredible spacecraft."

After working through their final checklists and powering down the shuttle, the STS-135 crew exited Atlantis and boarded the Crew Transport Vehicle. They were greeted on the runway by NASA Administrator Charles Bolden and other NASA officials before returning to the crew quarters. Atlantis was later towed back to Orbiter Processing Facility 2 (OPF-2), where a ceremonial walk-around was held for NASA employees before the shuttle was retired as a museum exhibit at the Kennedy Space Center Visitor Complex.

No specific contingency mission was designated for STS-135, the final Shuttle mission. Instead, NASA planned to rely on Russian Soyuz spacecraft for any required rescue operations. This decision reflected the evolving partnership between NASA and international space agencies and the transition toward new spacecraft and methods of ensuring crew safety.

On July 21, 2011, NASA hosted a special employee appreciation event outside OPF-2, with Atlantis parked on display. The ceremony featured remarks from NASA Administrator Charles Bolden, Kennedy Space Center Director Robert Cabana, and other officials, celebrating the shuttle's legacy and the people who made the program possible. Two NASA employees, Rita Wilcoxson and Patricia Stratton, were honored with the Distinguished Service Medal and the Distinguished Public Service Medal, recognizing their leadership and contributions to the Space Shuttle Program.

The following day, a public "welcome home" ceremony was held for the crew at Ellington Field in Houston, where thousands gathered to celebrate the conclusion of the Space Shuttle Program and honor the crew of STS-135.

Canceled Missions

One of the earliest cancellations occurred during the initial development phase of the Shuttle program. An emergency flight abort mission, designed as a Return to Launch Site (RTLS) sub-orbital test, was canceled due to the high risk involved. This decision underscored the inherent dangers of spaceflight and the cautious approach NASA adopted in the early days of Shuttle operations.

As the Shuttle program progressed, several other planned missions were canceled. Delays in the Shuttle's development pushed back numerous flights, and the devastating accidents of Challenger in 1986 and Columbia in 2003 led to the indefinite suspension of others. These tragedies forced NASA to re-evaluate the Shuttle program's safety protocols and mission priorities, resulting in a more conservative approach to mission planning.

On July 13, Flight Day 6 of the STS-135 mission, the crew of *Atlantis* began their day with a special wake-up message from Sir Elton John.

The message followed the morning's wake-up song, which played at 6:29 UTC, signaling the start of a busy day focused on cargo transfer operations.

The primary objective for the day was to continue unpacking supplies from the Raffaello Multi-Purpose Logistics Module (MPLM). By this point, the crew had already transferred 26% of the total 15,069 pounds of cargo, which included 9,403 pounds brought to the station and 5,666 pounds to be returned to Earth aboard *Atlantis*. The supplies delivered to the International Space Station (ISS) were essential for sustaining the orbiting outpost through 2012.

The *Atlantis* crew, consisting of Commander Chris Ferguson, Pilot Doug Hurley, and Mission Specialists Sandra Magnus and Rex Walheim, received assistance in these cargo transfer operations from the ISS Expedition 28 crew—Andrey Borisenko, Sergei Volkov, and Satoshi Furukawa. Together, they worked to unload the contents of Raffaello and store some of the material in the Pressurized Mating Adapter-3 (PMA-3), attached to the Tranquility module.

By the afternoon, *Atlantis*'s lead flight director, Chris Edelen, reported that approximately 50% of the cargo had been moved from both the Raffaello MPLM and the shuttle's middeck to the ISS. Despite the rigorous work schedule, all four shuttle crew members paused briefly at 16:54 UTC to participate in interviews with reporters from WBNG-TV and WICZ-TV in Binghamton, New York—near Pilot Doug Hurley's hometown of Apalachin—as well as KGO-TV in San Francisco.

Meanwhile, back on Earth, NASA was making preparations for the end of the shuttle program. On Flight Day 6, Space Shuttle *Discovery* was moved from the Orbiter Processing Facility (OPF-2) to the nearby Vehicle Assembly Building (VAB) for storage. This move was part of the plan to clear OPF-2 for *Atlantis* after its final landing.

On July 14, Flight Day 7, the crew continued their cargo transfer operations. Throughout the day, more supplies were unpacked from Raffaello and distributed around the ISS. However, the crew also enjoyed some much-deserved downtime. At 10:59 UTC, Commander Ferguson and Mission Specialist Magnus took part in interviews with Fox News Radio and local TV stations in St. Louis, and later the entire crew participated in interviews with media outlets from Chicago, Oakland, and Philadelphia.

After their media engagements, the crew had the afternoon off. In a moment of camaraderie, both the *Atlantis* and ISS crews gathered for a special "All-American Meal" that included barbecue brisket or grilled chicken, southwestern corn, baked beans, and apple pie—a symbolic celebration of the mission. NASA even invited the public to join virtually, sharing the details of the meal online.

As the day came to a close, the *Atlantis* crew encountered an unexpected issue. At 22:07 GMT, a master alarm sounded aboard the shuttle, indicating a failure in one of the spacecraft's five IBM AP-101 General Purpose Computers (GPC-4). Commander Ferguson responded quickly, heading to *Atlantis* to assess the problem. With guidance from ground control, the crew successfully transferred the failed GPC's programs to GPC-2. The operation, which took about 45 minutes, was completed by utilizing communication links at White Sands, New Mexico. Once the issue was resolved and the shuttle was confirmed to be in good condition, the crew returned to sleep. CAPCOM Shannon Lucid reassured the crew, telling them, "You all have done an absolutely fabulous job. Everyone is ready for you to go back to sleep." To compensate for the lost sleep, mission control extended the crew's sleep period by 30 minutes.

On July 15, Flight Day 8, the *Atlantis* crew was awakened by a special message and song from Sir Paul McCartney at 4:59 UTC—half an hour later than their usual wake-up time, allowing them to catch up on the rest they missed due to the GPC-4 failure. Commander Ferguson and Pilot Hurley spent the early part of the day reloading software and successfully restarting GPC-4, while flight controllers in Houston closely monitored the computer's performance through data downloads.

As Ferguson and Hurley addressed the computer issue, Mission Specialists Magnus and Walheim, along with the station crew, continued the critical task of transferring cargo between

Atlantis and the ISS. In addition to moving supplies, Walheim also transferred Extravehicular Mobility Unit (EMU) and airlock equipment to *Atlantis* for return, as these items would no longer be needed on the station following the shuttle's retirement.

Throughout the day, the crew participated in a series of media interviews, with Ferguson and Hurley speaking to CBS Radio, KYW-TV, and the Associated Press, followed by all four crew members speaking to WPVI-TV, KYW Radio, and Reuters. During a 45-minute news conference from the Japanese Kibo Laboratory, *Atlantis* crew members and their station colleagues fielded questions from reporters at multiple NASA centers and in Japan.

A notable moment occurred at 16:30 UTC when President Barack Obama called the combined Expedition 28 and *Atlantis* crews. During the call, he thanked everyone involved in the shuttle program and expressed pride in the achievements of the crew. Commander Ferguson responded, saying that all the international partners on the station were honored to represent their home countries in this collaborative effort. Flight Engineer Sergei Volkov added that the station and shuttle crews, hailing from three different nations, had become "one big family."

During the Mission Status Briefing, STS-135 lead flight director Kwatsi Alibaruho reported that 70% of the cargo transfer was complete. Although the crew had been ahead of schedule earlier in the mission, the GPC-4 issue had slowed progress. Alibaruho also noted a minor issue with a latch on *Atlantis*'s middeck locker, which contained lithium hydroxide (LiOH) canisters used to scrub carbon dioxide from the shuttle's cabin air. The issue was temporarily resolved by fastening the entire locker panel to the floor with fasteners. While docked to the ISS, *Atlantis* relied on the station's Carbon Dioxide Removal Assembly (CDRA) for air revitalization, but the LiOH canisters would be required once the shuttle undocked.

The GPC-4 failure, though rare, had occurred previously on *Atlantis* during the STS-71 mission and once on STS-9. Despite the setback, the crew remained on track as they prepared for the final stages of their mission.

Missions Cut Short

Even among the missions successfully launched, some were forced to return earlier than planned. Four missions were notably cut short while in orbit:

STS-2: The second Shuttle flight was reduced by a day due to a technical failure with one of the Shuttle's fuel cells. The mission, launched in November 1981, was the first time a Shuttle had been reused, but the early termination highlighted the challenges of the program's early years.

STS-35: This mission, launched in December 1990, was cut short by a day due to adverse weather conditions at the landing site. Despite the shortened mission, the crew successfully conducted significant astronomical observations using the ASTRO-1 observatory.

STS-44: In November 1991, this mission was reduced by three days after a malfunction in the Inertial Upper Stage (IUS) rocket that was supposed to deploy a Defense Support Program satellite. The crew managed the situation effectively, but the incident underscored the complexities of Shuttle operations.

STS-83: Launched in April 1997, this mission was shortened due to a problem with one of the Shuttle's fuel cells. The mission was relaunched as STS-94 in July 1997, allowing the crew to complete their microgravity experiments without further issues.

Contingency Missions

In the later years of the Shuttle program, NASA developed a series of contingency plans known as Launch on Need (LON) missions. These missions, designated STS-3xx, were designed to rescue crews in the event that a Shuttle became disabled or damaged and could not safely return to Earth.

The LON missions were prepared to be launched on short notice, with a crew of four astronauts who would execute the rescue. The rescued Shuttle's crew, combined with the LON crew, would return to Earth aboard the rescue Shuttle, making the return journey with ten or eleven crew members. These missions were planned to last approximately eleven days.

Notable LON missions included:

STS-300: Designated as the rescue mission for both STS-114 and STS-121, Atlantis would have launched this mission.

STS-400: Planned as the rescue mission for STS-125, Endeavor would have launched this mission if Atlantis encountered issues during its service mission to the Hubble Space Telescope.

Fortunately, these contingency missions were optional, as all planned Shuttle flights completed their missions and returned safely to Earth.

In summary, while the Space Shuttle program was a triumph of engineering and human perseverance, it was also marked by the realities of space exploration, where the risks and uncertainties often necessitated cautious planning and adaptability. The canceled missions, shortened flights, and contingency plans all played a part in shaping the legacy of the Shuttle era, ensuring that every mission was approached with the utmost care and consideration for the astronauts' safety.

Epilogue

NASA's Space Shuttle Program represents a defining chapter in the history of human spaceflight. Over 30 years, from its first mission in 1981 to its final flight in 2011, the Shuttle Program accomplished the extraordinary. It launched 135 missions, flew over 500 astronauts, and became the first reusable spacecraft capable of traveling to and from low Earth orbit, making it a cornerstone of the United States' space exploration endeavors. The program was a triumph of engineering and a demonstration of international collaboration, scientific achievement, and human resilience.

Each of NASA's five Space Shuttles—Columbia, Challenger, Discovery, Endeavour, and Atlantis—played a unique role in advancing space exploration. Columbia, the first Shuttle to fly, carried out 28 missions, helping to refine spaceflight science before its tragic loss in 2003. Challenger, lost in 1986, nonetheless flew 10 missions, contributing to early Shuttle successes. Discovery, the most-flown orbiter, completed 39 missions and was crucial in building the International Space Station (ISS) and launching significant payloads. Endeavour, constructed as a replacement for Challenger, carried out 25 missions, including vital repairs to the Hubble Space Telescope. Atlantis, which closed the Shuttle era with its final flight, completed 32 missions, contributing to both ISS assembly and international cooperative efforts, including missions to Russia's Mir space station.

Across these missions, 355 astronauts from 16 countries ventured into space aboard the Shuttle, including many who reached historic milestones. Sally Ride became the first American woman in space during Challenger's STS-7 mission in 1983. Prince Sultan bin Salman of Saudi Arabia, who flew aboard Discovery during STS-51G, became the first Arab and Muslim astronaut, as well as the youngest at just under 29 years of age. In contrast, John Glenn, who had first orbited Earth in 1962, became the oldest person to fly in space at age 77 when he returned aboard Discovery's STS-95 mission in 1998.

The Space Shuttle Program was instrumental in both assembling the ISS and advancing space science. Shuttles delivered crucial components to the ISS, conducted 36 dockings with the station, and provided the ability to conduct scientific experiments in microgravity. The fleet also achieved nine dockings with Mir, a precursor to the collaborative efforts that would define the ISS.

Throughout its operational life, the Shuttle Program faced challenges, including the devastating losses of Challenger and Columbia. These tragedies led to intense reflection and reevaluation at NASA, resulting in heightened safety protocols and greater determination to advance human space exploration. The lessons learned from these missions, successes and the losses, forged a path for safer and more reliable space travel in the future.

When Atlantis made its final landing at Kennedy Space Center on July 21, 2011, it marked not only the end of the Shuttle era but the beginning of a new phase in NASA's journey. With the Shuttle retired, NASA turned its focus toward deep space, developing the Space Launch System (SLS) and Orion spacecraft to pursue missions to the Moon, Mars, and beyond. At the same time, the Shuttle's legacy of collaboration laid the foundation for a new era of spaceflight. Commercial companies like SpaceX and Boeing have taken up the mantle of transporting astronauts and cargo to low Earth orbit, building on the technological advancements and lessons learned from the Shuttle Program.

The impact of the Shuttle Program cannot be overstated. In total, NASA's Shuttles traveled over 537 million miles, completing nearly 21,000 orbits around Earth. These missions helped revolutionize space science, enabled international cooperation, and opened space to a diverse group of astronauts. They also demonstrated humanity's capacity for resilience and ambition—qualities that continue to drive exploration efforts today.

As humanity looks to the future of space exploration, the legacy of the Space Shuttle Program endures. The knowledge and experience gained from three decades of Shuttle flights influence every facet of modern spaceflight, from spacecraft design and astronaut training to international space policy. The program continues to inspire new generations of scientists, engineers, and dreamers who carry the torch forward in NASA's ongoing quest to push the boundaries of human achievement. The Shuttles, once the workhorses of space exploration, are now museum pieces, but their story remains a symbol of what can be accomplished when we dare to venture beyond the Earth.

In the decades to come, as humans set foot on new worlds and journey farther into the cosmos, the Shuttle's legacy will serve as a reminder of where we began—and of all that we are capable of when we unite in the pursuit of knowledge and discovery. The Space Shuttle was more than just a vehicle; it was a bridge to the future, a testament to human ingenuity and perseverance, and a foundation for the next great adventures in space.

About the Author

Thornton D. "TD" Barnes is a distinguished author, entrepreneur, and former military intelligence specialist. Born in Dalhart, Texas and raised on a ranch near Clayton, New Mexico and Dalhart, Texas, he cultivated a passion for exploration. After high school in Oklahoma, Barnes embarked on a ten-year military journey, initially serving in Korea as an intelligence specialist. While in the Army, he also specialized in missile and radar electronics, defending against Soviet threats and later attending the Artillery Officer Candidate School. An injury ended his military career, but Barnes soon transitioned to aerospace endeavors. He worked on pivotal projects at NASA's High Range in Nevada, including the X-15, the NASA NERVA nuclear rocket project, and atomic bomb testing at the Nevada Test Site. Furthermore, he participated in the CIA's Mach 3 A-12 Project OXCART and stealth projects at Area 51.

Barnes founded and led an oil and gas exploration company outside the aerospace sphere for over 40 years, delving into uranium and gold mining ventures. In retirement, he's dedicated to preserving Area 51's history, serving as president of Roadrunners Internationale and the Nevada Aerospace Hall of Fame Director Emeritus. His contributions have been spotlighted in documentaries on National Geographic, the History Channel, and other major networks. Barnes has authored several books, including "The Secret Genesis of Area 51" and "The CIA Area 51 Chronicles." He currently resides in Henderson, Nevada, continuing to influence aerospace, exploration, and literature, focusing on the formally highly classified of the CIA's era at Area 51.

Bibliography

NASA Archives
2011.07.05-shuttle-era-facts.pdf
wings-ch3a-pgs53-73.pdf
https://www.scribd.com/document/753355368/SP-4221
Wikipedia